Earthquake Science and Engineering

I0036536

Earthquakes form one of the categories of natural disasters that sometimes result in huge loss of human life as well as destruction of (infra)structures, as experienced during recent great earthquakes. This book addresses scientific and engineering aspects of earthquakes, which are generally taught and published separately. This book intends to fill the gap between these two fields associated with earthquakes and help seismologists and earthquake engineers better communicate with and understand each other. This will foster the development of new techniques for dealing with various aspects of earthquakes and earthquake-associated issues, to safeguard the security and welfare of societies worldwide.

Because this work covers both scientific and engineering aspects in a unified way, it offers a complete overview of earthquakes, their mechanics, their effects on (infra)structures and secondary associated events. As such, this book is aimed at engineering professionals with an earth sciences background (geology, seismology, geophysics) or those with an engineering background (civil, architecture, mining, geological engineering) or with both, and it can also serve as a reference work for academics and (under)graduate students.

Earthquake Science and Engineering

Ömer Aydan

Emeritus Professor, University of the Ryukyus, Okinawa, Japan

CRC Press
Taylor & Francis Group
Boca Raton   London   New York   Leiden

CRC Press is an imprint of the
Taylor & Francis Group, an **informa** business

A BALKEMA BOOK

Cover image: Ömer Aydan

First published 2023
by CRC Press/Balkema
Schipholweg 107C, 2316 XC Leiden, The Netherlands
e-mail: enquiries@taylorandfrancis.com
www.routledge.com – www.taylorandfrancis.com

CRC Press/Balkema is an imprint of the Taylor & Francis Group, an informa business

Library of Congress Cataloging-in-Publication Data
Names: Aydan, Ömer, author.
Title: Earthquake science and engineering / Ömer Aydan, Emeritus
Professor, University of the Ryukyus, Okinawa, Japan.
Description: Boca Raton : CRC Press, Taylor & Francis Group, 2022. |
"A Balkema Book." | Includes bibliographical references and index.
Subjects: LCSH: Earthquake engineering. | Seismology. | Earthquake
damage.
Classification: LCC TA654.6 .A95 2022 (print) | LCC TA654.6 (ebook)
| DDC 624.1/762—dc23/eng/20220309
LC record available at https://lccn.loc.gov/2021062703
LC ebook record available at https://lccn.loc.gov/2021062704

ISBN: 978-0-367-75877-6 (hbk)
ISBN: 978-0-367-75878-3 (pbk)
ISBN: 978-1-003-16437-1 (ebk)

DOI: 10.1201/9781003164371

Typeset in Times New Roman
by codeMantra

Contents

Preface

This book is intended for readers interested in earthquakes and their scientific and engineering aspects. Most books cover either scientific or engineering aspects, and there is no book that treats both aspects. This book aims to combine both aspects so that readers can have a clear understanding of the phenomenon of earthquakes. Therefore, this book addresses readers having an earth science background (geology, seismology, geophysics) or an engineering background (civil, architecture, mining, geological engineering) or both, and simultaneously it can be a reference book for undergraduate and graduate students.

The contents of the book evolved from earthquake engineering classes taught by the author in the Marine Civil Engineering Department of Tokai University (Shizuoka, Japan) and in the Civil Engineering Department of the University of the Ryukyus (Okinawa, Japan) when the author was asked to prepare the syllabus and to teach undergraduate students in association with the re-organization and re-formation of the curriculum of Civil Engineering Classes at both universities. Furthermore, some of the contents of this book were developed when the author was involved in the establishment of the laboratory for the simulation of natural disasters as the director of the Disaster Prevention Research Center for Islands Region (DPRCIR) of the University of the Ryukyus. The establishment of the laboratory was intended to demonstrate the physics of natural disaster phenomena such as earthquakes, tsunami, tornado, and geotechnical disasters and the response of structures during ground motions to an audience encompassing primary schools, the general public, and undergraduate and graduate university students.

Readers with a background in earth sciences will be able to understand what earthquake engineering is all about, while readers with an engineering background will understand the physics of earthquakes and associated secondary natural phenomena. Therefore, the goal of this book is to bring together both scientific and engineering aspects of earthquakes into a single entity so that readers can have a clear and complete understanding of the phenomenon of earthquakes and to adequately address issues associated with them.

Acknowledgements

The involvement of the author with the science and engineering of earthquakes started with his participation in the reconnaissance of the 1992 Erzincan earthquake in Turkey together with Emeritus Professor M. Hamada of Waseda University (formerly, Tokai University) and Emeritus Professor Z. Hasgür (Istanbul Technical University) on behalf of the Association for the Development for Earthquake Prediction (ADEP) and supported by Emeritus Prof. Dr. R. Yarar (Istanbul Technical University) of Turkish Earthquake Foundation (TDV) in April 1992. This reconnaissance provided the chances for the author to meet many eminent scholars of Turkey, Japan and the United States in person, and such connections provided the author opportunities to participate in the reconnaissance of many great earthquakes in Turkey, Japan and other countries. Furthermore, these connections led the author to be a member, and later the chairman, of the Earthquake Disaster Investigation Sub-Committee of the Earthquake Engineering Committee of the Japan Society of Civil Engineers since 1999, and he has become an advisor of the Earthquake Engineering Committee since May 2021.

This involvement of the author led him to specialize in and to teach earthquake engineering subject at Tokai University and later at the University of the Ryukyus until his retirement, besides his main field research in Rock Mechanics and Rock Engineering. He still teaches Disaster Prevention Engineering at Tokoha University as a part-time lecturer since April 2021. A further involvement of the author in the topic evolved when the author was asked to establish the demonstration facility for the simulation of natural disasters as the director of the Disaster Prevention Research Center for Islands Region (DPRCIR) of the University of the Ryukyus. The establishment of the facility was intended to demonstrate the physics of natural disaster phenomena such as earthquakes, tsunami, tornado, geotechnical disasters and response of structures during ground motions to audiences ranging from primary school students, the public to undergraduate and graduate students.

When the author contemplated this book, he received great encouragements from Prof. Z. Hasgür, Prof. K. Kawashima and Prof. R. Ulusay. The author sincerely thanks Emeritus Prof. Dr. M. Hamada of Waseda University, Emeritus Prof. K. Kawashima of Tokyo University of Technology, Prof. J. Kiyono of Kyoto University and Prof. Dr. M. Miyajima of Kanazawa University for collaboration, help and guidance for about four decades. He extends special thanks to Emeritus Prof. Dr. Zeki Hasgür for his continuous support, guidance and leading the author to study the key elements of earthquake engineering since April 1992. He sincerely thanks Prof. Dr. R. Ulusay of Hacettepe University and Prof. Dr. Halil Kumsar for joining the author in the reconnaissance of earthquakes in Turkey, Italy and New Zeeland. He also extends sincere thanks to Prof. Dr. J. Tomiyama for his help and collaboration with the author during the earthquake engineering classes at the University of the Ryukyus. The author would also like to thank his former students at Tokai University and the University of the Ryukyus for attending and shaping up the content of this book. In particular, special thanks are to his former students Mr. Mitsuo Daido and Dr. Yoshimi Ohta of Tokai University and K. Horiuchi, T. Aikawa, Y. Murayama and M. Tamashiro of University of the Ryukyus.

Finally, he thanks his wife Reiko, daughter Ay and son Turan Miray, and his parents for their continuous help and understanding throughout, without which this book could not have been completed.

The author heartfully dedicates this book to the profound memory of the late Emeritus Prof. Dr. Rıfat YARAR, who continuously supported and guided the author since 1992, and the people who lost their lives or suffered in the earthquakes in the reconnaissance of which the author was involved.

About the author

Ömer Aydan was born in 1955 and studied Mining Engineering at the Technical University of Istanbul, Turkey (B.Sc., 1979), Rock Mechanics and Excavation Engineering at the University of Newcastle upon Tyne, UK (M.Sc., 1982), and received his Ph.D. in Geotechnical Engineering from Nagoya University, Japan in 1989. Prof. Aydan worked at Nagoya University as a research associate (1987–1991), and then in the Department of Marine Civil Engineering at Tokai University, first as Assistant Professor (1991–1993), then as Associate Professor (1993–2001), and finally as Professor (2001–2010). He then became Professor of the Institute of Oceanic Research and Development at Tokai University, and is currently Professor at the University of the Ryukyus, Department of Civil Engineering & Architecture, Nishihara, Okinawa, Japan. He is also the director of the Disaster Prevention Research Center for Island Region of the University of the Ryukyus. Prof. Aydan has played an active role in numerous organizations such as ISRM, JSCE, JGS, SRI and Rock Mechanics and other National Group of Japan committees, and has organized several national and international symposia and conferences.

Chapter 1

Introduction

Earthquakes are well known as the natural disasters resulting in huge loss of human lives as well as of properties as experienced in the recent great earthquakes such as the 1999 Kocaeli, Düzce, Chi-chi; the 1995 Kobe earthquakes; the 2004, 2005, 2007 and 2009 off-Sumatra earthquakes; the 2008 Wenchuan earthquake; and the 2011 Great East Japan Earthquake. It is well known that ground motion characteristics, deformation and surface breaks of earthquakes depend upon the causative faults. While many large earthquakes occur along the subduction zones, which are far from the land, and their effects appear as severe ground shaking, the large in-land earthquakes may occur just beneath or nearby urban and industrial zones as observed in the recent great earthquakes. Earthquakes are due to the temporary instability of Earth's crust resulting from stress state changes. While the accumulation of stress takes for very long period from seconds to thousands of years, which is called stick phase, stress release occurs within a few seconds to 500–600 seconds and it is called the slip phase.

The location, magnitude, occurrence time, the mechanism of earthquakes and the possibility of tsunami are determined by national and international seismological centres, while strong motions are recorded by strong-motion networks in each country and publicized in social networks, TVs and radios. In other words, the earthquakes and related effects are a part of the daily lives of humankind. Earthquakes and related events will continue to occur as long as Earth continues its motion around the Sun in the solar system. Humankind should take counter-measures and design structures against earthquakes and resulting tsunamis for their safe living against their devastating effects.

Most books are intended for either scientific or engineering aspects, and there is no book addressing both aspects. This book addresses scientific and engineering aspects of earthquakes, which are generally taught separately. This book intends to fill the gap between these two fields associated with earthquakes and to make seismologists and earthquake engineers understand and communicate with each other for developing new techniques to deal with various aspects of earthquakes and earthquake-induced issues for better and safe societies worldwide.

This book specifically attempts to provide

- A unified treatise of scientific and engineering aspects of earthquakes
- Clear definitions and explanation of fundamental concepts
- Explanation of procedures for determining the fundamental parameters essential to the science and engineering of earthquakes. Specifically, location, magnitude,

DOI: 10.1201/9781003164371-1

wave propagation, focal mechanism related to scientific aspects and analyses of strong motions, measurements, their utilization in earthquake-proof design, tsunami-resistant structures, the design of structures against permanent ground movements, etc.

• In-depth view of procedures and solution techniques

Therefore, this book is expected to address readers having an earth-science background (geology, seismology, geophysics), an engineering background (civil, architecture, mining, geological engineering) or both, and simultaneously it can be a reference book for undergraduate and graduate students.

Chapter 2 is concerned with the physics of earthquakes and explains the background of earthquakes starting from laboratory experiments to real-world occurrences. Laboratory experiments involve fracturing of intact rock, slippage of artificial and natural discontinuities. These experiments explain various physical variations such as wave velocity, electrical resistivity and acoustic emission (AE) activities before and during fracturing. Furthermore, acceleration responses of mobile and stationary sides are presented and discussed in relation to strong-motion observations. Finally, some observations of the differences between earthquakes and volcanic eruptions are explained and discussed.

Earthquakes are always associated with motions that propagate as waves. Chapter 3 presents the fundamental law of equation of motion, which establishes the equations of wave propagation. The wave propagation induced by earthquakes is also utilized to infer the interior structure of Earth as well as the hypocenter data and magnitude of earthquakes.

Faults and faulting mechanisms are the most important elements in the science and engineering of earthquakes. Faults are known to be geological discontinuities, and their temporary instability resulting in slippage causes disastrous effects on structures, environment, and life. The accumulation of stress causes their instability, and evaluation of the critical conditions is essential. In Chapter 4, the stress state in the close vicinity of various types of faults, the asperities and their stress states during stable and unstable conditions are explained through actual experiments. For this purpose, results of some photoelasticity experiments, which enable us to see the stress state visually, are presented for some typical faulting conditions and the same conditions are also analysed through the finite element method. Some empirical procedures are described to evaluate some interrelations for estimating potential earthquakes from fault length and relative slip amount. In the final part, procedures for inferring focal mechanisms from waves and fault striations are explained.

Every earthquake causes vibrations and temporary and/or permanent movement of ground. The ground motions caused by earthquakes in/around their epicentral area having a dominant frequency >1 Hz would be categorized strong ground motions. These ground motions recorded as acceleration and/or velocity decrease in amplitude in relation to the distance, while the motions recorded as displacements having dominant frequency <1 Hz are called broad-band ground motions, which are commonly used to infer the hypocenter and magnitude of earthquakes. Chapter 5 describes ground motions and their characteristics, and techniques for measuring and recording them. As it is difficult to measure ground motions at every location, empirical, semi-analytical or numerical procedures are described to estimate ground motions.

In addition, this chapter also explains some procedures developed to infer the strong ground motion parameters from the collapse, failure or slippage of simple structures and simplified reinforced concrete structures. These methods are also of great importance to assess the magnitudes of past earthquakes as well as modern earthquakes where instrumental data is scarce.

Chapter 6 is concerned with response analyses of structures, which is of great importance in the evaluation of behaviour of structures under earthquake motions. First, the fundamental procedures for response analyses are described and the fundamental concepts for the vibration of structures are explained utilizing some simplified analyses. Then the techniques used for measuring and analysing the vibrational characteristics of structures such as Fourier spectra and response spectra methods are presented. Some specific examples of vibration measurements for model structures as well as actual structures are given. Furthermore, some empirical procedures for estimating the natural frequencies of some structures are presented and compared with actual data. In addition, the response and vibrational characteristics of large structures such as dams, tanks, wind turbines, underground openings, caverns and shafts are explained.

The effects of past earthquakes and associated surface ruptures on engineering structures should be well understood for their earthquake-proof or earthquake-resistance design and performance. Chapter 7 explains the observed effects of past earthquakes and associated surface ruptures on various engineering structures such as bridges and viaducts, dams, tunnels, failures of slopes and rockfalls, embankments and foundations and deformation of underground caverns and shafts. Ground liquefaction is also an important geotechnical phenomenon. The mechanism and estimation of ground liquefaction and associated lateral spreading and their effects on various structures are explained.

Chapter 8 is concerned with the seismic design of various structures. First, the fundamental procedures of seismic design such as seismic coefficient method, modified spectral seismic coefficient method, modal pushover and response analysis methods are described. Then some specific seismic design procedures for dams, bridges and viaducts, tunnels, pipelines, embankments, slopes, foundations of critical structures, buildings, minarets and dome-type structures are explained. As ground liquefaction is also a very critical issue, the techniques for assessing ground liquefaction and its effects such as uplift, settlement and lateral spreading of structures are also presented.

The devastating 2011 Great East Japan Earthquake (GEJE) and the 2004 off-Sumatra (Aceh) mega earthquakes showed the tremendous extent of structural damage tsunamis can cause. In Chapter 9, the mechanism of tsunamis and their effects on various engineering structures are explained through some specific examples the author observed during the reconnaissance of the 2011 GEJE, and the 2004 Aceh mega earthquakes are explained. Laboratory-scale tsunami model tests are also described to illustrate the forces resulting from tsunamis. Then, procedures for evaluating the effects of tsunamis on structures and the principles of tsunami-proof structural design are explained. Finally, an empirical procedure for estimating the magnitude of earthquakes from tsunami boulders is presented, and some specific examples are given.

Earthquake prediction, which requires the prediction of the time of occurrence, magnitude and location, is always the most desirable yet most difficult issue in earthquake science. Chapter 10 is concerned with this issue. First, the physical background on anomalous phenomena observed in earthquakes is presented, and the implications of responses of rocks and discontinuities during fracturing and slippage are discussed. Then, the available methods of earthquake prediction are described together with some attempts of prediction. In recent years, the GPS method has become widely used. A section is dedicated to earthquake prediction using the GPS method, and it is shown that prediction of the location and time of earthquakes may be possible. Furthermore, the possibility for estimating the magnitude is discussed. Several specific examples of anomalous phenomena during earthquakes in Turkey are given, and the application of the multi-parameter monitoring system for earthquake prediction in Denizli Basin of Turkey is described. In the final part, some principles of earthquake prediction using various techniques are established.

Chapter 2

Physics of earthquakes

2.1 CAUSES OF EARTHQUAKES

The main cause of earthquakes are changes in stresses of the Earth's crust and its temporary mechanical instability. The plate tectonic theory based on Wegener's theory on continental drift proposed in 1912 is often used to explain the occurrence of quakes on the Earth, called "earthquakes." However, Wegener's theory was not well received among mainstream geologists and geophysicians until the discovery of palaeomagnetism and modern seismology. Figure 2.1 shows the plots of earthquake epicentres and major plates of the Earth.

It is often assumed that seismic regions coincide with plate boundaries. There are three major plate boundaries, namely, convergent, divergent and transform plate boundaries. High mountains are found at/along convergent boundaries and plate thickness increases up to 70 km. Furthermore, large earthquakes occur along convergent plate boundaries, while the magnitude of earthquake along divergent plate boundaries is generally smaller. Transform plate boundaries may have a dextral or sinistral sense of relative slip, and the magnitude of earthquakes generally ranges between those of earthquakes caused by convergent and divergent plate boundaries (Figure 2.2).

The driving force of plates is often said to be due to mantle convection as illustrated in Figure 2.3. However, it is well known that the convection in the Earth cannot take place if the Earth consists of uniformly distributed co-centric layers of materials. The subducting plate along the convergent plate boundaries and the divergent plate boundary constitutes the mantle convection cell. This subducting plate is likely to be caused by the plastic yielding of the upper mantle and the Earth's crust as explained in the next section.

The Earth is rotating around the Sun in a given orbit, and this orbit has an elliptical shape so that the distance between the Earth and the Sun varies during a year. As a result, the acceleration and velocity of the Earth are not the same throughout a year as shown in Figure 2.4. The stress acting in various parts of the Earth would be changing with different periods such as 3 months, 28 days, 24 hours and 6 hours as a result of the motion of the Earth around the Sun, the Earth's rotation and the Moon's revolution around the Earth besides other cycles, such as the wobbling of the rotation axis of the Earth. Figure 2.5 shows the effect of the Sun and the Moon on the Earth's velocity and acceleration during a day.

DOI: 10.1201/9781003164371-2

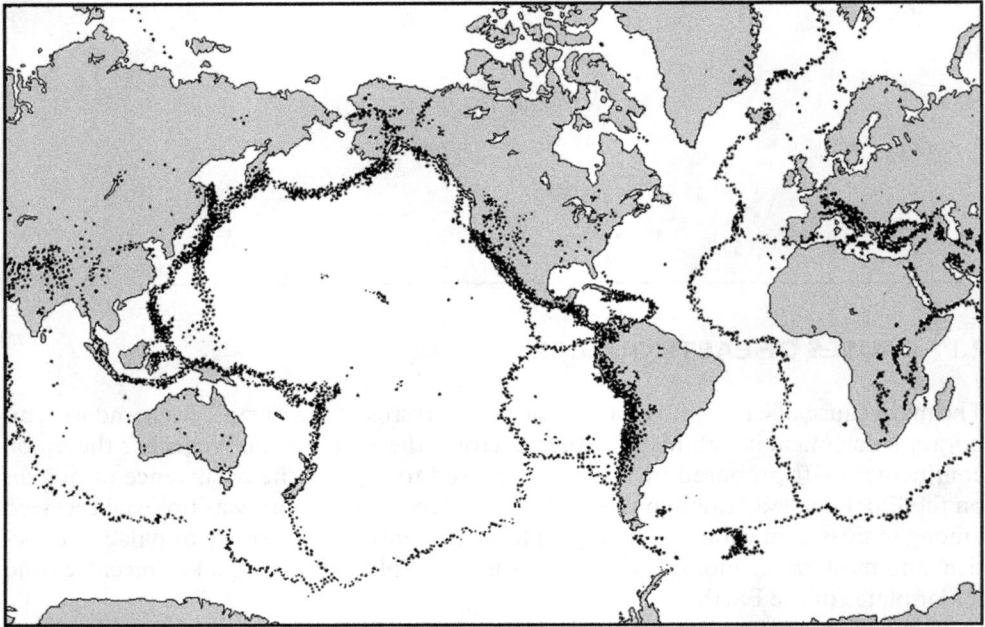

(a)

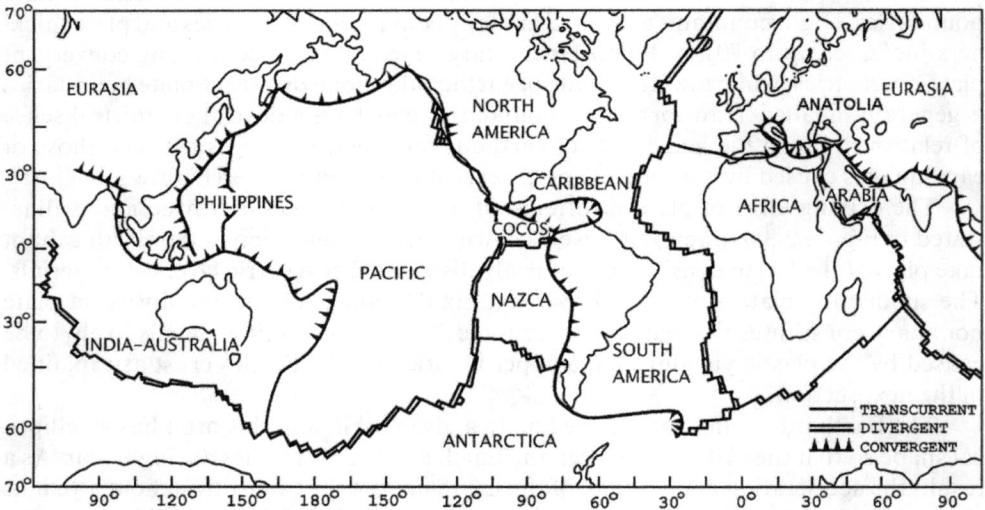

(b)

Figure 2.1 (a) World seismicity and (b) the major plates of the Earth. (Modified from Udias (1999).)

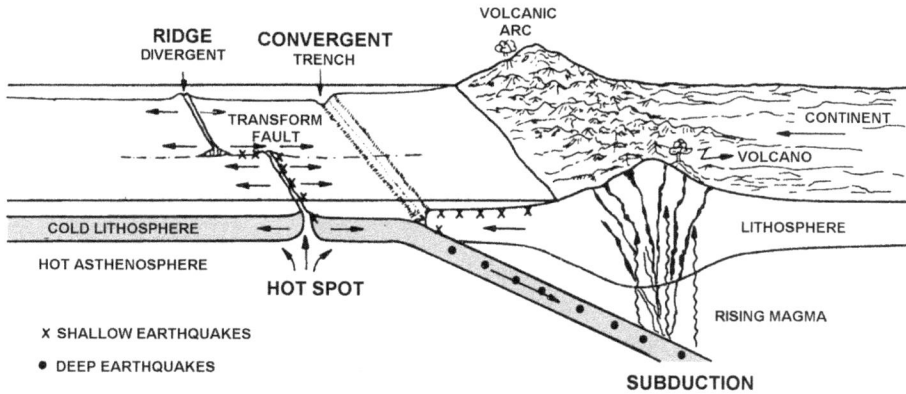

Figure 2.2 Types of plate boundaries.

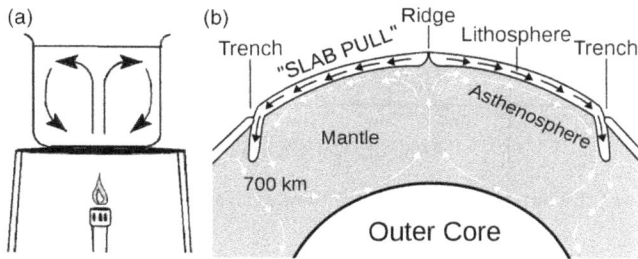

Figure 2.3 Illustration of (a) a simple convection cell and (b) mantle convection cells in the Earth.

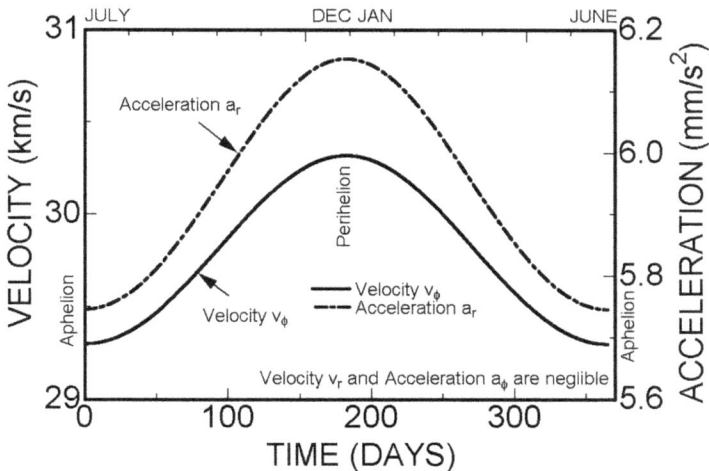

Figure 2.4 Acceleration and velocity variations of the Earth during a year.

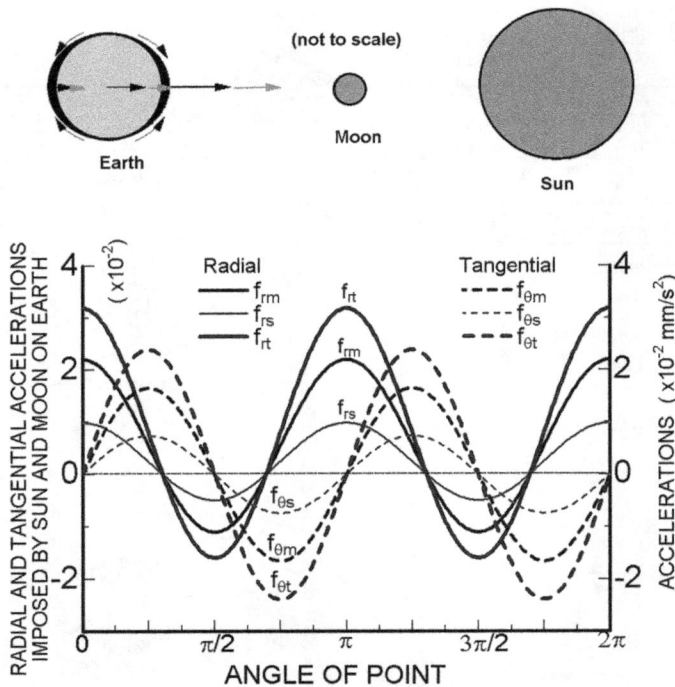

Figure 2.5 Effects of the Sun and the Moon on the Earth's radial and tangential accelerations.

2.2 THE STRESS STATE OF THE EARTH AND THE EARTH'S CRUST

In situ stresses are of great importance for the design and stability assessment of rock engineering structures as well as for understanding and predicting earthquakes. Several researchers investigated the stress state of the Earth and its crust in the past. Jeffreys and Bullen (1940) considered that the stress state of the Earth was hydrostatic and they calculated the pressure of the Earth by considering the variations in density and gravitational acceleration. Anderson and Hart (1976) modified this approach by considering recent findings. Nadai (1950) also derived a formula for radial and tangential stresses by assuming that the density of Earth is constant and the gravitational acceleration varies linearly with depth.

Aydan (1995a) analysed the stress state of the Earth by modelling it as a spherical body consisting of layers exhibiting thermo-elasto-plastic behaviour under pure gravitational acceleration using the finite element method (Figure 2.6). The computational results of Aydan (1995a) are briefly explained in this section. The co-centric layers constituting the Earth are isotropic while the overall structure is anisotropic. Considering the distribution of material properties (elastic modulus, Poisson's ratio, density) and the variation in gravity through the Earth (Fowler 1990), the rock mass is assumed to behave thermo-elasto-plastically, obeying the thermoplastic yield criterion of Aydan (1995a).

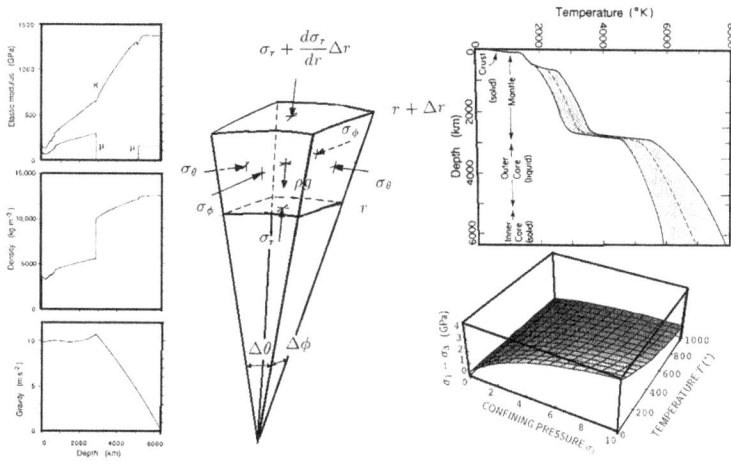

Figure 2.6 Illustration of a spherical model of the Earth and its material properties. (a) Variation in material properties and gravitational acceleration with depth; (b) control-volume for a spherical Earth; and (c) temperature distribution and thermoplastic yield surface. (Arranged from Aydan (1993, 1995a) and Fowler (1990).)

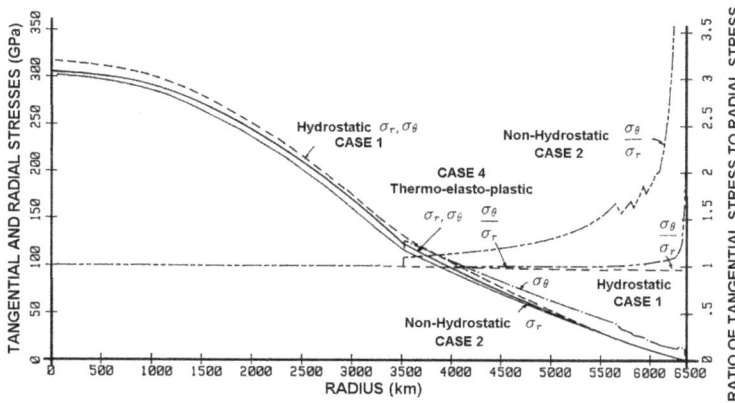

Figure 2.7 The stress state of the Earth. (From Aydan (1995a).)

Figure 2.7 shows the distribution of radial and tangential stresses on the Earth. From the study, the following conclusions were drawn for the stress state of the Earth:

a. If the spherical structure of the Earth is taken into account, it is possible to explain why the horizontal stress is larger than the vertical stress near the ground surface. In other words, the large horizontal stress is due to gravity, and not due to presumed tectonic forces resulting from unknown sources.

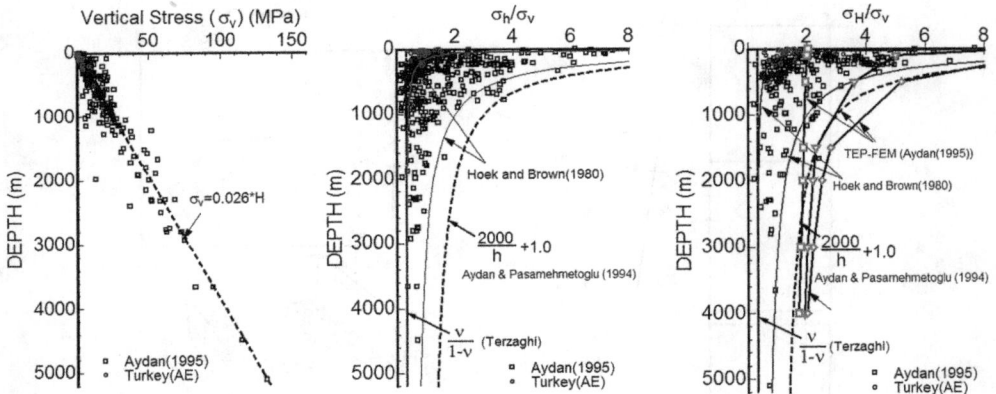

Figure 2.8 Comparison of in situ stress measurements with various methods. (From Aydan (2016a, 2019a).)

b. For a spherically symmetric Earth, the tangential stress (lateral stress) is always the maximum principal stress and the radial stress (vertical stress) is the minimum principal stress irrespective of the mechanical behaviour of rocks.

c. The crust and the mantle are in plastic state, which may have some important implications in geoscience. In other words, the Earth is not an elastic body as presumed in many different studies, and the plastic behaviour of the crust together with the upper mantle could be the main cause of nonuniformity resulting in the right conditions for mantle convection to occur.

Aydan's approach provided very valuable information, and it is a first approximation to the stress state of the Earth. Although the gravitational pull is the governing element in shaping the stress state of the Earth, slight variations in the computed stress state from the gravitational model are caused by rotation around its axis as well as the Sun and mantle convection due to the nonuniformity in the thermal field resulting from the subduction of the cooler plates into the hot mantle.

Figure 2.8 compares the stress measurements together with available in situ stress measurements worldwide (Aydan 1995a, 2016a) and some empirical relations for horizontal stress ratio over the overburden pressure developed by Brown and Hoek (1978), Hoek and Brown (1980) and Aydan and Paşamehmetoğlu (1994) as well as those from the thermo-elasto-plastic finite element analyses by Aydan (1995a). It is interesting to note that the stress measurements are similar to theoretical and empirical formulas (Aydan and Kawamoto 1997), while some discrepancies occur in terms of quantitative values.

2.3   THE STRESS STATE OF A FAULT AND ITS CHANGES DURING EARTHQUAKES

Aydan (2000a, 2016a) proposed a stress inference method based on the striation of faults. Let's first consider the average stress state on a fault plane. The stress state and geometrical parameters of the faults can be illustrated as shown in Figure 2.9. Vectors

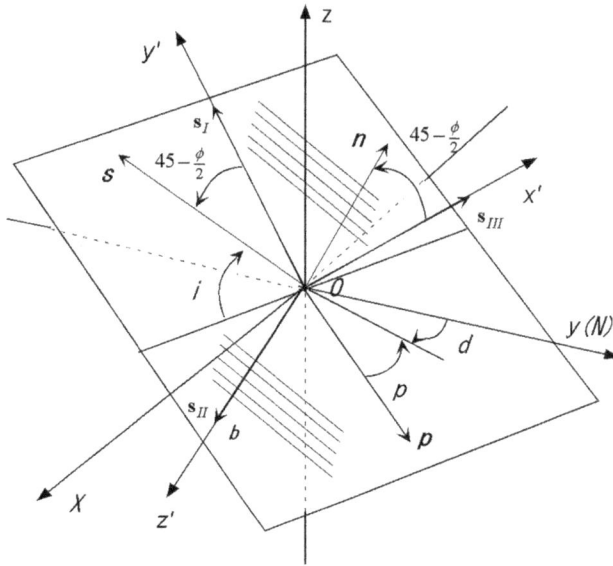

Figure 2.9 Illustration of notation for a fault plane with directions of principal stresses, slip, normal and neutral vectors.

n, s and **b** are normal, sliding and neutral vectors with respect to the fault plane, respectively. Neutral vector **b** is perpendicular to the plane defined by normal and sliding vectors **n, s**. Parameters p, d and i stand for dip (plunge), dip direction and sliding direction on the fault plane. When the value of i is 0–180, it corresponds to the faults with normal component. On the other hand, if the value of i is 180–360, it corresponds to faults with reverse components (Figure 2.10).

The normal and sliding vectors in terms of their components can be specifically written as shown below:

$$\mathbf{n} = \left\{ \begin{array}{ccc} n_x & n_y & n_z \end{array} \right\}; \quad \mathbf{s} = \left\{ \begin{array}{ccc} s_x & s_y & s_z \end{array} \right\} \tag{2.1}$$

where

$$n_x = \sin p \sin d; \quad n_y = \sin p \cos d; \quad n_z = \cos p; \quad s_x = -\cos p \sin d \sin i + \cos d \cos i;$$
$$s_y = -\cos p \cos d \sin i - \sin d \cos i; s_z = \cos p \sin i.$$

As the neutral vector **b** is perpendicular to the plane of normal and sliding vectors, it can be mathematically expressed as given below:

$$\mathbf{b} = \mathbf{s} \times \mathbf{n} \tag{2.2}$$

The traction vector **t** acting on the fault plane can be related to stress tensor $\boldsymbol{\sigma}$ and the normal **n** of the fault plane through the Cauchy equation (e.g. Mase 1970; Eringen 1980; Aydan 2021), as below:

$$\mathbf{t} = \boldsymbol{\sigma} \cdot \mathbf{n}$$

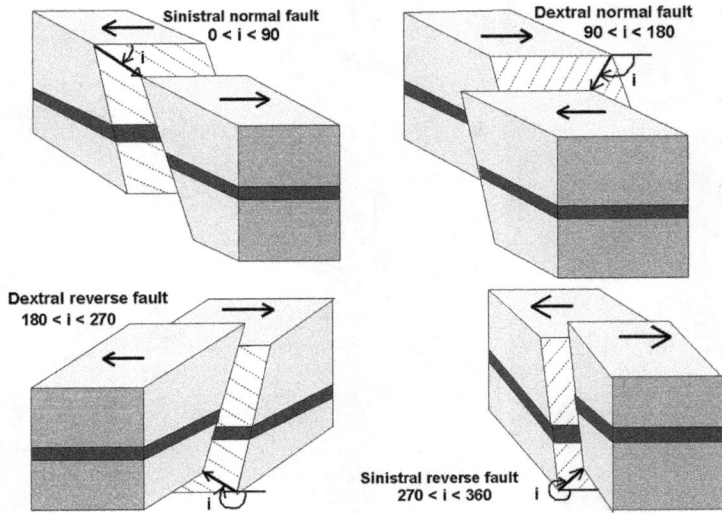

Figure 2.10 Illustration of the definition of a striation angle on a fault surface.

The stress tensor can be written in matrix form as given below:

$$\sigma = \begin{bmatrix} \sigma_{xx} & \sigma_{xy} & \sigma_{xz} \\ \sigma_{xy} & \sigma_{yy} & \sigma_{yz} \\ \sigma_{xz} & \sigma_{yz} & \sigma_{zz} \end{bmatrix} \tag{2.3}$$

Furthermore, the following relations can be written for the normal, shear and neutral vectors:

$$\sigma_N = \mathbf{n} \cdot \mathbf{t} \tag{2.4a}$$

$$\sigma_S = \mathbf{s} \cdot \mathbf{t} \quad \text{and} \quad \sigma_S = \mathbf{s} \cdot \sigma \cdot \mathbf{n} \tag{2.4b}$$

$$\sigma_B = \mathbf{b} \cdot \mathbf{t} \quad \text{and} \quad \sigma_B = \mathbf{b} \cdot \sigma \cdot \mathbf{n} \tag{2.4c}$$

Aydan (1995a) both theoretically and numerically showed that the vertical component σ_{zz} of the stress tensor can be taken as a quantity obtained by the multiplication of depth h and unit weight γ of rock as given below, by taking into account the sphericity and gravitational acceleration of the Earth.

$$\sigma_{zz} = \gamma h \tag{2.5}$$

Let us introduce a normalised stress obtained by dividing every component of the stress tensor by its vertical component:

$$\mathbf{N} = \begin{bmatrix} N_{xx} & N_{xy} & N_{xz} \\ N_{xy} & N_{yy} & N_{yz} \\ N_{xz} & N_{yz} & N_{zz} \end{bmatrix} \tag{2.6}$$

where N_{zz} is equal to 1. Let us consider a coordinate system $ox'y'z'$ whose axes are aligned with the principal stress components, as shown in Figure 2.9.

$$\sigma' = \begin{bmatrix} \sigma_I & 0 & 0 \\ 0 & \sigma_{II} & 0 \\ 0 & 0 & \sigma_{III} \end{bmatrix} \tag{2.7}$$

Similarly the normalised principal stress tensor by the vertical stress can be written as follows:

$$\mathbf{N}' = \begin{bmatrix} N_I & 0 & 0 \\ 0 & N_{II} & 0 \\ 0 & 0 & N_{III} \end{bmatrix} \tag{2.8}$$

The shearing of rock takes place along the direction of the maximum shear stress according to the *least work principle of the mechanics*. Therefore, the shear stress on the plane of normal and sliding vectors must be nil, which implies that the direction of the neutral vector must coincide with that of the intermediate principal stress. Thus, this can be mathematically expressed as

$$\sigma_B \equiv \sigma_{II} \tag{2.9}$$

and

$$\mathbf{b} \equiv \mathbf{s}_{II} \quad \text{and} \quad \mathbf{b} \equiv \left\{ \begin{array}{ccc} b_x & b_y & b_z \end{array} \right\}; \ \mathbf{s}_{II} = \left\{ \begin{array}{ccc} l_2 & m_2 & n_2 \end{array} \right\} \tag{2.10}$$

In the lights of experimental facts on rocks, the following relations may be written among normal, sliding and neutral vectors and the maximum and minimum principal stresses:

$$\mathbf{s} \cdot \mathbf{s}_I = \cos\left(45 - \frac{\phi}{2} \right) \ \text{ve} \ \mathbf{s} \cdot \mathbf{s}_{III} = \cos\left(135 - \frac{\phi}{2} \right) \tag{2.11a}$$

$$\mathbf{n} \cdot \mathbf{s}_I = \cos\left(45 + \frac{\phi}{2} \right) \ \text{ve} \ \mathbf{n} \cdot \mathbf{s}_{III} = \cos\left(45 - \frac{\phi}{2} \right) \tag{2.11b}$$

$$\mathbf{b} \cdot \mathbf{s}_I = 0 \ \text{ and } \ \mathbf{s} \cdot \mathbf{s}_{III} = 0 \tag{2.11c}$$

Therefore the direction vectors $\mathbf{s}_I$ and $\mathbf{s}_{II}$ of the maximum and minimum principal stresses can be easily obtained from the above relations as

$$\mathbf{s}_I = \left\{ \begin{array}{c} l_1 \\ m_1 \\ n_1 \end{array} \right\} = \begin{bmatrix} s_x & s_y & s_z \\ n_x & n_y & n_z \\ b_x & b_y & b_z \end{bmatrix}^{-1} \left\{ \begin{array}{c} \cos\left(45 - \frac{\phi}{2} \right) \\ \cos\left(45 + \frac{\phi}{2} \right) \\ 0 \end{array} \right\} \tag{2.12a}$$

$$\mathbf{s}_{\mathrm{III}} = \left\{ \begin{array}{c} l_3 \\ m_3 \\ n_3 \end{array} \right\} = \left[\begin{array}{ccc} s_x & s_y & s_z \\ n_x & n_y & n_z \\ b_x & b_y & b_z \end{array} \right]^{-1} \left\{ \begin{array}{c} \cos\left(135 - \dfrac{\phi}{2}\right) \\[2mm] \cos\left(45 - \dfrac{\phi}{2}\right) \\[2mm] 0 \end{array} \right\} \qquad (2.12\mathrm{b})$$

When the Mohr–Coulomb yield criterion is used, the value of intermediate principal stress becomes indeterminate although its value is bounded by the maximum and minimum principal stresses. Aydan (2000a) established a relation through the use of Mohr–Coulomb criterion and Drucker–Prager yield criterion for frictional condition, which is commonly used in numerical analysis of structures in geomaterials, and he derived the following inequality relation to obtain the value of intermediate principal stress:

$$\beta^2 - \frac{(1+6\alpha^2)(q+1)}{(1-3\alpha^2)}\beta + \frac{(1-3\alpha^2)(q^2+1)-q(1+6\alpha^2)}{(1-3\alpha^2)} = 0 \qquad (2.13)$$

where

$$\beta = \frac{\sigma_{\mathrm{II}}}{\sigma_{\mathrm{III}}} \qquad (2.14)$$

The inequality relation yields two roots, and one of the root is chosen such that the intermediate stress would have a value between the maximum and minimum principal stresses. When the peak friction angle is utilized, the stress state would correspond to the stress state at the time slip. On the other hand, if the residual friction is utilized, it should correspond to the stress state at the equilibrium following the termination of the earthquake.

An example of this concept has been applied to the 1984 Nagano Prefecture Seibu earthquake with a moment magnitude of 6.2. The normalized stress states by the vertical stress for peak (30°) and residual (26°) friction angles are obtained and shown in lower-hemisphere stereo-net projections in Figure 2.11. The computed normalized stress changes are given in Table 2.1. The average shear stress change on the fault plane would be about 14.88 MPa for a linear distribution of vertical stress.

Aydan (2000b, 2003a) developed an interpolation technique based on finite element method, and it is used to compute the strain rate and consequently stress rate in the tangential plane to the surface of the Earth's crust from crustal deformation measurements by GPS. This approach was applied to the 2011 Great East Japan Earthquake (GEJE) using the area shown in Figure 2.12 and an element consisting of Oshika, Wakuya and Rifu GPS stations of the GEONET. Figure 2.13 shows the computed variations in principal stress, maximum shear stress and disturbing stress and the orientation of the principal stress change from east together with another definition of magnitude-distance intensity (MRI), which was suggested by Aydan et al. (2015c) as follows:

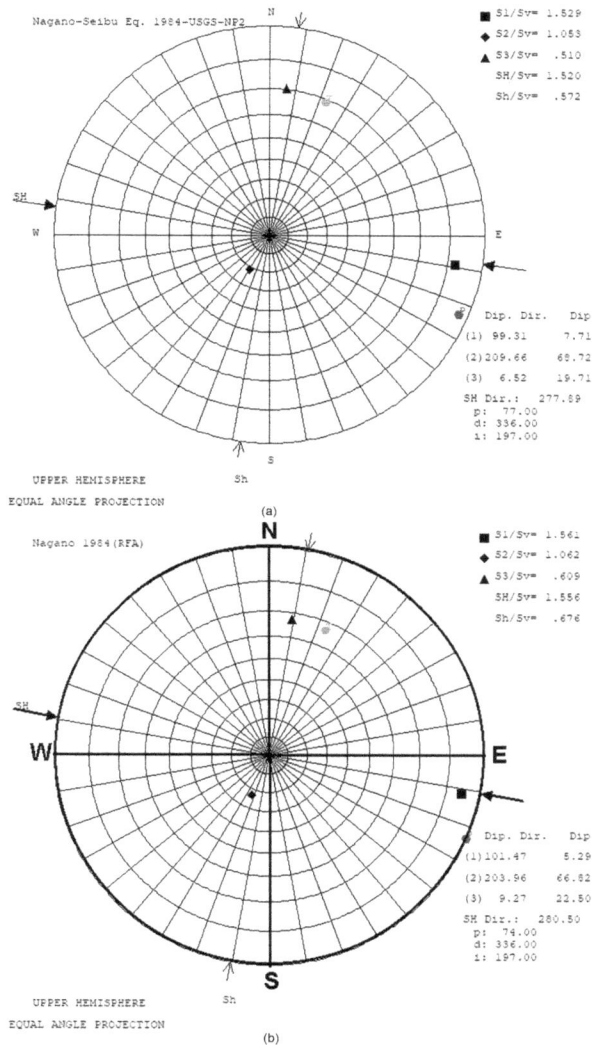

Figure 2.11 Normalized stress states obtained for (a) peak and (b) residual friction angles for the 1984 Nagano Prefecture Seibu earthquake.

Table 2.1 Stress and orientation changes for the 1984 Nagano Prefecture Seibu earthquake

Normal stress $\Delta\sigma_n / \sigma_v$	Shear stress $\Delta\tau_d / \sigma_v$	Orientation $\Delta\theta$
0.14	0.114534	2.11

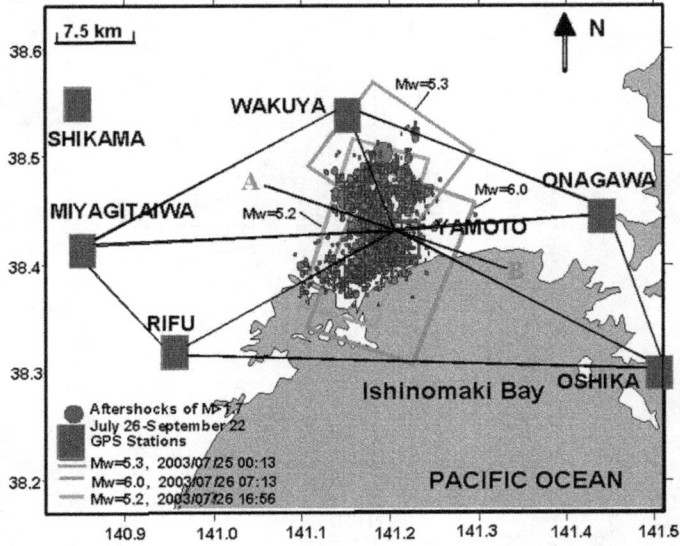

Figure 2.12 GPS stations and configuration of GPS mesh. (From Aydan (2004b).)

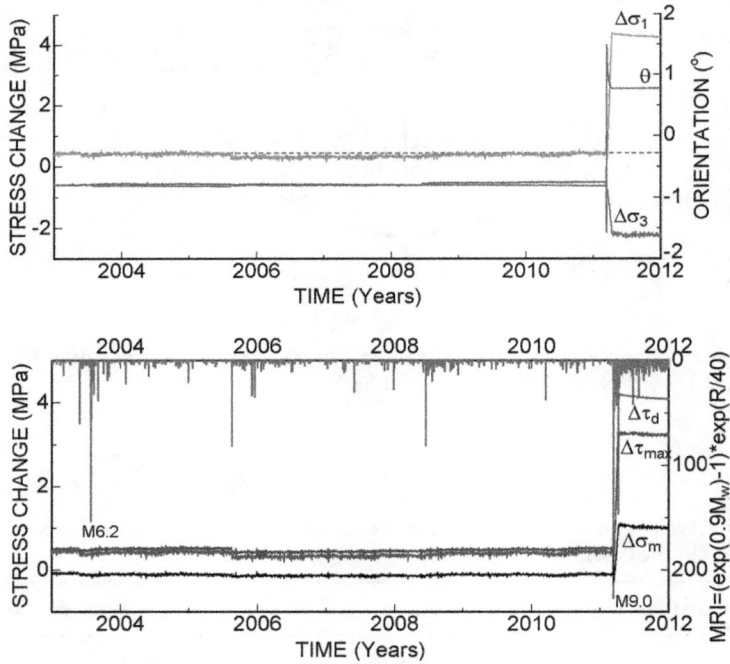

Figure 2.13 Stress changes between the beginning of 2003 and the end of 2011.

$$\text{MRI} = \left(e^{0.9M_w} - 1\right)e^{-R/40} \tag{2.15}$$

where the unit of hypocentral distance R is given in km. As seen from Figure 2.13, stress changes can occur in a wide area, and co-seismic stress changes could be evaluated.

2.4 LABORATORY EXPERIMENTS

2.4.1 Uniaxial compression experiments in relation to earthquakes

Tests on instrumented samples of various rocks such as Ryukyu limestone, tuff, granite, porphyrite, andesite, sandstone, etc. were performed (e.g. Aydan et al. 2007c, 2011; Ohta 2011; Ohta and Aydan 2010; Ohta et al. 2008). Two examples, named Fuji-TV No. 1 and Mitake Sandstone MS2, are selected for discussions herein. Fuji-TV No. 1 is a prismatic granite sample ($100 \times 100 \times 200\,\text{mm}^3$). The acceleration responses start to develop when the applied stress exceeds the peak strength and it attains the largest value just before the residual state is achieved as seen in Figure 2.14a. This pattern was observed in all experiments. Another important aspect is that the acceleration of the upper platen is much larger than that of the lower platen. This is also a common feature in all experiments. In other words, the amplitude of accelerations of the mobile part of the loading system is higher than that of the stationary part.

Mitake Sandstone MS2 sample (Height: 93 mm; Diameter: 45 mm) is a soft rock and an accelerometer was attached to the sample at the mid height. Figure 2.14b shows

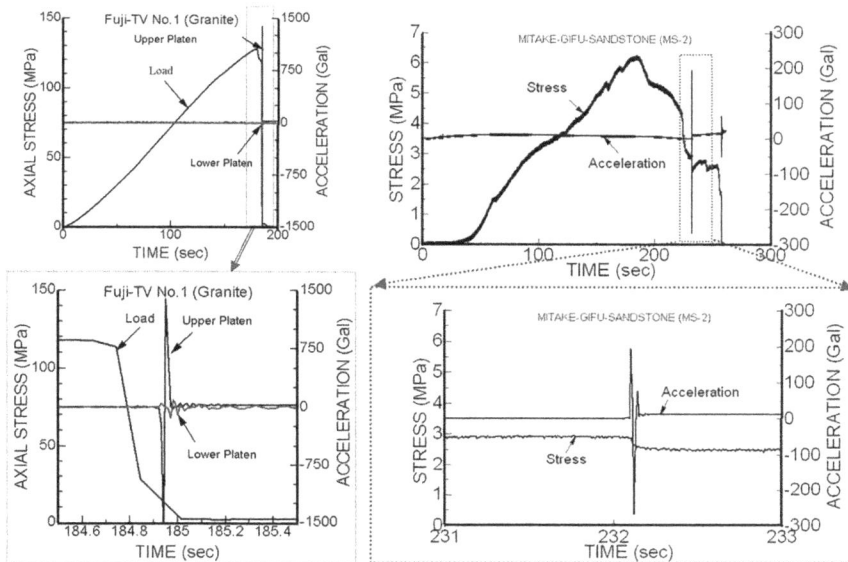

Figure 2.14 Acceleration and axial responses of a granite sample denoted (a) Fuji-TV No. 1 and (b) Mitake-Gifu sandstone (MS-2).

the axial stress and acceleration response as a function of time. The failure of this rock sample is ductile and the maximum acceleration is much less than that for the granite sample. Nevertheless, the maximum acceleration occurs just before the residual state, which is very similar to that observed during the fracturing of the hard brittle granite sample. Fundamentally, the observed acceleration responses during the fracturing of various rocks are similar to each other except their absolute values. The most striking feature is the chaotic acceleration response during the initiation and propagation of the macroscopic fracture of the sample. This chaotic response is very remarkable for the radial acceleration component in particular, and this phase is probably associated with the small fragment detachments before the final burst of rock samples. The small fragments result from splitting cracks aligned along the direction of loading before they coalesce into a large shear band. Furthermore, the audible sounds of fracturing are emitted from the rock during this phase. As shown by Aydan (2004a) and his co-workers (Aydan et al. 2007c, 2011), work done (according the definition in Continuum Mechanics) on tested samples and maximum acceleration increases proportionally to the maximum acceleration and it is always higher on the mobile part compared with that of the stationary side of loading system (Aydan 2004a). This result should probably have very important implications in many disciplines of geoscience and earthquake engineering for inferring and understanding ground motions.

Aydan (2018a,b) reported experiments on several crystals such as quartz, orthoclase, tourmaline, calcite, aragonite and gypsum utilizing a multi-parameter monitoring system, including infrared imaging techniques. During experiments, high-strength minerals such as quartz and tourmaline failed in a mode of explosion while gypsum, rock salt and aragonite failed in a very ductile manner. The experiments on quartz are only presented herein.

Responses of several measurable parameters such as stress, strain, electric potential and cumulative AE are shown in Figure 2.15 together with views of quartz crystal before and after the failure. The failure mode of the quartz crystal was in the form of an explosion and the remains of the sample after the experiment was like a powder, as seen in the same figure. The maximum acceleration was 13 times the gravitational acceleration. Distinct variations of various measurable parameters such as electric potential, acoustic emission besides load and displacement were observed during the deformation and fracturing processes.

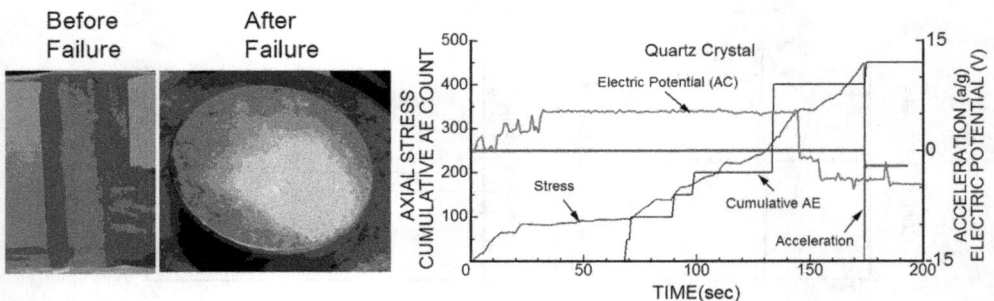

Figure 2.15 Multi-parameter responses of quartz crystal sample and its view before and after failure.

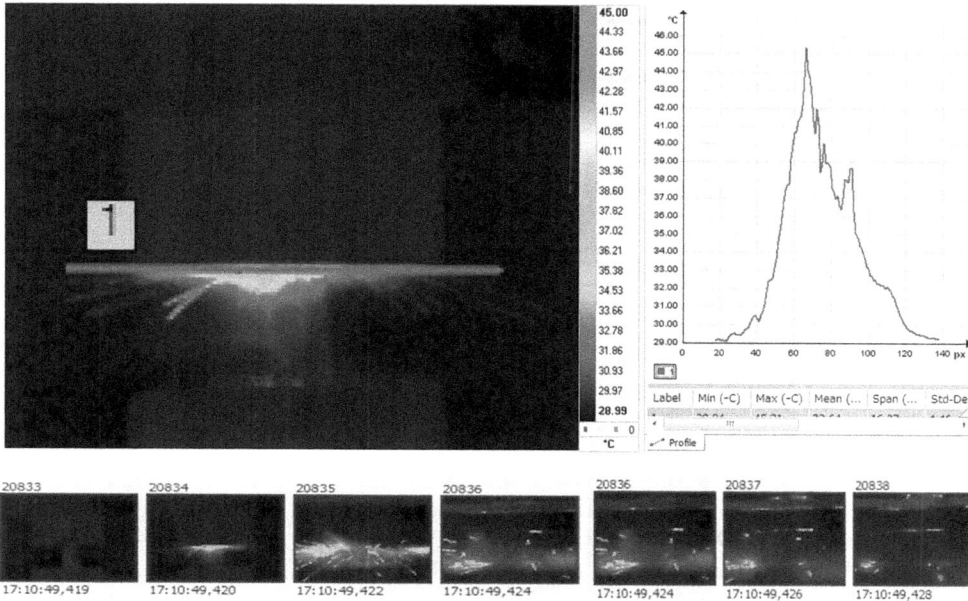

Figure 2.16 (a) An infrared thermograph image of the quartz crystal during failure and temperature rise along the selected line, and (b) sequential infrared thermographic images during fracture process.

An infrared thermographic image and temperature response along the selected line of the quartz crystal sample during the initiation of failure are shown in Figure 2.16. The maximum temperature difference was observed almost at the centre of the top surface of the crystal and the temperature difference was about 16°C from the ambient temperature. Furthermore, one can easily notice the ejection trajectories of some fragments from the failing quartz sample.

2.4.2 Stick-slip phenomenon for simple mechanical explanation of earthquakes, and some experiments

Brace and Byerlee (1966) and Byerlee (1970) were first to suggest the stick-slip phenomenon as a possible explanation of the mechanics of earthquake occurrence. Nevertheless, the stick-slip phenomenon is well known in the field of tribology (e.g. Bowden and Leben 1939; Bowden and Tabor 1950; Jaeger and Cook 1979).

2.4.2.1 A simple theory of the stick-slip phenomenon

In this model, the basal plate is assumed to be moving with a constant velocity v_m, and an overriding block is assumed to be elastically supported by the surrounding medium, as illustrated in Figure 2.17. The basic concept of modelling assumes that

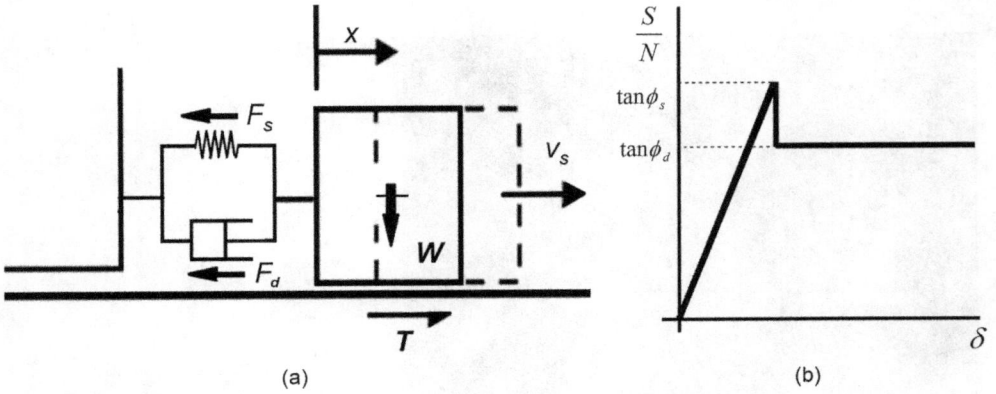

Figure 2.17 (a) Mechanical modelling of stick-slip phenomenon and (b) constitutive model.

the relative motion between the basal plate and overriding block is divergent. Let us assume that the motion of the plate can be modelled as a stick-slip phenomenon (e.g. Bowden and Leben 1939; Jaeger and Cook 1979; Ohta and Aydan 2010; Aydan 2017; Aydan et al. 2021). The governing equation and its closed form solution are described as follow

During the stick phase, the following holds

$$\dot{x} = v_s, \quad F_s = k \cdot x \tag{2.16}$$

where v_s is the velocity of the moving base and k is the stiffness of the system. The initiation of slip obeys the friction law as given below (Figure 2.18):

$$F_y = \mu_s N \tag{2.17}$$

where μ_s is the static friction coefficient and N is the normal force. For the block shown in Figure 2.16, the normal force is equal to block weight W and it is related to its mass m and gravitational acceleration g through m_g. During slip phase, the force equilibrium is

$$-kx + \mu_k W = m\frac{d^2 x}{dt^2} \tag{2.18}$$

where μ_k is the dynamic friction angle. The solution of the equation above can be obtained as

$$x = A_1 \cos\Omega t + A_2 \sin\Omega t + \mu_k \frac{W}{k} \tag{2.19}$$

If initial conditions ($t = t_s$, $x = x_s$ and $\dot{x} = v_s$) are introduced in Eq. (2.19), integration constants A_1 and A_2 are obtained as follows:

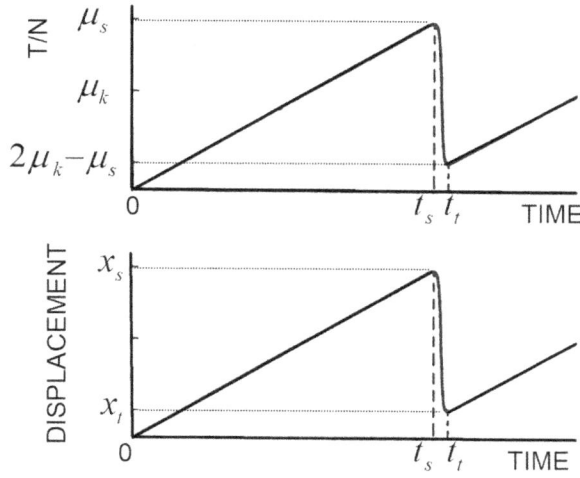

Figure 2.18 Frictional forces during a stick-slip cycle.

$$x = \frac{W}{k}(\mu_s - \mu_k)\cos\Omega(t - t_s) + \frac{v_s}{\Omega}\sin\Omega(t - t_s) + \mu_k\frac{W}{k}$$
$$\dot{x} = -\frac{W}{k}(\mu_s - \mu_k)\Omega\sin\Omega(t - t_s) + v_s\cos\Omega(t - t_s) \quad (2.20)$$
$$\ddot{x} = -\frac{W}{k}(\mu_s - \mu_k)\Omega^2\cos\Omega(t - t_s) - v_s\Omega\sin\Omega(t - t_s)$$

where $\Omega = \sqrt{k/m}$ and $x_s = \mu_s\dfrac{W}{k}$.

At $t = t_t$, velocity becomes equal to the velocity of the moving base, which is given as $\dot{x} = v_s$. This yields the period of slip phase as follows:

$$t_t = \frac{2}{\Omega}\left(\pi - \tan^{-1}\left(\frac{(\mu_s - \mu_k)W\Omega}{k \cdot v_s}\right)\right) + t_s \quad (2.21)$$

where $x_s = v_s \cdot t_s$. The rise time, which is slip period, is given by

$$t_r = t_t - t_s \quad (2.22)$$

Rise time can be specifically obtained from Eqs. (2.21) and (2.22) as

$$t_r = \frac{2}{\Omega}\left(\pi - \tan^{-1}\left(\frac{(\mu_s - \mu_k)W\Omega}{k \cdot v_s}\right)\right) \quad (2.23)$$

If the velocity of the moving base is negligible, that is, $v_s \approx 0$, the rise time (t_p) reduces to the following form:

$$t_r = \pi \sqrt{\frac{m}{k}} \qquad (2.24)$$

The amount of slip is obtained as

$$x_r = |x_t - x_s| = 2\frac{W}{k}(\mu_s - \mu_k) \qquad (2.25)$$

The force drop during slip is given by

$$F_d = 2(\mu_s - \mu_k)W \qquad (2.26)$$

It should be noted that the formulation given above does not consider the damping associated with slip velocity. If the damping resistance is linear, the governing Eq. (2.18) takes the following form:

$$-kx - \eta\dot{x} + \mu_k W = m\frac{d^2 x}{dt^2} \qquad (2.27)$$

Static and dynamic friction angles from a typical slip phase as shown in Figure 2.18 can be determined from Eq. (2.26). Rearranging Eq. (2.26) yields the following relation for dynamic friction angle as

$$\phi_k = \tan^{-1}\left(\mu_s - \frac{1}{2}\frac{F_d}{N}\right) \qquad (2.28)$$

It should be noted that the increase in normal load on the blocks would increase the amount of slip, and decrease the velocity during slip and prolong the period of stick-slip cycles.

2.4.2.2   Device of stick-slip tests

Figure 2.19 shows a view of the experimental device. The experimental device consists of an endless conveyor belt and a fixed frame. The inclination of the conveyor belt can be varied so that tangential and normal forces can be easily imposed on the sample as desired. To study the actual frictional resistance of interfaces of rock blocks, the lower block is stuck to a rubber belt while the upper block is attached to the fixed frame through a spring as illustrated in Figure 2.17a. Some experiments were conducted using the rock samples of granite with planes having different surface morphologies. The base blocks were 200–400 mm long, 100–150 mm wide and 40–100 mm thick. The upper block was 100–200 mm long, 100 mm wide and 50–100 mm thick.

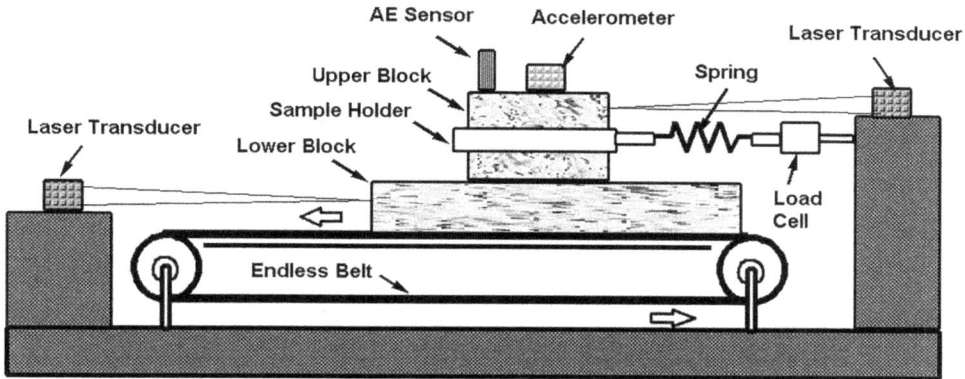

Figure 2.19 Stick-slip experimental setup.

When the upper block moves together on the base block at a constant velocity (stick phase), the spring is stretched at a constant velocity. The shear force increases to a critical value and then a sudden slip occurs with an associated spring force drop. Because the instability sliding of the upper block occurs periodically, the upper block slips violently over the base block. Normal loads can also be easily increased in experiments.

To measure the frictional force acting on the upper block, the load cell (KYOWA LUR-A-200NSA1) is installed between spring and fixed frame. During experiments, the displacement of the block is measured through a laser displacement transducer produced by KEYENCE or OMRON and a contact type displacement transducer with a measuring range of 70 mm, while the acceleration responses parallel and perpendicular to the belt movement are measured by a three-component accelerometer (TOKYO SOKKI) attached to the upper block. The measured displacement, acceleration and force are recorded on laptop computers.

2.4.2.3   Stick-slip experiment

Many stick-slip experiments were performed on various natural rock blocks as well as other types of blocks made of foam, plastic, wood and aluminium (e.g. Aydan 2003a; Aydan et al. 2011, 2019a; Ohta and Aydan 2010). Here some experimental results on discontinuities in granite are quoted as examples. Three different combinations of the surface roughness conditions of granite blocks were investigated while keeping the system stiffness, upper block weight and base velocity constant. These combinations are rough to rough (tension joint), rough to smooth (saw-cut) and smooth to smooth interfaces. Measured responses of a discontinuity with the rough to rough combination are shown in Figures 2.20. As noted from Figure 2.20, the velocity of the upper block starts to change before the slippage.

Figure 2.21 shows a series of stick-slip experiments on discontinuities in Ryukyu limestone together with interpretations of peak and residual friction angles. The peak (static) friction angle can be evaluated from the T/N response while the residual (kinetic) friction angle is obtained from the theoretical relation (2.26).

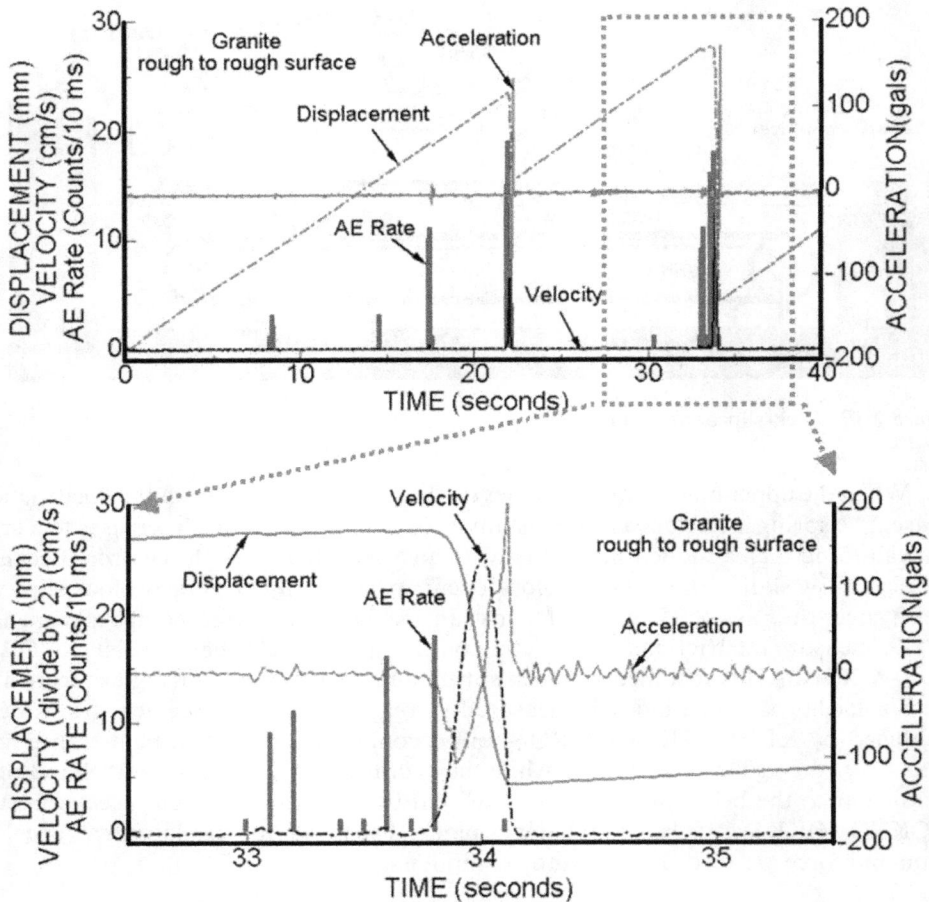

Figure 2.20 The response of a discontinuity surface during a stick-slip experiment.

Some tilting experiments were carried out on the same discontinuity planes (Aydan et al. 2019a, 2021). Peak (static) friction angle for both discontinuity plane obtained from tilting tests and stick-slip experiments were very close to each other. The residual or kinetic friction angles for rough discontinuity plane of granite are also very close to each other. Similarly, the residual (kinetic) friction angle of saw-cut discontinuity plane of Ryukyu limestone obtained from stick-slip experiments are very close to those obtained from tilting experiments. Nevertheless, the kinetic or residual friction angle is generally lower than those obtained from the tilting experiments. As expected from the theoretical formulas (2.23 and 2.24) derived in the previous section, the phenomenon would be periodic. If the peak and residual friction angles are the same, the slip would be continuous with a given velocity, which may be the fundamental explanation of fault-creep observed in some segments of North Anatolian Fault in Turkey and San Andreas Fault in the United States.

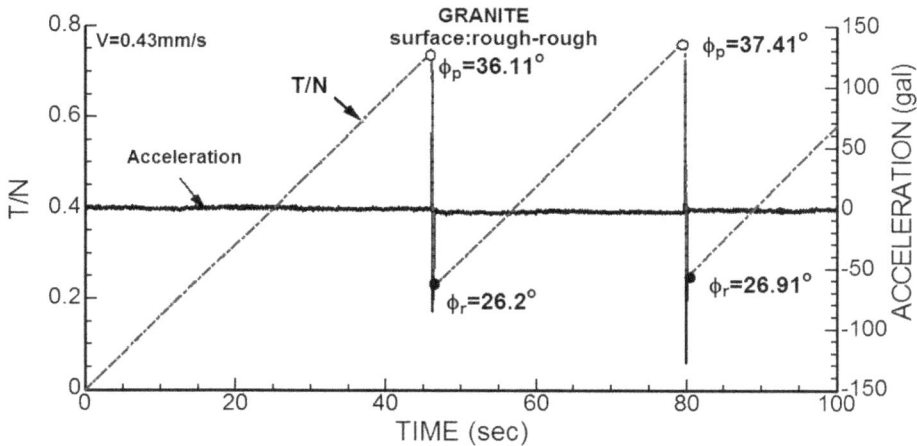

Figure 2.21 Stick-slip response of saw-cut plane of Ryukyu limestone.

2.5 RELATIONS BETWEEN EARTHQUAKES AND VOLCANIC ERUPTIONS

It is often speculated that volcanic eruptions cause earthquakes or vice versa. Without any doubt, earthquakes are temporary instabilities in the Earth's crust, and the dissipation of accumulated mechanical energy occurs as earthquakes. Furthermore, there are some areas where creep-like deformation of the Earth's crust takes place. On the other hand, the intrusion of magma into the crust decreases the strength of the crust as well as the stress state in the Earth's crust is altered. This issue is discussed herein.

2.5.1 Observations

It is always speculated that volcanic eruptions are triggered following large earthquakes or vice versa. Three examples are selected for illustrative purposes. The first example is related to Mt. Fuji, which erupted in 864 AD (Aokihara Lava Plain), and in 864 AD (M8.4–9.0) the Jogan earthquake occurred, whose epicentre is almost the same as of the GEJE. Before the 1707 Hoei eruption (December 16, 1707) of Mt. Fuji, the Genroku (Mw 8.2) (epicentre is about 100 km east) earthquake occurred on December 31, 1703, and the Hoei (Mw 8.7) earthquake occurred on October 28, 1707, and this earthquake caused a huge slope failure, which is known as the third largest slope failure in Japan, named Oya-kuzure. The Hoei volcanic eruption occurred about 49 days after the Hoei earthquake.

The second example is related to the Valdivia earthquake on May 22, 1960, and the Cordón Caulle volcano that erupted on May 24, 1960. There are some volcanic eruptions of Planchón-Peteroa and Tupungatito in July 1960, the Calbuco volcano in February 1961 (8 months after the earthquake) and the Villarrica volcano (July 2, 1960, 1961 and February 25, 1963) following the 1960 earthquake.

The third example is related to the December 26, 2005, Aceh (M9.0–9.3) and the March 28, 2005, Nias (M8.5–8.7) earthquakes along Sumatra subduction in Indonesia. Before the earthquakes, the Kerinci volcano (about 500–900 km SE) erupted during June 16–22, 2004. Following the earthquakes, the Talang volcano (450–790 km SE) erupted April 10, 2005, and the Barren Island Volcano erupted on May 28, 2005. Sinabung volcano, which is the nearest volcano to the 2004 and 2005 off-Sumatra earthquakes, erupted on September 15, 2013, and since then its activity has been continuing.

A common feature of such associations between volcanoes and earthquakes are strato-volcanoes on volcanic arcs, where the upper plate ruptures to allow access to the rising magma. Interplate earthquakes that occur along the subduction zones are associated with the release of stored strain and allow more than 10 m relative slip probably causing expansive volumetric strain, which may be quite large along volcanic arcs and it may cause additional rupturing in the volcanic arc. Besides these mechanical deformations, it must be noted that the release of the stored mechanical energy would be converted to heat as explained in the next section and it is experimentally confirmed (e.g. Aydan et al. 2011, 2015). The heat production inducing high temperatures would decrease the strength of the Earth's crust as well as cause the melting of surrounding rock.

2.5.2 Mechanical background of heat emission during crustal deformation

2.5.2.1 Fundamental governing equation for energy conservation law

The well-known governing equation of energy conservation in media is given in the following form (e.g. Aydan 2016b, 2021; Eringen 1980):

$$\rho c \frac{\partial T}{\partial t} = -\nabla \cdot \left(k \nabla T \right) + \sigma \cdot \dot{\varepsilon} + Q_h \tag{2.29}$$

where $\dot{\varepsilon}$ is strain tensor, k is thermal conductivity, σ is stress tensor, Q is energy generated per unit volume per unit time, c is specific heat capacity and T is temperature.

2.5.2.2 Temperature distribution in the vicinity of geological active faults

As a first case, a geological fault is assumed to be sandwiched between two nonconductive rock slabs, and closed form solutions are derived for temperature rises within the fault. Then, a more general case is considered such that a seismic energy release is assumed to occur within the fault, and adjacent rock is conductive. The solution of the governing equation for this case is solved with the use of the finite element method. Several examples were solved by considering hypothetical energy release functions, and their implications are discussed.

A geological fault and its close vicinity may be simplified to a one-dimensional situation as shown in Figure 2.22 by assuming that mechanical energy release is due purely to shearing with no heat production source. Thus, Eq. (2.29) may be reduced to the following form:

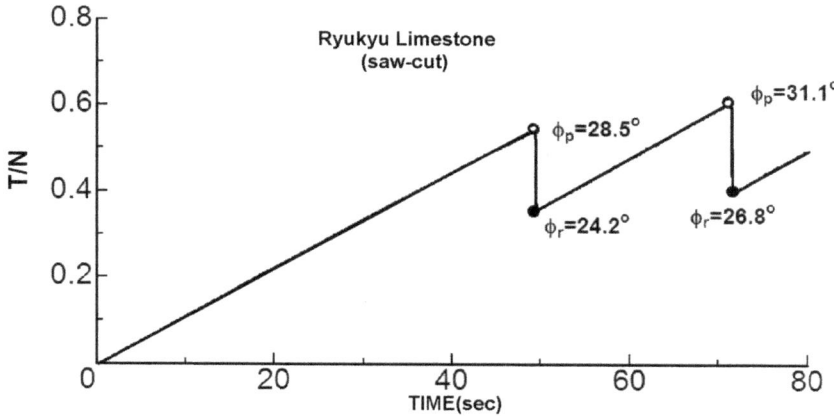

Figure 2.22 Fault model.

$$\rho c \frac{\partial T}{\partial t} = -\nabla q + \tau \dot{\gamma} \tag{2.30}$$

Let us assume that the heat flux obeys Fourier's law, which is given by

$$q = -k \frac{\partial T}{\partial x} \tag{2.31}$$

Inserting Eq. (2.31) into Eq. (2.30) yields the following equation:

$$\rho c \frac{\partial T}{\partial t} = k \frac{\partial^2 T}{\partial x^2} + \tau \dot{\gamma} \tag{2.32}$$

The solution of the above equation will yield the temperature variation with time.

Energy release during earthquakes is a very complex phenomenon. Nevertheless, some simple forms relevant for the overall behaviour may be assumed in order to have some insight to the phenomenon. Two energy release rate functions of the following form are assumed as given below:

$$\dot{E} = \tau \dot{\gamma} = A t e^{-\frac{t}{\theta}} \tag{2.33}$$

$$\dot{E} = \tau \dot{\gamma} = A^* e^{-\frac{t}{\theta^*}} \tag{2.34}$$

Constants A and A^* depend on the shear stress and shear strain rate history with time and fault thickness. Constants θ and θ^* are time history constants. For a situation illustrated in Figure 2.23, constants A and A^* will take the following forms:

$$\text{For Eq. (2.33)} \quad A = \frac{\tau_o u_f}{h \theta^2} \tag{2.35}$$

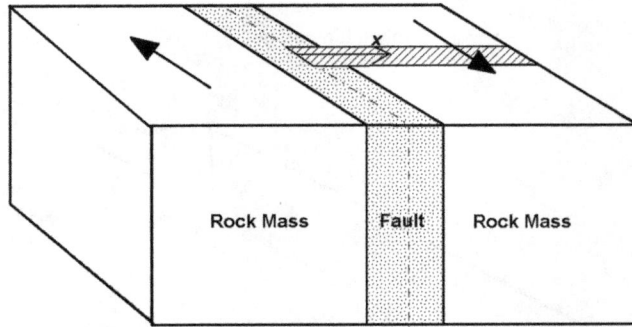

Figure 2.23 Faulting models and energy release types.

For Eq. (2.34) $A^* = \dfrac{\tau_o u_f}{h \theta^*}$ (2.36)

where u_f, h are final relative displacement and thickness of the fault. τ_o is the shear stress acting on the fault and it is assumed to be constant during the motion.

Two specific situations are analysed, namely:

1. Creeping Fault
2. Fault with hill-shaped seismic energy release rate

In the case of creeping fault, the energy release rate is almost constant with time. The geometry of the fault is assumed to be one-dimensional as shown in Figure 2.23. Figures 2.24 and 2.25 show the computed temperature differences at selected locations with time and temperature difference distribution throughout the whole domain at selected time steps. In the computations, the energy release is assumed to be taking place within the fault zone only. The increase of temperature difference is parabolic, and it keeps increasing with time. Nevertheless, the temperature difference increases are about 1/10 of those of the fault sandwiched between nonconductive rock mass slabs.

Figures 2.26 and 2.27 show the computed temperature differences at selected locations with time and temperature difference distribution throughout the whole domain at selected time steps for a fault with a hill-like energy release rate. In the computations, the energy release rate is assumed to be taking place within the fault zone only. The increase in temperature difference is parabolic. Temperature difference increases first, and then it tends to decay with time in a similar manner to the assumed seismic energy release rate function. This situation is probably quite similar to the actual situation in nature. The temperature difference increases are about 1/10 of those of the fault sandwiched between nonconductive rock mass slabs. These results indicate that the observation of ground temperatures may be a highly valuable source of information in the predictions of earthquakes. Atmospheric temperature measurements near the ground surface may be quite problematic in interpreting the observations. However, the observation of hot-spring temperature, which reflects the actual ground temperature, may be a very good indicative tool for such measurements without any deep boring.

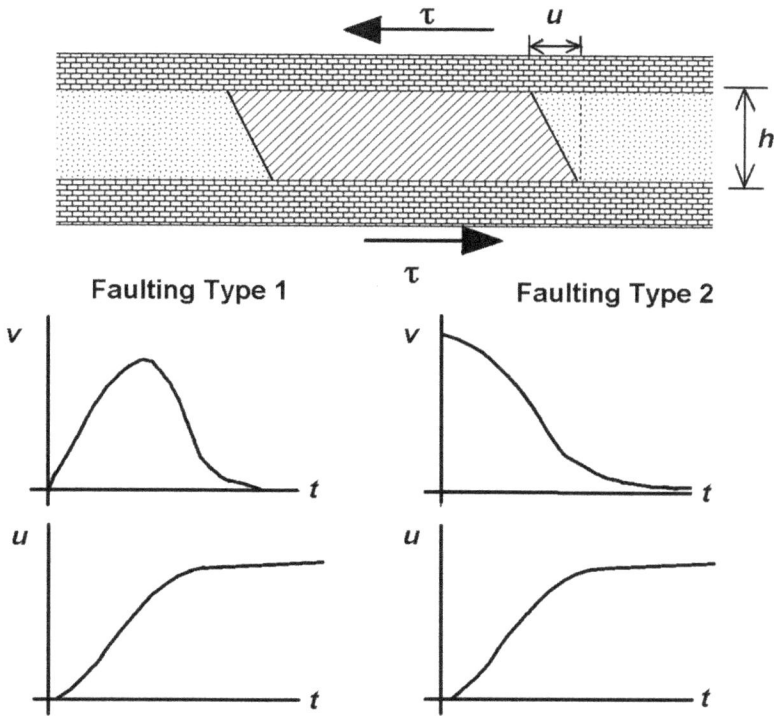

Figure 2.24 Temperature difference variations for a fault sandwiched between conductive rock mass slabs for creeping condition.

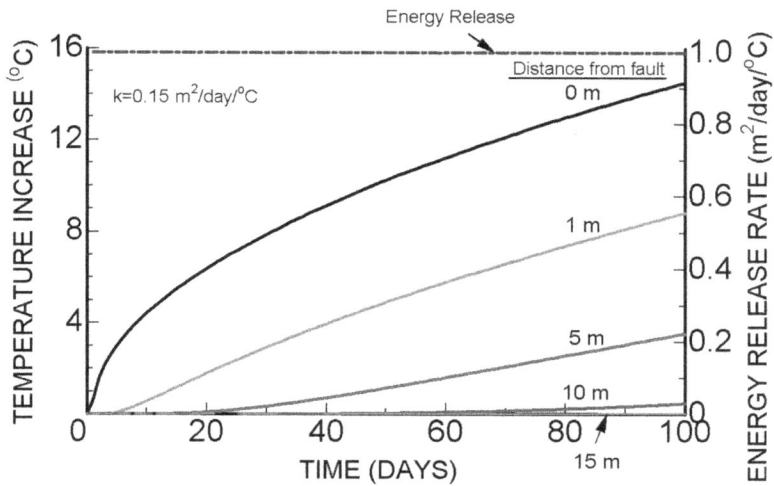

Figure 2.25 Temperature distributions at different time steps for a fault sandwiched between conductive rock mass slabs for creeping condition.

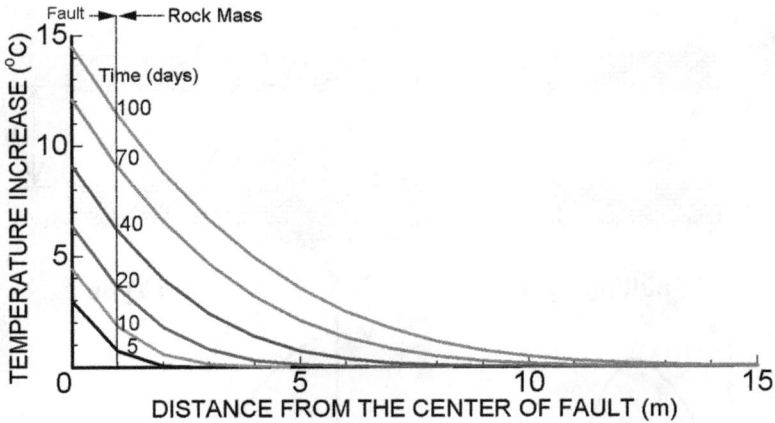

Figure 2.26 Temperature difference variations for a fault sandwiched between conductive rock mass slabs for hill-shaped energy release function.

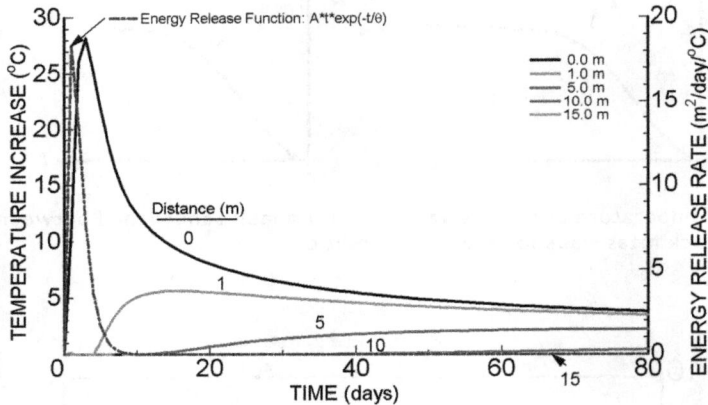

Figure 2.27 Temperature distributions at different time steps for a fault sandwiched between conductive rock mass slabs for hill-shaped energy release function.

2.5.3   Strength reduction due to temperature increase

In geomechanics, there is almost no yield (failure) criterion incorporating the effect of temperature on yield (failure) properties of rocks, although there are some experimental researches (Hirth and Tullis 1994). Aydan (1995a) proposed a yield function for the thermoplasticity yielding of rock as given by

$$\sigma_1 = \sigma_3 + \left[S_\infty - \left(S_\infty - \sigma_c \right) e^{-b_1 \sigma_3} \right] e^{-b_2 T} \tag{2.37}$$

where S_∞ is the ultimate deviatoric strength, while coefficients b_1, b_2 are empirical constants. This yield criterion implies that the strength of rock and rockmass decrease with increase of temperature. Aydan's criterion is the only criterion known to incorporate

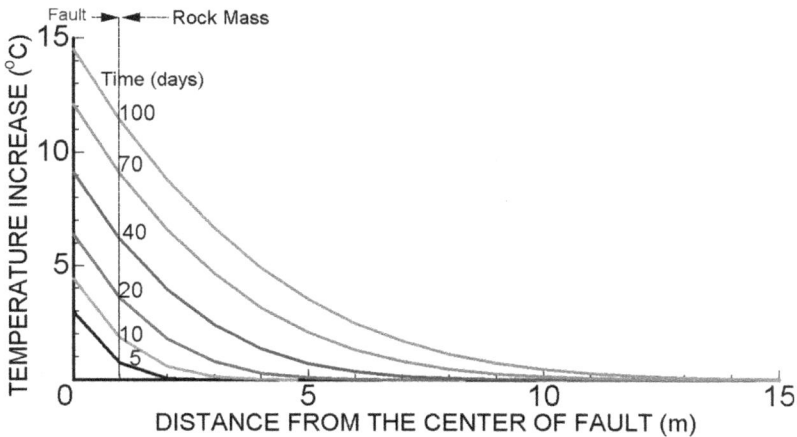

Figure 2.28 Experimental results of Hirth and Tullis (1994) on quartz for three different ambient temperatures.

the temperature, and it was used to study the stress state of the Earth (Aydan 1995a). As noted from the previous section, temperature increases in the close vicinity faults and the Earth's crust.

Figure 2.28 shows the experimental results for three different values of ambient temperature reported by Hirth and Tullis (1994), while Figure 2.29 shows the reduction in strength with temperature for a given confining pressure of 1.17–1.2 GPa.

Aydan's yield (failure) criterion is applied to experimental results shown in Figures 2.28 and 2.29, and the results are shown in Figure 2.30.

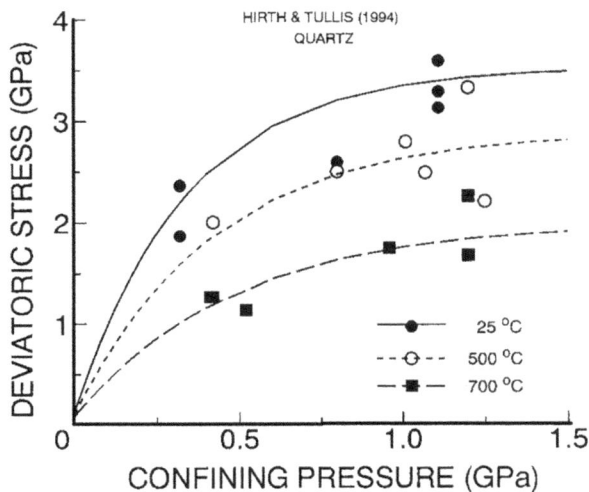

Figure 2.29 The reduction of deviatoric strength of quartz as a function of temperature for a confining pressure of 1.17–1.2 GPa.

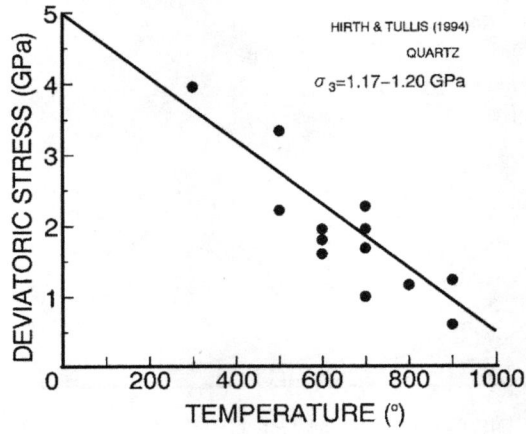

Figure 2.30 A three-dimensional representation of Aydan's failure criterion for the experimental results of Hirth and Tullis (1994).

Chapter 3

Waves and theory of wave propagation

3.1 MOMENTUM CONSERVATION LAW

For any structure or continuum body subjected to motion, the momentum conservation law should hold. Momentum is defined in integral form as given below (e.g. Eringen 1980; Mase 1970; Aydan 2017, 2021):

$$\mathbf{p} = \int_{\Omega} \rho \mathbf{v} d\Omega \tag{3.1}$$

Preliminary relations

$$\int_{\Omega} \nabla \cdot \sigma d\Omega = \int_{\Gamma} \sigma \cdot \mathbf{n} d\Gamma; \quad \frac{d(d\Omega)}{dt} = (\nabla \cdot \mathbf{x}) d\Omega; \quad t = \sigma \cdot \mathbf{n} \tag{3.2}$$

Conservation of momentum is written in the following form in view of Eq. (3.2), which is also known as Reynolds transport theorem (Figure 3.1):

$$\frac{d}{dt} \int_{\Omega} \rho \mathbf{v} d\Omega = \int_{\Gamma} \mathbf{t} d\Gamma + \int_{\Omega} \mathbf{b} d\Omega \tag{3.3}$$

Equation (3.3) may be rewritten as

$$\int_{\Omega} \left(\frac{d(\rho \mathbf{v})}{dt} + (\rho \mathbf{v}) \nabla \cdot \mathbf{v} \right) d\Omega = \int_{\Omega} \nabla \cdot \sigma d\Omega + \int_{\Omega} \mathbf{b} d\Omega \tag{3.4}$$

Carrying out the derivation in Eq. (3.4), we have the following:

$$\int_{\Omega} \left(\left(\frac{d\rho}{dt} + \rho(\nabla \cdot \mathbf{v}) \right) \mathbf{v} \right) d\Omega + \int_{\Omega} \rho \frac{d\mathbf{v}}{dt} d\Omega = \int_{\Omega} (\nabla \cdot \sigma + \mathbf{b}) d\Omega \tag{3.5}$$

The first term on the left-hand side disappears by virtue of mass conservation law and takes the following form:

$$\int_{\Omega} \rho \frac{d\mathbf{v}}{dt} d\Omega = \int_{\Omega} (\nabla \cdot \sigma + \mathbf{b}) d\Omega \tag{3.6}$$

DOI: 10.1201/9781003164371-3

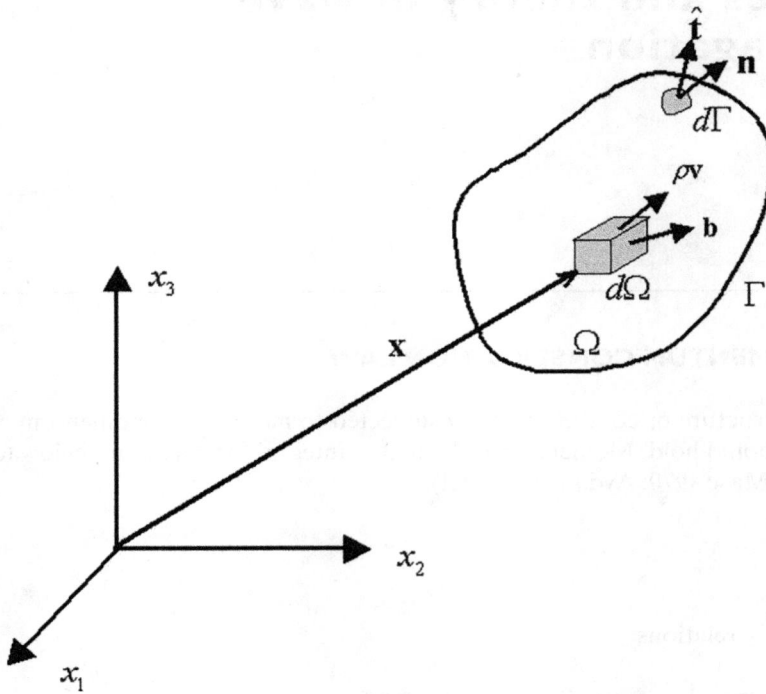

Figure 3.1 Illustration of momentum conservation law. (From Aydan (2017).)

Equation (3.6) may be rewritten as

$$\int_{\Omega}\left[\rho\frac{d\mathbf{v}}{dt}-(\nabla\cdot\boldsymbol{\sigma}+\mathbf{b})\right]d\Omega = 0 \tag{3.7}$$

To satisfy Eq. (3.7), the integrand should be zero so that we have the following relation:

$$\rho\frac{d\mathbf{v}}{dt}=\nabla\cdot\boldsymbol{\sigma}+\mathbf{b} \tag{3.8}$$

Furthermore, the derivation on the left-hand side may be related to acceleration or displacement vectors as follows:

$$\frac{d\mathbf{v}}{dt}=\mathbf{a} \qquad \frac{d\mathbf{v}}{dt}=\frac{d^2\mathbf{u}}{dt^2} \tag{3.9}$$

3.2 EARTHQUAKE-INDUCED WAVES

It is known that earthquakes cause fundamentally two types of waves (Figure 3.2). The first type of waves are body waves, and they are named P-waves and S-waves. P-waves

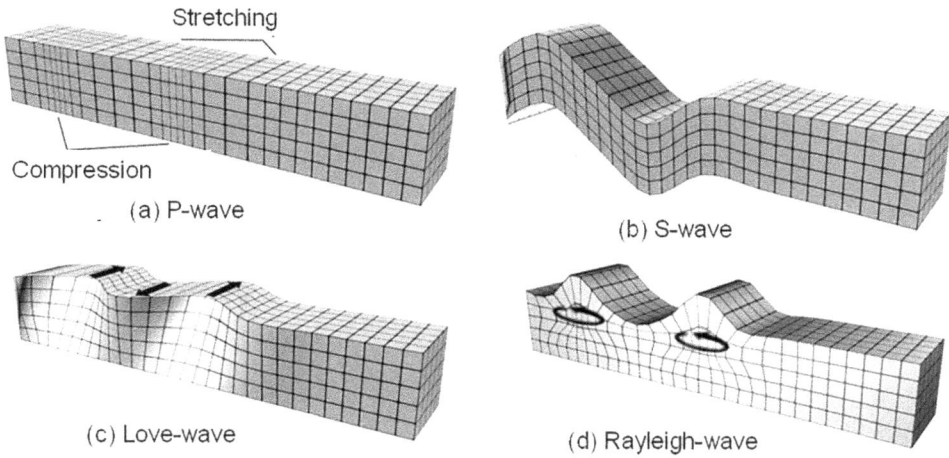

Figure 3.2 Illustration of wave types. (Modified from Sayısal Grafik (2004).)

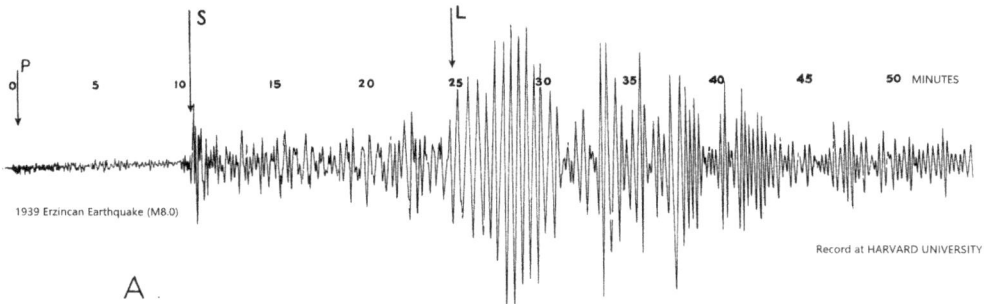

Figure 3.3 The seismogram of the 1939 Erzincan earthquake from Harvard University. (Modified from Ketin (1973).)

or primary waves pass through every material. S-waves or secondary waves arrive at a given observation point after the P-waves. The second type of waves are surface waves and they are further subdivided into Rayleigh and Love waves. Figure 3.3 shows a record of the 1939 Erzincan earthquake taken from Harvard University. It is known that shear waves are not transmitted through materials in liquid phase. Surface waves (Rayleigh and Love) are observed near Earth's surface and they disappear as the depth increases.

The equation of motion given by Eq. (3.8) can be rewritten in the following form:

$$\frac{\partial \sigma_{ij}}{\partial x_j} + b_i = \rho \frac{\partial^2 u_i}{\partial t^2}$$

(3.10a)

or specifically

$$\frac{\partial \sigma_{11}}{\partial x_1} + \frac{\partial \sigma_{12}}{\partial x_2} + \frac{\partial \sigma_{13}}{\partial x_3} + b_1 = \rho \frac{\partial^2 u_1}{\partial t^2} \tag{3.10b}$$

$$\frac{\partial \sigma_{12}}{\partial x_1} + \frac{\partial \sigma_{22}}{\partial x_2} + \frac{\partial \sigma_{23}}{\partial x_3} + b_2 = \rho \frac{\partial^2 u_2}{\partial t^2} \tag{3.10c}$$

$$\frac{\partial \sigma_{13}}{\partial x_1} + \frac{\partial \sigma_{23}}{\partial x_2} + \frac{\partial \sigma_{33}}{\partial x_3} + b_3 = \rho \frac{\partial^2 u_3}{\partial t^2} \tag{3.10d}$$

Normal strain components are related to components of displacement vector if infinitesimal strain approach is adopted

$$\varepsilon_{ij} = \frac{1}{2} \left(\frac{\partial u_i}{\partial x_j} + \frac{\partial u_j}{\partial x_i} \right) \tag{3.11a}$$

or specifically

$$\varepsilon_{11} = \frac{\partial u_1}{\partial x_1}; \quad \varepsilon_{22} = \frac{\partial u_2}{\partial x_2}; \quad \varepsilon_{33} = \frac{\partial u_3}{\partial x_3} \tag{3.11b}$$

Engineering shear strains are related to the strain tensor components as given below:

$$\gamma_{ij} = 2\varepsilon_{ij} \quad \text{with} \quad i \neq j \tag{3.11c}$$

or specifically

$$\gamma_{23} = \frac{\partial u_2}{\partial x_3} + \frac{\partial u_3}{\partial x_2}; \quad \gamma_{12} = \frac{\partial u_1}{\partial x_2} + \frac{\partial u_2}{\partial x_1}; \quad \gamma_{13} = \frac{\partial u_1}{\partial x_3} + \frac{\partial u_3}{\partial x_1} \tag{3.11d}$$

Rotational strains are defined as

$$\omega_1 = \frac{1}{2} \left(\frac{\partial u_3}{\partial x_2} - \frac{\partial u_2}{\partial x_3} \right); \quad \omega_2 = \frac{1}{2} \left(\frac{\partial u_1}{\partial x_3} - \frac{\partial u_3}{\partial x_1} \right); \quad \omega_3 = \frac{1}{2} \left(\frac{\partial u_2}{\partial x_1} - \frac{\partial u_1}{\partial x_2} \right) \tag{3.11e}$$

The constitute law between stress and strain, if the material is an isotopic elastic body (e.g. Mase 1970; Eringen 1980; Aydan 2017, 2019a, 2021), can be expressed as

$$\sigma_{ij} = \lambda \delta_{ij} \varepsilon_{kk} + 2\mu \varepsilon_{ij}; \quad \varepsilon_{kk} = \varepsilon_{11} + \varepsilon_{22} + \varepsilon_{33} \tag{3.12a}$$

or specifically

$$\begin{Bmatrix} \sigma_{11} \\ \sigma_{22} \\ \sigma_{33} \\ \sigma_{12} \\ \sigma_{23} \\ \sigma_{13} \end{Bmatrix} = \begin{bmatrix} \lambda + 2\mu & \lambda & \lambda & 0 & 0 & 0 \\ \lambda & \lambda + 2\mu & \lambda & 0 & 0 & 0 \\ \lambda & \lambda & \lambda + 2\mu & 0 & 0 & 0 \\ 0 & 0 & 0 & \mu & 0 & 0 \\ 0 & 0 & 0 & 0 & \mu & 0 \\ 0 & 0 & 0 & 0 & 0 & \mu \end{bmatrix} \begin{Bmatrix} \varepsilon_{11} \\ \varepsilon_{22} \\ \varepsilon_{33} \\ \gamma_{12} \\ \gamma_{23} \\ \gamma_{13} \end{Bmatrix} \tag{3.12b}$$

where λ and μ are the Lame coefficients specifically given in the following form:

$$\lambda = \frac{E\upsilon}{(1+\upsilon)(1-2\upsilon)} \quad \text{and} \quad \mu = \frac{E}{2(1+\upsilon)} \tag{3.13}$$

Let us introduce the following:

$$\Delta = \frac{\partial u_1}{\partial x_1} + \frac{\partial u_2}{\partial x_2} + \frac{\partial u_3}{\partial x_3} \tag{3.14a}$$

$$\nabla^2 = \nabla \cdot \nabla = \frac{\partial^2}{\partial x_1^2} + \frac{\partial^2}{\partial x_2^2} + \frac{\partial^2}{\partial x_3^2} \tag{3.14b}$$

Equation (3.14a) corresponds to volumetric strain while Eq. (3.14b) is called Laplacian operator. Inserting constitutive law given by Eq. (3.12) together with relations between strain and displacement components given by Eq. (3.11) into the equation of motion and differentiating Eqs. (3.11b), (3.11c) and (3.11d) with respect to x_1, x_2 and x_3, respectively, yields for each respective direction provided that elastic coefficients, density and body forces are constant as follows (e.g. Jaeger and Cook 1976):

$$(\lambda+\mu)\frac{\partial^2 \Delta}{\partial x_1^2} + \mu\nabla^2\frac{\partial u_1}{\partial x_1} = \rho\frac{\partial}{\partial x_1}\left(\frac{\partial^2 u_1}{\partial t^2}\right) \tag{3.15a}$$

$$(\lambda+\mu)\frac{\partial^2 \Delta}{\partial x_2^2} + \mu\nabla^2\frac{\partial u_2}{\partial x_2} = \rho\frac{\partial}{\partial x_2}\left(\frac{\partial^2 u_2}{\partial t^2}\right) \tag{3.15b}$$

$$(\lambda+\mu)\frac{\partial^2 \Delta}{\partial x_3^2} + \mu\nabla^2\frac{\partial u_3}{\partial x_3} = \rho\frac{\partial}{\partial x_3}\left(\frac{\partial^2 u_3}{\partial t^2}\right) \tag{3.15c}$$

Summing up relations given by Eqs. (3.15a), (3.15b) and (3.15c) results in the following partial differential equation:

$$(\lambda+\mu)\nabla^2\Delta + \mu\nabla^2\Delta = \rho\frac{\partial^2 \Delta}{\partial t^2} \quad or \quad V_p^2\nabla^2\Delta = \frac{\partial^2 \Delta}{\partial t^2} \quad or \quad V_p^2\nabla^2\varepsilon_v = \frac{\partial^2 \varepsilon_v}{\partial t^2} \tag{3.16}$$

where

$$\lambda + 2\mu = \frac{E\upsilon}{(1+\upsilon)(1-2\upsilon)} + \frac{E}{1+\upsilon} = \frac{E(1-\upsilon)}{(1+\upsilon)(1-2\upsilon)}$$

$$V_p = \sqrt{\frac{E(1-\upsilon)}{\rho(1+\upsilon)(1-2\upsilon)}}$$

$$\varepsilon_v = \Delta$$

Equation (3.16) is known as the governing equation of P-wave propagation in solids. As noted from this equation, P-wave propagation is directly related to volumetric straining. During the propagation of P-wave, solids undergo dilatational and compressive volumetric straining.

Similarly inserting constitutive law given by Eq. (3.12) together with relations between strain and displacement components given by Eq. (3.11) into the equation of motion takes the following form specifically for each respective direction:

$$(\lambda + \mu)\frac{\partial \Delta}{\partial x_1} + \mu\nabla^2 u_1 + b_1 = \rho\frac{\partial^2 u_1}{\partial t^2} \tag{3.17a}$$

$$(\lambda + \mu)\frac{\partial \Delta}{\partial x_2} + \mu\nabla^2 u_2 + b_2 = \rho\frac{\partial^2 u_2}{\partial t^2} \tag{3.17b}$$

$$(\lambda + \mu)\frac{\partial \Delta}{\partial x_3} + \mu\nabla^2 u_3 + b_3 = \rho\frac{\partial^2 u_3}{\partial t^2} \tag{3.17c}$$

The differentiation of Eqs. (3.17b) and (3.17c) with respect to x_3 and x_2 yields the following provided that elastic coefficients, density and body forces remain constant:

$$(\lambda + \mu)\frac{\partial^2 \Delta}{\partial x_3 \partial x_2} + \mu\nabla^2 \frac{\partial u_2}{\partial x_3} = \rho\frac{\partial^2}{\partial t^2}\left(\frac{\partial u_2}{\partial x_3}\right) \tag{3.18a}$$

$$(\lambda + \mu)\frac{\partial^2 \Delta}{\partial x_2 \partial x_3} + \mu\nabla^2 \frac{\partial u_3}{\partial x_2} = \rho\frac{\partial^2}{\partial t^2}\left(\frac{\partial u_3}{\partial x_2}\right) \tag{3.18b}$$

Subtracting Eq. (3.18a) from Eq. (3.18b) results in

$$(\lambda + \mu)\left(\frac{\partial^2 \Delta}{\partial x_2 \partial x_3} - \frac{\partial^2 \Delta}{\partial x_3 \partial x_2}\right) + \mu\nabla^2\left(\frac{\partial u_3}{\partial x_2} + \frac{\partial u_2}{\partial x_3}\right) = \rho\frac{\partial^2}{\partial t^2}\left(\frac{\partial u_3}{\partial x_2} - \frac{\partial u_2}{\partial x_3}\right) \tag{3.19}$$

Using the rotational strain definition given by Eq. (3.11e) and dividing Eq. (3.19) by density yields

$$\frac{\mu}{\rho}\nabla^2 \omega_1 = \frac{\partial^2 \omega_1}{\partial t^2} \tag{3.20a}$$

Using the same procedure for other directions together with rotation strain components given by Equation (3.11e), one can easily derive the following relations:

$$\frac{\mu}{\rho}\nabla^2 \omega_2 = \frac{\partial^2 \omega_2}{\partial t^2} \tag{3.20b}$$

$$\frac{\mu}{\rho}\nabla^2 \omega_3 = \frac{\partial^2 \omega_3}{\partial t^2} \tag{3.20c}$$

Equation (3.20) is the governing equation of Rayleigh waves. The coefficient in Eq. (3.20) is interpreted as the propagation velocity of rotational waves:

$$V_s = \sqrt{\frac{\mu}{\rho}} \quad \text{or} \quad V_s = \sqrt{\frac{E}{2\rho(1+\upsilon)}} \tag{3.21}$$

If volumetric strain (Δ) is nil and body force is negligible, one easily gets the following expressions from Eq. (3.17):

$$\frac{\mu}{\rho}\nabla^2 u_1 = \frac{\partial^2 u_1}{\partial t^2} \quad \text{or} \quad V_s\nabla^2 u_1 = \frac{\partial^2 u_1}{\partial t^2} \tag{3.22a}$$

$$\frac{\mu}{\rho}\nabla^2 u_2 = \frac{\partial^2 u_2}{\partial t^2} \quad \text{or} \quad V_s\nabla^2 u_2 = \frac{\partial^2 u_2}{\partial t^2} \tag{3.22b}$$

$$\frac{\mu}{\rho}\nabla^2 u_3 = \frac{\partial^2 u_3}{\partial t^2} \quad \text{or} \quad V_s\nabla^2 u_2 = \frac{\partial^2 u_2}{\partial t^2} \tag{3.22c}$$

Equation (3.22) is the fundamental equation of distortion (shear) waves known as S-waves. It should be noted that the propagation velocity of S-waves is the same as that of Rayleigh waves.

3.3   WAVE PROPAGATION IN A POND

Figure 3.4 shows wave propagation in a pond with a single source and two separated sources of disturbances. In this particular case, the wave propagation obeys Eq. (3.16) without shear wave propagation. As noted from Figure 3.4, the induced waves propagate co-centrically. Regarding the wave propagation associated with the double sources of disturbances, the waves start to interfere with each other after a short period of time and then the wave propagation obeys the superposed waves emanating from two sources thereafter and looks elliptical. The P-waves induced by the rupture of a single asperity and the simultaneous rupture of double asperities during an earthquake would be quite similar to those seen in Figure 3.4.

3.4   WAVE REFRACTION

Let us consider a two-layered medium as shown in Figure 3.5. S denotes the source geophone and R denotes the receiver geophone. It is assumed that the wave velocity V_1 of layer 1 is smaller than that of layer 2 ($V_2 > V_1$). For this particular situation, there will be numerous wave paths. Among them, three wave paths would be of particular importance. These wave paths are called direct wave path (S-R), reflected wave path (S-C-R) and refracted wave path (S-A-B-R).

If the medium is assumed to be elastic and its density remains to be the same within the layer, Snell's law holds for incidence angle and refraction angle as given below:

$$\frac{\sin i}{\sin r} = \frac{V_1}{V_2} \tag{3.23a}$$

The refraction angle r of the refracted wave path shown in Figure 3.5 is 90°. Therefore, the critical incidence wave angle i_c can be easily obtained from Eq. (3.23a) as follows:

$$\sin i_c = \frac{V_1}{V_2} \tag{3.23b}$$

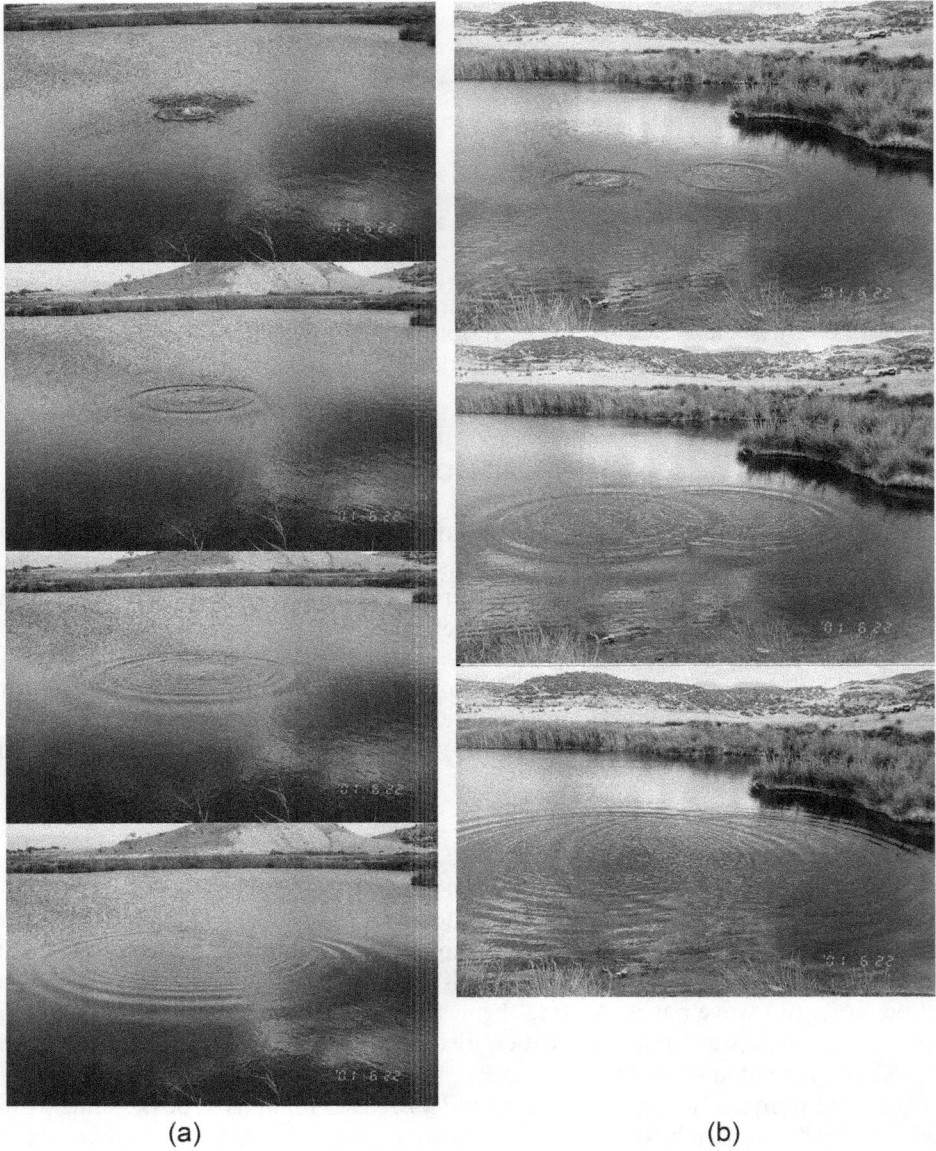

Figure 3.4 Wave propagations in a pond. (a) Single source disturbance and (b) double sources of disturbance.

One can easily write the following relation between distance and arrival time for the direct wave path as

$$t = \frac{X}{V_1}$$

(3.24)

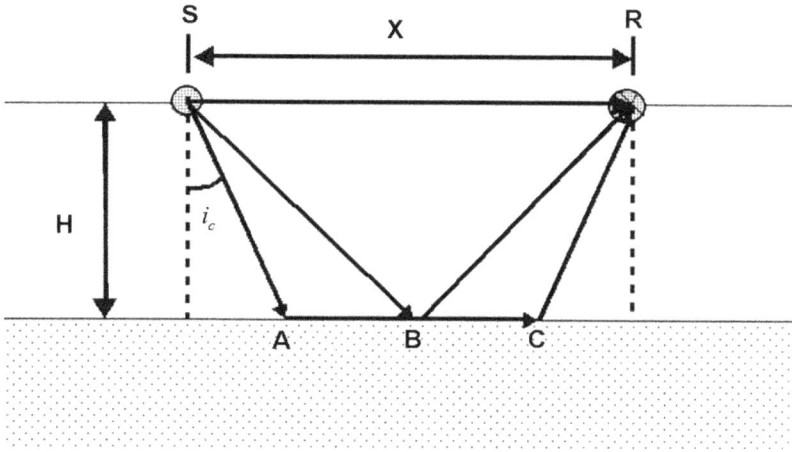

Figure 3.5 Wave paths in a two-layered medium.

As for the reflected wave path, the relations between distance and arrival time are given by

$$t = \frac{SC}{V_1} + \frac{CR}{V_1} \qquad (3.25)$$

As $SC = CR$ and is given by

$$SC = CR = \sqrt{H^2 + \frac{X^2}{4}} \qquad (3.26)$$

equation (3.25) becomes

$$t = \frac{2}{V_1} \sqrt{H^2 + \frac{X^2}{4}} \qquad (3.27)$$

The relation between distance and arrival time for the refracted wave path shown in Figure 3.5 can be written as follows:

$$t = \frac{SA}{V_1} + \frac{AB}{V_2} + \frac{BR}{V_1} \qquad (3.28)$$

From the geometry of the path, one can write the following relations:

$$SA = BR = \frac{1}{V_1} \cdot \frac{H}{\cos i_c}; \quad AB = X - 2H \tan i_c \qquad (3.29)$$

As $\quad \cos i_c = \sqrt{1 - \left(\frac{V_1}{V_2}\right)^2}$

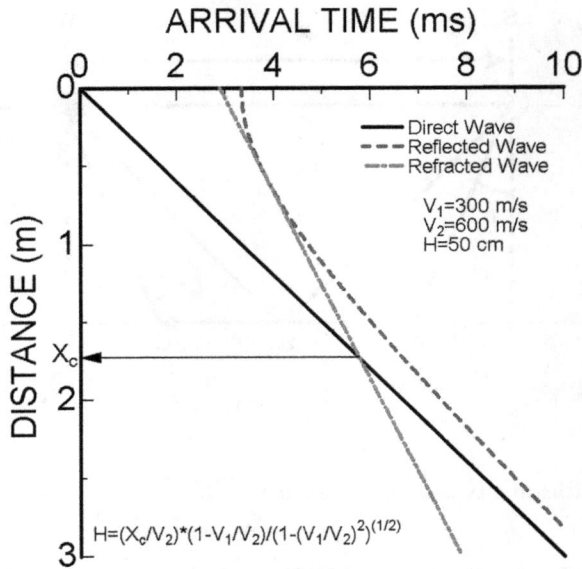

Figure 3.6 Relations between distance and arrival time for different wave paths.

Equation (3.28) takes the following form:

$$t = \frac{2H}{V_1}\sqrt{1 - \left(\frac{V_1}{V_2}\right)^2} + \frac{X}{V_2} \tag{3.30}$$

If critical distance X_c and wave velocities V_1 and V_2 are obtained from the records, the thickness of layer 1 can be obtained by equating the arrival times of the direct wave and the refracted wave as follows:

$$H = \frac{X_c}{V_2}\frac{1 - \dfrac{V_1}{V_2}}{\sqrt{1 - \left(\dfrac{V_1}{V_2}\right)^2}} \tag{3.31}$$

As an application of the theory presented above, an example computation was carried out, and the results are shown in Figure 3.6 together with the assumed wave velocities and the thickness of layer 1.

The fundamental principle of this technique is generally used to identify the layered structures as well as the interior of the Earth. Several applications of this method to the interior structure of the Earth are given in the following section.

3.5   WAVE PROPAGATION THROUGH THE EARTH AND INFERENCE OF THE EARTH'S INTERIOR

As said previously, waves are categorized into body and surface waves induced by earthquakes, which travel throughout the Earth. The propagation of surface waves is restricted to the observations near the surface of the Earth. P-waves or dilatational waves can propagate any medium while shear waves cannot propagate through fluids. This feature of waves is very important for the interpretation of wave observations. In a uniform homogenous medium, the wave ray would be a straight line, while the variation of density and elastic constants of medium would change the ray path, and this abrupt change of medium would cause refraction. Utilizing the wave propagation theory and Snell's law, analyses of records at seismographs installed at various locations on the Earth revealed the interior of the Earth as illustrated in Figure 3.7. If the Earth is considered as a simple three-layered structure consisting of mantle-crust, outer core and inner core with different densities and wave velocity characteristics and outer core is fluid, there will be zones of shadow for P-wave and S-wave. P-wave shadow occurs between 103° and 143° due to refraction caused by the characteristics of mantle-crust and outer core while S-wave shadow is due to no S-wave propagation through the outer core. As the outer core of the Earth is in liquid phase, no shear waves are observed beyond 103° from the focus of an earthquake.

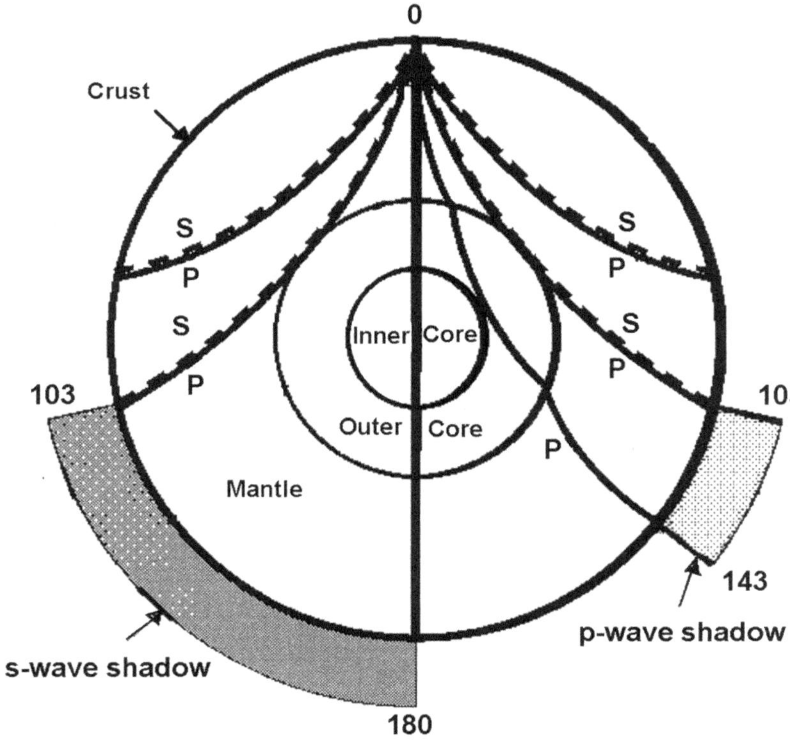

Figure 3.7 Wave propagation and interior structure of the Earth.

As shown in Figure 3.3, various waves induced by an earthquake would be observed and the waves will continue to appear on the seismic records for a certain period of time and they will eventually disappear due to volumetric and viscous dispersion. In seismology, the observed waves are distinguished according to their path and the media they propagated through, as illustrated in Figure 3.8. For example, P-waves would correspond to the first arrival of P-wave while PP would correspond to a reflected P-wave. Waves denoted PKP would involve a path through mantle–outer core–mantle, while PKIKP would involve a path through mantle (P)–outer core (K)–inner core (I)–outer core (K)–mantle (P). P-wave can be reflected as S-wave. In such cases, the observed wave would be denoted as PS. Figure 3.9 shows an example of records observed at different locations on the Earth. The distance is measured as the angle from the focus of an earthquake.

Milne (1898), who is originally a mining engineer from UK, together with T. Gray was the first to develop the modern age seismographs following the 1880 Yokohama earthquake. Seismographs were updated continuously since then and there are many seismic networks, internationally, country-wise and locally, these days. With the analyses of seismic records and several techniques developed for interpretation and

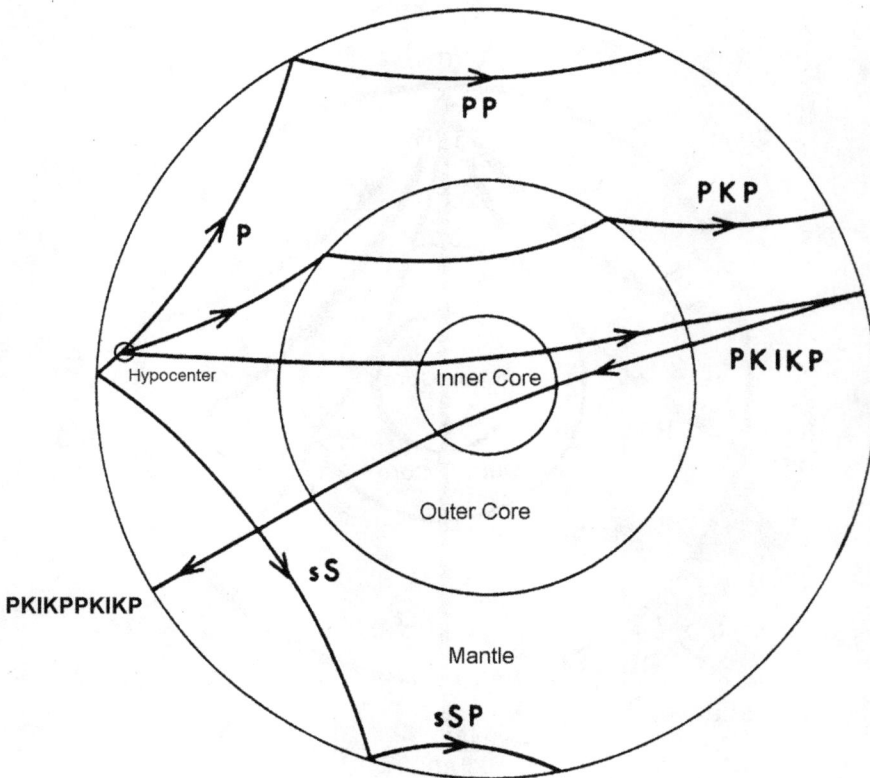

Figure 3.8 Illustration of various arrival waves.

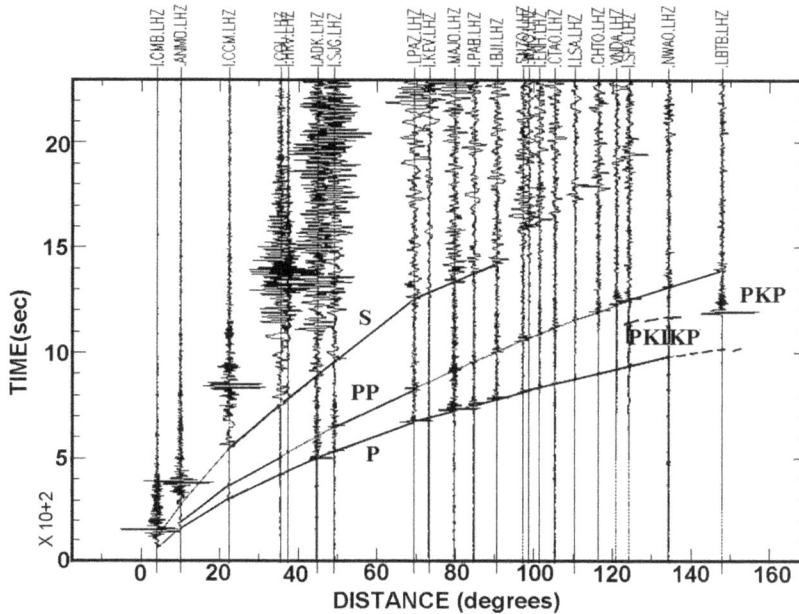

Figure 3.9 Observed wave forms and various phase arrivals.

inferences (e.g. Jeffreys and Bullen 1940), the interior structure of the Earth was contemplated, as shown in Figure 3.10. From the density, P-wave and S-wave distributions and some thermo-mechanical tests on rocks and materials existing on the Earth, it is shown that the Earth broadly consists of core, mantle and crust. While the outer core is assumed to be in fluid state in view of its shear wave velocity, the inner core is presumed to be solid or solid-like. The mantle is divided into two layers, namely the upper mantle and the lower mantle. Particularly, the upper mantle seems to be highly disturbed. The Earth's crust varies from place to place. While it is about 33 km thick in average, it becomes quite thick in the close vicinity of subduction zones and thinner near the divergent plate boundaries.

3.6   DETERMINATION OF OCCURRENCE TIME

The motions induced by earthquakes can be recorded using strong motion equipment (acceleration, velocity) with short periods and broadband seismographs. Particularly, broadband seismometers are used to record long distance earthquakes. However, they may become saturated in the close vicinity of earthquake epicentres with large magnitudes. The occurrence time of an earthquake fundamentally utilizes the arrival time differences of P-wave and S-wave. Omori (1894) proposed the following relation for the distance between the hypocentre of an earthquake and time lag of wave arrivals at a given seismic observation point for P-wave and S-wave arrivals (Figure 3.11). Fundamentally, the earthquake occurrence time is unknown.

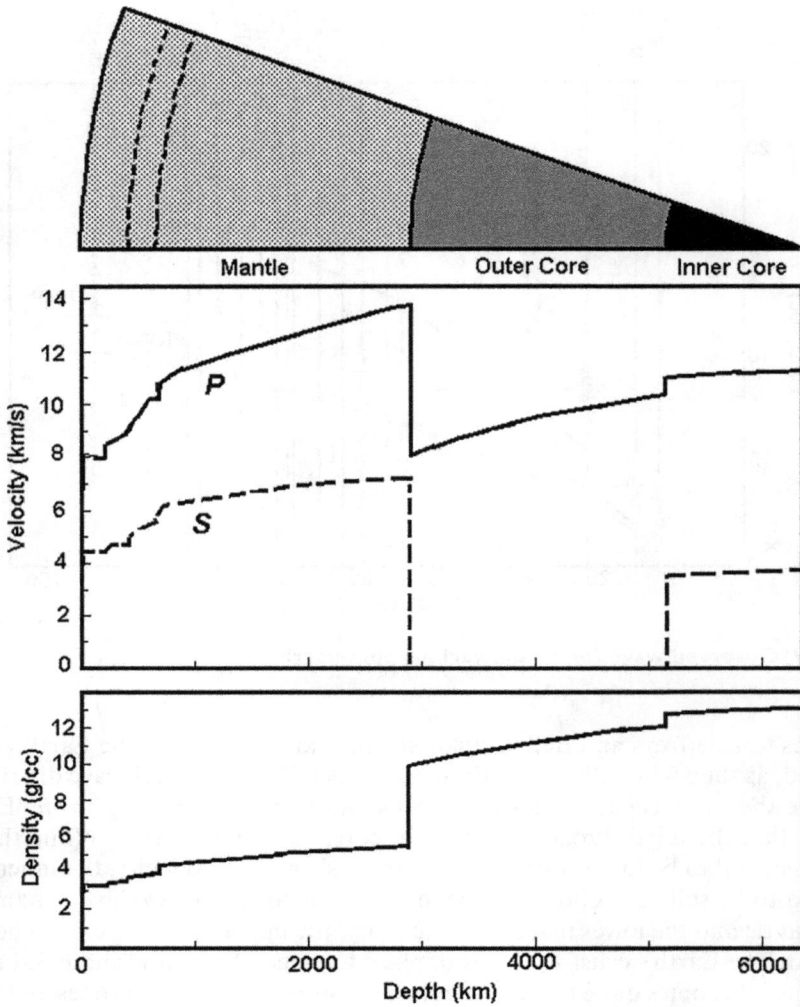

Figure 3.10 Inferred interior structure of the Earth.

$$t_p - t_o = \frac{R}{V_p} \quad \text{and} \quad t_s - t_o = \frac{R}{V_s} \tag{3.32}$$

Subtracting S-wave arrival time lag from the P-wave arrival time lag yields the following equation:

$$t_s - t_p = \left(\frac{1}{V_s} - \frac{1}{V_p}\right) R = \frac{R}{k} \tag{3.33}$$

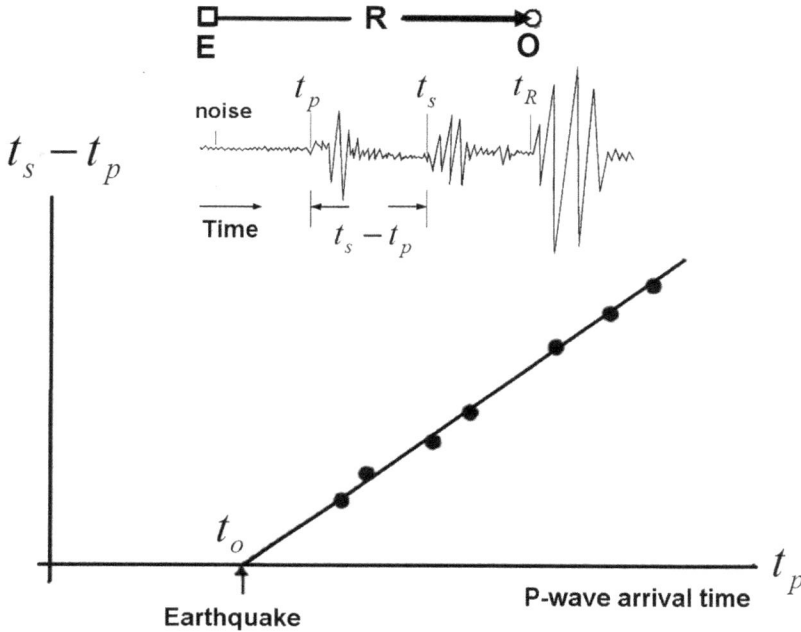

Figure 3.11 Illustration of arrival times of P-wave, S-wave and surface wave.

where k, which is called Omori coefficient, takes the following form:

$$k = \frac{V_s V_p}{V_p - V_s} = \frac{1}{\dfrac{V_p}{V_s} - 1} V_p \tag{3.34}$$

where V_p and V_s are the P-wave and S-wave velocities of the Earth's crust, which are not known. However, the following relation exists between these wave velocities in view of Eqs. (3.16) and (3.21):

$$\frac{V_s}{V_p} = \sqrt{\frac{1 - 2v}{2(1 - v)}} \tag{3.35}$$

As Poisson's ratio of the Earth's crust ranges between 0.25 and 0.3, the following values for V_s / V_p ratio may be assumed:

$$\frac{V_s}{V_p} = 0.53 - 0.58 \tag{3.36}$$

Thus, the value of the Omori coefficient would take the following value in terms of P-wave velocity as

$$\frac{k}{V_p} = 0.724 - 0.887 \tag{3.37}$$

The average P-wave velocity of the Earth's crust ranges from 6.5 to 7.2 km/s for upper-crust earthquakes and from 7.5 to 8.2 km/s for lower-crust earthquakes.

3.7   DETERMINATION OF HYPOCENTRE AND EPICENTRE

It is well known that earthquakes are initiated at a certain depth. The initiation location of earthquakes is called the hypocentre. In this section, the determination of hypocentre for two-dimensional and three-dimensional situations is explained. The theory again utilizes Omori's method.

3.7.1   Two-dimensional determination of hypocentre and epicentre

Let us consider two observation points denoted by A and B and an earthquake that occurred along the straight line at a point denoted H, which will be assumed to be corresponding to the hypocentre, and its projection on the ground surface is E, the epicentre (Figure 3.12).

Using Omori's formula, the hypocentral distance to the earthquake's focus is given by

$$R = V_p\left(t_p - t_o\right) \tag{3.38}$$

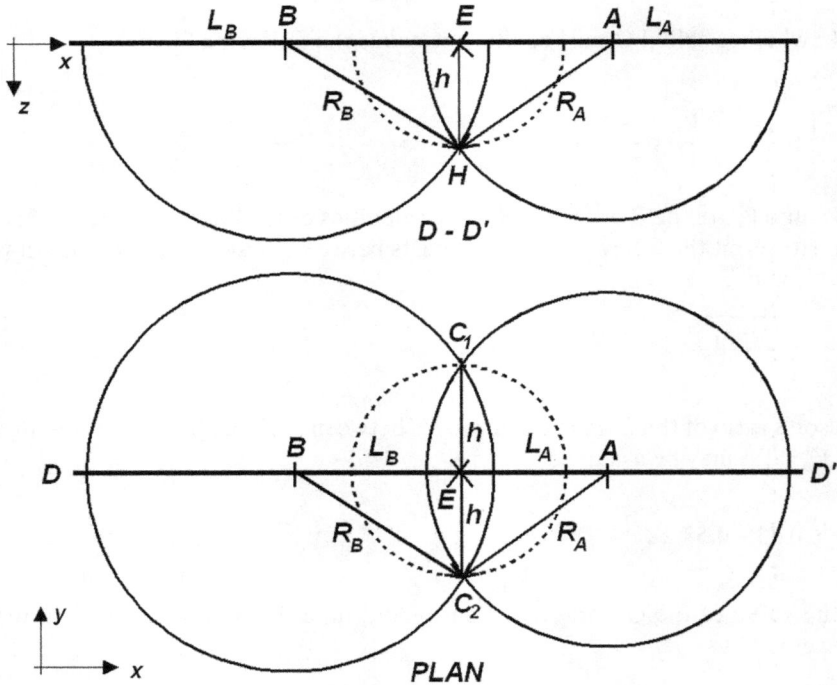

Figure 3.12 Illustration of hypocentre and epicentre for a two-dimensional situation.

In this particular problem, besides the distance between points A and B, the hypocentre distances R_A, R_B can be computed using Omori's formula by assigning a value to Omori's coefficient. However, the results will not be generally affected by the value of Omori's coefficient. Thus, the known parameters are

$$R_A, R_B, L \tag{3.39}$$

On the other hand, the unknown parameters are

$$h, L_A, L_B \tag{3.40}$$

However, unknown distances L_A, L_B are related to the total distance between points A and B though the following relation:

$$L_A + L_B = L \tag{3.41}$$

Let's introduce parameter α to relate the distances L_A, L_B to the total distance L as

$$\frac{L_B}{L} = (1 - \alpha); \frac{L_A}{L} = \alpha \tag{3.42}$$

From the geometry one can easily determine parameter α as

$$\alpha = \frac{1}{2L^2}\left(R_A^2 + L^2 - R_B^2\right) \tag{3.43}$$

Once parameter α is determined, first L_A, L_B is determined and then the focal depth is computed from one of following relations:

$$h = \sqrt{R_A^2 - L_A^2} \tag{3.44a}$$

or

$$h = \sqrt{R_B^2 - L_B^2} \tag{3.44b}$$

3.7.2 Three-dimensional determination of hypocentre and epicentre

Three-dimensional determination of hypocentre and epicentre can be carried out using the same procedure and/or graphically. Let us consider that there are three seismic stations with given coordinates and S-wave and P-wave arrival time differences. In graphical method, the following procedure is adopted:

1. Using Omori's coefficient, circles with diameters obtained from Omori's method are drawn first and then three common chords among circles are drawn.
2. The intersection of three chords is denoted by E, which would correspond to the epicentre.

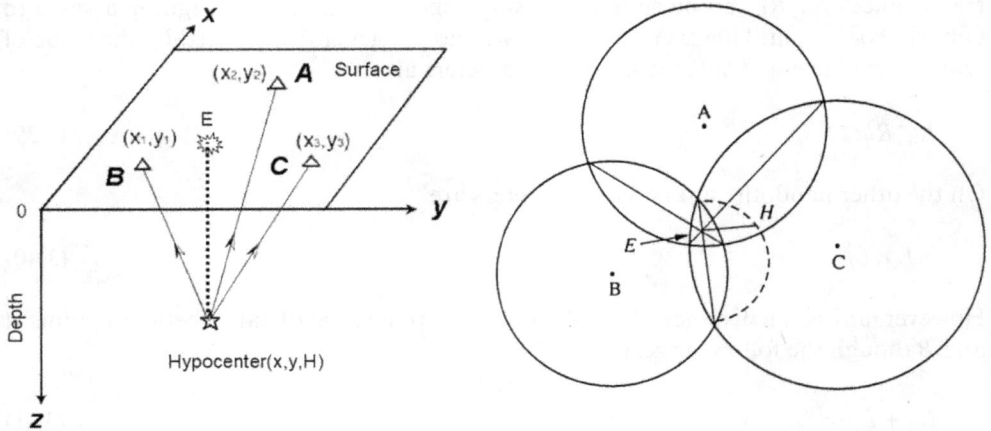

Figure 3.13 Illustration of a three-dimensional hypocentre and epicentre determination procedure.

3. Then a semi-circle having the shortest common chord length as its diameter is drawn and a straight line from E is drawn perpendicularly to the shortest chord and the intersection point H is obtained. The length between E and H would correspond to the depth of the hypocentre.

The concepts explained in the previous section and the graphical method explained above can be combined. As the geometry becomes somewhat cumbersome, computational schemes would be necessary. For example, if the number of station is assumed to be three, the radius for each station is obtained and then common chords for each pair circles of stations and the intersection point of the chords are obtained. Then the shortest chord length is searched, and the hypocentre is obtained. Figure 3.14 shows such a computational example of the procedure. The data used in this example are given in Table 3.1.

In practice, there are many observation stations. The procedure is extended to observational data, but the fundamental procedure would remain the same. For observational data, the results would not always be the same, and some differences would occur. Furthermore, the velocity and seismic rays may be path dependent as the Earth is spherical. In general, such effects are considered. For this reason, some minimization procedure for differences is adopted. For this reason, some commonly used software is available in literature.

3.7.3   Specific application: the 1998 Adana-Ceyhan earthquake

A specific example of the procedure explained in the previous section is given here. The 1998 Adana-Ceyhan earthquake is chosen as an example. This earthquake, presumed to be a lower-crust earthquake, occurred on June 27, 1998, with a moment magnitude of 6.3. In this particular example, P-wave and S-wave arrival time differences are obtained from strong motions records (Turkish National Strong Motion Network, DAD-ERD 1998). Omori's coefficient k was assigned to be 7.12. Table 3.2 compares

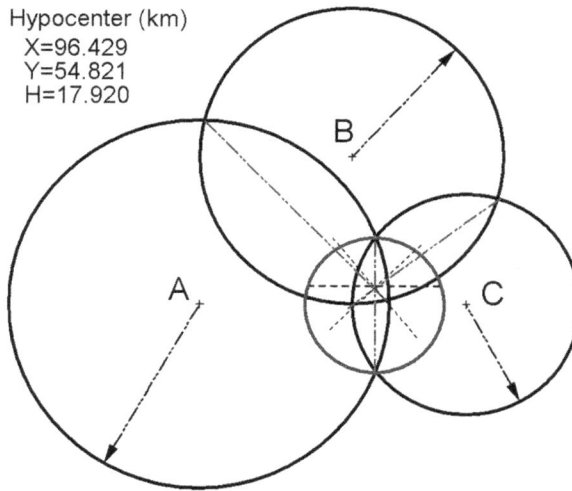

Figure 3.14 A sample solution.

Table 3.1 Locations of stations and arrival time difference

Station name	X (km)	Y (km)	Δt_{sp} (s)	$R = k\Delta t_{sp}$
A	50	50	6.25	50
B	120	50	3.125	25
C	90	90	3.75	30
$k = 8$ km/s				

Table 3.2 Comparison of computed hypocentre parameters by different institutes

Station name	Lat (km)	Lon (km)	Depth (km)	Time of occurrence
Ö.A.	36.813	35.583	14.85	13:55:50.89
DAD-ERD	36.85	35.55	23.0	13:55:53
USGS	36.95	35.31	14.0	13:55:52

the computed hypocentre parameters by different institutes and the method described herein (Aydan et al. 1998). As noted from Table 3.2, the results are quite similar to each other (Figure 3.15).

3.8   DETERMINATION OF MAGNITUDE

The energy E released during an earthquake can be obtained from the following relation:

$$E = \int \Delta t \cdot u \, dS \qquad (3.45)$$

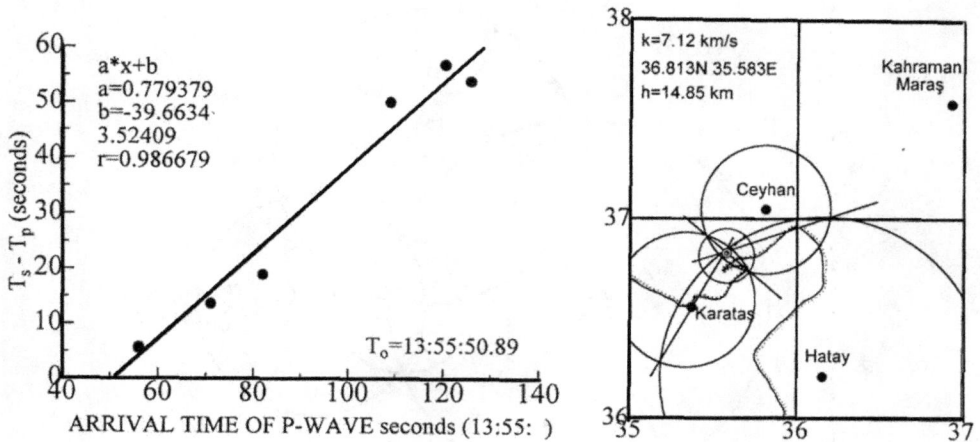

Figure 3.15 (a) Determination of time of occurrence and (b) hypocentre determination.

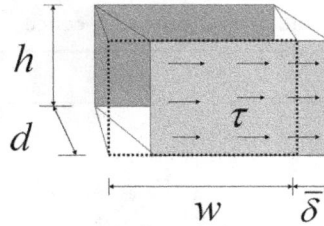

Figure 3.16 Illustration of a seismic moment.

where Δt is the released traction on fault surface, u is the relative slip and S is the area of fault. If average values are used, the equation above takes the following form:

$$E = \bar{t} S \cdot \bar{u} \tag{3.46}$$

Earthquake moment (M_0) is first defined by Aki (1966), and it is given in the following form (Figure 13.16):

$$M_o = \mu S \bar{u} \tag{3.37}$$

where μ is the shear modulus of Earth's crust, S is the rupture area and $\bar{u}$ is the average relative slip. The average relative slip is generally related to the maximum relative slip u_{max} as

$$u_{max} = 2\bar{u} \tag{3.48}$$

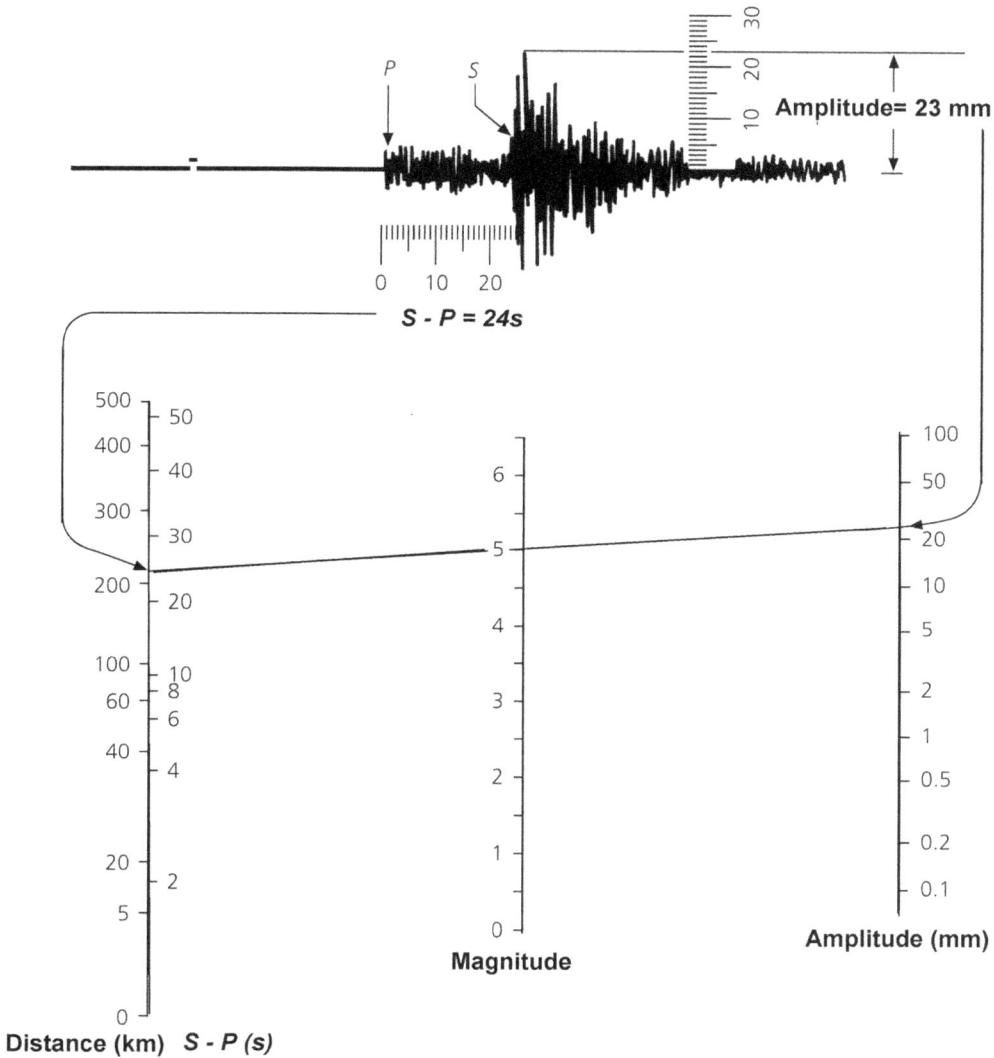

Figure 3.17 Manual chart of Richter to determine the magnitude of earthquakes. (From Richter (1935).)

Although reconnaissance, GPS and DInSAR can be used to obtain the relative slip of the causative earthquake, the seismic records are used to obtain the seismic moment. It should be noted that the seismic moment obtained from various seismological institutes would generally differ from each other. Hanks and Kanamori (1979) defined moment magnitude utilizing seismic moment in the following form:

$$M_w = \frac{2}{3}\log_{10} M_0 - 10.7 \tag{3.49}$$

The concept of earthquake magnitude was first introduced by Richter (1935) and it is known as local earthquake magnitude in modern times. Richter (1935) determined his magnitude from seismographs measured by Wood-Anderson torsion seismograph located at a distance of 100 km. Richter (1935) prepared a scale to determine the magnitude of earthquakes as shown in Figure 13.17.

All earthquake magnitude scales utilize the maximum amplitude of seismogram produced by a seismograph and its period together with some correction factors considering local conditions and distance, as given below:

$$M = \log_{10}\left(\frac{A}{T}\right) + q(\Delta, h) + a \qquad (3.50)$$

A is the amplitude, T is the period, $q(\Delta, h)$ is a correction factor depending upon the distance, and a is an empirical factor.

There are different magnitude scales of earthquakes depending upon the utilized wave type as given below:

M_b: body wave magnitude (P-wave or S-wave)

M_s: surface wave magnitude (surface wave)

M_d: duration magnitude (duration of seismogram for small earthquake)

M_L: local magnitude (Richter type)

M_w: moment magnitude

The correction factor generally depends upon the local conditions of seismographs and the distance from the hypocentre. When earthquakes are quite large, seismograms may be saturated. On the other hand, deep earthquakes cannot be determined by Richter-type and surface-wave-based seismographs.

Chapter 4

Faults and faulting mechanism of earthquakes

4.1 CHARACTERISTICS OF EARTHQUAKE FAULTS

A fault is geologically defined as a discontinuity in the geological medium along which a relative displacement takes place. Faults are broadly classified into three major groups: normal faults, thrust (reverse) faults and strike-slip faults, as seen in Figures 4.1 and 4.2. The fault is geologically defined active if a relative movement along the fault took place within a period <2 million years.

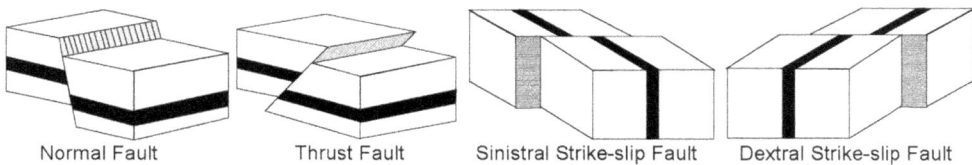

| Normal Fault | Thrust Fault | Sinistral Strike-slip Fault | Dextral Strike-slip Fault |

Figure 4.1 Fault types. (From Aydan (2003b, 2012a).)

Figure 4.2 Some examples of actual faults. (a) Normal faulting, (b) strike-slip faulting, and (c) thrust faulting.

DOI: 10.1201/9781003164371-4

It is well known that a fault/fracture zone may involve various kinds of fractures with crushed clayey core, as illustrated in Figure 4.3a, and it is a zone having a finite volume (Aydan et al. 1999b; Aydan 2003a, 2012a,b; Ulusay et al. 2002). In other words, it is not a single plane. Furthermore, the faults may have a negative or a positive flower structure as a result of their trans-tensional or trans-pressional nature and the reduction of vertical stress near the Earth's surface (Aydan et al. 1999b). For example, even a fault having a narrow thickness at depth may cause a broader rupture zone and numerous fractures on the ground surface during earthquakes (Figure 4.3b). Furthermore, the movements of a fault zone may be diluted if a thick alluvial deposit is found on the top of the fault, e.g. 1992 Erzincan earthquake (Hamada and Aydan 1992; Aydan 2003b; Aydan et al. 1999b). The appearance of ground breaks is closely related to the geological structure, characteristics of sedimentary deposits, their geometry, the magnitude of earthquakes and fault movements.

4.2   PHYSICAL MODELS ON FAULTING

4.2.1   Photo-elasticity tests

Aydan (2019c) and Aydan et al. (2020) performed some photo-elasticity tests on the stress state of faults under different loading regimes. In this section, the results of those photo-elasticity tests are discussed.

4.2.1.1   Material properties

Finite element analyses require data on mechanical properties of physical models. Some compression experiments were carried out to determine the elastic modulus and Poisson's ratio. Figure 4.4 shows the stress–strain response of several polyurethane samples under uniaxial compression condition. Table 4.1 summarizes the material

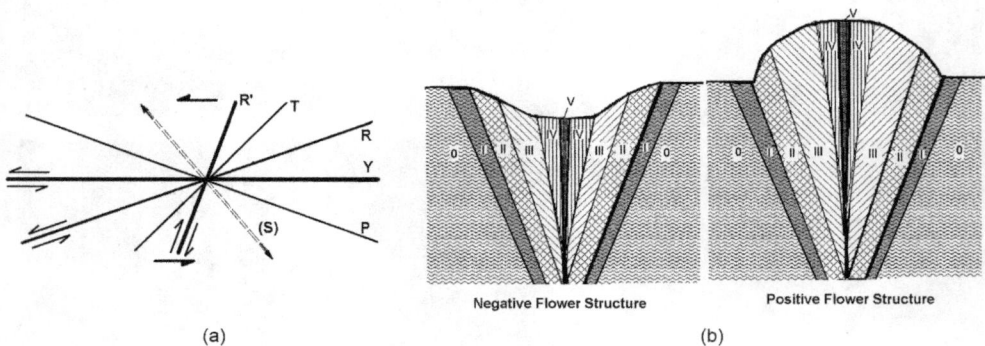

Negative Flower Structure   Positive Flower Structure

(a)   (b)

Figure 4.3 (a) Fractures in a shear zone or fault; (b) negative and positive flower structures due to trans-tension or trans-pression faulting and zoning. (From Aydan (2012a).)

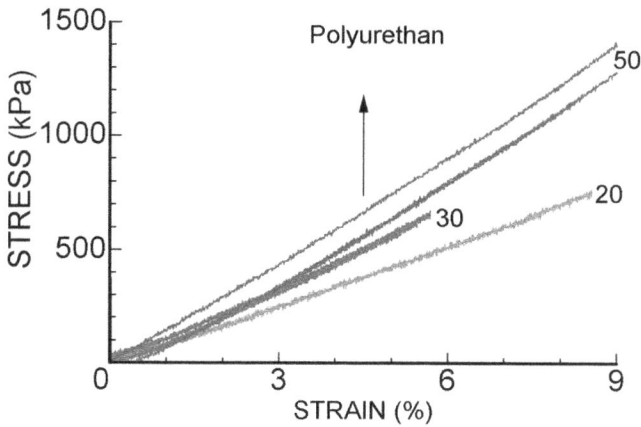

Figure 4.4 The stress–strain response of several polyurethane samples under uniaxial compression condition.

Table 4.1 The elastic modulus and Poisson's ratio of several polyurethane samples

Material	Elastic modulus (MPa)	Poisson's ratio	Models used
Polyurethane-20	7.3	0.38	Faults, samples with slits

properties of polyurethane-20 samples. Polyurethane-20 is used for testing faults, layers and blocks.

4.2.1.2   *Photo-elasticity tests on the stress state of faults*

4.2.1.2.1   SLIT-LIKE FAULTS

Samples having slit-like faults are often used to study the effect of distributed cracks on the overall behaviour of faults. For this purpose, samples having a slit with its longitudinal axis oriented at 0°, 45°, 90° from the horizontal were prepared and subjected to different loading regimes, as shown in Figures 4.5 and 4.6. As noted from the figures, the stress distributions in samples are quite different depending upon the orientation of the longitudinal axis of the slit. While the effect of a 90° slit on stress state is less, the stress concentrations occur at the tip of the slit and tensile stresses occur near the central part of the slits for samples having 0 and 45 slits.

Next, the same slit-like fault samples are subjected to normal and shear loads, as illustrated in Figure 4.7. Figures 4.8 and 4.9 show the maximum stress distributions in the vicinity of slit-like faults under normal load only and normal and shear loads. As noted from the figures, the stress concentrations occur at the tips of the slits.

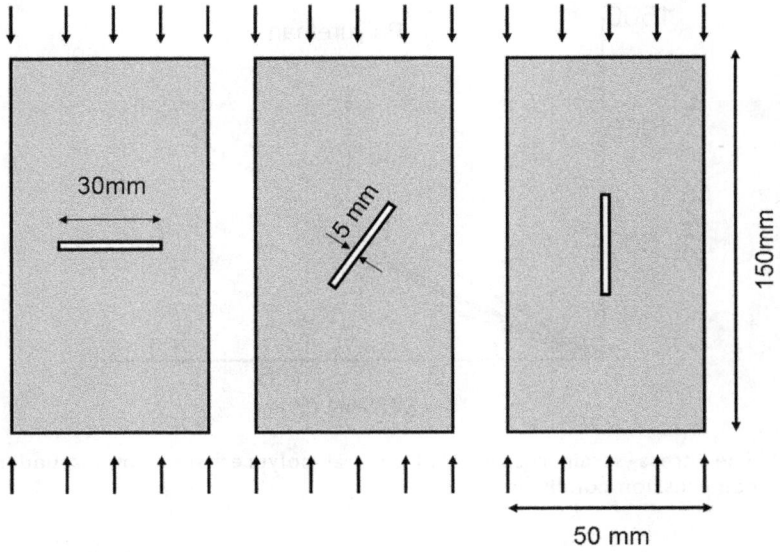

Figure 4.5 Dimensions and loading conditions of slit-like fault models.

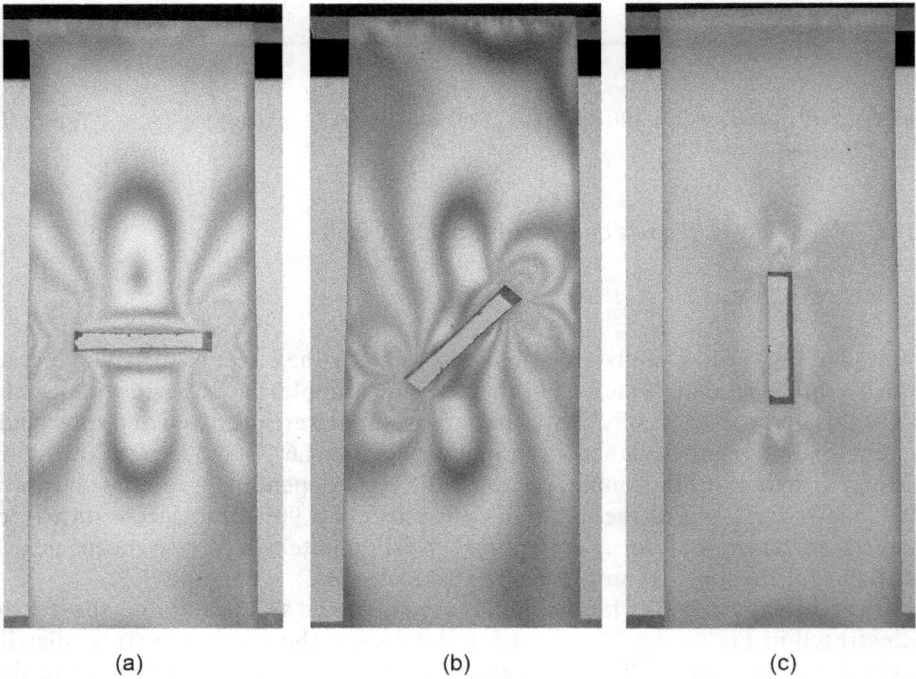

(a)   (b)   (c)

Figure 4.6 Stress distribution in samples having slits with different orientations. (a) 0°
slit, (b) 45° slit, and (c) 90° slit.

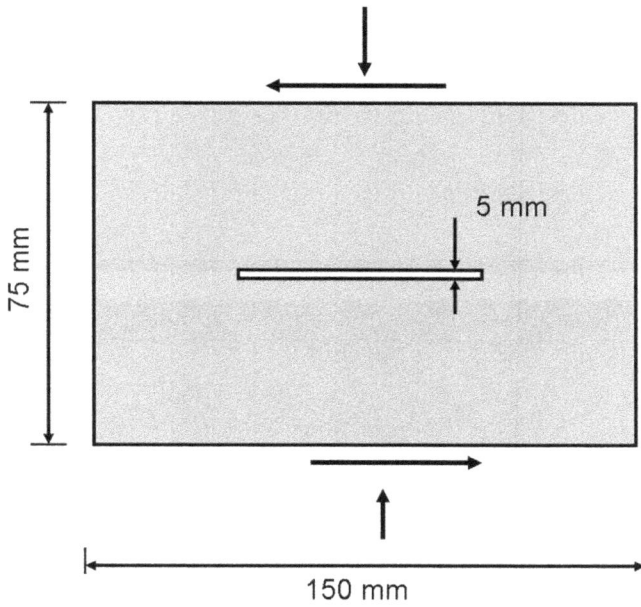

Figure 4.7 Illustration of normal and shear loads on the slit-like faults samples.

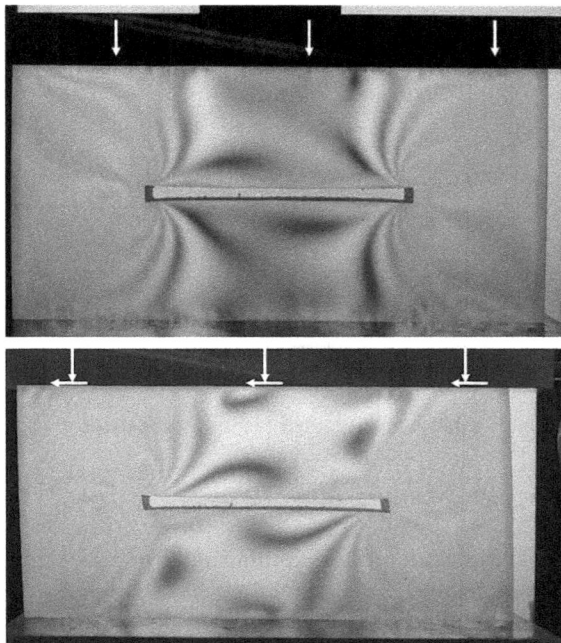

Figure 4.8 Maximum shear stress distributions under different loading regimes (0).

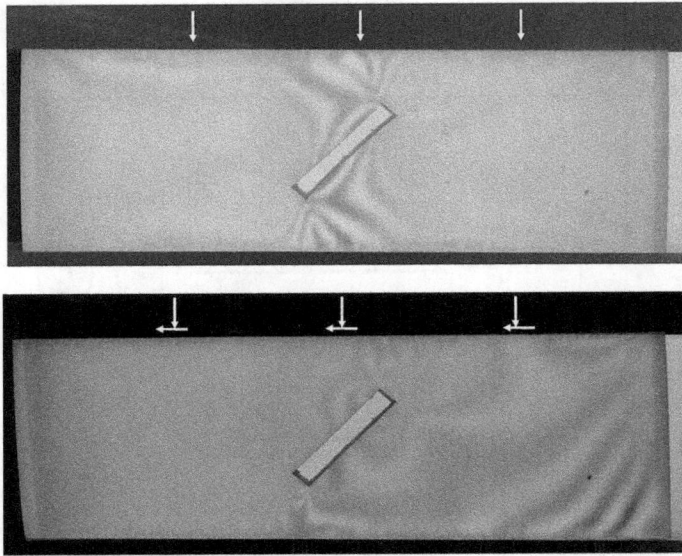

Figure 4.9 Maximum shear stress distributions under different loading regimes (45).

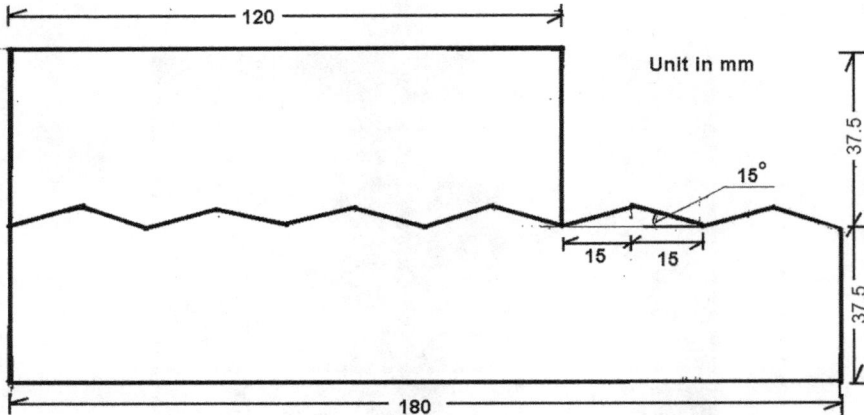

Figure 4.10 Dimensions of a fault model with regular asperities.

4.2.1.3   Faults with regular asperities

Faults with asperities are considered for simulating strong motions in earthquake science. Such models assume that fault surfaces are flat. However, they are actually not flat and mostly undulating. The sample used has a discontinuity plane consisting of regular asperities with an inclination of 15°, as illustrated in Figure 4.10. Figures 4.11 and 4.12 show the maximum stress distributions subjected to normal load only and a

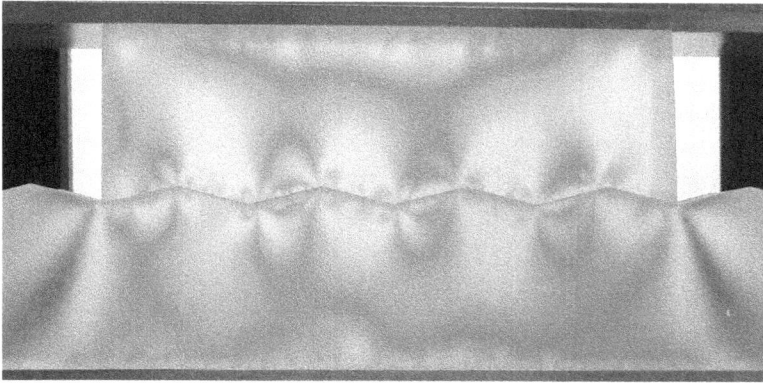

Figure 4.11 Stress distribution of a sample with regularly spaced asperities under normal loading only.

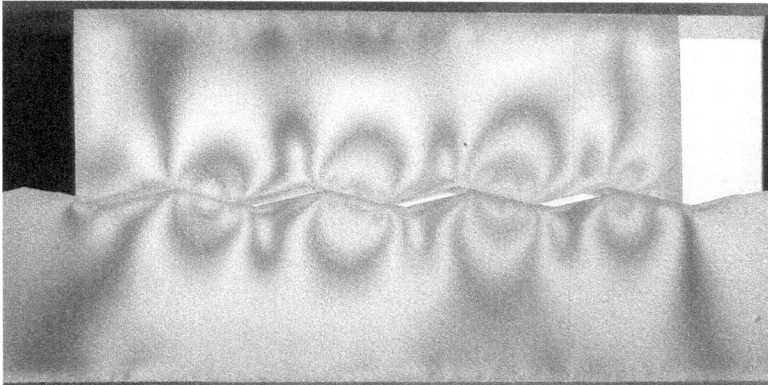

Figure 4.12 Stress distribution of a sample with regularly spaced asperities under combined normal and shear loading with relative slip.

combined shear and normal loading together with a relative slip, respectively. As noted from Figures 4.11 and 4.12, the existence of the discontinuity plane plays a great role in the overall stress distribution. Furthermore, the slip results in a high compression on the contact side and tensile stresses on the separated side of asperities. This result may also have important implications on the visualization of asperity models commonly used in earthquake science for simulating strong motions.

4.2.1.4 Faults with irregular rough asperities

Stress distribution in the vicinity of a fault-like discontinuity plane in a continuum model subjected to dextral-fault-type boundary conditions is studied, as shown in Figure 4.13. The surface configuration of the fault plane was relatively rough. As noted from the figure, stress concentrations occur at the tips of the fault plane while symmetric stress

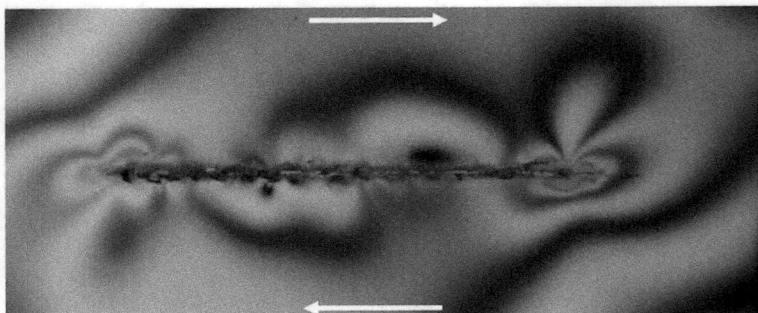

Figure 4.13 Stress distribution in the vicinity of a model fault plane subjected to dextral-type fault boundary conditions.

shadows occur in the central parts on both sides of the fault plane. Nevertheless, the stress distributions are quite complex compared to those of well-shaped fault models.

4.2.1.5   Finite element analyses of fault models

4.2.1.5.1   SLIT-LIKE FAULT MODELS UNDER UNIAXIAL COMPRESSION

Finite element analyses for slit-like faults with inclinations of 0°, 45° and 90° subjected to uniaxial loading conditions were carried out. The bottom side of the models was fully fixed. Figures 4.14–4.16 show the maximum shear stress distributions. As noted from the figures, stress concentrations are quite high at the end of slits. However, the amplitude of maximum shear stress is much higher for the 0° slit-like fault.

4.2.1.5.2   FAULT MODELS UNDER SHEARING

Next, finite faults with zero thickness were analysed under no-contact and frictional contact conditions. Figures 4.17–4.20 show minimum principal stress and maximum shear stress distributions for the given contact conditions. The bottom of the models was fixed and subjected to uniform shear traction at the top of the model. The high stress concentrations occur in the vicinity of the tips of fault. The stresses are much higher under no-contact or frictionless fault compared with frictional contact. Distributions of minimum principal stress are quite similar to those observed in the photo-elasticity tests.

4.2.2   Physical model tests

4.2.2.1   Experimental device, materials and procedure

An experimental device, as shown in Figure 4.21, was developed for model tests on faulting under dynamic conditions (Ohta 2011). One side of the device is movable

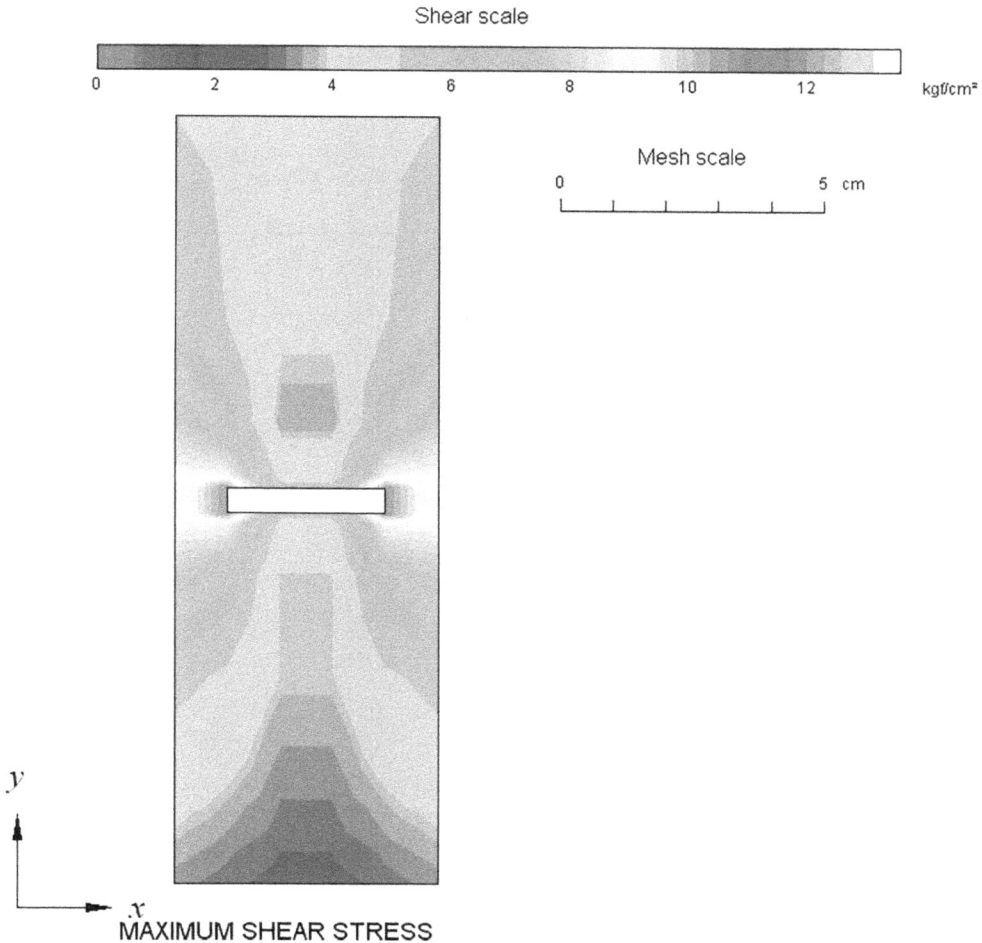

Figure 4.14 Maximum shear stress distribution for 0° slit-like fault.

in a chosen direction to induce base movements similar to normal or thrust fault-ing with a different inclination. The device can simulate from 45° normal faulting movement to 135° thrust faulting. The box is 780 mm long, 300 mm high and 300 mm wide. The length of the movable side is 400 mm. The motion of the movable side of the device is achieved through its own weight by removing a stopper. The amount of the vertical movement of the moving base can be up to 200 mm and it can be set to a certain level as desired. The device equipped with non-contact laser displacement transducers and contact type accelerometers with three components. Besides the continuous monitoring of movements of the model ground through laser displace-ment transducers and accelerometers, the experiments were recorded using digital video cameras.

Two kinds of ground material were considered: granular material simulating soft ground on a rigid base and a rock-like weak material to model layered and jointed

Shear scale

Mesh scale

MAXIMUM SHEAR STRESS

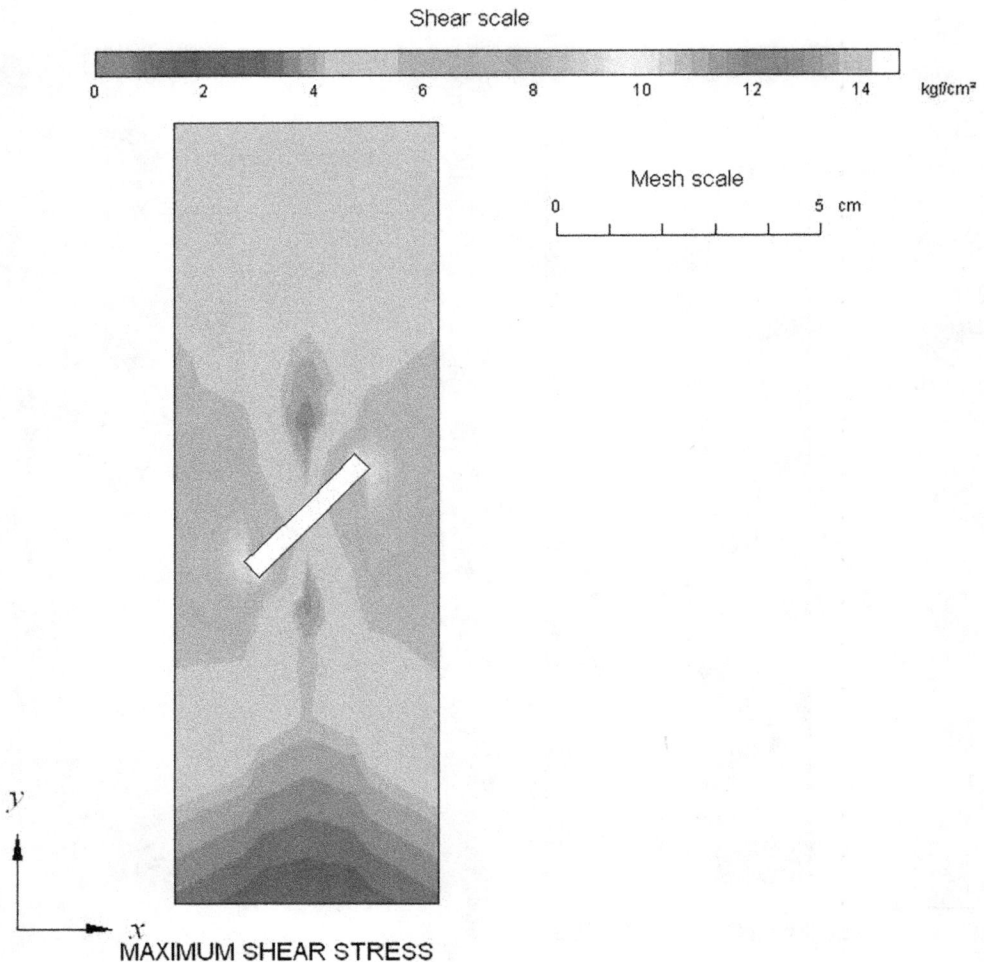

Figure 4.15 Maximum shear stress distribution for 45° slit-like fault.

rock mass. Granular ground material on the rigid movable base is dry quartz sand and its grain-size distribution is shown in Figure 4.22a. Direct shear experiments on ground material were carried out. Figure 4.22b shows some of the displacement versus shear stress responses. The behaviour of sand is elasto-plastic without any softening. Figure 4.22c shows some yield criteria for experimental results and the friction angle of sand ranges between 32° and 34°.

Layers or blocks used for physical models of layered or jointed rock mass were created through the compaction of a special mixture consisting of $BaSO_4$, ZnO and Vaseline oil under a chosen pressure. Various researchers determined the properties of this solid-like material (see Aydan and Amini 2009 for details). Frictional characteristics of discontinuities, unit weight, tensile and compressive strength of these samples have been measured in the laboratory as seen in Figure 4.23.

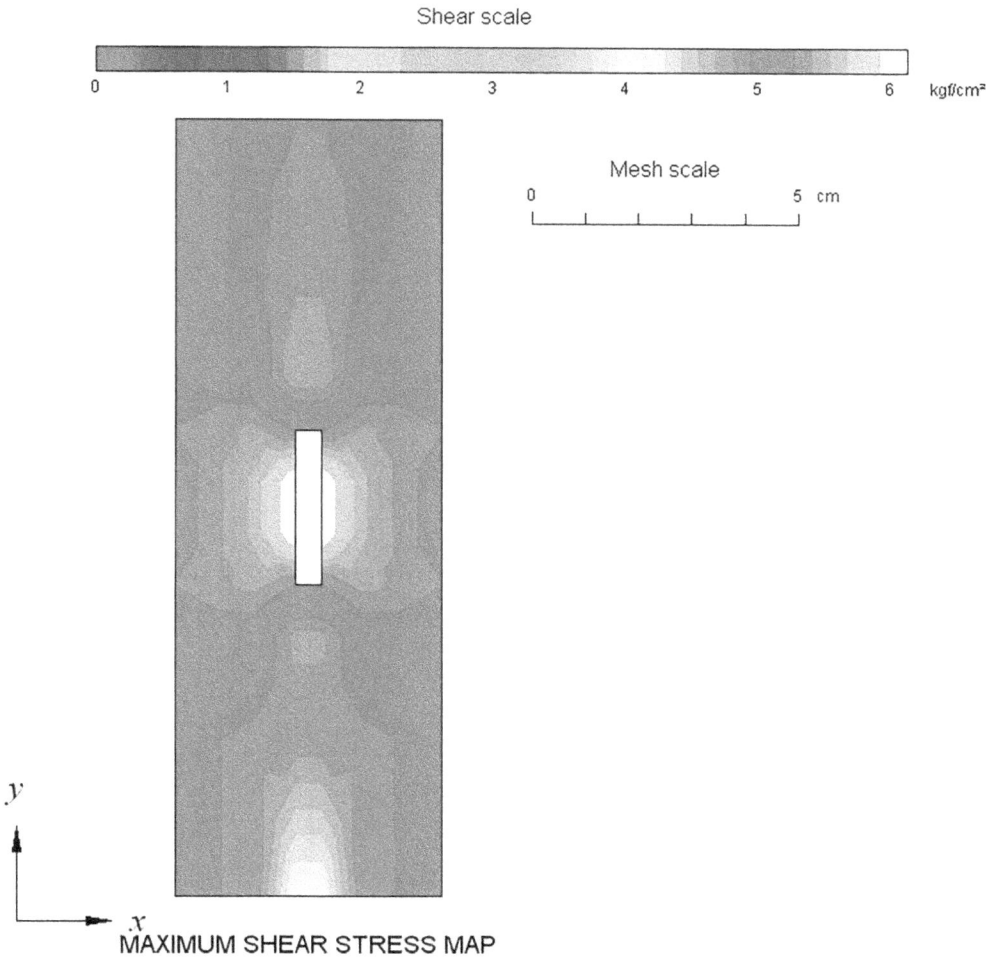

Figure 4.16 Maximum shear stress distribution for 90° slit-like fault.

4.2.2.2   Experiments on granular ground

The sand was poured into the soil box without any compaction. Therefore, the relative density of the sand ranges between 35% and 45% with an average of 40%. Black dyed sand marker lines were set at an interval of about 50 mm. However, the procedure for marker lines is extremely tedious and it is generally difficult to get perfectly straight lines.

Once the soil model was prepared, the monitoring devices consisting of laser displacement transducers and accelerometers to observe ground motions on ground surface were set at three locations, specifically, movable and stationary blocks and just above the fault (see Figure 4.21). Each experiment was recorded through digital video recorders and pictures were taken during the experiment. Furthermore, variations of

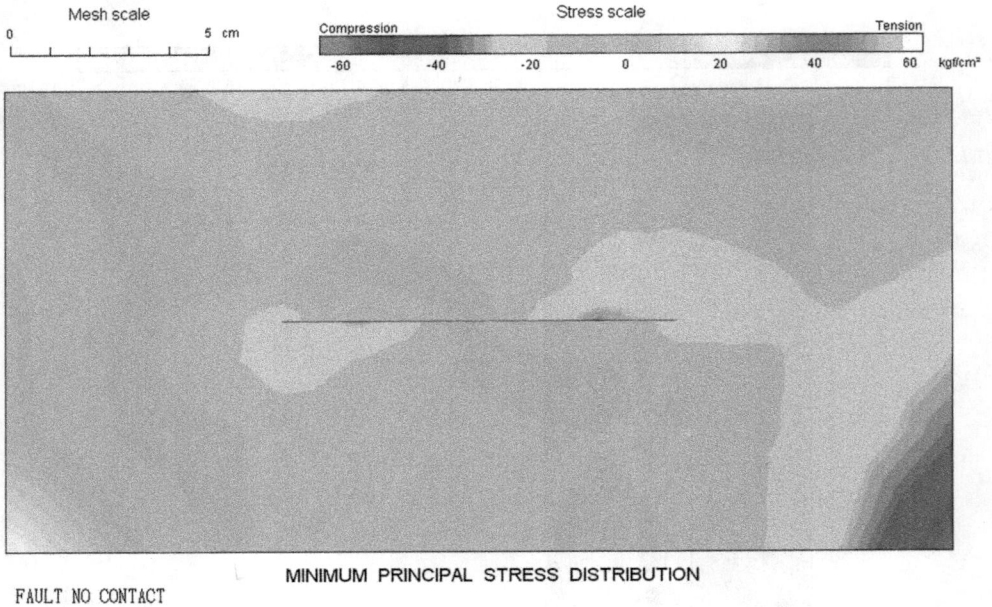

Figure 4.17 Minimum principal stress distribution (no-contact fault).

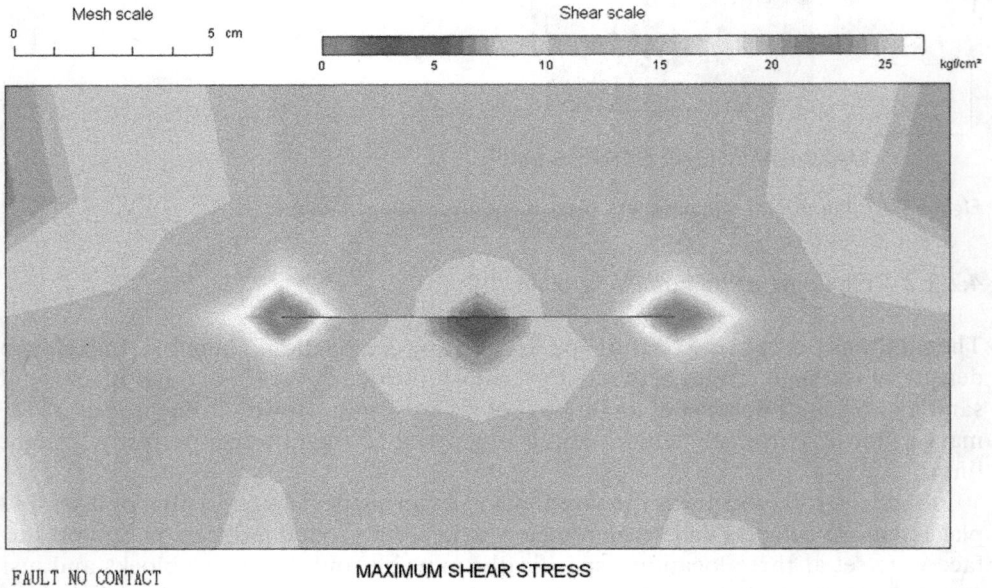

Figure 4.18 Maximum shear stress distribution (no-contact fault).

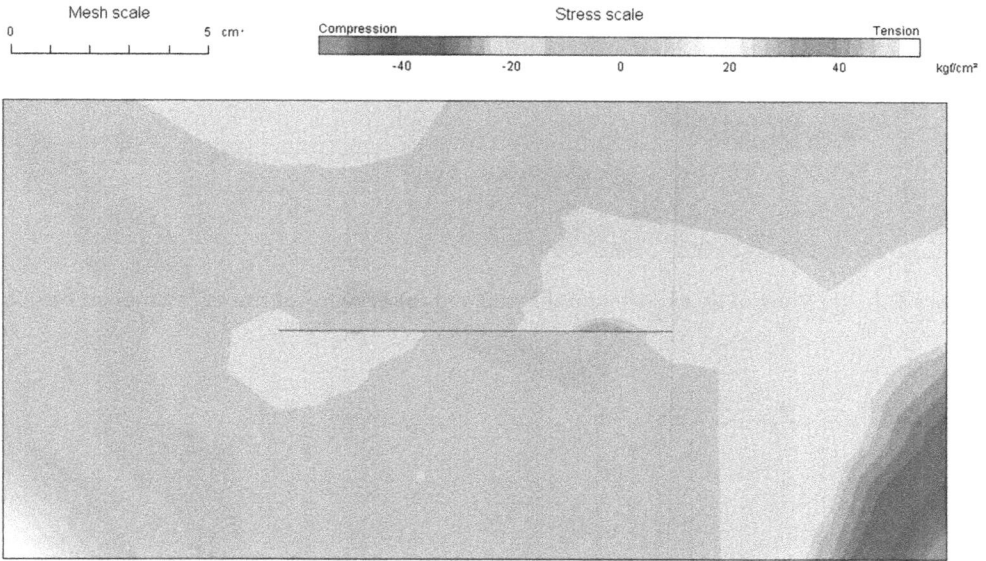

Mesh scale

0   5   cm·

Stress scale

Compression   Tension

-40   -20   0   20   40   kgf/cm²

MINIMUM PRINCIPAL STRESS DISTRIBUTION

FAULT FRICTIONAL CONTACT

Figure 4.19 Minimum principal stress distribution (frictional fault).

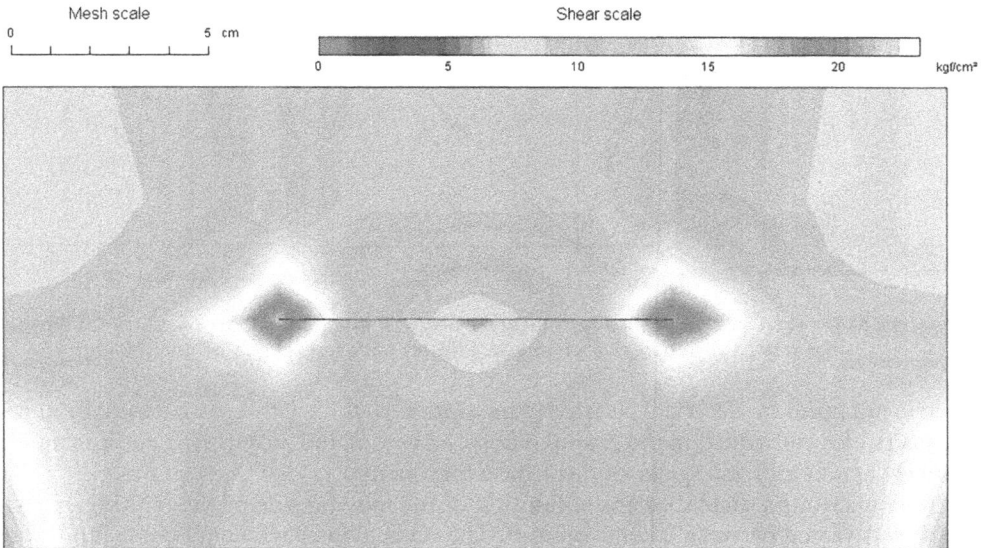

Mesh scale

0   5   cm

Shear scale

0   5   10   15   20   kgf/cm²

FAULT FRICTIONAL CONTACT   MAXIMUM SHEAR STRESS

Figure 4.20 Maximum shear stress distribution (frictional contact fault).

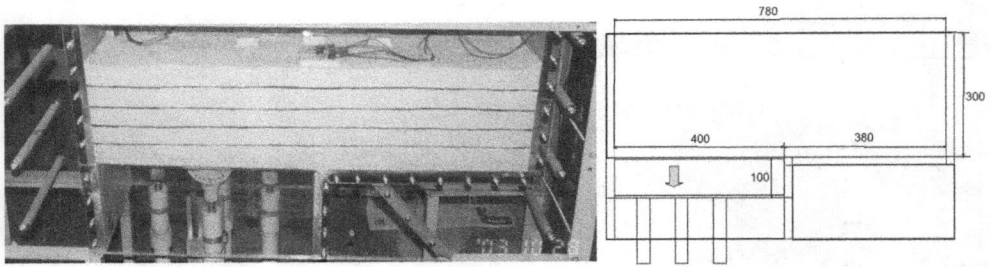

Figure 4.21 (a) View of an experimental setup and (b) drawing of the experimental device.

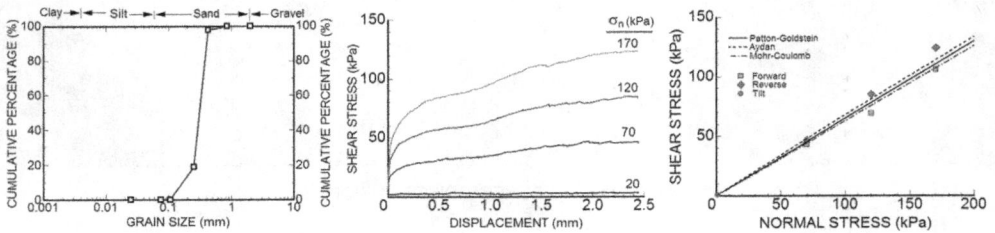

Figure 4.22 Grain-size distribution and shear strength properties of granular material. (a) Grain-size distribution, (b) displacement-shear stress responses, and (c) failure criteria.

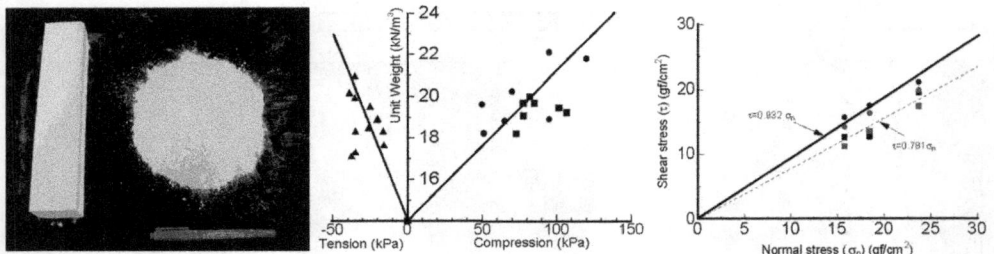

Figure 4.23 View of rock-like model material and its mechanical properties. (a) Model material, (b) density vs strength, and (c) shear strength of discontinuity.

electric potential or electrical resistivity in relation to faulting were also measured using electrodes embedded in the granular soil and associated monitoring equipment for electric current and voltage in some of the experiments.

The maximum displacement of faulting of the moving side of the faulting experiments was varied between 25 and 100 mm. The vertical displacements of the fault were 25, 50, 75 and 100 mm. Due to the nature of the problem, the vertical component of accelerations becomes maximum among other components. Figure 4.24a shows the vertical acceleration and displacement measured simultaneously in the experiment

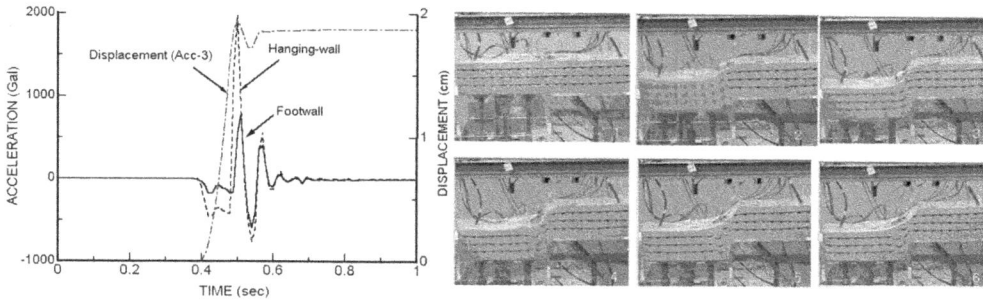

Figure 4.24 Acceleration responses and motions of 200 mm thick soil deposit for 90° normal faulting. (a) Acceleration responses and (b) the motion of the model at several time steps.

Figure 4.25 Views of faulting experiments.

with 200 mm thick soil deposit and 90° normal faulting, and Figure 4.24b shows its motion at several time steps. As seen in Figure 4.23a, the maximum acceleration of the movable side is greater than that of the stationary side. Furthermore, the maximum acceleration is observed when the movement of the movable side is restrained and the acceleration response is entirely unsymmetric while the acceleration response of the stationary side is almost symmetric. Although it is not the purpose of this section to discuss and compare with observations in actual earthquakes, the responses measured during these experiments are quite similar to the observations in actual earthquakes as well as in rock fracture experiments (i.e. Ohta 2011; Ohta and Aydan 2010). The variation in soil deposit thickness has a certain effect on the resulting accelerations. However, if the displacement of the fault is the same, its effect would be small compared to that of the variation of the fault displacement.

The observations on deformation and slip-lines in experiments carried out for inclinations of 45°, 90° and 135° are shown in Figure 4.25. Comparisons were made for three different inclinations of faulting for the same amount of vertical displacements. The most extensive studies are carried out on experiments with the faulting inclination of 90°, in which the effects of allowable vertical displacements and soil thickness were investigated (Figure 4.26). The top soil deposit on the hanging-wall (mobile) side is highly deformed while the soil deposit on the footwall (stationary) side is much less.

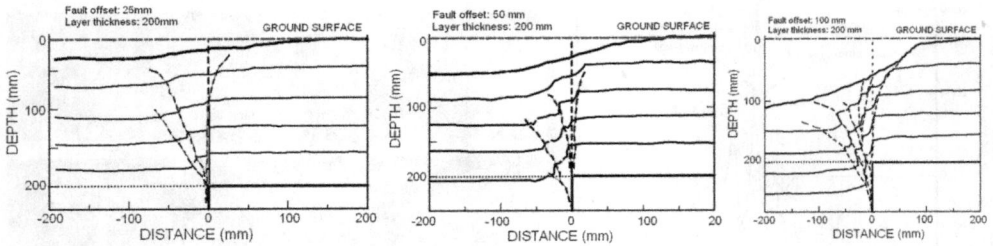

Figure 4.26 Slip-lines in experiments with varying vertical displacements for faulting inclination of 90°. (a) 45° (b) 90°, and (c) 135°.

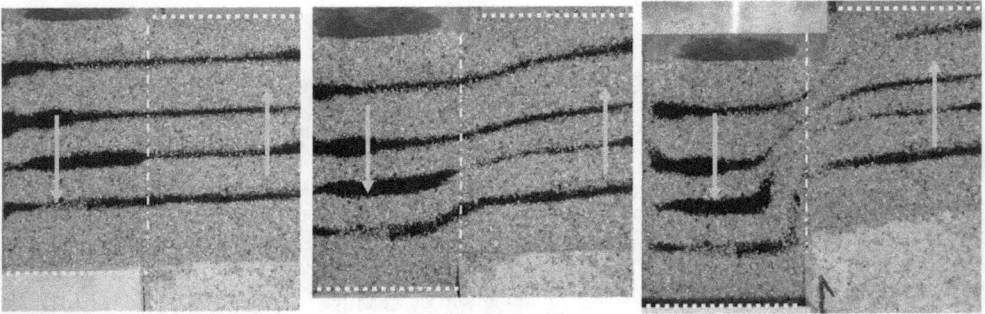

Figure 4.27 Negative views of several stages in a strike-slip faulting experiment in a granular medium.

Furthermore, the number of slip-lines on the hanging wall (mobile part) of the soil
layer is greater than that in the footwall side. This is probably associated with the
amount of displacement of the mobilized soil on the hanging-wall side. It is also interesting to note that thrust-type slip-lines occur at the hanging-wall side while normal
fault type slip-lines develop in the footwall side (see Figure 4.26). Such slip-lines may
be of great significance when ground deformation and slip-lines interpreted for faults
in situ.

Some experiments were carried out by the author and his group using an experimental faulting device of constant velocity type. Although the basic pattern of ground
deformation and slip-line formation is similar, the experimental results showed that
the volume of the deformed wedge-like body is smaller and the inclination of slip-lines
is less and slightly steeper. Furthermore, the ground surface profiles are not sharp as
they are in the experiments shown here. Similar conclusions are also valid for normal
faulting experiments.

In addition, some strike-slip experiments are carried out under constant velocity
conditions. Figure 4.27 shows negative images of several stages of a strike-slip faulting
experiment on granular media. Although slip-lines were not apparent from the figure,
a wide deformation band with a thickness equivalent to the amount of relative displacement developed on both sides of the projected fault-line.

4.3 CHARACTERIZATION OF EARTHQUAKES FROM FAULT RUPTURES

Turkey is one of the well-known earthquake-prone countries in the world and most of its large earthquakes involve ground surface rupturing. The data from the past and present earthquakes of Turkey as well as other countries may be quite useful to establish and/or to revise empirical relations among the characteristics of earthquakes accompanying ground surface rupturing. The data compiled by the author are from the Turkish earthquake data-base (TEDBAS) developed by Aydan (1997) and additional inputs are from publications on recent earthquakes (Ambraseys 1988; Aydan 1996; 2002; Aydan and Hamada 1992; Aydan and Kumsar 1997a; Aydan et al. 1996c, 1998, 1999a,b, 2000a,b, 2005a–c, 2006a, 2007a,b, 2008a,b, 2009a–f, 2010a–c, 2012a; Ergin et al. 1967; Emre et al. 1999; Soysal et al. 1981; Eyidogan et al. 1991; Gencoglu et al. 1990; Ulusay and Aydan 2005; Ulusay et al. 2001, 2002, 2003a,b). The data for other countries are those compiled by Wells and Coppersmith (1994), Matsuda (1975, 1981) and Sato (1989). The data on source properties of earthquakes are gathered from well-known seismological institutes such as USGS, Harvard, ERI of Tokyo University and the Swiss Seismological Institute. The number of data sets varies depending upon the studied empirical relations. For example, the number of data sets for the relation between M_w and M_s is 206. The following items are chosen as the characteristics of earthquakes:

a. Magnitude (moment and surface wave magnitudes, M_w, M_s).
b. Length of earthquake fault (L): L denotes the length of source fault or that estimated by the ground surface trace observed in the field or aftershock distribution if the surface rupture is hindered by the thick sedimentary deposits (e.g. 1992 Erzincan 1998; Adana-Ceyhan earthquake).
c. Depth of earthquake hypocenter (D).
d. Rupture area (S): S denotes the ruptured area of the earthquake fault inferred from aftershock distribution or the multiplication of surface rupture length produced by the earthquake by its hypocenter depth with the assumption of a rectangular source area.
e. Net slip of the earthquake fault (U_{max}): U_{max} denotes the maximum slip along the slip direction.
f. Maximum ground acceleration and velocity (AMAX, VMAX) (hypocenter distance is mostly in the range of 15–25 km).
g. Rupture mode – striation orientation.
h. Ratio of vertical maximum acceleration to the horizontal maximum acceleration (RVAHA).

It should be noted that the minimum value of M_w is assumed to be 0 in all empirical relations presented hereafter although some events may have negative values and such events have no physical significance in earthquake science and engineering.

4.3.1 Relation between surface wave magnitude and moment magnitude

Aydan (Aydan 1997; Aydan et al. 1996c) originally selected surface wave magnitude M_s in developing his empirical relations for earthquakes in Turkey since a lot of data based on surface wave magnitude M_s were available and the magnitude of earthquakes did not exceed 8 so far. It is pointed out that surface wave magnitude M_s becomes unreliable if it exceeds the value of 8 (i.e. Fowler 1990). Furthermore, it is becoming more popular to use the moment magnitude M_w in place of surface wave magnitude M_s as many seismological institutes release moment magnitude data rather than surface wave magnitude data, recently. Nevertheless, the moment magnitude data determined by various institutes for the same earthquake are not always the same. Furthermore, the moment magnitude data must be assigned to previous earthquakes before the development of a moment magnitude concept. Kanamori (1983) suggested that the surface wave and moment magnitudes of earthquakes can be taken as equal to each other within the range of 5–7.5. Aydan (Aydan 1997; Aydan et al. 1996c) proposed the following relation between surface and moment magnitudes of earthquakes for earthquakes in Turkey.

$$M_w = 1.044 M_s \quad \text{or} \quad M_s = 0.958 M_w \tag{4.1}$$

However, Ulusay et al. (2004) recently suggested the following formula for their data set on earthquakes in Turkey:

$$M_w = 0.6798 M_s + 2.0402 \tag{4.2}$$

Figure 4.28 compares the data set of earthquakes compiled by the author, including all recent data on earthquakes in Turkey and worldwide, which are fitted to the following empirical relation:

$$M_w = 1.2 M_s e^{-0.028 M_s} \tag{4.3}$$

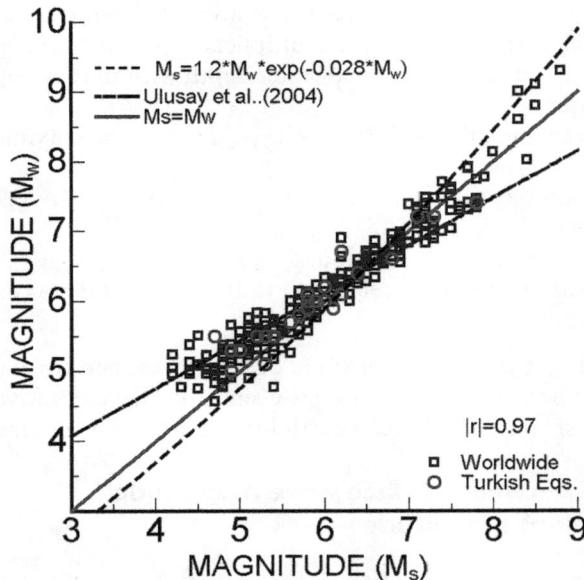

Figure 4.28 Relation between moment magnitude and surface wave magnitude.

As noted from the figure, all data set generally support the suggestion of Kanamori (1983) for relating surface wave magnitude to moment magnitude for the magnitude range of 4–8. Therefore, it is safe to adopt the previous empirical relations proposed by Aydan (1997, 2001a, 2012a) based on surface wave magnitude for the seismic characteristics of earthquakes in Turkey together with the replacement of surface wave magnitude with moment magnitude. However, the constants of functions should be re-calculated if the independent variable is chosen as moment magnitude.

4.3.2   Relation between MMI and moment magnitude

The magnitude of historical earthquakes is mainly inferred from the Modified Mercalli Intensity (MMI). Aydan et al. (1996c) and Aydan (1997) proposed the following empirical relation between MMI intensity and surface wave magnitude M_s (see also Aydan 2001a):

$$I_o = 1.317 M_s \tag{4.4}$$

In this study, the following relation between moment magnitude and MMI intensity is proposed with the consideration of recent large earthquakes in Turkey and worldwide.

$$I_o = 1.32 M_w \tag{4.5}$$

Yarar et al. (1980) and Gürpinar et al. (1979) also studied the relation between Mercalli-Karnik-Sponheuer (MKS) intensity and earthquake magnitude for earthquakes in Turkey. Their proposed relations were essentially similar to those given by Eqs. (4.4) and (4.5), except the constant with the minus sign. However, the magnitude in their formula is local magnitude. Furthermore, the coefficient for magnitude is about 1.6 times the one given in Eqs. (4.4) and (4.5). Kudo (1983) also studied the relation between the MKS intensity and the intensity of the Japan Meteorological Agency. As the maximum values of intensity were different from each other, two intensity values of MKS would be designated as one intensity value in the intensity scale of the Japan Meteorological Agency. However, the intensity scales of 4, 5 and 6 of the Japan Meteorological Agency are recently revised and subdivided as weak and strong intensity levels. Nowadays, the intensity scale of the Japan Meteorological Agency discussed by Kudo (1983) is no longer used in Japan.

4.3.3   Relation between moment magnitude and rupture length, area and net slip of fault

It is natural to expect that the sense of faulting of earthquakes greatly influence their seismic characteristics. The faulting sense is considered in empirical relations between moment magnitude and rupture length (L), rupture area (A) and net slip (U_{max}) of fault given below as an extension of the previous works of Aydan et al. (1966) and Aydan (1997, 2007a, 2012a) and tested with the recent mega earthquake data (Aydan 2015c):

$$L, S \text{ or } U_{max} = A \cdot M_w e^{M_w/B} \tag{4.6}$$

Table 4.2 Values of constants for Eq. (4.6) for each fault parameter

	Rupture length L (km)		Rupture area S (km²)		Maximum displacement U_max (cm)	
	A	B	A	B	A	B
Normal faulting	0.0014525	1.21	0.003	1.5	0.0003	1.6
Strike-slip faulting	0.0014525	1.25	0.001	1.7	0.00035	1.6
Thrust faulting	0.0014525	1.19	0.0032	1.5	0.0014	1.4

The functional form of the empirical relations is same while constants A and B differ depending upon the faulting sense, which are given in Table 4.2. Kudo (1983) and Ambraseys (1988) proposed similar empirical relations between rupture length and earthquake magnitude for earthquakes in Turkey. Most of the earthquakes were due to strike-slip faulting. Although their relations are good fits to the data set they used, they have limited applicability for the range of data set used in his study. Figure 4.29 shows the plot of data for several parameters listed above together with empirical functions given by Eq. (4.6); Aydan (1997) and Well and Coppersmith (1994). The horizontal axis of the plots is the moment magnitude of earthquakes. As seen in the figure, the data are somewhat scattered. Nevertheless, the function together with constants for each seismic parameter fits best to observational data. The standard deviations of fitted equations to observational data were obtained and are shown in Figure 4.29. Furthermore, the relation proposed by Aydan (1997) without considering faulting sense is still valid as noted from Figure 4.29 provided that the conversion of magnitudes is taken into account.

Another important source parameter is the duration of fault rupture. The duration is longer if the rupture propagation is unilateral. However, it is shortened if bi-lateral rupture propagation takes place. Figure 4.30 shows a compilation of data on earthquakes in Turkey and worldwide including the most recent events such as the 2004 and 2005 Sumatra earthquakes and the 2005 Pakistan earthquake. The functional form of the empirical equation has the form of Eq. (4.6) and it is shown in Figure 4.30 together with empirical relations proposed by Dobry et al. (1978). The empirical relation shown in the figure holds for the data of earthquakes in Turkey as well as worldwide data.

4.4   INFERENCE OF FAULTING MECHANISM AND EARTHQUAKES

4.4.1   Inference from striations of earthquake faults

The striations and internal structure of these faults are just evidences of what type of stress state caused them, and they may also indicate what type of earthquake they produced. The methodology for the inference of the possible stress state and focal plane solutions of earthquakes from the faults require data on dip, dip direction and striation orientation (Aydan 2000a; Aydan and Kim 2002; Aydan et al. 2002a). Figure 4.31 shows an illustration of how striation angle is measured. Figure 4.32 shows the focal mechanism of earthquakes estimated from the fault striations obtained for the 1891 Kiso-Beya earthquake in Japan and the 1999 Kocaeli earthquake in Turkey.

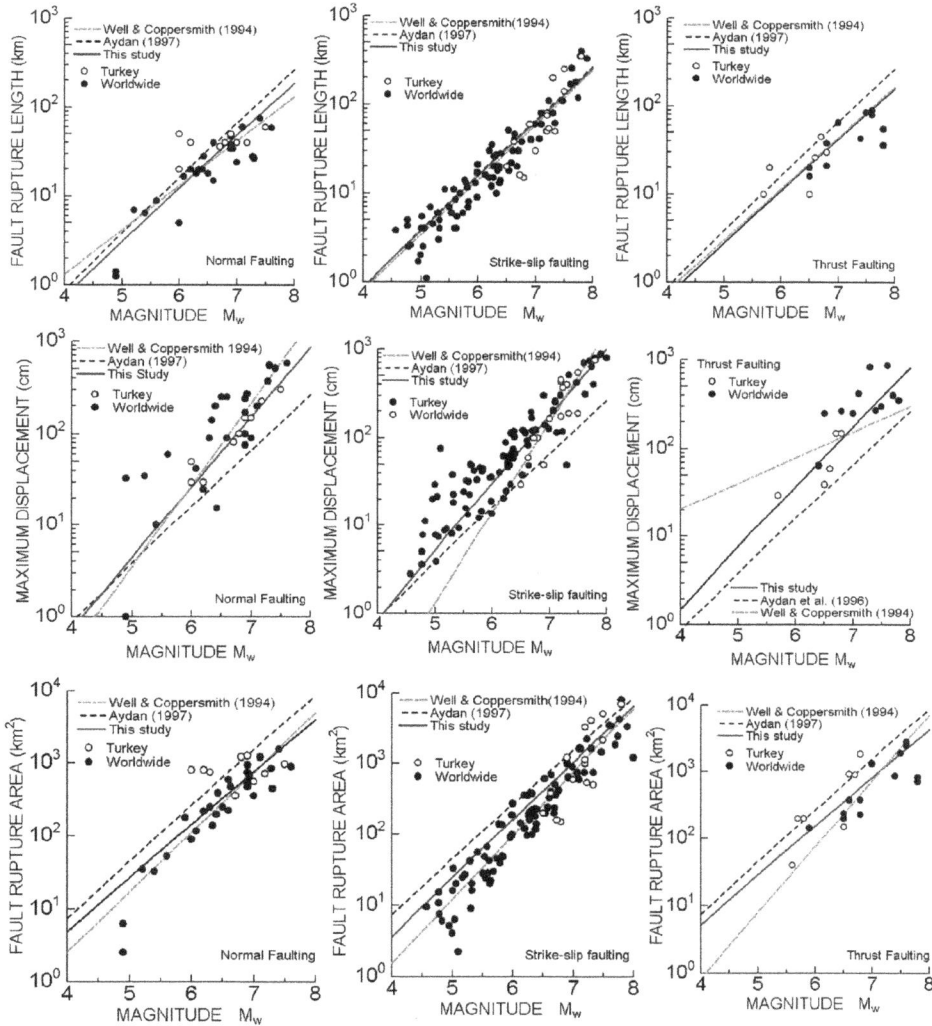

Figure 4.29 Comparison of various relations with observed data.

4.4.2   Inference from wave propagation characteristics

The focal mechanism of the earthquakes may also be inferred from the P-waves and S-waves induced by earthquakes. Figure 4.33 illustrates the concept of obtaining focal mechanism of a vertical strike-slip and associated wave responses at the observation points around the focus of the earthquake. Figure 4.34b shows the focal mechanism solution for the motion of fault plane shown in Figure 4.34a and the seismograms for this solution would look like those shown in Figure 4.34. Figure 4.35 shows the distribution of push-up and pull-down distributions to infer the faulting mechanism of the 1943 Fukui earthquake as an example.

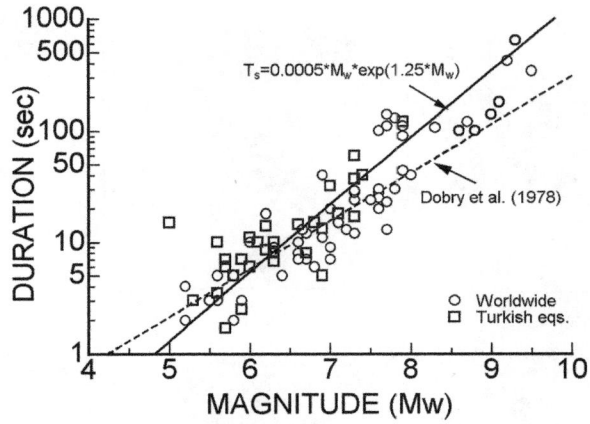

Figure 4.30 Comparison of various relations with observed data.

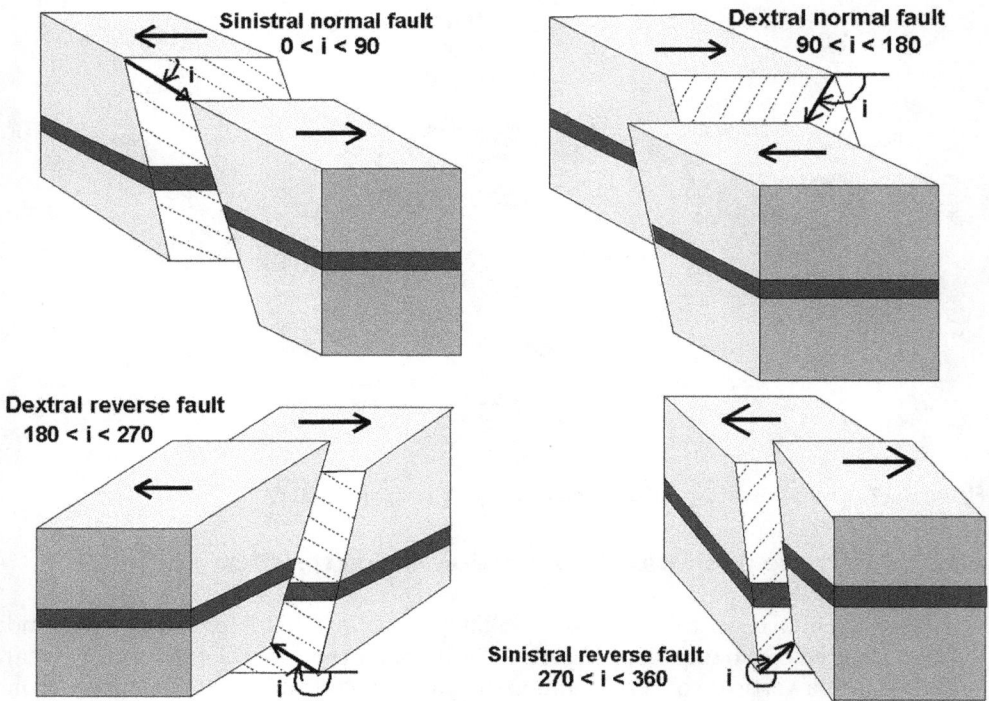

Figure 4.31 Definition of striation in structural geology.

NEODANI

N

W — P

E

T

S

LOWER HEMISPHERE
EQUAL ANGLE PROJECTION

BAŞISKELE

N

P

T

W

E

S

LOWER HEMISPHERE
EQUAL ANGLE PROJECTION

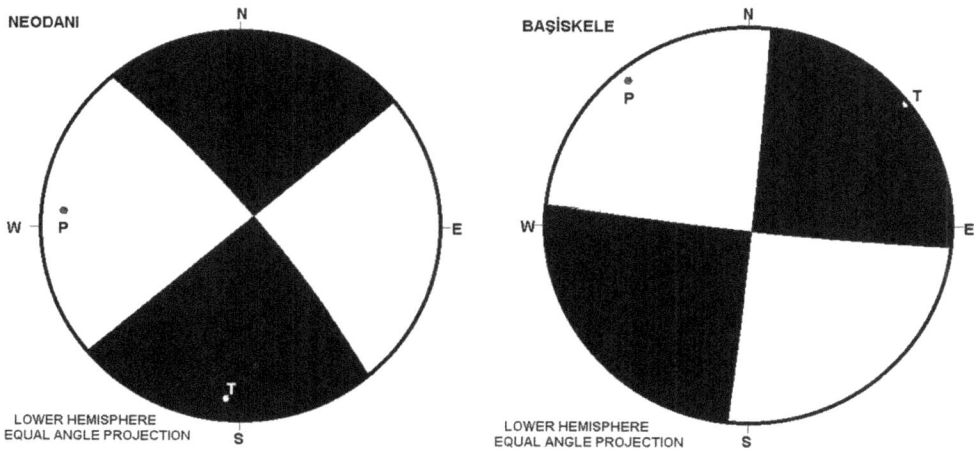

Figure 4.32 Inferred faulting mechanism for some earthquakes from fault striations. (a) 1881 Kiso-Beya earthquake and (b) 1999 Kocaeli earthquake.

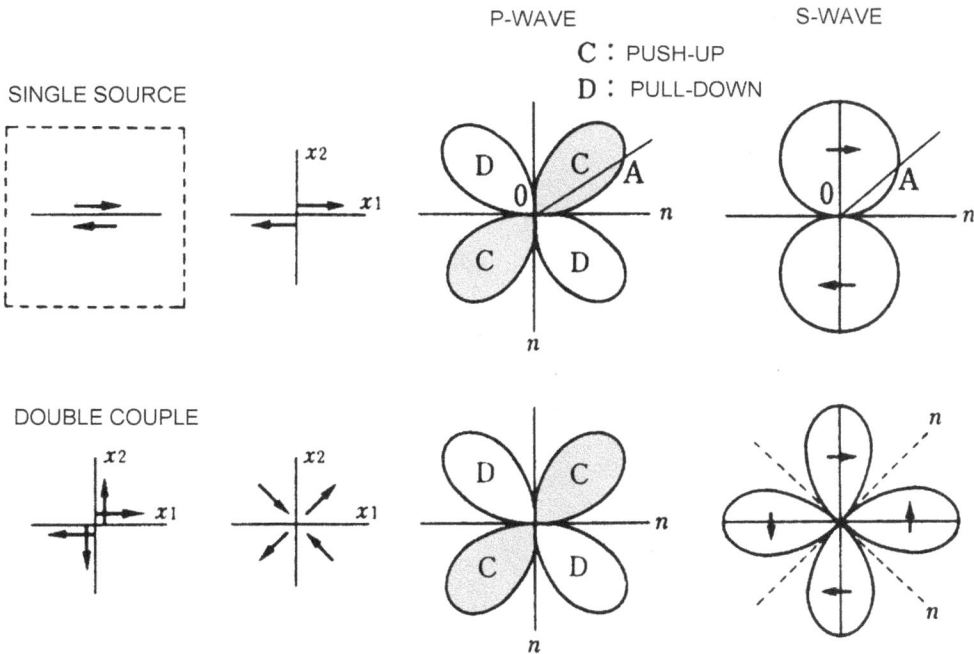

P-WAVE

S-WAVE

C : PUSH-UP

D : PULL-DOWN

SINGLE SOURCE

x_2

x_1

D   C
0   A
        n
C   D

n

0   A
        n

DOUBLE COUPLE

x_2

x_1

x_2

x_1

D   C
        n
C   D

n

n

n

Figure 4.33 Illustration of the fundamental concept to obtain focal mechanism solutions from seismic waves.

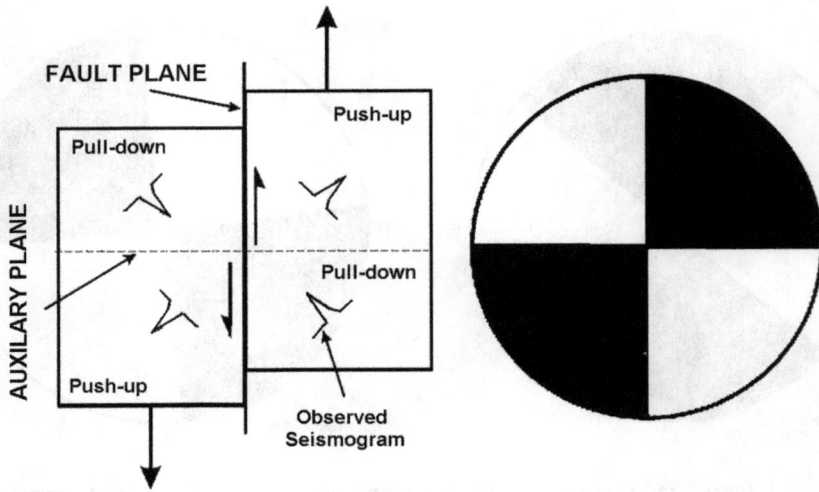

Figure 4.34 (a) Fault motion and (b) inferred focal mechanism.

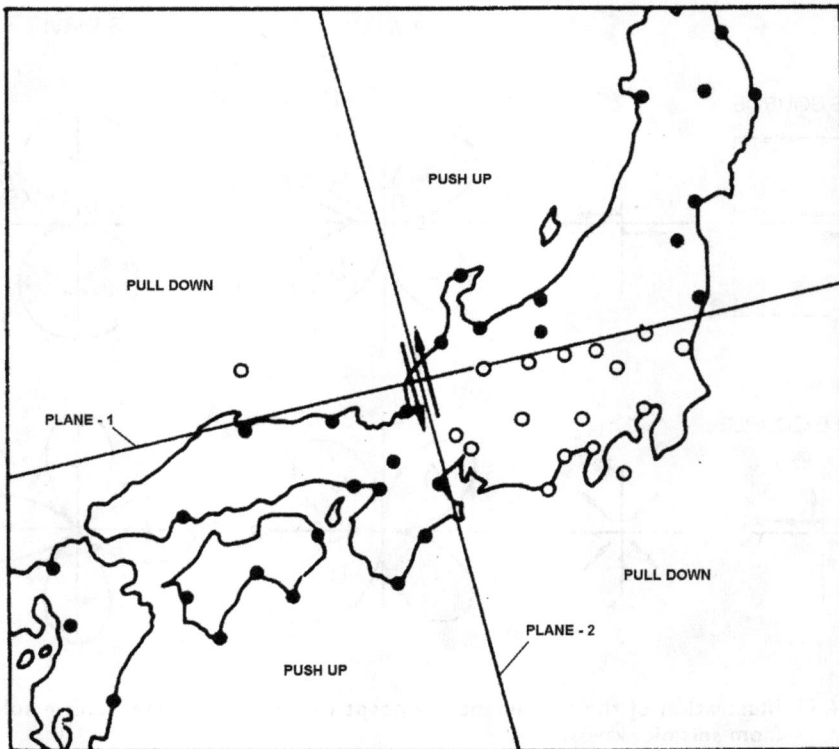

Figure 4.35 Focal mechanism of the 1948 Fukui earthquake from pull-down and push-up distributions.

The inclinations of fault planes have different orientations and data-processing requires the depth, dip direction and take-off angle of waves. For this purpose, a focal sphere fixed to the hypocenter of the earthquake, the character of recorded waves and the location of seismometer are considered. Figure 4.36 illustrates such an example for sign and polarities of P-wave first-motion seismic rays while Figure 4.37 shows focal mechanism of the 1984 Nagano-Seibu earthquake. Many seismological institutes provide take-off angle and the sense of motion data, which are used in obtaining focal plane solutions. However, it should be noted that a minimization procedure for observed data is necessary to determine focal plane solutions.

The focal plane solutions used in geo-science for inferring the faulting mechanism of earthquakes are derived by assuming that the pure-shear condition holds. As a result of this assumption, one of the principal stresses is always compressive while the

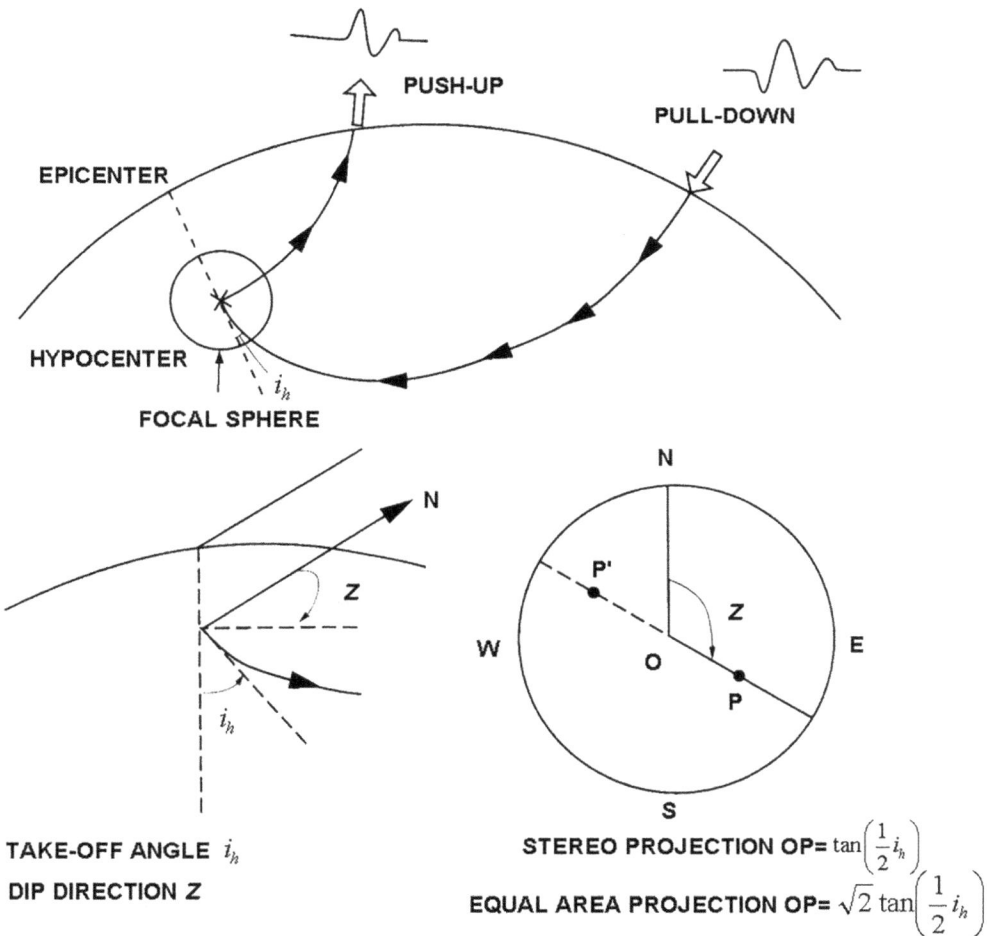

TAKE-OFF ANGLE i_h

DIP DIRECTION Z

STEREO PROJECTION OP= $\tan\left(\dfrac{1}{2}i_h\right)$

EQUAL AREA PROJECTION OP= $\sqrt{2}\,\tan\left(\dfrac{1}{2}i_h\right)$

Figure 4.36 Sign and polarities of P-wave first-motion seismic rays. (Modified from Ando et al. (1996).)

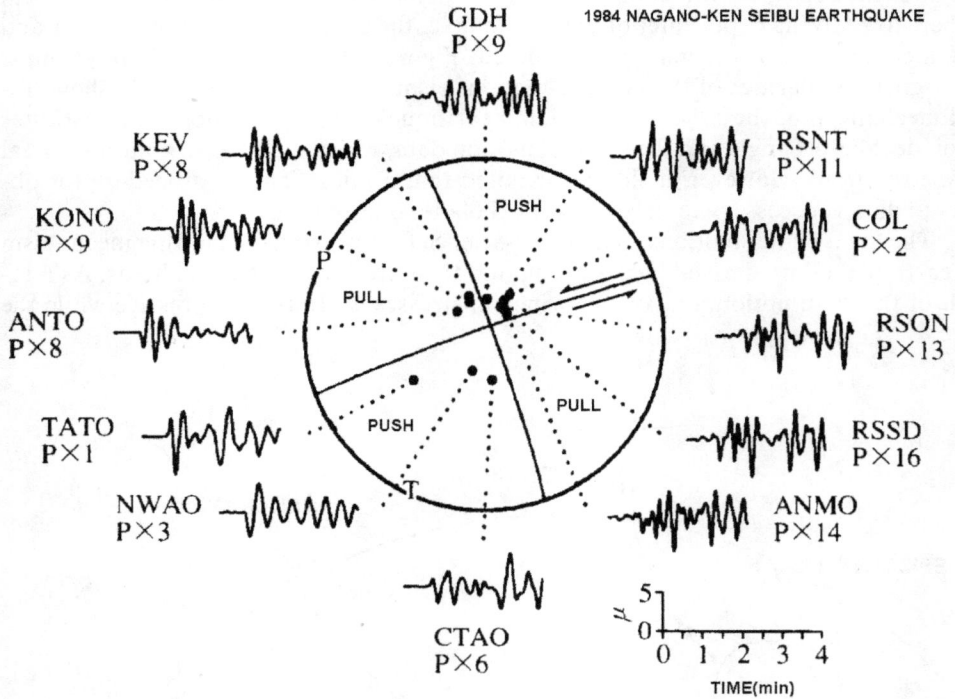

Figure 4.37 Focal mechanism solution for the 1984 Nagano-Seibu earthquake determined from P-wave first-motion seismic rays. (Modified from Ando et al. (1996).)

other one is tensile in focal plane solutions. This condition may also imply that the friction angle of the fault is assumed to be nil. Therefore, the principal stresses are inclined at an angle of 45° with respect to the normal of slip direction. This condition is used to determine the P and the T axes in focal plane solutions. Each focal plane solution involves the fault plane, on which the sliding takes place, and the auxiliary plane (Figure 4.38). The normal of the auxiliary plane corresponds to the slip vector, and it is orthogonal to the neutral plane on which the P and T axes exist.

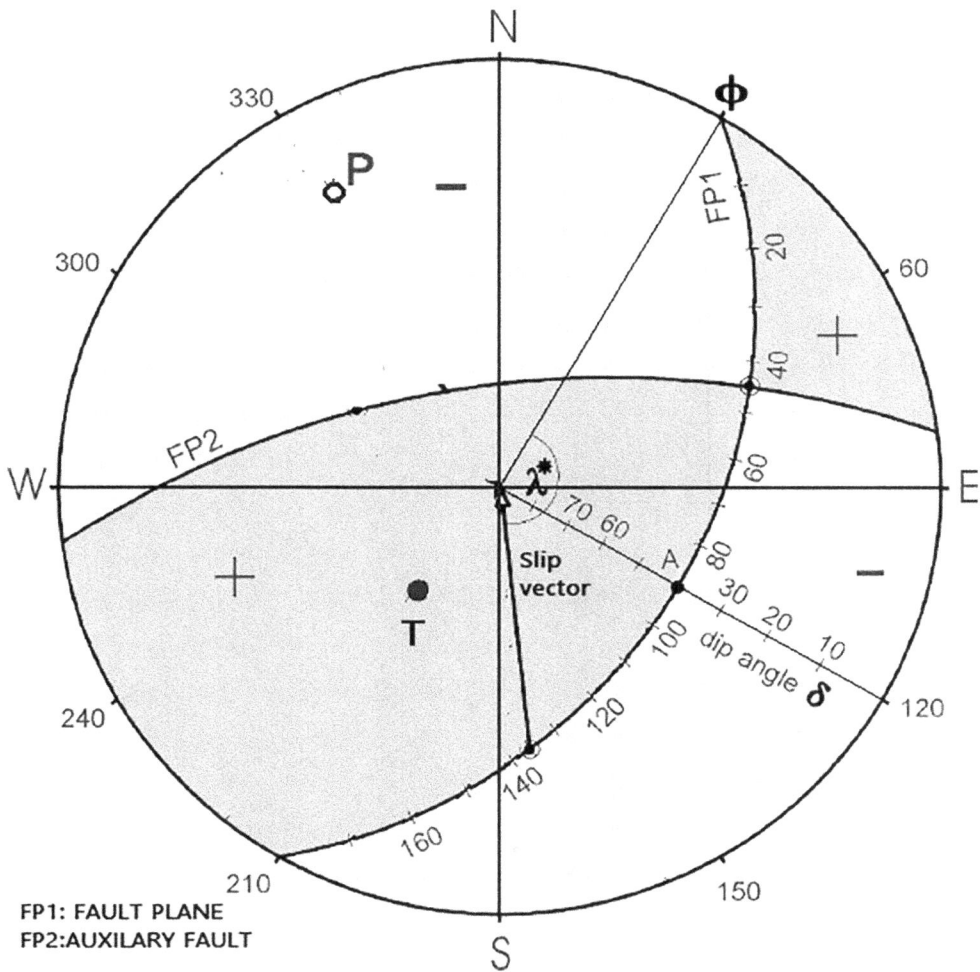

Figure 4.38 Illustration of fault plane, auxiliary fault and slip-vector on a lower-hemi-sphere stereo-net.

Chapter 5

Strong ground motions and permanent ground deformations

5.1 OBSERVATIONS ON STRONG MOTIONS AND PERMANENT DEFORMATIONS

5.1.1 Observations on maximum ground accelerations

It is observationally known that the ground motions induced by earthquakes could be much higher in the hanging-wall block or mobile side of the causative fault as observed in recent earthquakes such as the 1999 Kocaeli earthquake (strike-slip faulting), the 1999 Chi-chi earthquake (thrust faulting), the 2004 Chuetsu earthquake (blind thrust faulting) and the 2000 Shizuoka earthquake and L'Aquila earthquake (normal faulting) (e.g. Ohta 2011; Chang et al. 2001; Abrahamson and Somerville 1996; Somerville et al. 1997; Aydan 2003b, 2012a; Aydan et al. 2007a–c; Aydan and Ohta 2011), as seen in Figures 5.1–5.3.

Figure 5.4 illustrates the effect of hanging-wall effect on the attenuation of maximum ground accelerations observed in the 1999 Chi-chi earthquake (Taiwan), the 1999 Düzce earthquake (Turkey) and the 2001 Geiyo earthquake (Japan) with different faulting mechanisms (Ohta and Aydan 2010).

Figure 5.5 shows the records of accelerations at the ground surface and at bedrock 260 m below the ground surface at Ichinoseki strong motion station (IWTH25) of KiK-NET (2008) strong motion network of Japan measured during the 2008 Iwate-Miyagi earthquake. The strong motion station was located on the hanging-wall side of the fault and it was very close to the surface rupture. As noted from the figure, the ground acceleration of the UD component was amplified by 5.67 times that at the bedrock and the acceleration records are not symmetric with respect to time axis. This record is also the highest strong motion recorded in the world so far.

5.1.2 Permanent ground deformation

Recent analysis of the GPS also showed that permanent deformations of the ground surface occur after each earthquake (Figures 5.6 and 5.7). The permanent ground deformation may result from different causes such as faulting, slope failure, liquefaction and plastic deformation induced by ground shaking (e.g. Aydan et al. 2010a–c; Kanibir et al. 2006). These types of ground deformations have limited effect on small

DOI: 10.1201/9781003164371-5

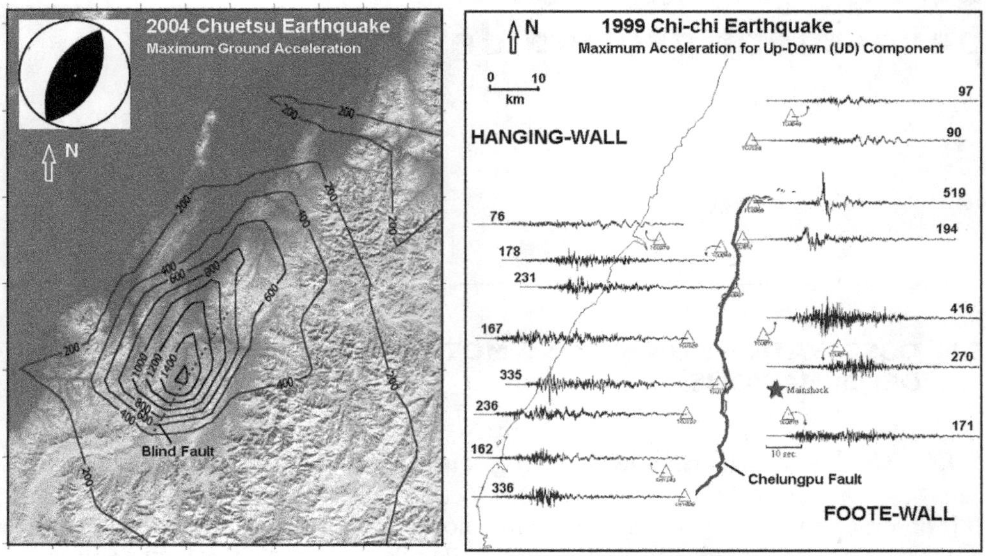

Figure 5.1 Footwall and hanging-wall effects on maximum ground accelerations (thrust faulting).

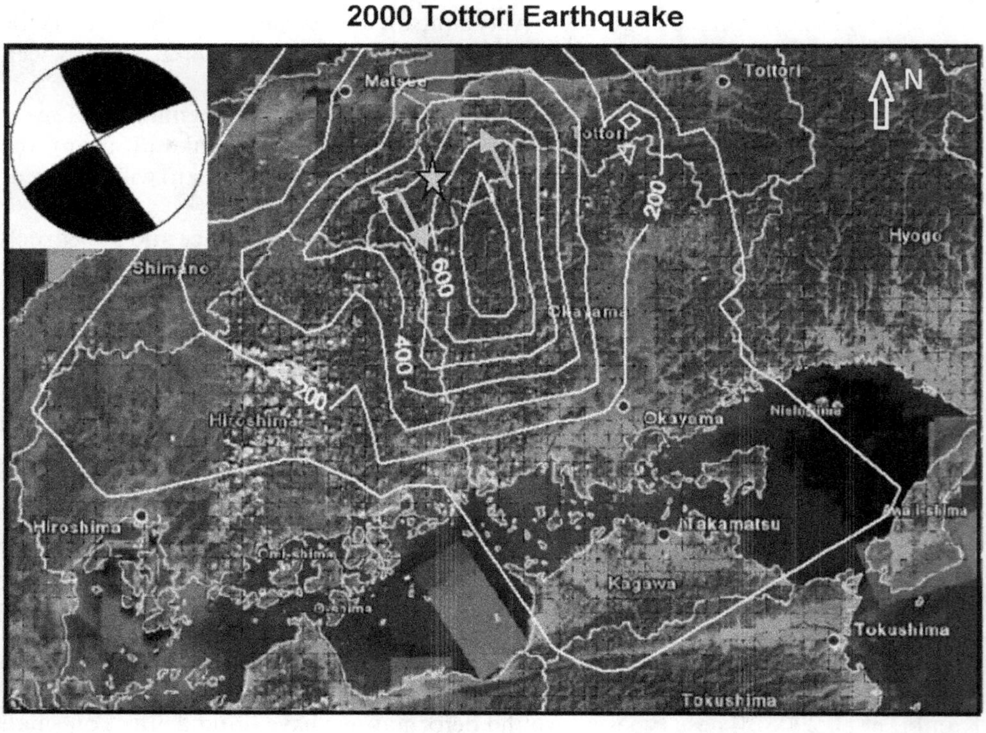

2000 Tottori Earthquake

Figure 5.2 Mobile and stationary block effects on the maximum ground accelerations observed in the 2000 Tottori Seibu earthquake (strike-slip faulting).

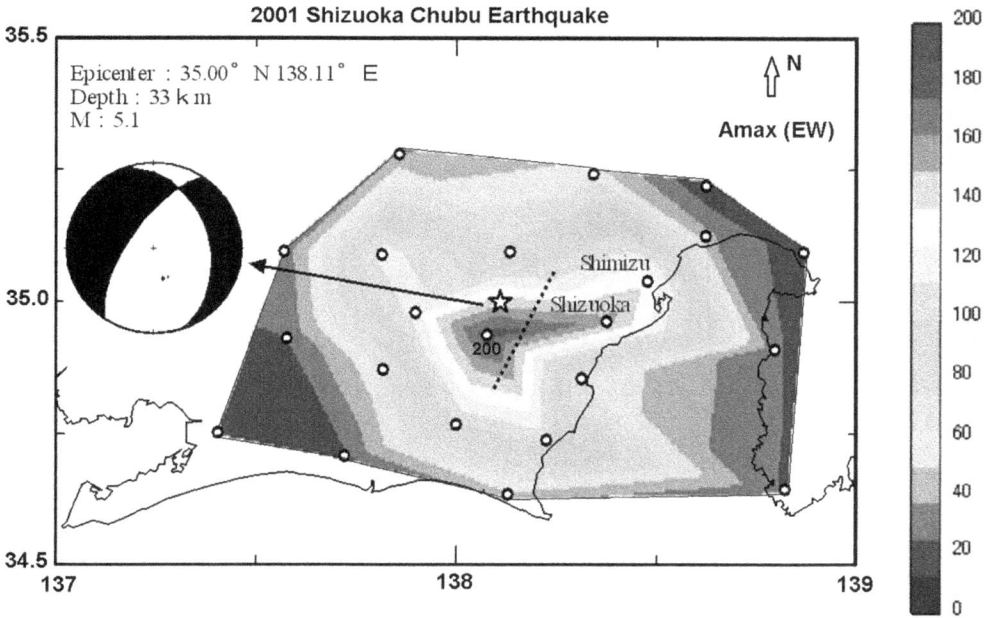

2001 Shizuoka Chubu Earthquake

Epicenter : 35.00° N 138.11° E
Depth : 33 km
M : 5.1

Amax (EW)

Shimizu
Shizuoka

Figure 5.3 Footwall and hanging-wall effects on maximum ground accelerations (normal faulting).

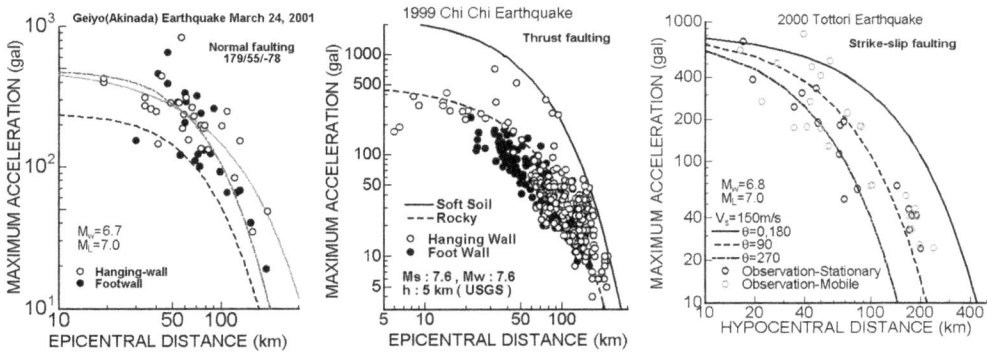

Figure 5.4 Attenuation of maximum ground accelerations for some earthquakes.

structures as long as the surface breaks do not pass beneath those structures. However, such deformations may cause tremendous forces on long and/or large structures, such as rock engineering structures. Ground deformation may induce large tensile or compression forces as well as bending stresses in structures depending upon the character of permanent ground deformations. Blind faults and folding processes may also induce some peculiar ground deformations and associated folding of overlaying soft

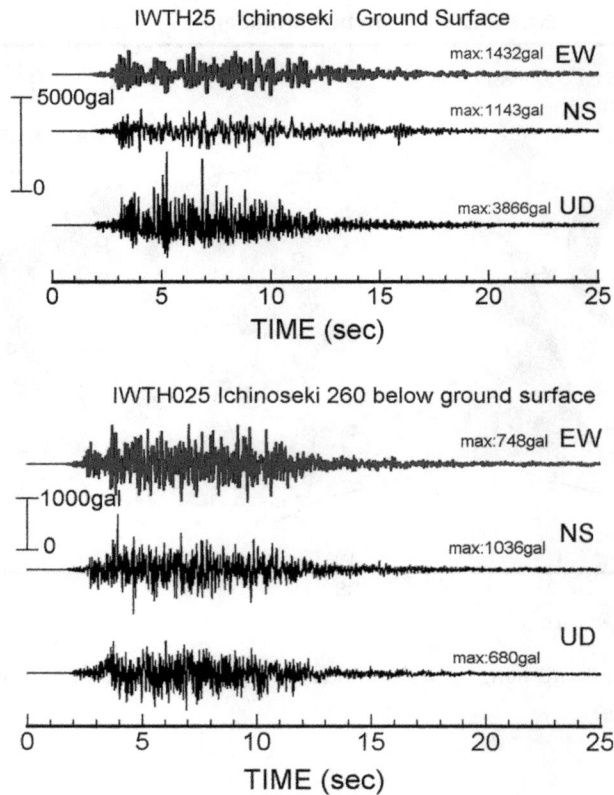

Figure 5.5 Acceleration records at ground surface and bedrock at the Ichinoseki strong motion station IWTH25 of KIK-NET for the Iwate-Miyagi earthquake.

sedimentary layers. Such deformations caused tremendous damage on tunnels during the 2004 Chuetsu earthquake although no distinct rupturing took place.

5.2 STRONG MOTION ESTIMATIONS

5.2.1 Empirical approach

There are many empirical attenuation relations for estimating ground motions in literature (i.e. Joyner and Boore 1981; Campbell 1981; Ambraseys 1988; Aydan et al. 1996c). Including so-called next-generation attenuation (NGA) relations, all these equations are essentially spherical or cylindrical attenuation relations and they cannot take into account the directivity effects. As discussed at the beginning of this section, ground motions such as maximum ground acceleration (AMAX) and maximum ground velocity (VMAX) have strong directivity effects in relation to fault orientation. Furthermore,

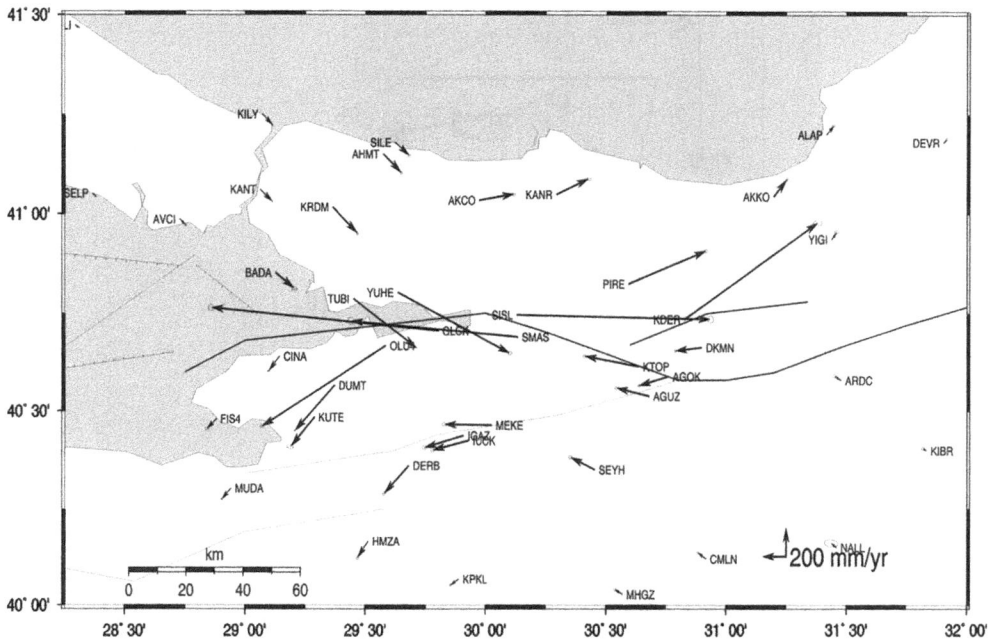

Figure 5.6 Permanent ground deformations and associated straining induced by the 1999 Kocaeli earthquake. (From Reilinger et al. (2000).)

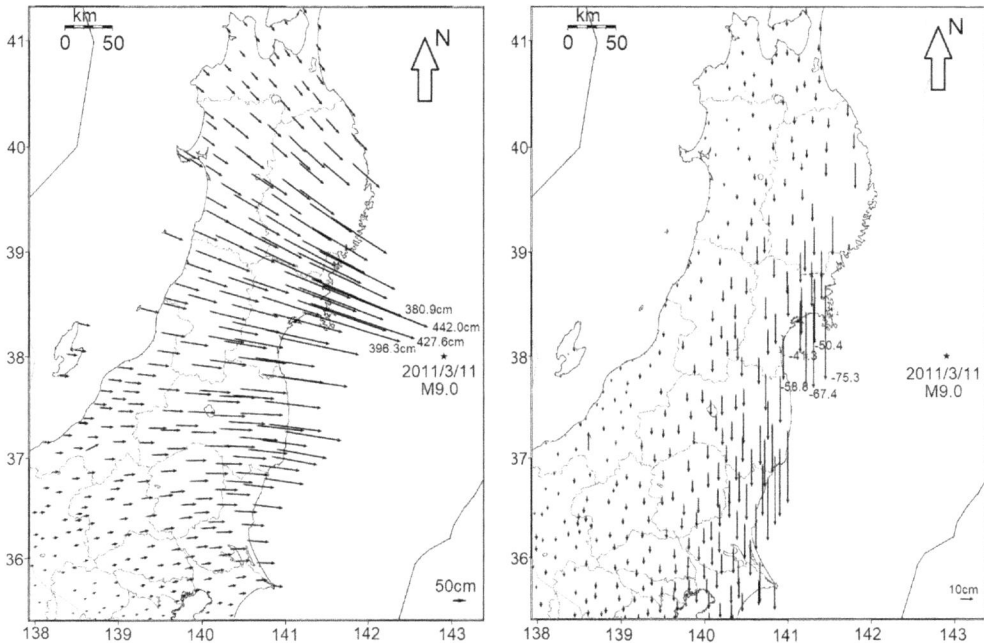

Figure 5.7 Ground deformation induced by the 2011 GEJE (measured using GPS by GSI, 2011).

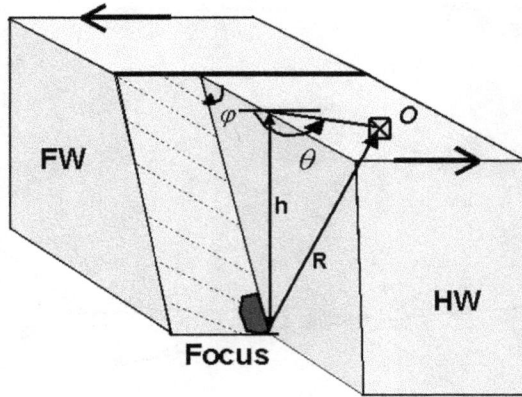

Figure 5.8 Illustration of geometrical fault parameters (R,θ,ϕ).

these relations are generally far below the maximum ground acceleration and they are incapable of obtaining AMAX or "peak ground acceleration (PGA)".

Aydan (2012a) proposed an attenuation relation by combining their previous proposals (Aydan et al. 1996c; Aydan 1997, 2001a, 2007b; Aydan and Ohta 2011) with the consideration of the inclination and length of earthquake fault using the following functional form (Figure 5.8):

$$\alpha_{\max} = F_1(V_s) * F_2(R,\theta,\phi,L^*) * F_3(M) \tag{5.1}$$

where V_s, $\theta,\phi,L*$ and M are the shear velocity of ground and the angle of the location from the strike and dip of the fault (measured anti-clockwise with the consideration of the mobile side of the fault) and earthquake magnitude. The following specific forms of functions in Eq. (5.1) were put forward:

$$F_1(V_s) = Ae^{-V_s/B} \tag{5.2a}$$

$$F_2(R,\theta,\phi,L^*) = e^{-R(1-D\sin\theta+E\sin^2\theta)(1+F\cos\phi)/L^*} \tag{5.2b}$$

$$F_3(M) = e^{M/G} - 1 \tag{5.2c}$$

The same form is also used for estimating the maximum ground velocity (VMAX). L^* (in km) is a parameter related to half of the fault length. And it is related to the moment magnitude in the following form:

$$L^* = a + be^{cM_w} \tag{5.3}$$

The specific values of constants of Eqs. (5.2) and (5.3) for this earthquake are given in Tables 5.1–5.3.

Table 5.1 Values of constants in Eq. (5.2) for inter-plate earthquakes

	A	B (m/s)	D	E	F	G (Mw)
AMAX	2.8	1,000	0.5	1.5	0.5	1.05
VMAX	0.4	1,000	0.5	1.5	0.5	1.05

Table 5.2 Values of constants in Eq. (5.2) for intra-plate earthquakes

	A	B (m/s)	D	E	F	G (Mw)
AMAX	2.8	1,000	0.5	1.5	0.5	1.16
VMAX	0.4	1,000	0.5	1.5	0.5	1.16

Table 5.3 Values of constants in Eq. (5.2) for earthquakes

Faulting type	a	b	c
Normal faulting	30	0.002	1.35
Strike-slip faulting	20	0.002	1.40
Thrust faulting	20	0.002	1.27

The most important parameter in this approach is the estimation of magnitude of the potential earthquake. If a very reliable database exists for a given region, one may estimate the magnitude of the most-likely earthquake from such a data base. Another approach may be an estimation from the characteristics (length, area, maximum relative slip) of active faults. Matsuda (1981), Sato (1989), Wells and Coppersmith (1994) and Aydan (1997) proposed empirical relations. Aydan (2007c, 2012a, 2015c) recently established several relations between moment magnitude and rupture length (L), rupture area (S) and net slip (U_{max}) of fault given below and checked their validity with available data as well as the data from the most recent event of the 2011 Great East Japan Earthquake (GEJE):

$$L, S \text{ or } U_{max} = A \cdot M_w e^{M_w / B} \tag{5.4}$$

The functional form of the empirical relations is same while their constants A and B differ depending upon the faulting sense, which are given in Table 5.4. If striation or sense of deformation of the potential active fault is known, it is also possible to infer its focal mechanism. Such a method is proposed by Aydan (2000a) and compared with the focal mechanism solutions inferred from fault striations or sense of deformation with those from telemetric wave solutions.

The attenuation relation given by Eq. (5.2) was used to evaluate the maximum ground acceleration and ground velocity of the GEJE and the 1999 Kocaeli earthquake and were compared with actual observation data in Figures 5.9 and 5.10. The same equation is used to evaluate the areal distribution of maximum ground

Table 5.4 Values of constants for Eq. (5.2) for each fault parameter

Fault type	Parameter	L (km)	S (km²)	U_{max} (cm)
Normal faulting	A	0.0014525	0.003	0.0003
	B	1.21	1.50	1.6
Strike-slip faulting	A	0.0014525	0.001	0.00035
	B	1.19	1.70	1.6
Thrust faulting	A	0.0014525	0.0032	0.0014
	B	1.25	1.50	1.6

L, rupture length; S, rupture area; U_{max}: maximum displacement.

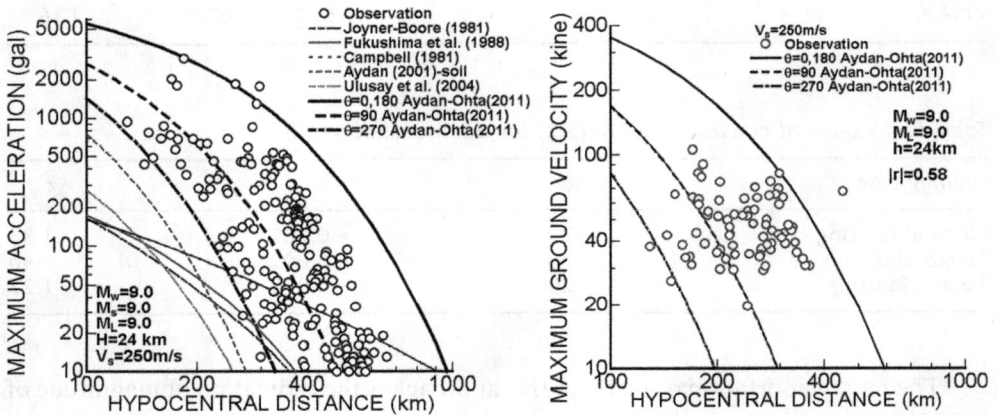

Figure 5.9 Comparison of estimated attenuation of maximum ground acceleration and ground velocity with observations for the 2011 GEJE.

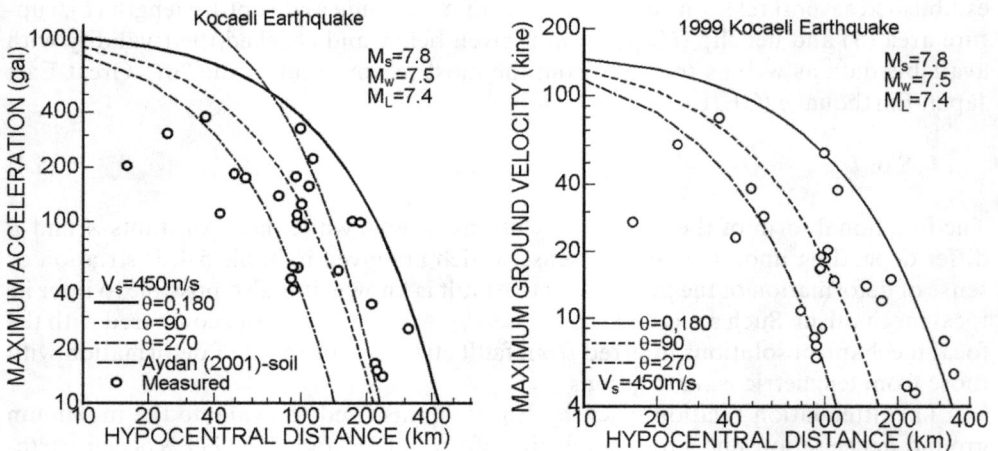

Figure 5.10 Comparison of estimated attenuation of maximum ground acceleration and ground velocity with observations for the 1999 Kocaeli earthquake.

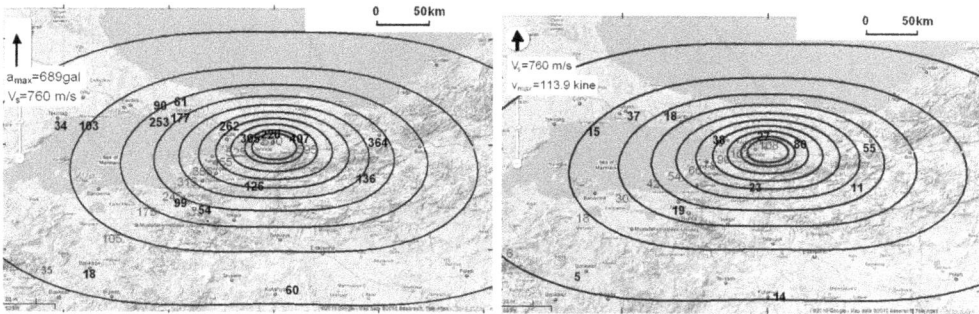

Figure 5.11 Comparison of estimated contours of maximum ground acceleration and ground velocity with observations for the 1999 Kocaeli earthquake.

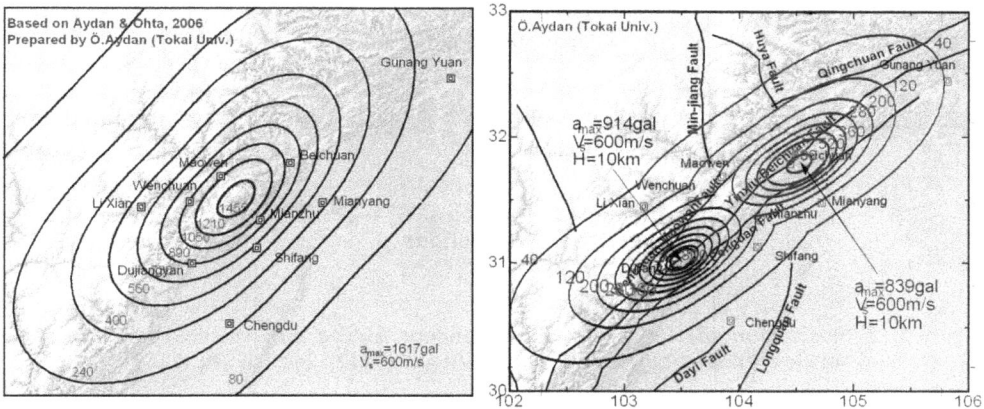

Figure 5.12 Comparison of single- and double-source models for maximum ground acceleration for the 2008 Wenchuan earthquake.

acceleration and velocity for Kocaeli earthquake and compared with observational data in Figure 5.11. When the Eq. (5.2) is applied to large earthquakes, the estimations based the segmentation of faults may be more appropriate. Figure 5.12 shows the single- and double-source models for the 2008 Wenchuan earthquake.

5.2.2 Green-function-based empirical waveform estimation

The empirical Green's function method was initially introduced by Hartzell (1978). Follow-up methods proposed by Irikura (1983) are modifications of Hartzell's method of summing empirical Green's functions. In empirical Green's function approach, rupture propagation and radiation pattern were specified deterministically and the source propagation and radiation effects were included empirically by assuming that the motions observed from aftershocks contained this information (Somerville et al. 1991). A

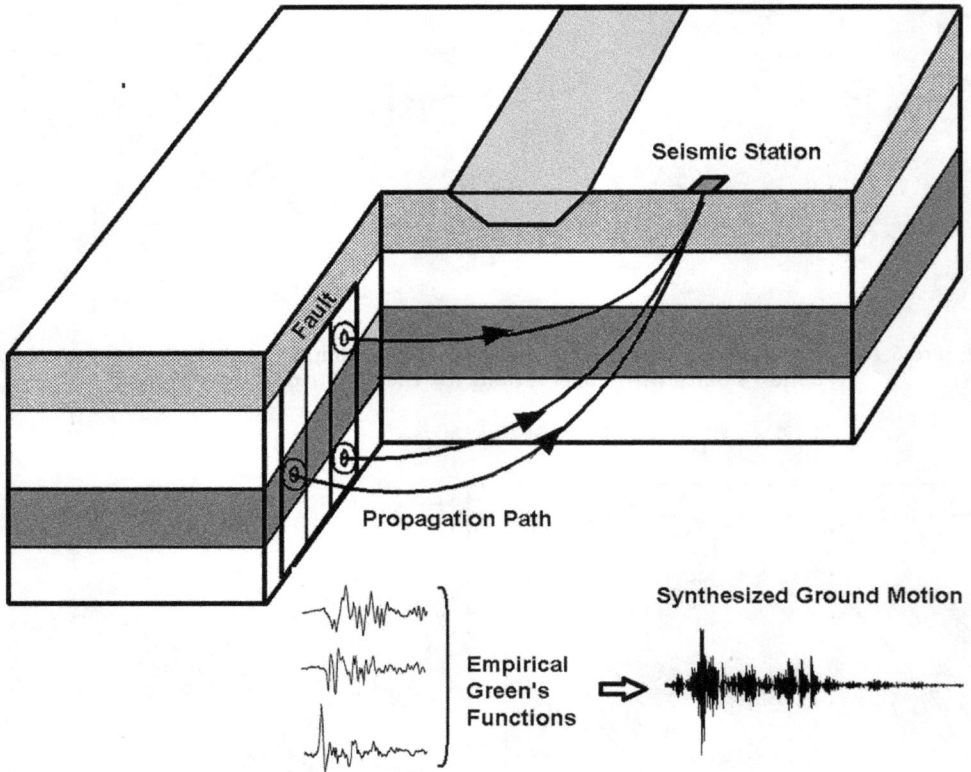

Figure 5.13 Illustration of the fundamental concept of the empirical Green's function method. (Based on Hutchings and Viegas (2012) and Irikura (1983).)

semi-empirical Green's function summation technique has been used by Wald et al. (1998), Cohen et al. (1991) and Somerville et al. (1991) which allows gross aspects of the source rupture process to be treated deterministically using a kinematic model based on first motion studies, tele-seismic modelling and distribution of aftershocks. Gross aspects of wave propagation are modelled using theoretical Green's functions calculated with generalized rays (Figure 5.13). Empirical Green's function method can be used only for a region where small events (i.e. aftershocks or foreshocks) of the target event are available.

Ikeda et al. (2016) recently performed an analysis of strong motion induced by the 2014 Nagano-ken-Hokubu earthquake using the empirical Green's function method. The earthquake fault was assumed to be 9.6 km long and 7.2 km wide with a dip angle of 50°. The stress drop was about 12.6 MPa and the rupture time was 2.7 km/s with a rise time of 0.6 s. Figure 5.14 shows the observed and simulated responses of acceleration, velocity and displacement in the NS direction for the K-NET Hakuba strong motion station. As seen from the figure, the simulated strong motions are close to the observations.

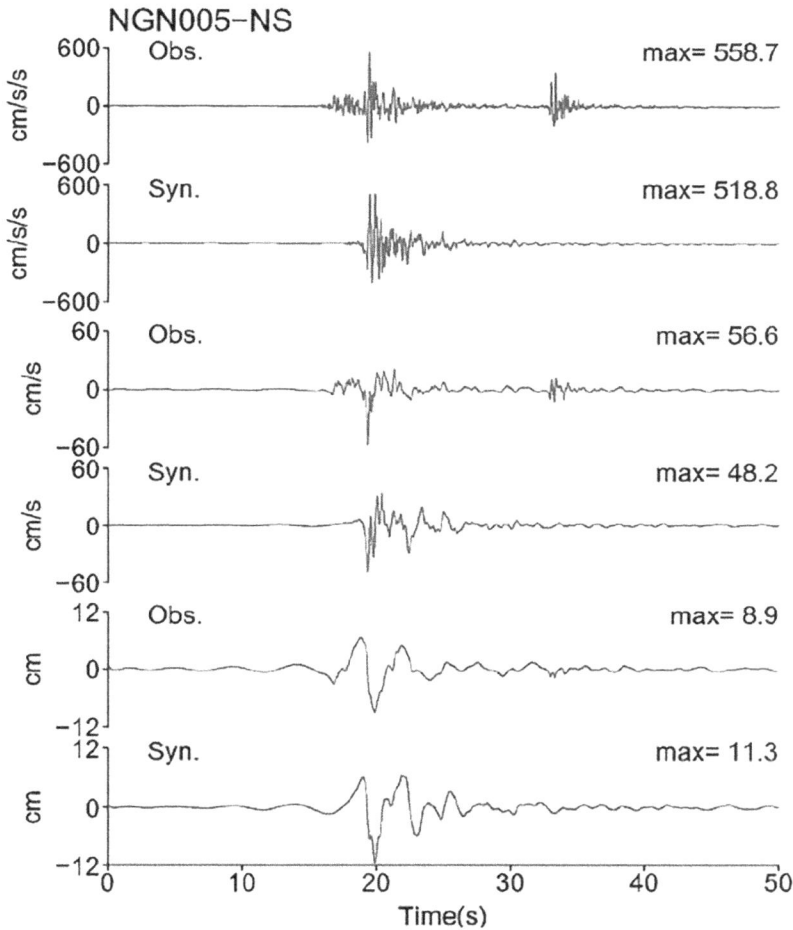

Figure 5.14 Comparison of simulations with observations for the NS component of Hakuba record. (From Ikeda et al. (2016).)

5.2.3 Numerical approaches

There are several numerical techniques, such as finite difference method (FDM), finite element method (FEM), and boundary element method (BEM). FDM is the earliest numerical model, while FEM and BEM became available during the 1960s and 1970s, respectively. Therefore, the first application of the numerical methods for strong motion estimation is related to FDM. When this method is applied for strong motion estimation, one needs to solve Eq. (3.8) together with appropriate constitutive laws for the medium and the assumption of a rupture plane. Particularly the geometrical definition of the rupture plane and its rupture velocity would also be the key parameters of the simulations. Furthermore, as both FDM and FEM consider finite size domain, the prevention of reflections of waves from the boundaries would be necessary. This issue is generally dealt with the introduction of the Lysmer-type viscous boundaries into the numerical model. Both FDM and FEM would evaluate the wave propagation without any assumption on how waves generated at the source are transferred to any point of particular interest, which is a major issue in Green's-function-based strong motion simulations. While FEM can easily handle the irregular boundaries such as the surface of the model with its topography as free boundary, FDM has a severe restriction in dealing with such boundaries with irregular geometry. Nevertheless, some procedures dealing with irregular surface topography have been proposed (i.e. Hestholm 1999; Gravers et al. 1996). For irregular surface topography, FDM and FEM are also combined (i.e. Ducellier and Aochi 2012). In addition, there are also some proposals to combine FDM or FEM with BEM in order to deal with newly developing ruptures.

5.2.3.1   Finite difference method

Constitutive law for the medium adjacent to rupture plane is generally assumed to be visco-elastic (i.e. Graves et al. 1996). The most difficult aspect is the simulation of the fault plane associated with the rupture process in FDM schemes. The most conventional technique is to assume that fault planes coincides with grid planes. Forced displacement field is introduced at domain where two points occupy the same space initially and can move relative to each other after the rupture. Many schemes also explore the incorporation of FEM or boundary integral model to simulate the fault plane, and the rest of the domain is discretized using FDM. Figure 5.15 shows a simulation of strong motion induced by the 1995 Kobe earthquake using FDM (Pitarka 1999).

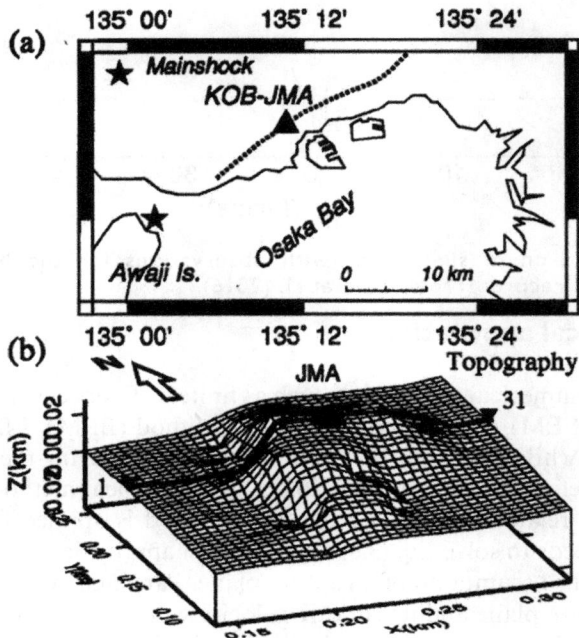

Figure 5.15 Fault normal ground velocity propagation induced by the 1995 Kobe earthquake. (From Pitarka (1999).)

(Continued)

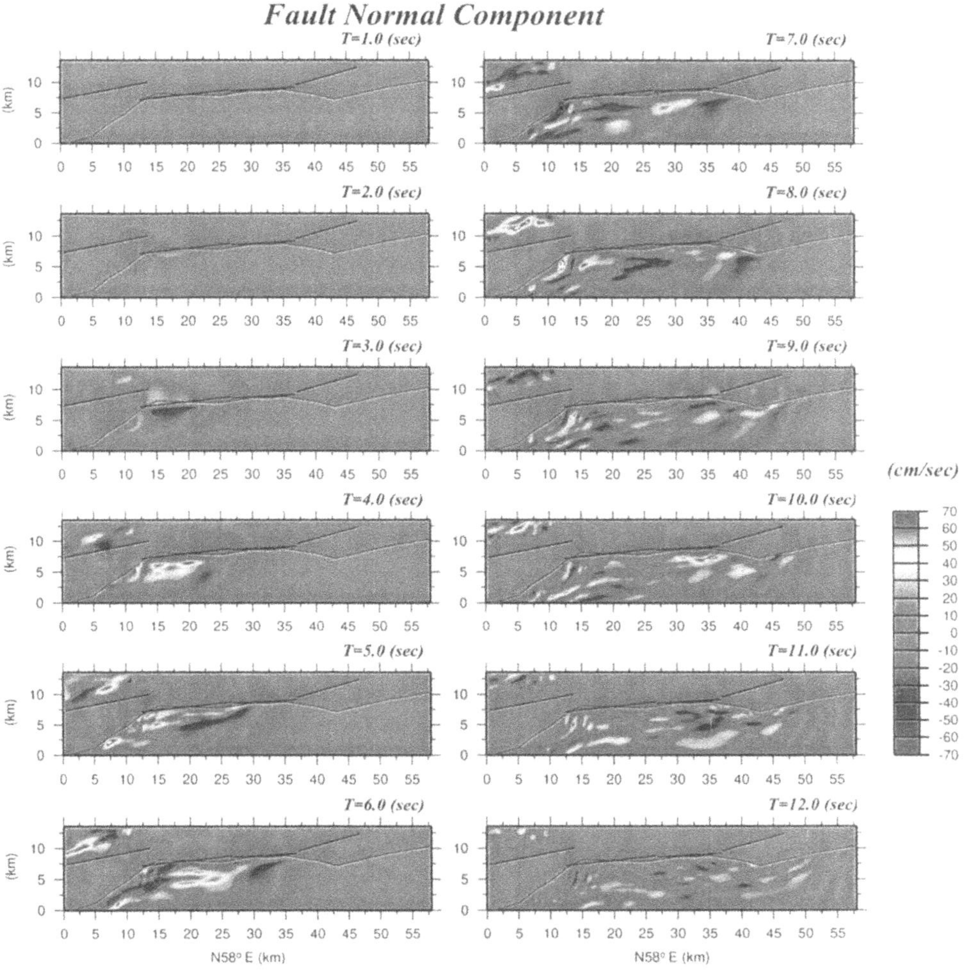

Fault Normal Component

Figure 5.15 (Continued) Fault normal ground velocity propagation induced by the 1995 Kobe earthquake. (From Pitarka (1999).)

5.2.3.2 Finite element method

Toki and Miura (1985) and Tsuboi and Muira (2000) used Goodman-type joint elements in 2D-FEM to simulate both rupture process and ground motions (Figure 5.16). Fukushima et al. (2010) used this method to simulate ground motions caused by the 2000 Tottori earthquake (Figure 5.17). Later, Mizumoto et al. (2005) extended the same method to 3D.

Iwata et al. (2016) recently investigated the strong motions induced by the 2014 Nagano-Hokubu earthquake. The model is based on a 3D FEM version. Figure 5.18a shows the fault parameters and Figure 5.18b shows the 3D mesh of the earthquake

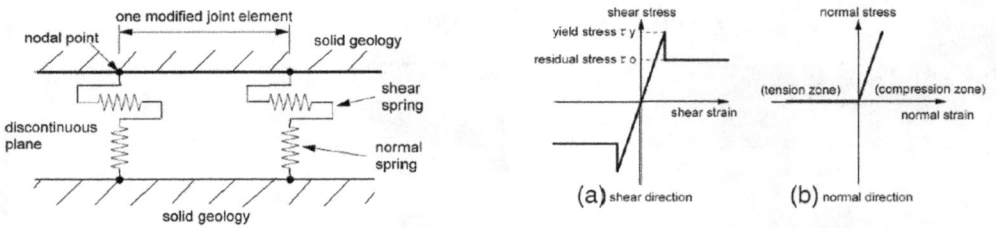

Figure 5.16 Representation of joint elements for faults and their constitutive law. (From Fukushima et al. (2010).)

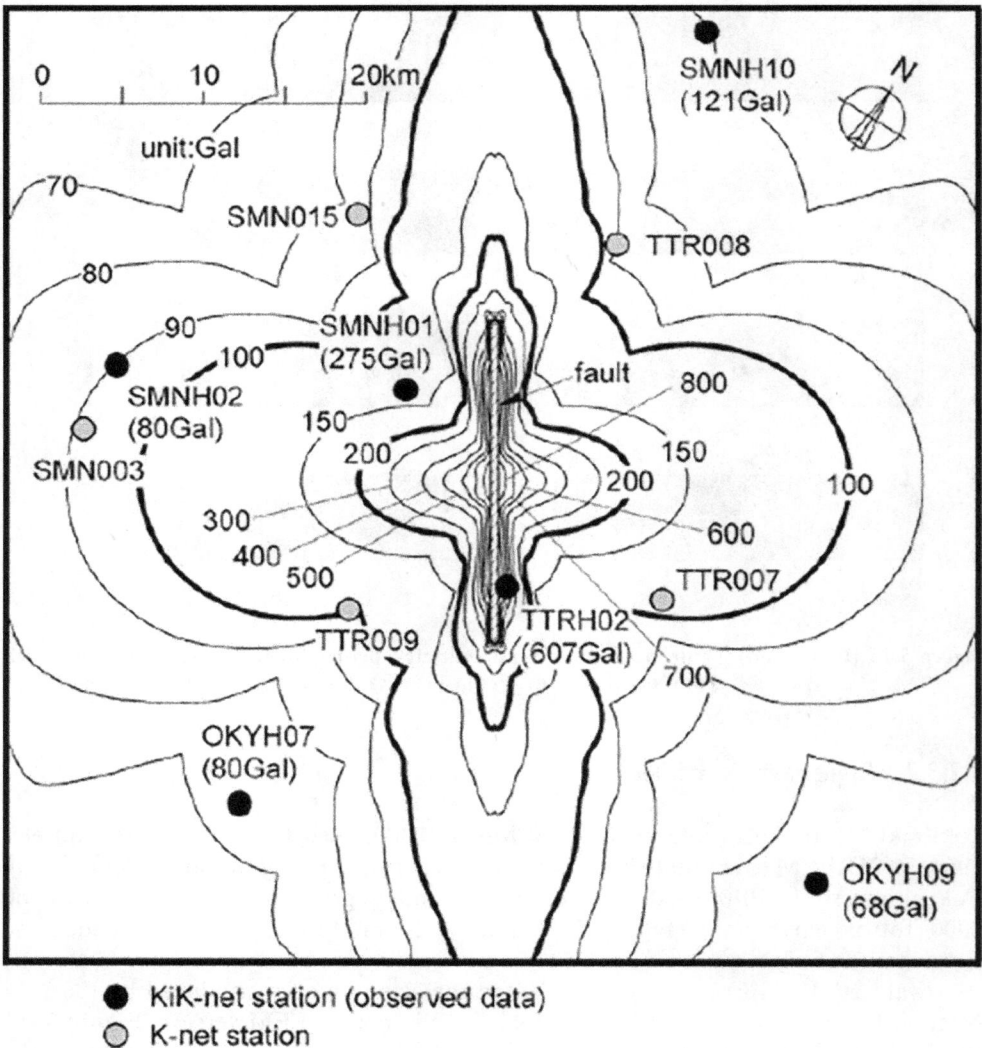

● KiK-net station (observed data)
○ K-net station

Figure 5.17 Comparison of computed and observed maximum ground acceleration for the 2000 Tottori earthquake. (From Fukushima et al. (2010).)

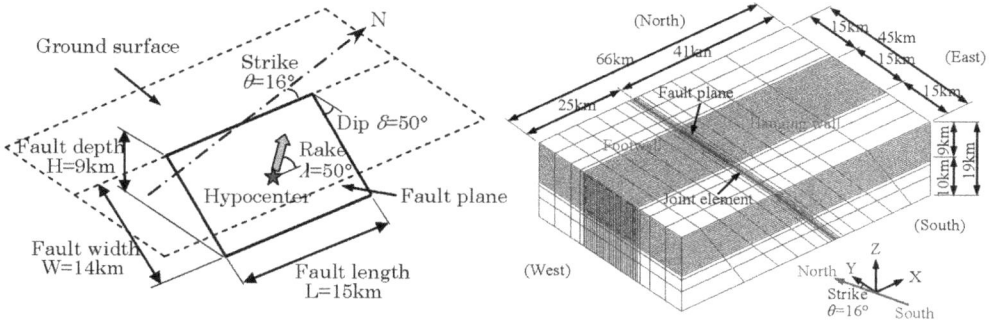

Figure 5.18 Fault model and 3D FEM mesh.

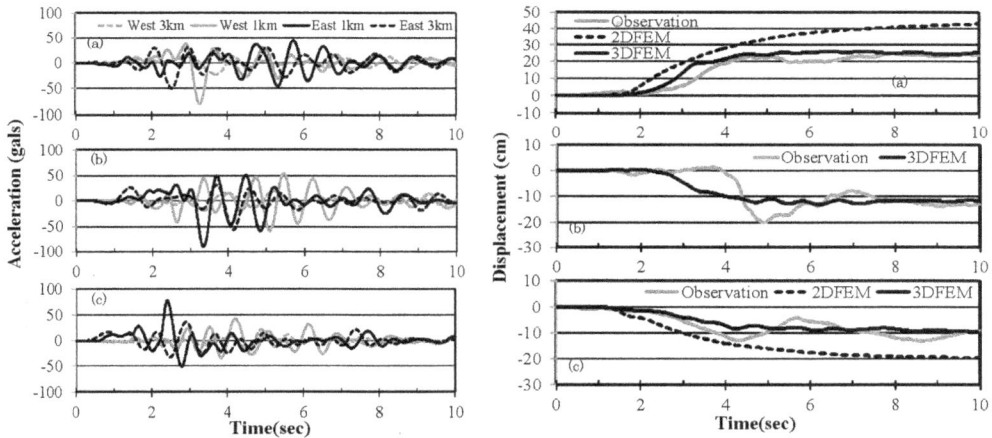

Figure 5.19 Computed acceleration (a) and displacement (b) responses.

fault and its vicinity. Figure 5.19a shows the time histories of surface acceleration at distances of 1 and 2 km from the surface rupture in 3D-FEM model. Rupture time is about 7–8 seconds. The maximum acceleration is higher on the east side (hanging-wall) than that on west side (footwall), which is close to the general trend observed in strong motion records. Nevertheless, the computed accelerations were less than the measured accelerations. Figure 5.19b shows the time histories of surface displacement at distances of 1 and 2 km from the surface rupture. The east side of the fault moves upward with respect to the footwall together with movement to the north direction and the vertical displacement of the east side is larger than that of the footwall, and the computed results are close to the observations. However, it is necessary to use finer meshes for better simulations of ground accelerations, which require the use of supercomputers.

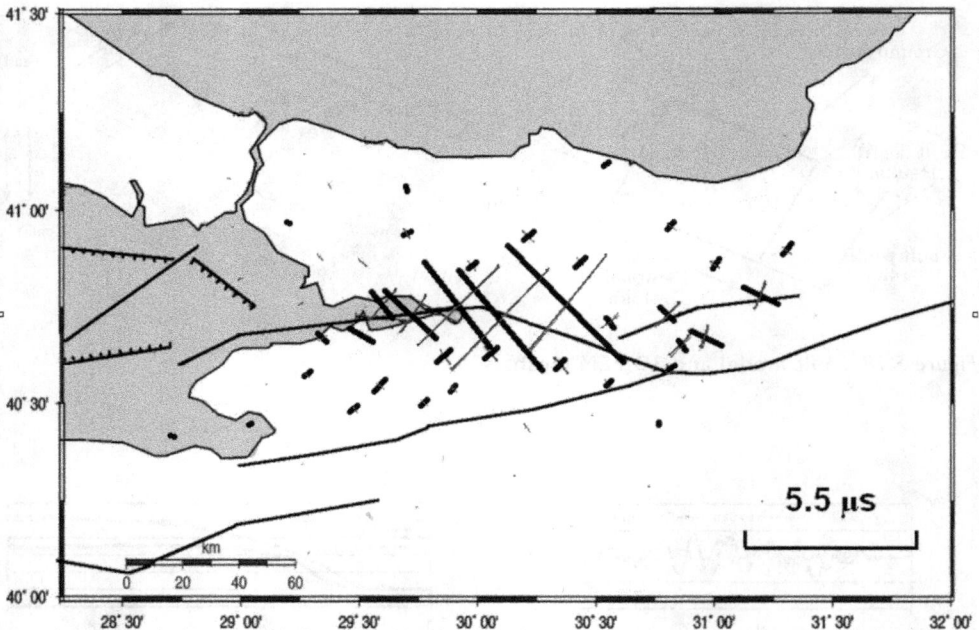

Figure 5.20 Computed ground straining from GPS measurements. (From Aydan et al. (2010b).)

5.2.3.3   GPS method

Permane and they may result from different causes such as faulting, slope failure, liquefaction and plastic deformation induced by ground shaking (Aydan et al. 2010). These types of ground deformations have limited effect on small structures as long as the surface breaks do not pass beneath those structures. However, such deformations may cause tremendous forces on long and/or large structures. Ground deformations may induce large tensile or compression forces as well as bending stresses in structures depending upon the character of permanent ground deformations. As an example, the ground straining induced by permanent ground deformations shown in Figure 5.6 are shown in Figure 5.20, which were caused by a strike-slip fault during the 1999 Kocaeli earthquake in Turkey. Blind faults and folding processes may also induce some peculiar ground deformations and associated folding of overlaying soft sedimentary layers. Such deformations caused tremendous damage on tunnels during the 2004 Chuetsu earthquake, although no distinct rupturing took place.

5.2.3.4   InSAR method

Interferometric synthetic aperture radar, abbreviated InSAR or IfSAR, is a radar technique used in geodesy and remote sensing. This geodetic method uses two or more synthetic aperture radar (SAR) images to generate maps of surface deformation or

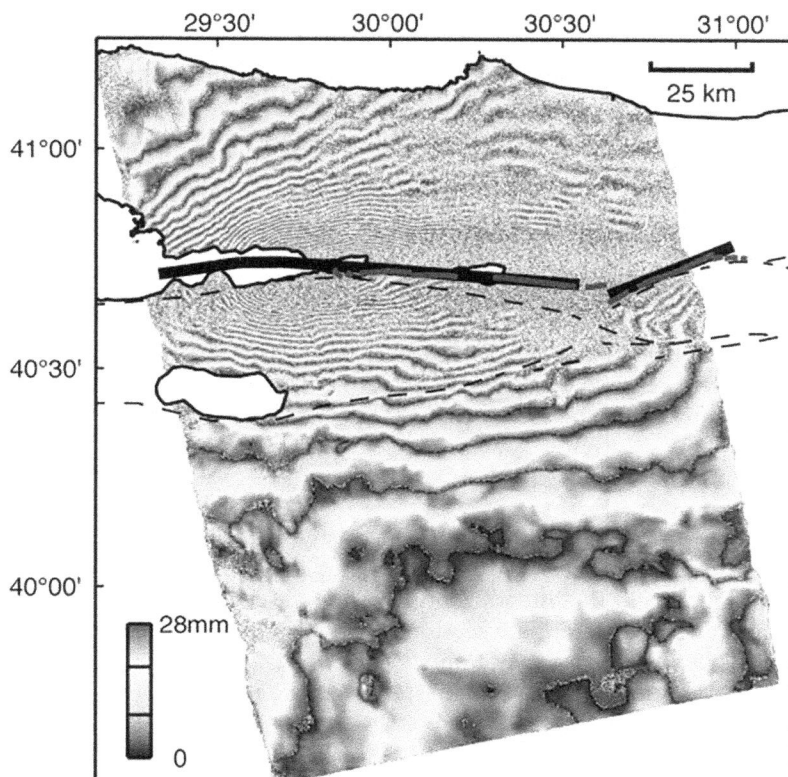

Figure 5.21 Interferogram produced using ERS-2 data from August 13 to September 17, 1999, for the 1999 Kocaeli earthquake (NASA/JPL-Caltech).

digital elevation, using differences in the phase of the waves returning to the satellite or aircraft. The technique can potentially measure centimetre-scale changes in deformation over spans of days to years. It has applications for geophysical monitoring of natural hazards, for example earthquakes, volcanoes and landslides, and in structural engineering, in particular monitoring of subsidence and structural stability. Figure 5.21 shows an application of the InSAR technique to estimate ground deformations induced by the 1999 Kocaeli earthquake.

5.2.3.5 EPS method

Ohta and Aydan (2007) have recently showed that permanent ground deformations may be obtained from the integration of acceleration records. The erratic pattern screening (EPS) method proposed by Ohta and Aydan (2004, 2007) and Aydan and Ohta (2011) can be used to obtain permanent ground displacement with the consideration of features associated with strong motion recording. The duration of shaking should be naturally related to the rupture time t_r. Depending on the arrival time difference of S-wave and P-wave, the shaking duration would be a sum of rupture duration and S-P arrival time difference Δt_{sp}. If ground exhibits plastic response due to yielding or ground liquefaction, the duration of shaking would be elongated. If co-seismic crustal deformations are obtained, the integration duration should be restricted the rupture duration with the consideration of S-P arrival time difference. The existence of plastic deformation can be assessed by comparing the effective shaking duration and the sum of rupture duration t_r and S-P arrival time difference (Δt_{sp}). If the effective shaking duration is longer than the sum of rupture duration t_r and S-P waves arrival time difference (Δt_{sp}), the integration can be carried for both durations and the difference can be interpreted as the plastic ground deformation. The most critical issue is the information of rupture duration. The data on rupture duration are generally available for earthquake with a moment magnitude >5.6 worldwide. The effective shaking duration may be obtained from the acceleration records using the procedure proposed by Housner (1961). When such data are not available, the empirical relation proposed by Aydan (2007b, 2012a) may be used.

$$t_r = 0.0005 M_w \exp(1.25 M_w) \tag{5.5}$$

S-P arrival time difference Δt_{sp} can also be easily evaluated from the acceleration record. They divided an acceleration record into three sections and apply filters in Sections 5.1 and 5.3 and the integration is directly carried in Section 5.2 without any filtering. The times to differentiate sections are t_1 and t_2. Time t_1 is associated with the arrival of P-wave on the strong motion station, while time t_2 is related to the arrival time of P-wave, rupture duration and S-P waves arrival time difference for the crustal deformations and it is given below:

$$t_2 = t_1 + t_r + \Delta t_{sp} \tag{5.6}$$

Any deformation after time t_2 must be associated with deformations related to local plastic behaviour of the ground at the instrument location. We show one example for defining times t_1 and t_2 on a record taken at the HDKH07 strong motion station of the KiKNET strong motion network for the 2003 Tokachi-oki earthquake, Japan (Figure 5.22). The estimated rupture time for this earthquake is about 40 seconds (Kikuchi 2003) with about 18 seconds of S-P waves arrival time difference (Δt_{sp}).

Another important issue is how to select filter values in Sections 5.1 and 5.2. This is somewhat a subjective issue and it depends upon the sensitivity of the accelerometers. The filter value ε_1 is generally small and this stage is associated with pre-trigger value of instruments. Our experience with the selection of ε_1 for K-NET and KIK-NET accelerometer records implies that its value be less than ±2 gals. As for the value of ε_2, higher values must be assigned. Again our experiences with the records of K-NET and KIK-NET accelerometers imply that its value should be ±6 gals. The experiences with acceleration records of Turkey and Italy, these values are much less than those high-sensitive accelerometers of networks in Japan.

This method is applied to results of laboratory faulting and shaking table tests, in which shaking-induced motions were recorded by using both accelerometers and laser displacement transducers, simultaneously. Furthermore, the method was applied to strong motion records of several large earthquakes with measurements of ground

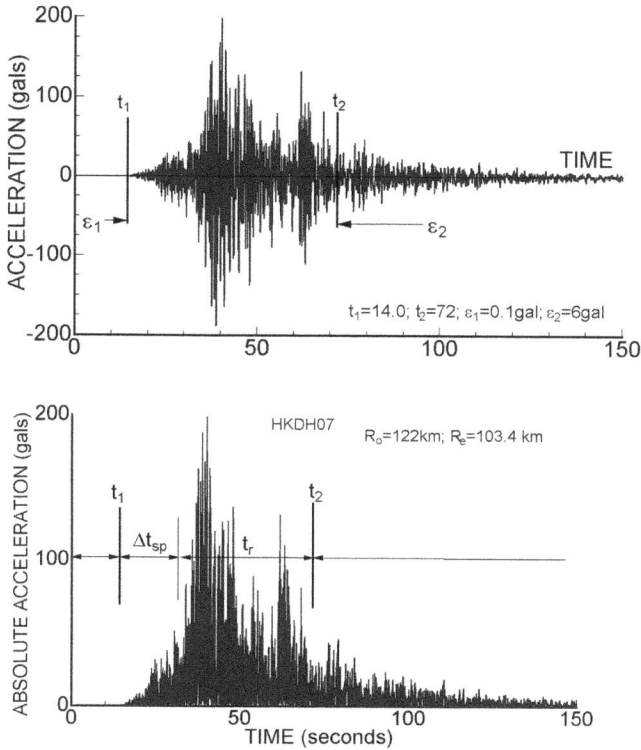

Figure 5.22 Definition of sections in the EPS method.

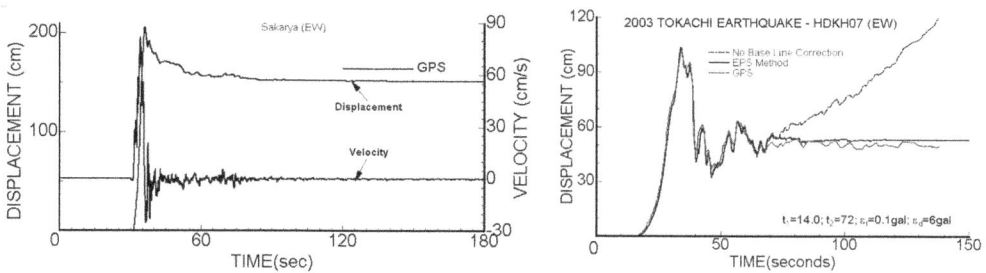

Figure 5.23 Comparison of permanent ground deformation by the EPS method with measured GPS recordings. (From Ohta and Aydan (2007).)

movements by GPS as seen in Figure 5.23. The comparison of computed responses with actual recordings was almost the same, implying that the proposed method can be used to obtain actual recoverable as well as permanent ground motions from acceleration recordings. Figure 5.24 shows the application of the EPS method to the strong motion records of the 2009 L'Aquila earthquake to estimate the co-seismic permanent

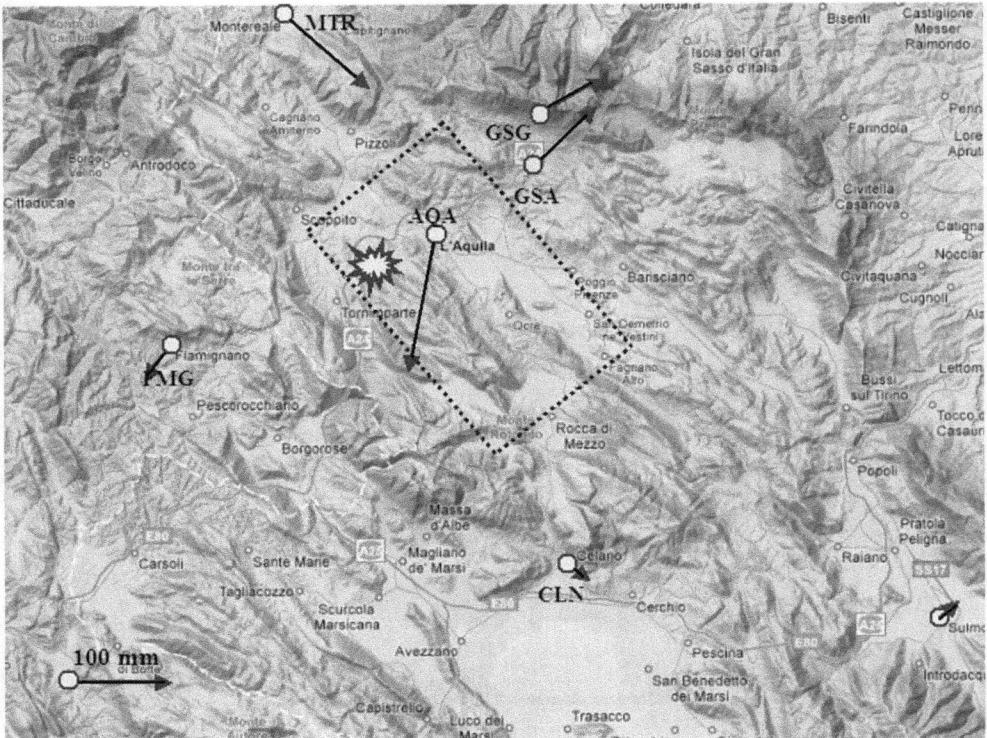

Figure 5.24 Estimated permanent ground displacements by the EPS method. (From Aydan et al. (2009a,b).)

ground displacements. These results are very consistent with the GPS observations. However, it should be noted that the permanent ground deformations recorded by the GPS do not necessarily correspond to those of the crustal deformation. Surface deformations may involve crustal deformation as well as those resulting from the plastic deformation of ground due to ground shaking. The records at ground surface and 260 m below the ground surface taken at IWTH25 during the 2008 Iwate-Miyagi earthquake clearly indicated the importance of this fact in the evaluation of GPS measurements (KiKNET 2008).

5.2.3.6 Okada's method

Okada (1992) proposed closed-form solutions for elastic dislocation in a half-space isotropic medium (Figure 5.25 and 5.26). Closed-form analytical solutions are presented in a unified manner for the internal displacements and strains due to shear and tensile faults in a half-space for both point and finite rectangular sources. These expressions evaluate deformations in an infinite medium, and a term related to surface deformation is obtained through multiplying by the depth of observation point. Stein

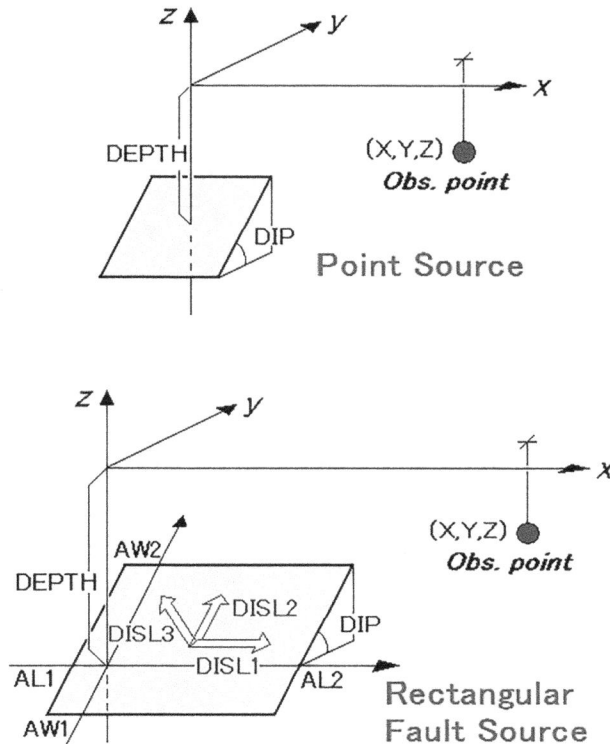

Figure 5.25 Geometrical illustration of assumed fault and relative displacements.

(2003) utilized the solutions of Okada's method in his software to compute permanent ground deformation and associated stress changes. This method is also used to forecast earthquakes with the introduction of a superposing displacement field and associated stress changes together with the use of the Mohr–Coulomb criterion (King et al. 1994). Stein's method is upgraded by his research group and applied to various earthquakes in recent years. Figure 5.27 shows an example of computations by Toda et al. (2002) for earthquake activity in Izu Islands.

5.2.3.7 Numerical methods

FDM, FEM, BEM or their combination produce permanent ground deformation as a natural output of computations provided that the rupture process is well simulated. Figure 5.28 shows an example of such a computation for the 2014 Nagano-Hokubu earthquake and the 2011 GEJE (or Tohoku earthquake). Although the displacement response could be more easily simulated, acceleration responses simulation requires fine meshes, which undoubtedly require the use of supercomputers. It should also be noted that FEM models could easily simulate both permanent displacements in addition to strong ground motions.

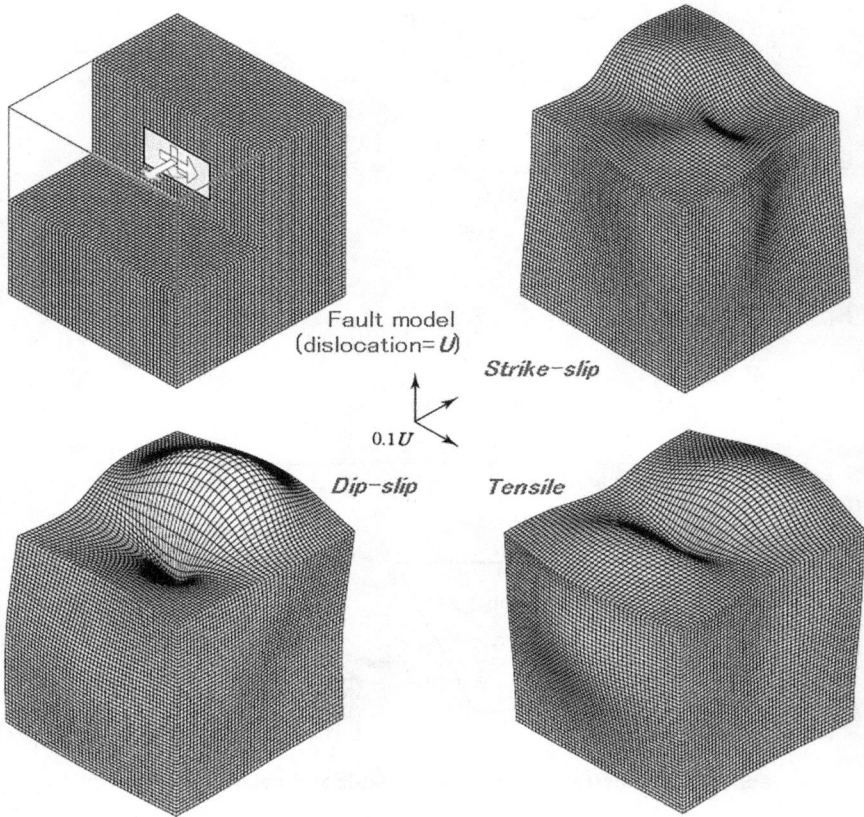

Figure 5.26 Illustration of ground deformation associated with faulting.

Figure 5.27 Relations between static stress changes and seismicity. (From Toda et al. (2002).)

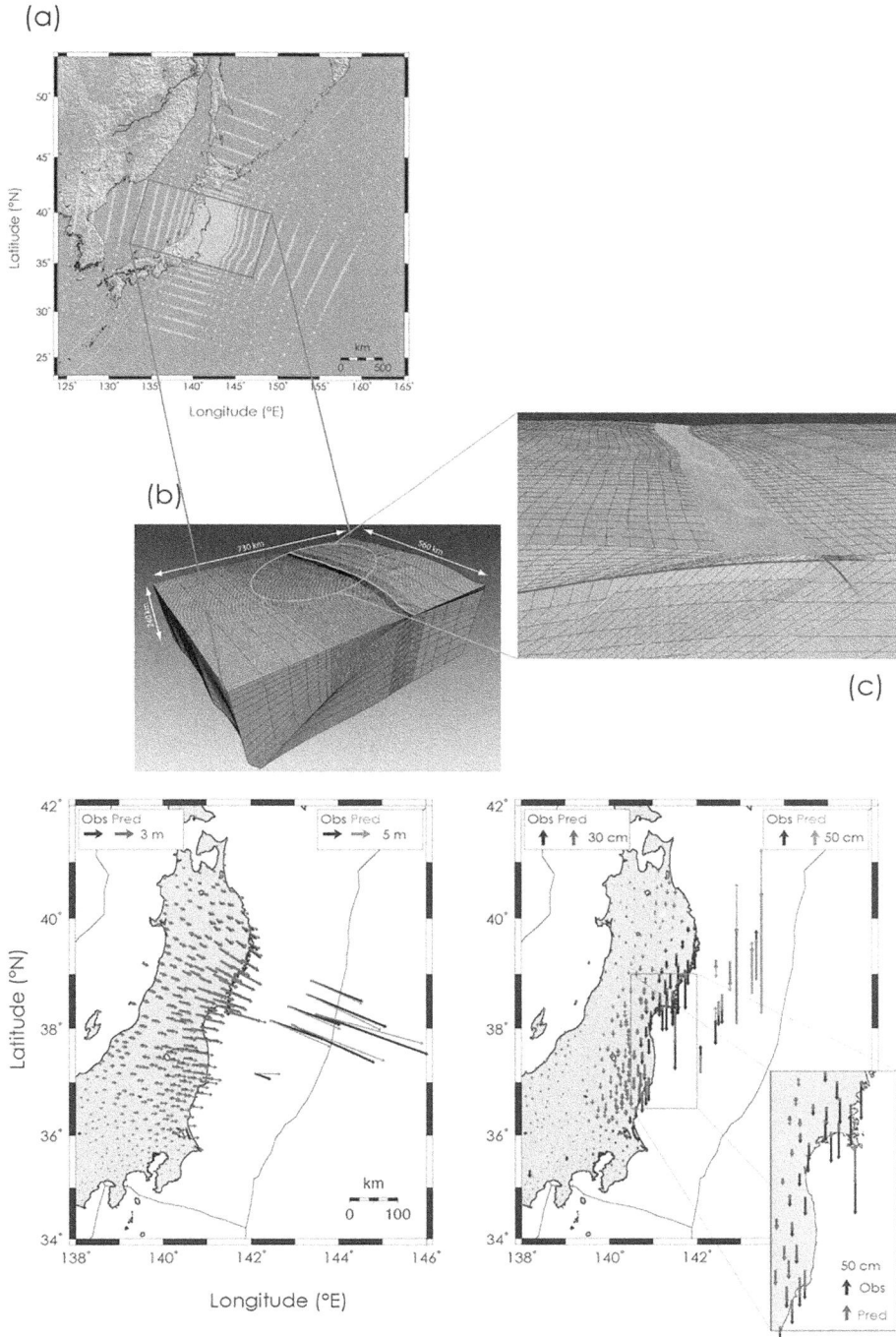

Figure 5.28 3D-FEM simulations of ground motions associated with the 2011 GEJE. (From Romano et al. (2014).)

5.3   ESTIMATIONS OF STRONG MOTION PARAMETERS FROM THE COLLAPSE, FAILURE AND SLIPPAGE OF SIMPLE STRUCTURES AND SIMPLIFIED REINFORCED CONCRETE STRUCTURES

5.3.1   Inference of strong motions from masonry walls

As indicated in the previous section, masonry buildings with mud mortar mostly collapsed during earthquakes. Since the bonding strength of mud mortar is very small, the seismic resistance of the masonry walls against toppling mostly depends upon the wall geometry while the seismic resistance against shear would be frictional. The conditions for different modes of failure, shown in Figures 5.29, derived for a horizontal seismic coefficient are as follows (Aydan et al. 1989):

Toppling condition

$$\frac{a}{g} > \frac{t}{h} \tag{5.7}$$

Sliding condition

$$\frac{a}{g} > \tan\phi \tag{5.8}$$

Toppling and sliding conditions

$$\frac{a}{g} > \frac{t}{h} \ \text{ and } \ \frac{a}{g} > \tan\phi \tag{5.9}$$

Figure 5.30 shows the relation between the ratio of wall height to width and lateral seismic coefficient. It seems that the walls of masonry buildings are vulnerable to both toppling (out of plane failure) and sliding failure (X-cracks) in the epicentral area. Such buildings away from the epicentral area may only suffer from toppling failure. The author plotted some observations and the results are shown in Figure 5.31 together

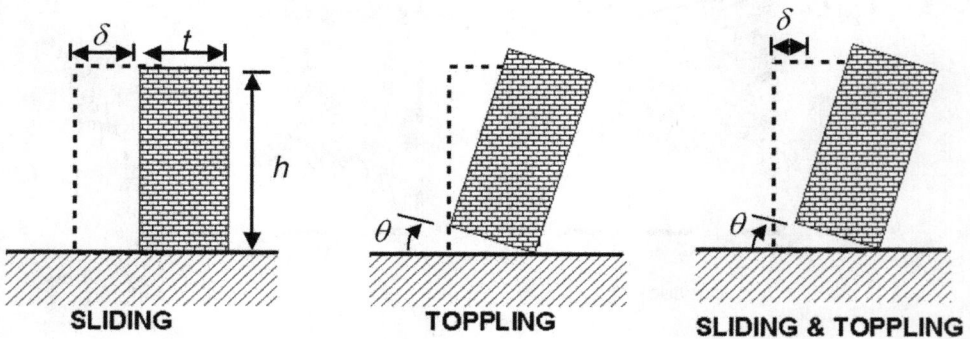

Figure 5.29 Failure modes of a wall.

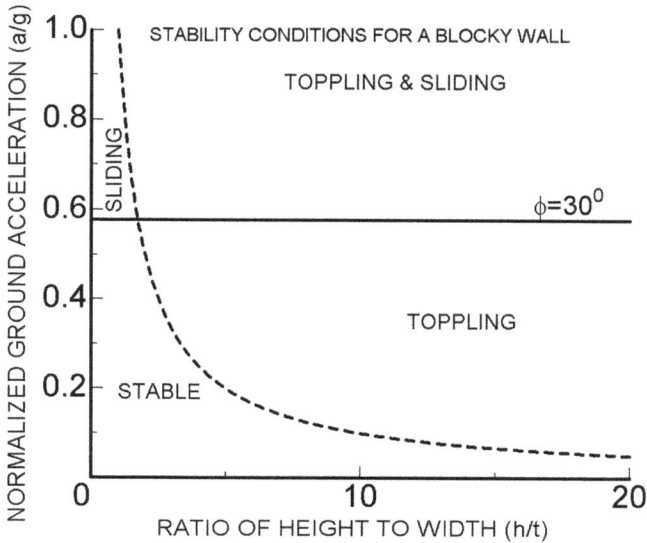

Figure 5.30 Relation between wall height to width ratio and lateral seismic coefficient.

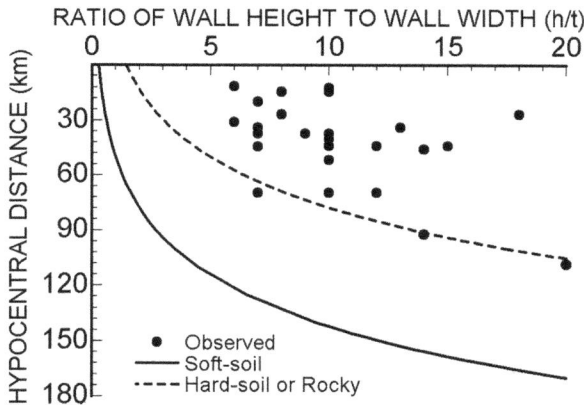

Figure 5.31 Relation between wall height to width ratio and hypocentral distance.

with the bounds for stability against toppling, computed from empirical attenuation relations proposed by Aydan (1997, 2001a, 2007a, 2012a) for various ground conditions. Since the ground conditions in many places are hard soil or rocky ground, the computed results are quite consistent with actual observations.

The maximum ground velocity may also be evaluated using the displaced and/ or toppled simple structures. This concept requires the equality of momentum to induce sliding, toppling or both to the imposed maximum kinetic energy associated with ground motions due to earthquakes. The final equations would take the following forms for the modes of sliding, toppling or both in terms of the block dimensions and the relative slip and/or rotation (Aydan et al. 2006b):

Toppling condition

$$v_{\max} = \theta \sqrt{gl} \qquad (5.10)$$

Sliding condition

$$v_{\max} = \sqrt{2gl\tan\phi} \qquad (5.11)$$

Toppling and sliding conditions

$$v_{\max} = \sqrt{gl(2\tan\phi + \theta^2)} \qquad (5.12)$$

where $l = \sqrt{h^2 + t^2}$, $\theta = \dfrac{\pi}{4}\tan^{-1}\left(\dfrac{t}{h}\right)$, ϕ: friction angle.

5.3.2   Inference of strong motions from reinforced concrete structures

Next, the relation between seismic coefficient and number of floors for reinforced concrete (RC) buildings, which collapsed as a result of soft-floor effect, was investigated. To derive the equations for seismic coefficient for the collapse of RC structures due to the soft-floor effect, the failure modes may involve the shearing of columns or bending as illustrated in Figures 5.32 and 5.33. If the load is assumed to be uniformly distributed over columns with a square cross section and the resistance of the columns against shearing and bending result from the shear and tensile strength of the reinforcing bars and column friction with the basement, one can easily derive the equations for a given seismic horizontal coefficient for the bending failure and for shear failure of the columns as follows:

Bending failure

$$\frac{a}{g} > \frac{t}{3h}\left(1 + \frac{\sigma_b}{\gamma_b h}i_b \cdot i_c \frac{1}{N}\right) \qquad (5.13)$$

Figure 5.32 Failure of RC buildings due to soft-floor effect in Bhuj.

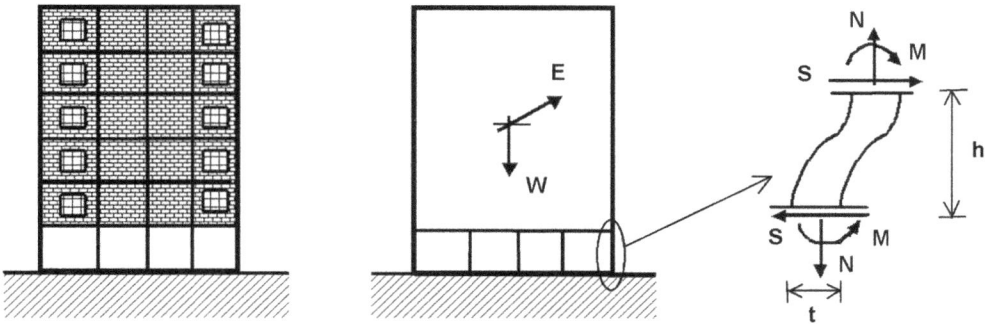

Figure 5.33 Illustration of the soft-floor effect and its mechanical modelling.

Shear failure

$$\frac{a}{g} > \tan\phi_c + \frac{\sigma_s}{\gamma_b h} i_b \cdot i_c \frac{1}{N} \qquad (5.14)$$

where

t is the column width
h is the floor height
N is the number of stories
σ_b is the reinforcing bar tensile strength
σ_s is the reinforcing bar shear strength
i_b is the areal ratio of reinforcing bar to the column area
i_c is the areal ratio of columns to the building area
γ_b is the average unit weight of the building
ϕ_c is the friction angle of columns with base floor

Figures 5.34 and 5.35 show the relation between the number of stories and lateral seismic coefficient for bending failure and shear failure modes. In computations, areal ratio coefficients are changed between 0.01 and 0.02, which are typical values in RC buildings in the earthquake region of the 2001 Kutch earthquake. The results indicate that the buildings should be more vulnerable to bending failure rather than the shear failure unless the ratio of column width to floor height is large, which should correspond to buildings having shear-wall-like columns. As the ratio of the floor height to column width is generally greater than six, the failure of RC buildings must have been due to bending failure. The site observations confirmed this conclusion. Furthermore, it is inferred that the damage to low-story RC buildings may also occur in the epicentral area due to the soft-floor effect, while high-rise buildings away from the epicentral area should be more vulnerable to tremors.

Some observations together with the bounds for stability against bending failure due to soft-floor effect, computed from empirical attenuation relations by Aydan (1997, 2001) for various ground conditions are plotted, and the results are shown in Figures 5.36 and 5.37. The plotted data clearly confirm the statement above. Furthermore, if the effect of vertical shaking is considered, it is expected that the damage to buildings due to soft-floor effect can be observed at distances far away from

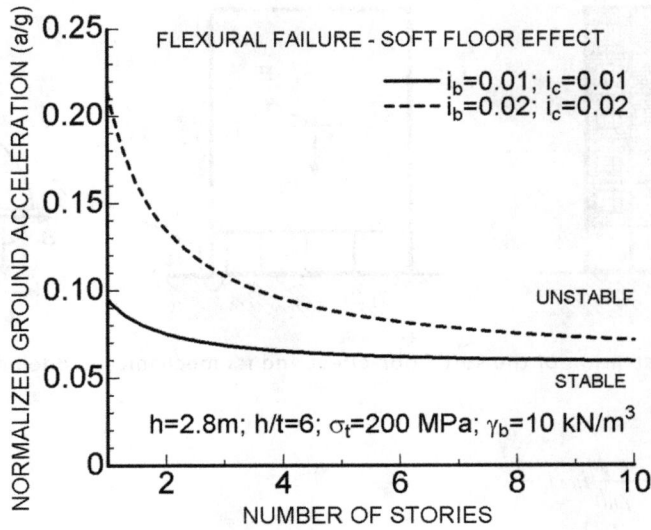

Figure 5.34 Relation between the number of stories and lateral seismic coefficient for bending failure due to soft-floor effect.

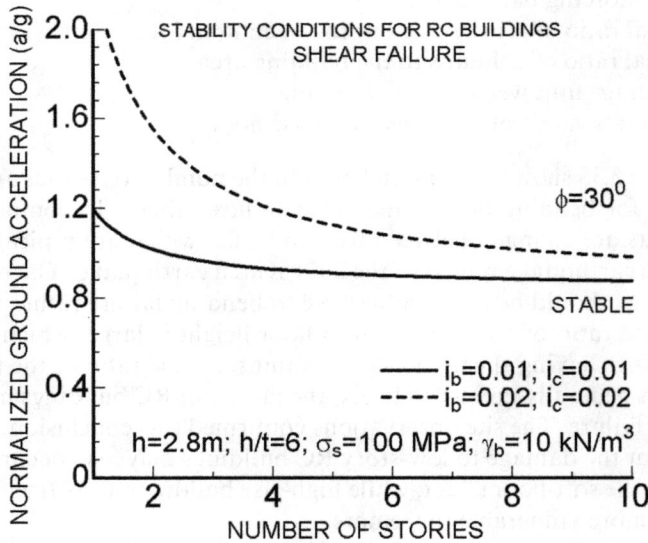

Figure 5.35 Relation between number of stories and lateral seismic coefficient for shear failure due to soft-floor effect.

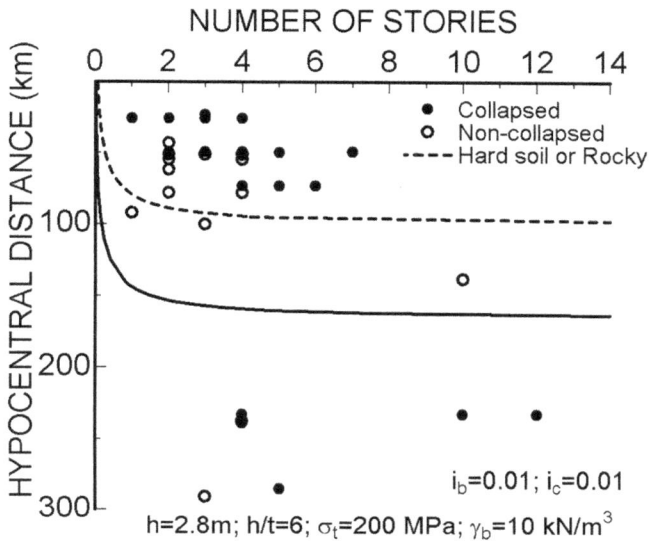

Figure 5.36 Relation between the number of stories and hypocentral distance.

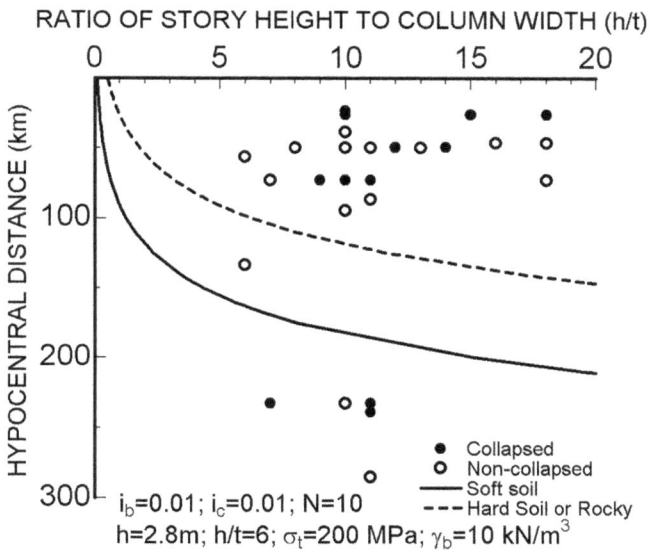

Figure 5.37 Relation between the ratio of story height to column width and hypocentral distance.

the epicentre. The damage to buildings as observed in Ahmedabad, Morbi, Surat in Gujarat state and Hyderabad in Pakistan far away from the epicentre may also be related to ground amplification and the resonance phenomenon in addition to poor construction quality.

5.3.3   Inference of strong motions from Mercalli Seismic Intensity

As pointed out in the previous section, if there is no strong motion record available for the epicentral area, a relation between Modified Mercalli Intensity and maximum ground acceleration may be utilized. For example, the relation proposed by Aydan (1997) is as follows:

$$a_{\max} = a\left(e^{0.6834*I} - 1\right) \tag{5.15}$$

where

I: Modified Mercalli Intensity at the particular location

$$a = A * B$$

A:2.8 for soft soil ground and 0.56 for stiff and rocky ground,

$B = e^{-0.025*h}$: h: hypocenter depth. If it is not known, it may be assumed to be 15–25 km.

Figure 5.38 compares the estimations obtained from Eq. (5.15) and other empirical relations proposed by Gutenberg–Richter (1958), Trifunac and Brady (1975) and Wald

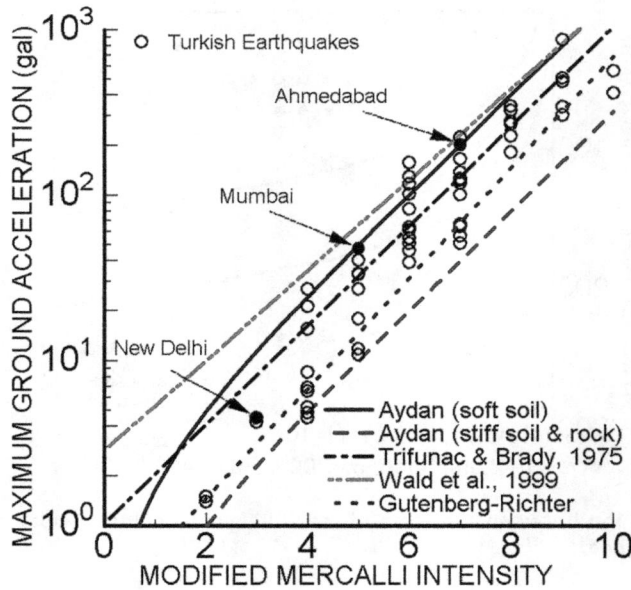

Figure 5.38 Comparison of empirical relations between Modified Mercalli Intensity and maximum ground acceleration with observations.

et al. (1999) in Figure 5.38. The actual data between maximum ground accelerations and Modified Mercalli Intensity (I) are from the measurements of earthquakes in Turkey, including the recent earthquakes. While the relation of Wald et al. (1999) is an upper-bound for the actual data, the relation of Gutenberg–Richter provides a lower bound for the data set. The empirical relation with the consideration of ground conditions provides better estimations and it may be used to infer the strong motions in the earthquake area where no strong motion data is available. In the same figure, the data for Ahmedabad, Mumbai (Bombay) and New Delhi associated with the 2001 Kutch earthquake in the Gujarat region of India are also plotted and the data from these sites almost coincide with the empirical relation proposed in this report for soft ground.

Chapter 6

Vibration analyses of structures

6.1 NUMERICAL METHODS

The final, discretized form of the equation of motion given in Chapter 3 (Eq. 3.8) irrespective of the solution method (FDM, FEM, BEM), depending upon the character of the governing equation, may be written in the following form (e.g. Aydan 2017, 2021):

$$[M]\{\ddot{\phi}\} + [C]\{\dot{\phi}\} + [K]\{\phi\} = \{F\} \tag{6.1}$$

The specific forms of matrices $[M]$, $[C]$, $[K]$ and vector $\{F\}$ in the equation above differ depending upon the method of solution chosen and dimensions of physical space. Viscosity matrix $[C]$ is associated with rate-dependency of geomaterials. However, in many dynamic solution schemes, viscosity matrix $[C]$ is expressed in the following form using the Rayleigh damping approach as follows:

$$[C] = \alpha[M] + \beta[K] \tag{6.2}$$

where α, β are called proportionality constants. This approach becomes highly convenient in large-scale problems if central finite difference technique and mass lumping are used.

Natural frequencies of any structure or continuum body can be obtained from the following equation system by assuming that the damping matrix $[C]$ is nil and the external force $\{F\}$ is zero in Eq. (6.2).

$$[M]\{\ddot{U}\} + [K]\{U\} = \{0\} \tag{6.3}$$

Equation (6.3) is known as the equation system of free vibration.

The variable vector $\{U\}$ can be assumed to be expressed as a harmonic function given as

$$\{U\} = \{\phi\}e^{iwt} \tag{6.4}$$

DOI: 10.1201/9781003164371-6

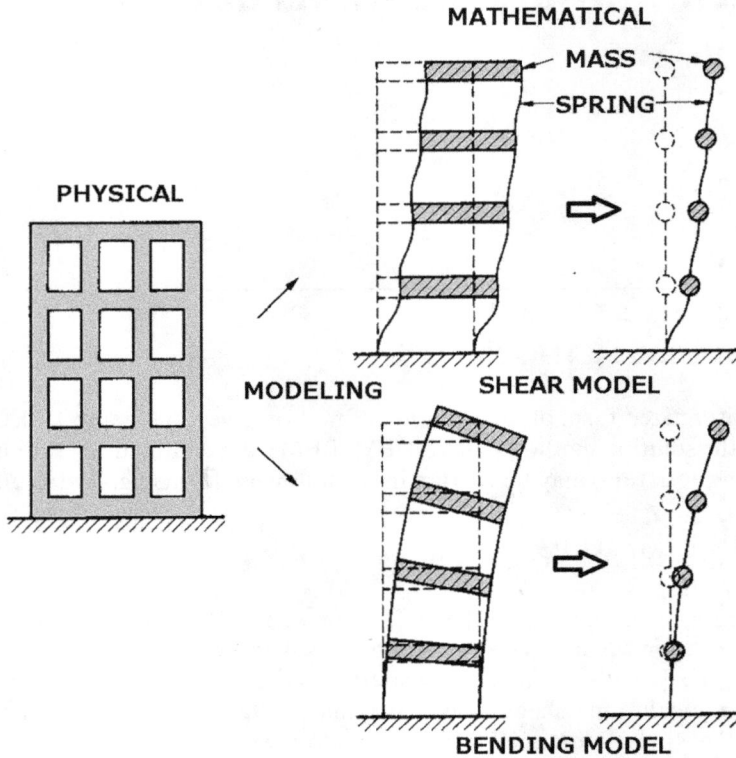

Figure 6.1 Mathematical models for representing building in earthquake engineering.

If Eq. (6.4) is inserted in Eq. (6.3), one can easily obtains the following equation system:

$$([K] - \omega^2 [M])\{\phi\} = \{0\} \tag{6.5}$$

The solution of Eq. (6.5) is known as Eigen value analyses yielding ω^2 and $\{\phi\}$. ω^2 is the known Eigen values and $\{\phi\}$ is the Eigen value vector.

Equation (6.1) is valid for any type of structure (e.g. 1D, 2D or 3D continuum with elastic, visco-elastic or elasto-visco-plastic constitutive law, single degree of freedom (SDOF) or multi-degree of freedom (MDOF) systems) for chosen seismic loading, and it is generally classified as response analyses. However, it is quite simplified for engineering purposes and one can find numerous techniques for various idealized loading and boundary conditions in some textbooks on earthquake engineering (e.g. Clough and Penzien 1975; Meirovitch 1986; Okamoto 1973; Osaki 1983; Usami 1990; Kawashima 2018) and mathematics (e.g. Kreyszig 1983). Figure 6.1 illustrates how a building is simplified for dynamic analyses, which is commonly used in building design (e.g. Usami 1980).

6.2 SIMPLIFIED ANALYSES OF STRUCTURES FOR THEIR VIBRATION CHARACTERISTICS

Although the most appropriate method would be the solution of equation of motions under given boundary and initial conditions, the structures are often simplified to obtain their approximate responses and vibration characteristics. They are mostly simplified to SDOF structures by assuming they may vibrate either in shear or bending mode.

The force equilibrium of the simplified structure shown in Figure 6.2 may be given as

$$\sum_{K=1}^{n} F_i = F_{eq} - F_d - F_e - F_I = 0 \tag{6.6}$$

where

$$F_d = c\frac{du}{dt}; \quad F_e = k\,u; \quad F_{eq} = -m\frac{d^2u_g}{dt^2}; \quad F_I = m\frac{d^2u}{dt^2} \tag{6.7}$$

m, c, k, t, u and u_g are the mass, damping, stiffness, time, displacement and ground displacement, respectively.

Equation (6.6) may be reduced to the following form by using relations given in Eq. (6.7) as

$$m\frac{d^2u}{dt^2} + c\frac{du}{dt} + k\,u = -m\frac{d^2u_g}{dt^2} \tag{6.8}$$

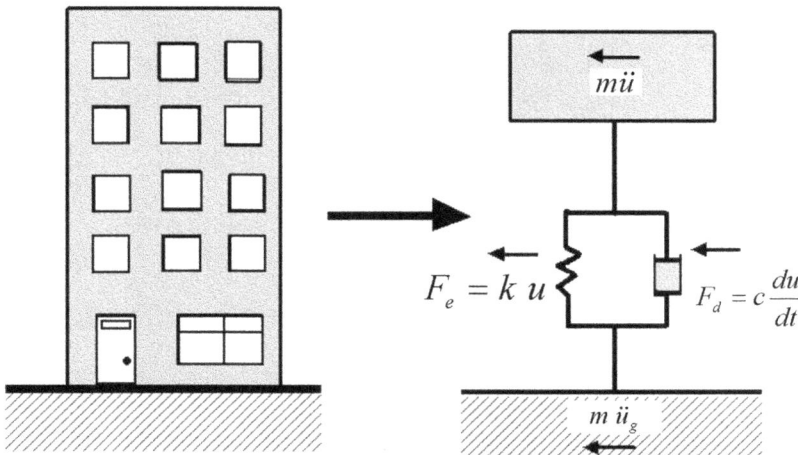

Figure 6.2 Simplification of structures as a single degree of freedom structure.

Introducing the following relations

$$\omega_0 = \sqrt{\frac{k}{m}}; \quad 2h\omega_0 = \frac{c}{m} \tag{6.9}$$

Equation (6.8) may be reduced to the following form:

$$\frac{d^2u}{dt^2} + 2h\omega_0 \frac{du}{dt} + \omega_0^2 u = -\frac{d^2 u_g}{dt^2} \tag{6.10}$$

where ω_0 is known as angular natural frequency and h is damping coefficient.

6.2.1   Free vibration

If damping coefficient and base acceleration are zero, Eq. (6.10) reduces to the following form:

$$\frac{d^2u}{dt^2} + \omega_0^2 u = 0 \tag{6.11}$$

The solution of Eq. (6.11) is obtained as follows:

$$u = C_1 e^{\lambda_1 t} + C_2 e^{\lambda_2 t} \tag{6.12}$$

where the roots are given specifically as

$$\lambda_1 = i\omega_0; \quad \lambda_2 = -i\omega_0; \quad i = \sqrt{-1} \tag{6.13}$$

Using the following relations

$$e^{i\omega_0} = \cos\omega_0 t + i\sin\omega_0 t \quad \text{and} \quad e^{-i\omega_0} = \cos\omega_0 t - i\sin\omega_0 t \tag{6.14}$$

the final form of solution is

$$u = A\cos\omega_0 t + B\sin\omega_0 t \tag{6.15}$$

Introducing the following initial conditions

$$u(0) = u_0; \quad \dot{u}(0) = v_0 \quad \text{at} \quad t = 0 \tag{6.16}$$

yields the integration coefficients as

$$A = u_0; \quad B = \frac{v_0}{\omega_0} \tag{6.17}$$

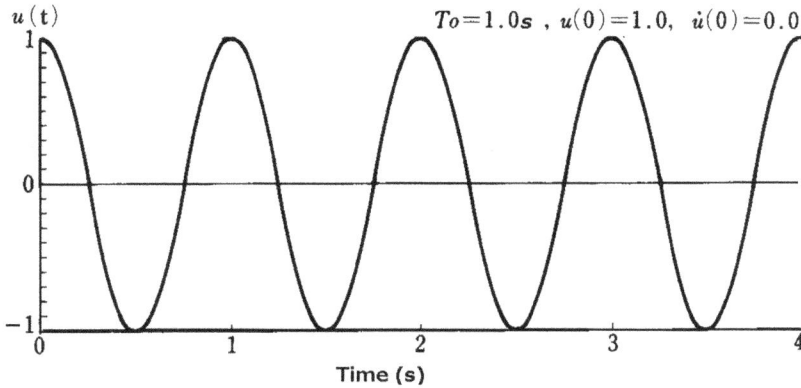

Figure 6.3 Displacement response.

Thus the exact solution takes the following form:

$$u = u_0 \cos\omega_0 t + \frac{v_0}{\omega_0}\sin\omega_0 t \tag{6.18}$$

Equation (6.18) may be rewritten by introducing the following identities:

$$R\cos\theta = u_0; \quad R\sin\theta = \frac{v_0}{\omega_0}; \quad R = \sqrt{\left(\frac{v_0}{\omega_0}\right)^2 + u_0^2} \tag{6.19}$$

as

$$u = R\cos(\omega_0 t - \theta) \tag{6.20}$$

Finally, one obtains the vibration characteristics of the structure, such as natural period, natural frequency in terms of natural angular frequency as

$$T_0 = \frac{1}{f_0} = \frac{2\pi}{\omega_0} \tag{6.21}$$

Figure 6.3 illustrates the displacement response for the given initial and boundary conditions shown in the figure. As noted from the figure, the selected structure vibrates with the period obtained from Eq. (6.21).

6.2.2   Damped free vibration

If base acceleration is zero, Eq. (6.10) reduces to the following form:

$$\frac{d^2u}{dt^2} + 2h\omega_0\frac{du}{dt} + \omega_0^2 u = 0 \tag{6.22}$$

The solution of Eq. (6.22) can be obtained as follows:

$$u = C_1 e^{\lambda_1 t} + C_2 e^{\lambda_2 t} \tag{6.23}$$

where roots are

$$\lambda_1 = -\omega_0 h + i\omega_d; \quad \lambda_2 = -\omega_0 h - i\omega_d; \quad i = \sqrt{-1}; \quad \omega_d = \omega_0 \sqrt{1 - h^2} \tag{6.24}$$

Thus the general form of solution takes the following form:

$$u = e^{-\omega_0 h t} \left(A\cos\omega_d t + B\sin\omega_d t \right) \tag{6.25}$$

Introducing the following initial conditions

$$u(0) = u_0; \quad \dot{u}(0) = v_0 \quad \text{at} \quad t = 0 \tag{6.26}$$

yields the integration constants as

$$A = u_0; \quad B = \frac{h\omega_0 u_0 + v_0}{\omega_d} \tag{6.27}$$

Inserting the integration constants given by Eq. (6.27) into Eq. (6.25) results in

$$u = e^{-\omega_0 h t} \left\{ u_0 \cos\omega_d t + \frac{h\omega_0 u_0 + v_0}{\omega_d} \sin\omega_d t \right\} \tag{6.28}$$

Again, using the following identities,

$$R\cos\theta = u_0; \quad R\sin\theta = \frac{h\omega_0 u_0 + v_0}{\omega_d}; \quad R = \sqrt{\left(\frac{h\omega_0 u_0 + v_0}{\omega_d} \right)^2 + u_0^2} \tag{6.29}$$

Equation (6.25) takes the following form:

$$u = e^{-\omega_0 h t} R\cos\left(\omega_d t - \theta\right) \tag{6.30}$$

where

$$T_d = \frac{1}{f_d} = \frac{2\pi}{\omega_d} = \frac{T_0}{\sqrt{1 - h^2}} \tag{6.31}$$

As noted from Eq. (6.31), damping results in a decrease in the period of shaking. Figure 6.4 shows the displacement response of the structure for chosen boundary and initial conditions. As noted from the figure, the damping coefficient has great influence on the vibration response.

Figure 6.5 shows the displacement response for a chosen damping coefficient. As noted from the figure, the amplitude decreases as time increases. The value of damping coefficient h is said to be having a value between 0.02 and 0.05 for linear behaviour.

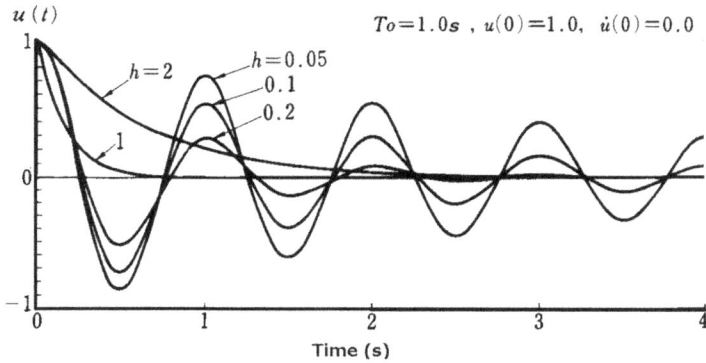

$u(t)$

$To=1.0s$, $u(0)=1.0$, $\dot{u}(0)=0.0$

$h=2$

$h=0.05$

0.1

0.2

1

Time (s)

Figure 6.4 Displacement responses for different damping coefficients.

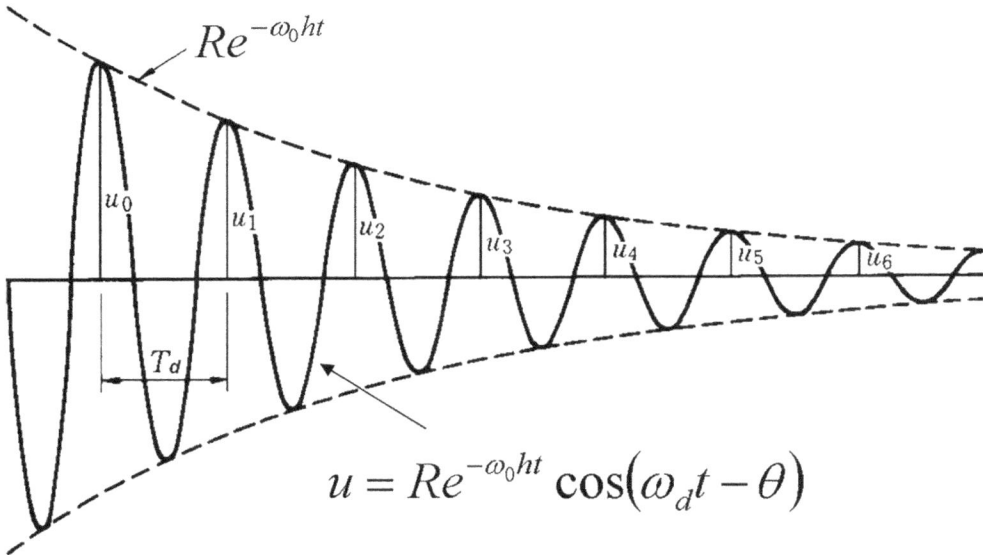

$Re^{-\omega_0 ht}$

u_0 u_1 u_2 u_3 u_4 u_5 u_6

T_d

$$u = Re^{-\omega_0 ht}\cos(\omega_d t - \theta)$$

Figure 6.5 Illustration of the decay of amplitude for damped vibration.

To determine the value of damping coefficient h, the following procedure may be followed. The value of displacement response at the given time step $t = t_{i+1}$ may be written as

$$u_{i+1} = e^{-\omega_0 h t_{i+1}} R\cos(\omega_d t_{i+1} - \theta) \tag{6.32}$$

Similarly at the time step $t = t_i$, the displacement response may also be given as

$$u_i = e^{-\omega_0 h t_i} R\cos(\omega_d t_i - \theta) \tag{6.33}$$

Diving Eq. (6.33) by Eq. (6.32) yields the following:

$$\frac{u_i}{u_{i+1}} = \frac{e^{-\omega_0 h t_i} R \cos(\omega_d t_i - \theta)}{e^{-\omega_0 h t_{i+1}} R \cos(\omega_d t_{i+1} - \theta)} \qquad (6.34)$$

Introducing the following relations using the peak to peak values

$$T_d = t_{i+1} - t_i; \quad \cos(\omega_d t_{i+1} - \theta) = 1; \quad \cos(\omega_d t_i - \theta) = 1 \qquad (6.35)$$

Equation (6.34) may be rewritten as

$$\ln\left(\frac{u_i}{u_{i+1}}\right) = \omega_0 h T_d = \frac{2\pi h}{\sqrt{1-h^2}} \qquad (6.36)$$

If $h \ll 1$, Eq. (6.36) may be further simplified so that the damping coefficient h can be obtained as follow:

$$h = \frac{1}{2\pi} \ln\left(\frac{u_i}{u_{i+1}}\right) \qquad (6.37)$$

However, there will be a number of peak values. Therefore, if peak values are plotted in the space of u_{i+1} and u_i, a curve-fitting procedure yields the value of damping coefficient h. Figure 6.6 shows an example of the procedure applied to an experiment.

6.2.3   Forced vibration subjected to sinusoidal vibration

Let us assume Eq. (6.10) is subjected to the harmonic base acceleration given as

$$\frac{d^2 u_g}{dt^2} = \alpha \sin \omega t \qquad (6.38)$$

The general solution of the resulting differential equations consists of homogenous and non-homogenous parts and it can be written as

$$u = u_h + u_p \qquad (6.39)$$

where

$$u_h = e^{-\omega_0 h t}\left(A\cos\omega_d t + B\sin\omega_d t\right); \quad u_p = C\sin\omega t + D\cos\omega t \qquad (6.40a)$$

$$C = -\frac{\alpha}{\omega_0^2} \cdot \frac{1-(\omega/\omega_0)^2}{\left[1-(\omega/\omega_0)^2\right]^2 + 4h^2(\omega/\omega_0)^2}; \qquad (6.40b)$$

$$D = \frac{\alpha}{\omega_0^2} \cdot \frac{2h\omega/\omega_0}{\left[1-(\omega/\omega_0)^2\right]^2 + 4h^2(\omega/\omega_0)^2} \qquad (6.40c)$$

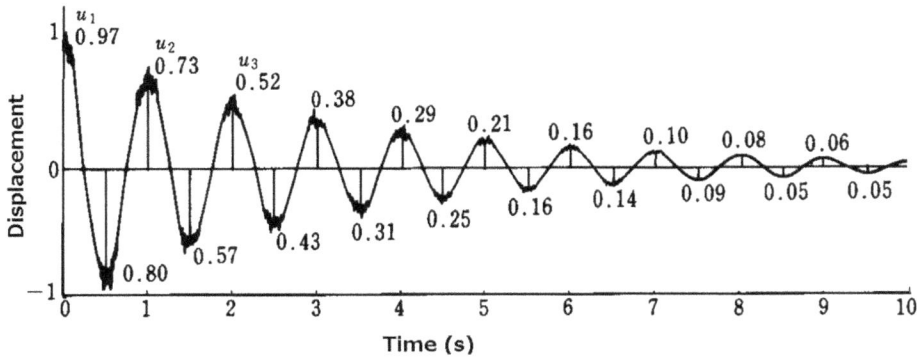

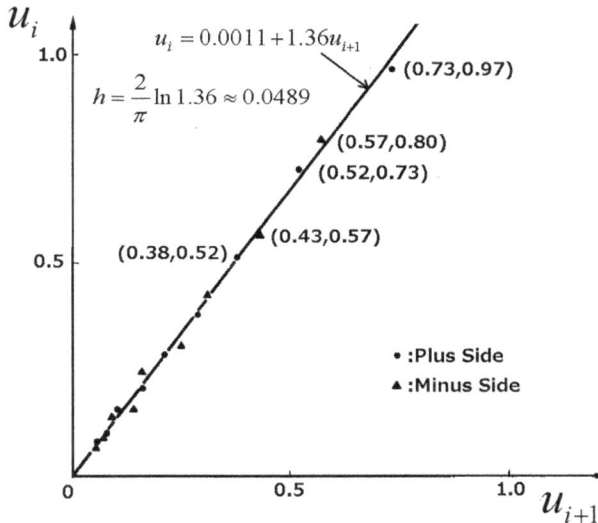

Figure 6.6 Determination of damping coefficient from displacement responses measured in an experiment. (a) Measured displacement response and (b) evaluation of amplitude of displacement responses. (Arranged from Usami (1990).)

The non-homogenous part of the solution may be rewritten as

$$u_p = R\sin(\omega_d t - \theta) \tag{6.41}$$

where

$$R = \frac{\alpha / \omega_0^2}{\sqrt{\left[1 - (\omega / \omega_0)^2\right]^2 + 4h^2(\omega / \omega_0)^2}}; \quad \theta = \tan^{-1}\left(\frac{2h\omega / \omega_0}{1 - (\omega / \omega_0)^2}\right) \tag{6.42}$$

Thus, the displacement response may be rewritten in terms of base displacement or base acceleration as follows:

$$\left|\frac{u}{u_g}\right| = \left|\frac{u}{\alpha/\omega^2}\right| = \frac{(\omega/\omega_0)^2}{\sqrt{\left[1-(\omega/\omega_0)^2\right]^2 + 4h^2(\omega/\omega_0)^2}}; \tag{6.43a}$$

$$\left|\frac{u}{\ddot{u}_g}\right| = \left|\frac{u}{\alpha}\right| = \frac{1/\omega_0^2}{\sqrt{\left[1-(\omega/\omega_0)^2\right]^2 + 4h^2(\omega/\omega_0)^2}} \tag{6.43b}$$

Figure 6.7 shows acceleration and displacement spectra from Eq. (6.43). As noted from the figure, both the acceleration and displacement responses become very large when $\omega = \omega_0$, and if there is no damping, the values tend to infinity. This is a very important observation from spectral values of acceleration and displacement responses.

If $\omega = \omega_0$, the displacement response of the non-homogenous part is

$$u_p = \frac{\alpha}{2h\omega_0^2}\sin\omega t \tag{6.44}$$

Introducing the initial conditions given as

$$u(0) = 0; \quad \dot{u}(0) = 0 \quad \text{at } t = 0 \tag{6.45}$$

yields the integration coefficients as

$$A = \frac{1}{2h\omega_0^2}; \quad B = \frac{1}{2\omega_0\omega_d} \tag{6.46}$$

Thus the final form of solution is

$$u = -\alpha e^{-h\omega_0 t}\left[\frac{1}{2h\omega_0^2}\cos\omega_d t + \frac{1}{2\omega_0\omega_d}\cos\omega_d t\right] + \frac{\alpha}{2h\omega_0^2}\sin\omega t \tag{6.47}$$

If $h \ll 1$, we have

$$\omega_0 \approx \omega_d \tag{6.48}$$

Thus Eq. (6.47) takes the following form:

$$u \approx \frac{\alpha}{2h\omega_0^2}\left(1 - e^{-h\omega_0 t}\right)\cos\omega_d t \tag{6.49}$$

Equation (6.49) is the displacement response of the structure in resonant mode and it is illustrated in Figure 6.8. As noted, the displacement response tends to infinity as time increases. In other words, if such a situation is prevalent, no structure can withstand against vibrations.

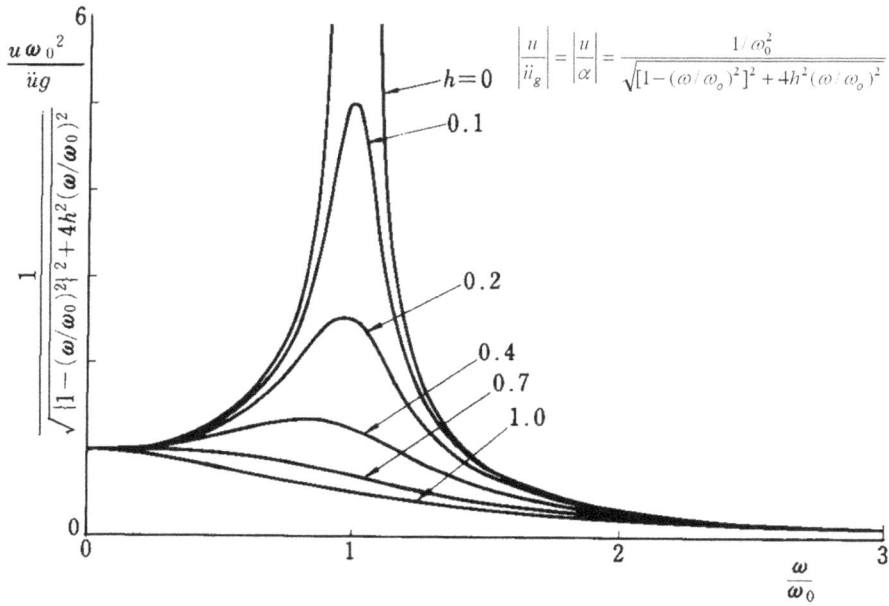

(a)

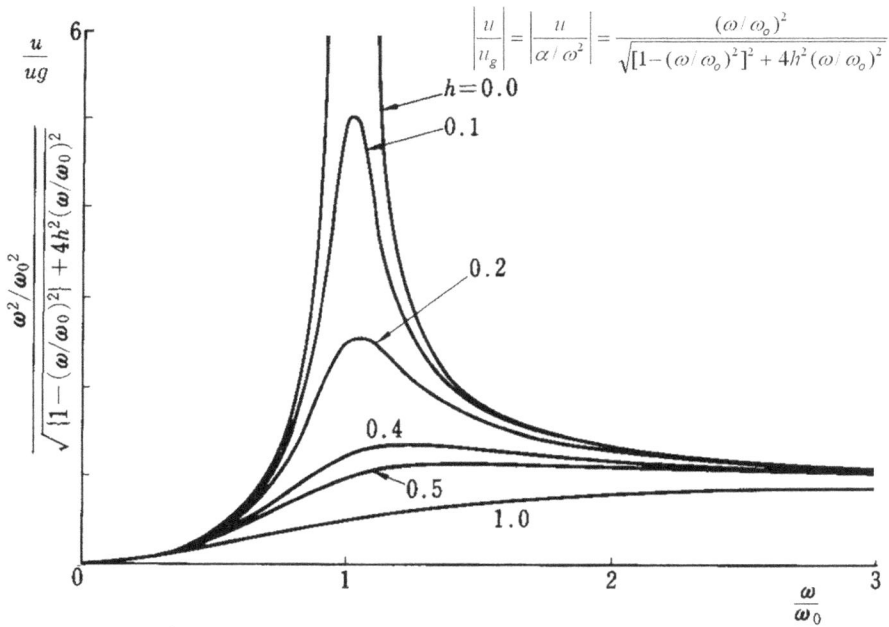

(b)

Figure 6.7 Acceleration and displacement spectra as a function of normalized angular frequency. (a) Acceleration spectra and (b) displacement spectra.

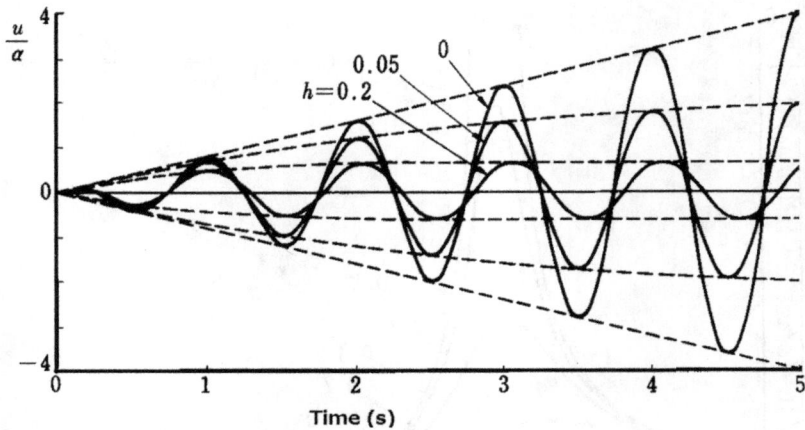

Figure 6.8 Displacement response in resonant mode.

6.2.4 Forced vibration subjected to arbitrary vibration

When base acceleration is arbitrary such as those caused by earthquakes, Eq. (6.10) may be rewritten for a given time step t_i as

$$\frac{d^2u_i}{dt^2} + 2h\omega_0\frac{du_i}{dt} + \omega_0^2 u_i = -\left(\frac{d^2u_g}{dt^2}\right)_i \tag{6.50}$$

Equation (6.50) can be solved using different integration techniques such as central difference technique, linear acceleration method or Newmark β method based on finite difference technique. For example, if the linear acceleration method is chosen, we have the following relations:

$$v_{n+1} = v_n + \frac{1}{2}(a_n + a_{n+1})\Delta t \tag{6.51a}$$

$$u_{n+1} = u_n + v_n\Delta t + \frac{1}{3}\left(a_n + \frac{1}{2}a_{n+1}\right)\Delta t^2 \tag{6.51b}$$

And they are subjected to the following initial conditions:

$$v_{t=0} = a_{t=0}\Delta t \quad \text{and} \quad u_{t=0} = u_0 \quad \text{at} \quad t = 0 \tag{6.52}$$

A specific application of the procedure described herein is shown in Figure 6.9 using the East-West (EW) acceleration component of the Erzincan strong motion station in Turkey with a moment magnitude of 6.8 that occurred on March 13, 1992, in Eastern Turkey (Aydan and Hamada 1992; Hamada and Aydan 1992). This earthquake was

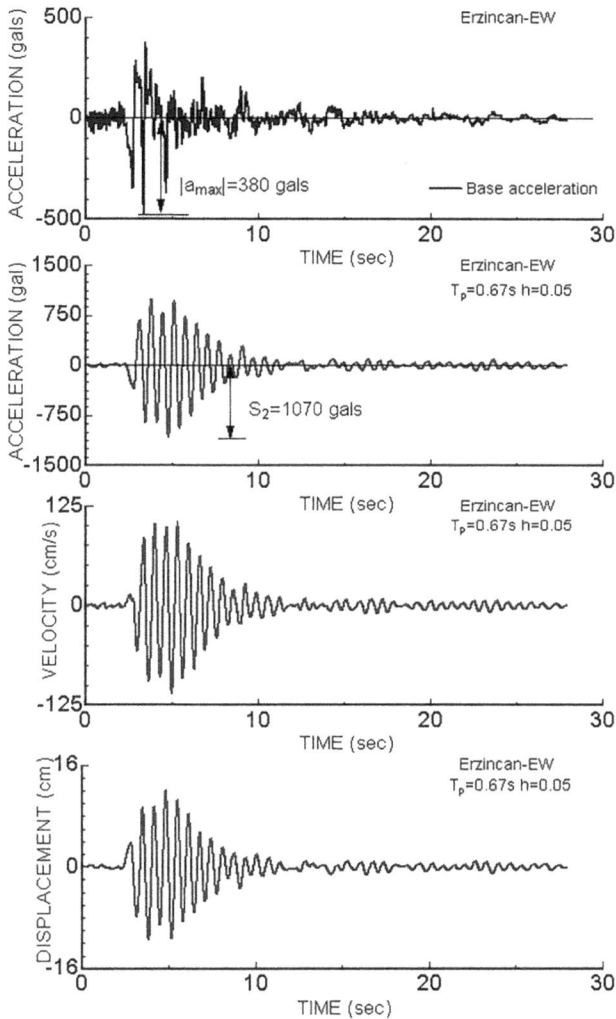

Figure 6.9 Computed acceleration, velocity and displacement of a SDOF structure with a natural period of 0.67 seconds and damping coefficient of 0.05 subjected to the EW component of the 1992 Erzincan earthquake.

associated with the well-known North Anadolu (Anatolian) Fault. The motion was greatly influenced by those observed in the close vicinity of nearby earthquake faulting. In this specific example, the period of the structure was selected as 0.67 seconds, which roughly corresponds to the natural period of 10–12-story reinforced concrete buildings. Despite the duration of the earthquake being <10 seconds, the responses of acceleration, velocity and displacement have very large amplitudes. Particularly, the amplitude of the velocity component is too high for any structure.

6.3 MEASUREMENT TECHNIQUES FOR VIBRATION CHARACTERISTICS

6.3.1 Free vibration

This technique is based on the measurement of acceleration responses of structures by subjecting to an initial displacement and velocity field. Such conditions may be introduced to the structure through some external forces such as passing vehicles, pulse-like impacts. Different accelerometers can be used for these purposes. Nowadays there are either wired or wireless accelerometers, which can be easily used for monitoring vibrations and storage, as well as for data-processing. Specific examples for free-vibration responses and the correlations between Fourier spectra analysis (FSA) and acceleration response analyses (ARA) are given in Section 6.6.

6.3.2 Forced vibration

This technique utilizes dynamic forces such as shaking table or other types of vibration sources. The acceleration responses are measured as a function of induced vibration source using accelerometers attached to the structure and vibrator. In this case, different accelerometers can also be used. They are either wired or wireless accelerometers, which can be easily used for monitoring vibrations and storage, as well as for data-processing. Specific examples for forced vibration responses and the correlations between FSA and ARA are given in Section 6.6.

6.3.3 Micro-tremor measurement technique

Micro-tremor is a low-amplitude ambient vibration of the ground caused by human-made or atmospheric disturbances. Micro-tremor measurements can give useful information on dynamic properties of the site as well as the structures. Micro-tremor measurements are easy to perform. There are different micro-tremor devices. Some comparisons of results obtained from micro-tremor measurements are given in Section 6.6.

6.4 FOURIER SPECTRA ANALYSIS

Fourier transformation is generally used to simulate real wave forms by numerical approximate functions (e.g. Aydan 2017). Let us consider the actual acceleration form given by the following function:

$$a = a(t) \tag{6.53}$$

Fourier transform of Function (6.53) is replaced by the following function

$$a(t) = \sum_{k=0}^{\infty} \left[A_k \cos(kt) + B_k \sin(kt) \right] \tag{6.54}$$

If Eq. (6.54) is approximated by a finite number (N) of data with a time interval (Δt), Eq. (6.54) can be rewritten as

$$\tilde{a}(t) = \frac{A_0}{2} + \sum_{k=1}^{N/2-1} \left[A_k \cos(2\pi f_k t) + B_k \sin(2\pi f_k t) \right] + \frac{A_{N/2}}{2} \cos 2\pi f_{N/2} t \tag{6.55}$$

where

$$f_k = \frac{k}{N\Delta t} \tag{6.56}$$

If time (t) is represented by $N\Delta t$, coefficients A_k and B_k in Eq. (6.55) would be expressed as

$$A_k = \frac{2}{N} \sum_{m=0}^{N-1} a_m \cos \frac{2\pi km}{N}, \quad k = 0,1,2,\ldots,\frac{N}{2}-1, \frac{N}{2} \tag{6.57a}$$

$$B_k = \frac{2}{N} \sum_{m=0}^{N-1} a_m \sin \frac{2\pi km}{N}, \quad k = 1,2,\ldots,\frac{N}{2}-1 \tag{6.57b}$$

For kth frequency, the maximum amplitude, phase angle and power would be obtained as

$$C_k = \sqrt{A_k^2 + B_k^2}; \quad \phi_k = \tan^{-1}\left(-\frac{B_k}{A_k}\right); \quad P_k = \frac{C_k^2}{2} \tag{6.58}$$

It should be noted that the amplitude of Fourier coefficient C_k is multiplied by half of the period of the record ($T/2$) in fast Fourier transformation (FFT). Therefore, the Fourier spectra explained above differ from the FFT spectra in the value of amplitudes and its unit.

6.5   RESPONSE SPECTRAL ANALYSES

Response spectra analyses (RSA) are often utilized for seismic design of structures as well as to assess the response of structures subjected to earthquake excitations. Figure 6.10 illustrates this concept utilizing the EW component of the acceleration record measured at Erzincan with a damping coefficient of 5%. Although the actual structures are complex, the modelled SDOF structures are often used to assess how they would behave during earthquakes with a given excitation wave form. The fundamental concept in RSA is based on the technique described in Section 6.2.4. The maximum acceleration, velocity and displacement are obtained for a given period (T) (frequency) and damping coefficient (h). Although the damping coefficient is unknown beforehand, it is generally assumed to be about 5% in view of various experimental investigations, as also shown in Figure 6.6. The spectral acceleration, velocity and displacement are defined as

$$S_a = \max\left| \frac{d^2u}{dt^2} + \frac{d^2u_g}{dt^2} \right| \tag{6.59}$$

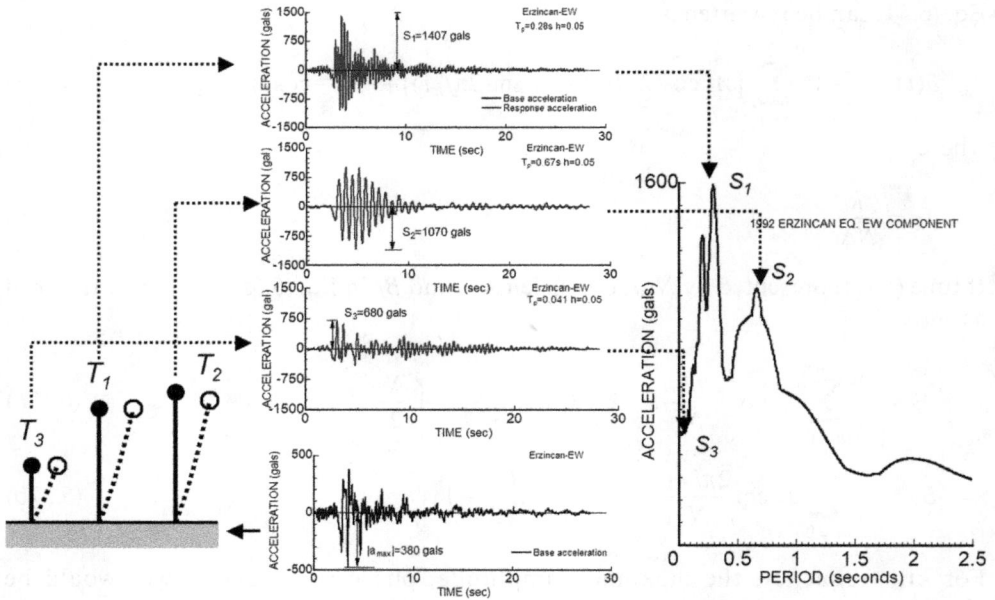

Figure 6.10 Illustration of the spectral analyses concept.

$$S_v = \max\left|\frac{du}{dt}\right| \tag{6.60}$$

$$S_v = \max|u(t)| \tag{6.61}$$

The spectral acceleration, velocity and displacement are plotted as a group of curves for different damping coefficient values.

6.6   APPLICATIONS

In this section, several applications to actual simple model structures are described.

6.6.1   Tower models

Three simple tower models, which are 6, 14 and 36 cm high and made of plastic material and fixed to the base with rigid 4 cm high clamps, were prepared and tested under free and forced vibration models. Figure 6.11 shows model towers on a shaking table equipped with wired or wireless accelerometers. Figures 6.12–6.14 show the acceleration response of towers subjected to the free-vibration condition.

Figure 6.15 shows the Fourier spectra of the records shown in Figures 6.12–6.14. As noted from the figure, the natural frequency of $(36 + 4)$ cm tower is much lower than the $(6 + 4)$ cm tower. This simple example illustrates that the natural frequency of the higher structures would be smaller than those of shorter structures.

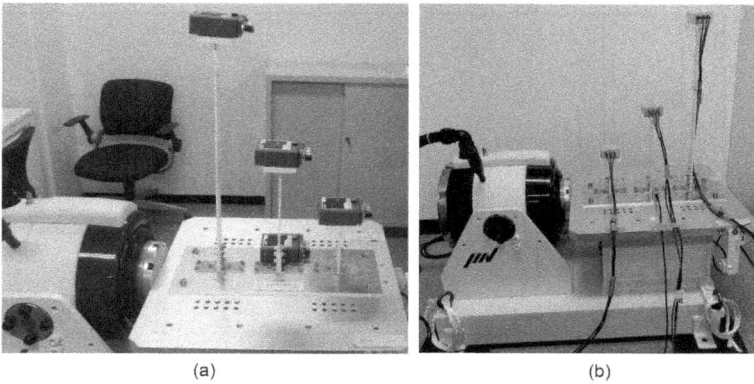

(a)   (b)

Figure 6.11 View of towers equipped with different types of accelerometers. (a) Towers with wireless accelerometers and (b) towers with wired accelerometers.

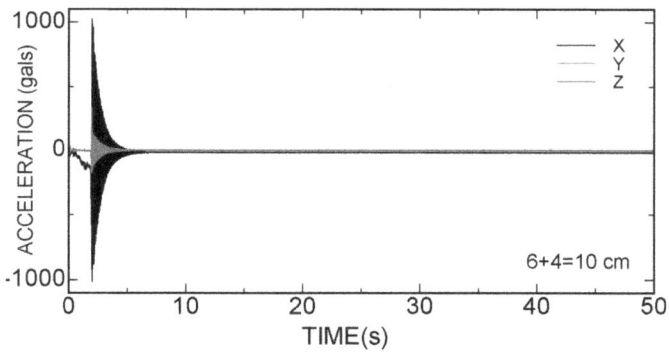

Figure 6.12 Free-vibration response of (6 + 4) cm tower model.

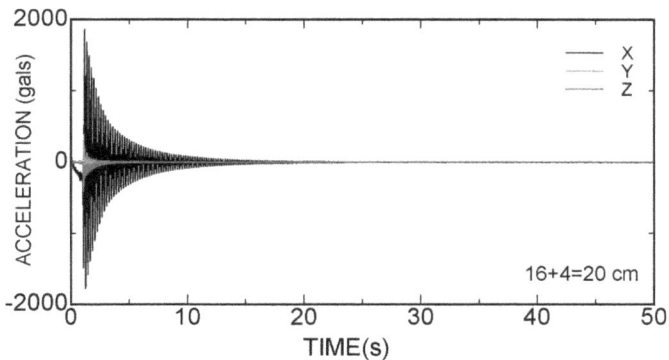

Figure 6.13 Free-vibration response of (16 + 4) cm tower model.

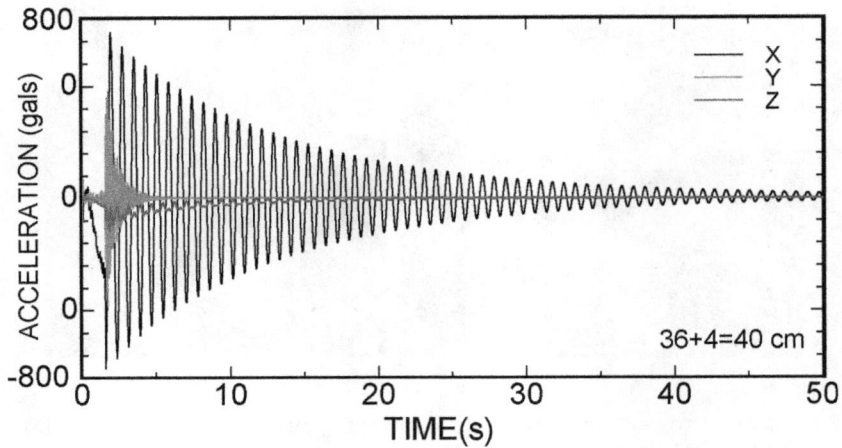

Figure 6.14 Free-vibration response of (36 + 4) cm tower model.

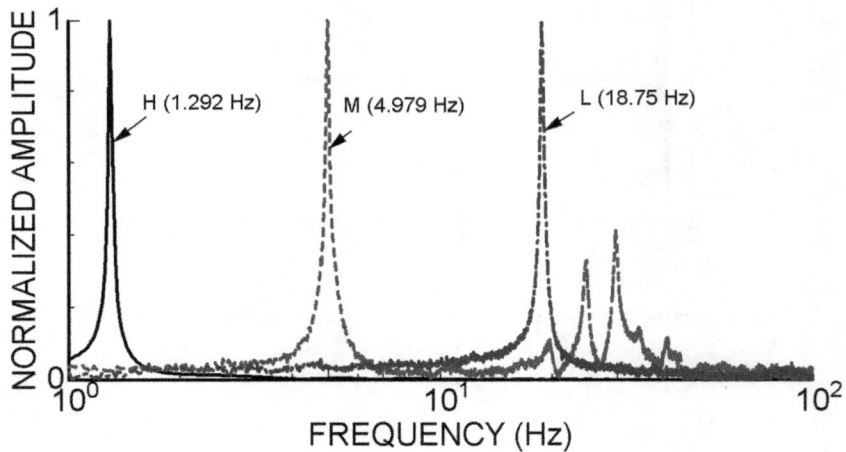

Figure 6.15 Normalized Fourier spectra of tower models subjected to free vibration.

The natural frequencies obtained from FSA are compared with ARA for the tower models in Figure 6.16. As noted from the figure, both analyses yield exactly the same results.

Next, the towers were subjected to a sweeping test condition and experiments were carried out and the records are shown in Figure 6.16. The amplitude of waves was about 100 gals with a frequency ranging between 1 and 70 Hz. As noted from the figure, the tall tower starts to shake at lower frequencies and its vibration is almost nil at higher frequencies.

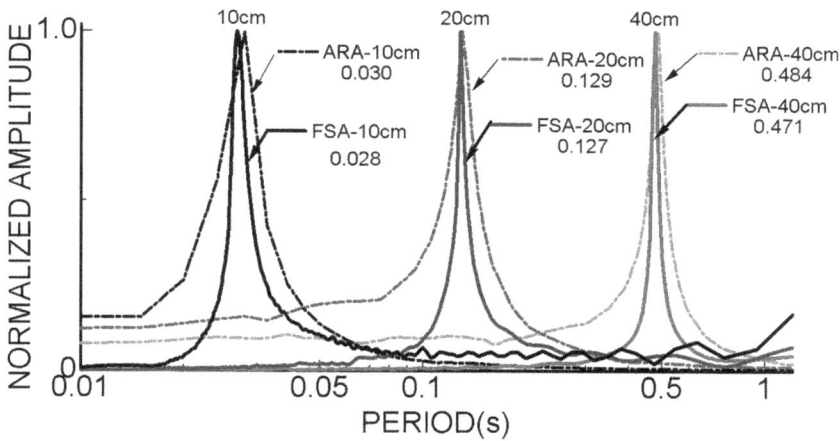

Figure 6.16 Normalized Fourier spectra and acceleration response spectra analyses of tower models subjected to free vibration.

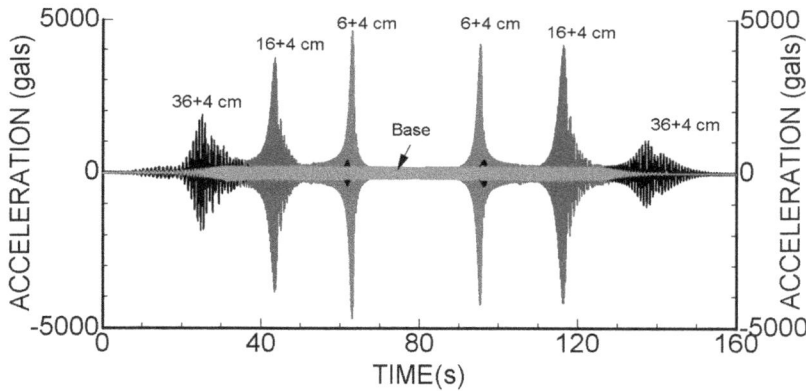

Figure 6.17 Acceleration responses of tower models during a sweeping test.

The Fourier spectra of the records shown in Figure 6.17 are shown in Figure 6.18. When Figure 6.18 is compared with Figures 6.15 and 6.16, the results are almost the same. It is interesting to notice that the vibration characteristics of towers can be easily obtained from free-vibration tests.

6.6.2 Building models

As the building models are more rigid than the tower models, the results of a sweeping experiment are explained herein. Three building models with 4, 8 and 12 stories made of plastic material and fixed to the base were prepared and tested under forced

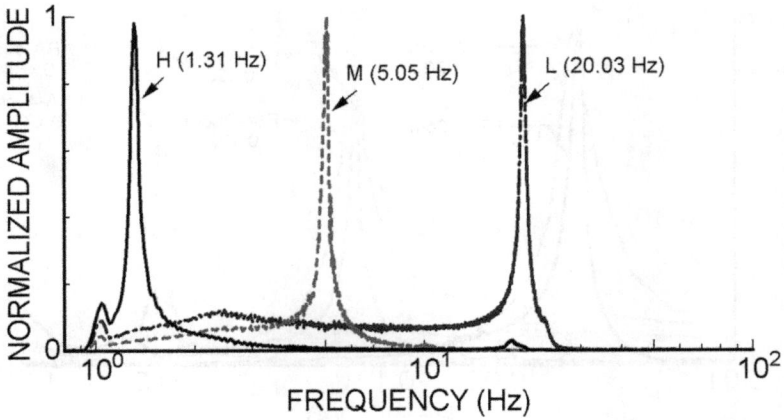

Figure 6.18 Normalized Fourier spectra of tower models subjected to forced vibration shown in Figure 6.17 ($H = 36 + 4$ cm; $M = 16 + 4$ cm; $L = 6 + 4$ cm).

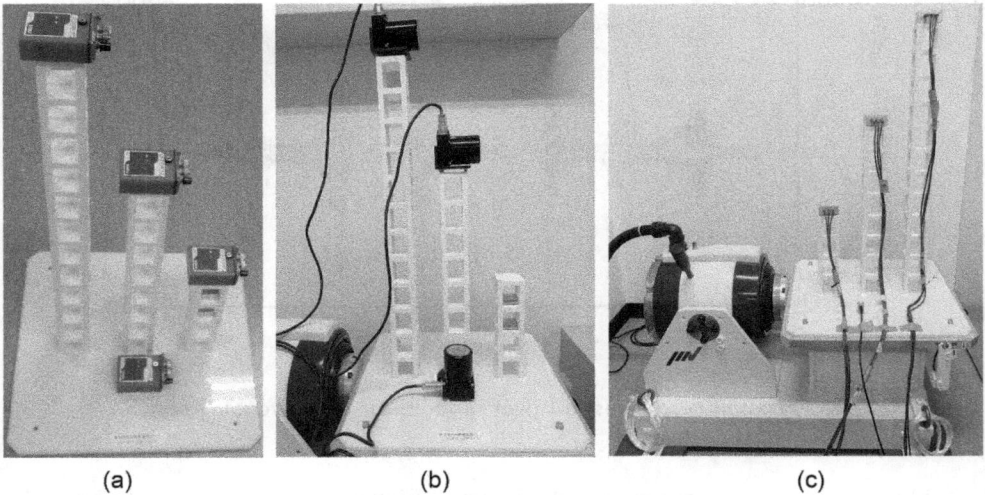

(a) (b) (c)

Figure 6.19 Instrumentation of building models. (a) Wireless accelerometers, (b) micro-tremor and (c) wired accelerometers.

vibration condition. Figure 6.19 shows model buildings on a shaking table equipped with wired or wireless accelerometers. Figure 6.20 shows the acceleration response of building models subjected to a sweeping test with a frequency ranging between 1 and 70 Hz, and Figure 6.21 shows the results of Fourier spectra analyses.

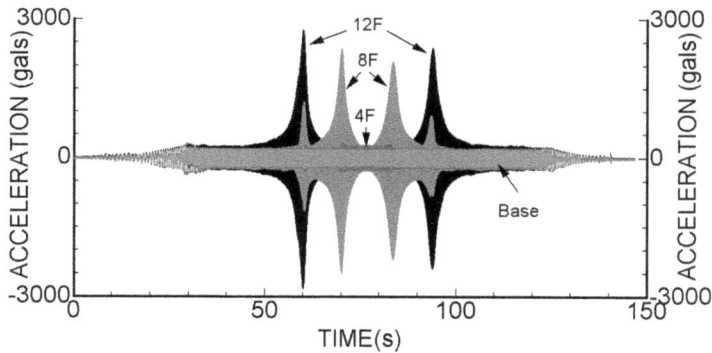

Figure 6.20 Acceleration responses of building models during sweeping tests.

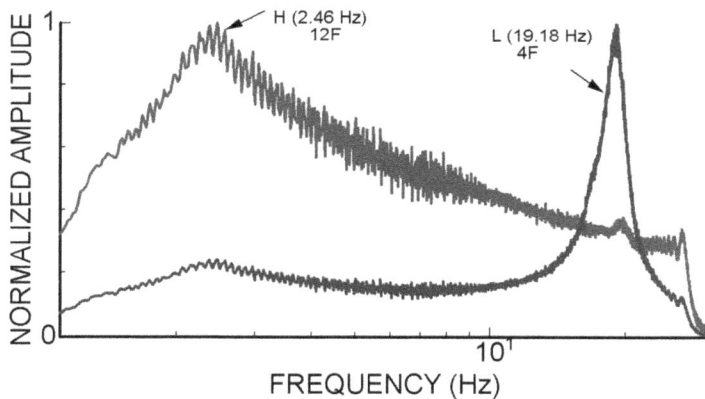

Figure 6.21 Normalized Fourier spectra of building models subjected to forced vibration.

6.6.3   Photo-elastic frame models and Eigen value analyses by FEM

6.6.3.1   Frame only

To understand the vibration characteristics of single frames, photo-elasticity tests were used and some free-vibration tests were carried out as shown in Figure 6.22. Wireless accelerometers set to triggering mode were used for this purpose.

Tests on frames with or without infill walls were carried out, and the Fourier spectra of records are shown in Figure 6.23. As expected, the dominant natural frequency of the frame with infill wall is greater than that for the frame without infill wall. This is due to the increased rigidity of the frame with infill wall.

Eigen value analyses of the single frames with/without infill wall using finite element method were carried out. Figure 6.24 shows the FEM models. Figure 6.25 shows

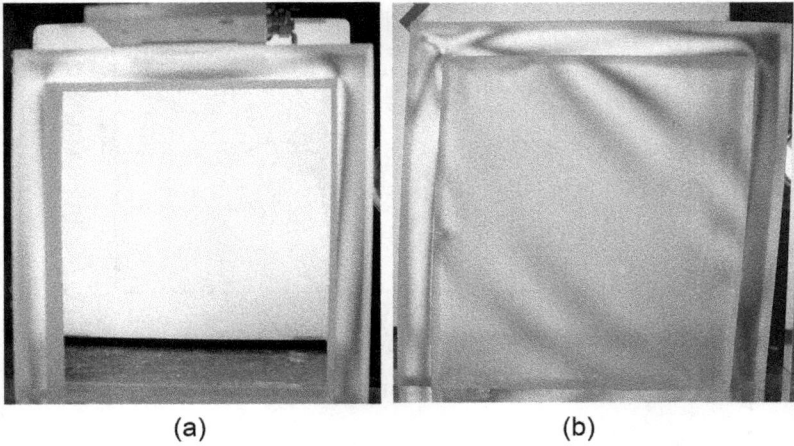

(a)   (b)

Figure 6.22 Views of setup of single frames with/without infill walls for free-vibration tests. (a) Single frame without infill wall (b) single frame with infill wall.

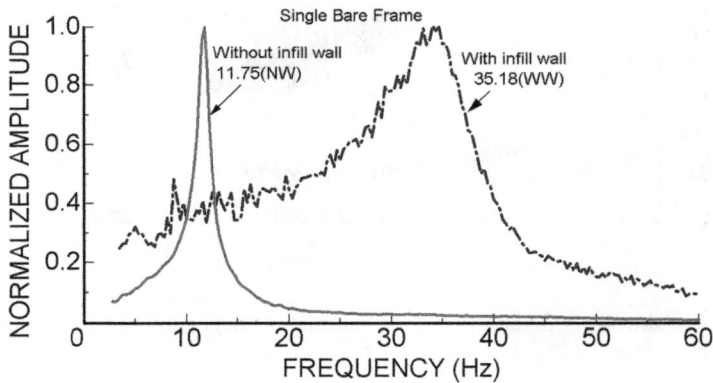

Figure 6.23 Fourier spectra of frames with/without infill wall subjected to free vibration.

the deformation response for Mode 1, and computed Eigen values for the single frame are given in Table 6.1. The computed results are somewhat different from measured values and they are higher than the measured results. One of the main reasons might be mechanical properties and constraint conditions. Nevertheless, the computational results confirm that the natural frequency of a single frame with infill wall would be greater than that of a bare single frame.

6.6.3.2   Four-story frame models

Photo-elasticity tests on four-story model framed structures are described in Chapter 7. They were subjected to free vibrations and their acceleration responses were measured

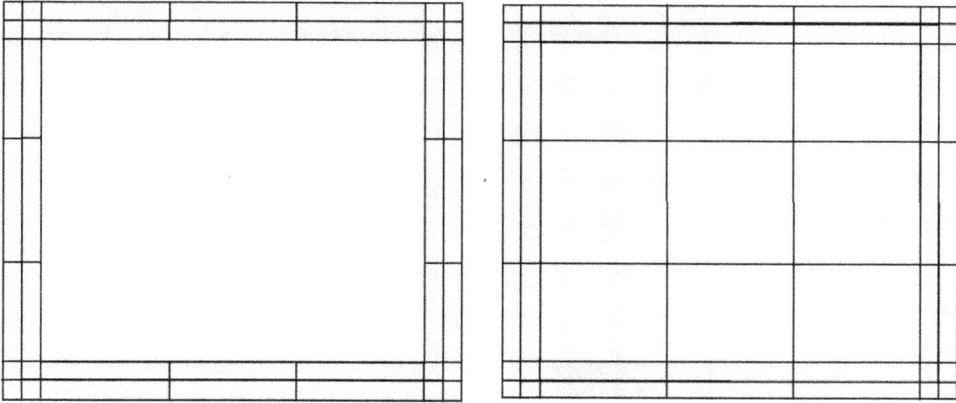

Figure 6.24 FEM models for Eigen value analyses. (a) Bare frame and (b) frame with infill wall.

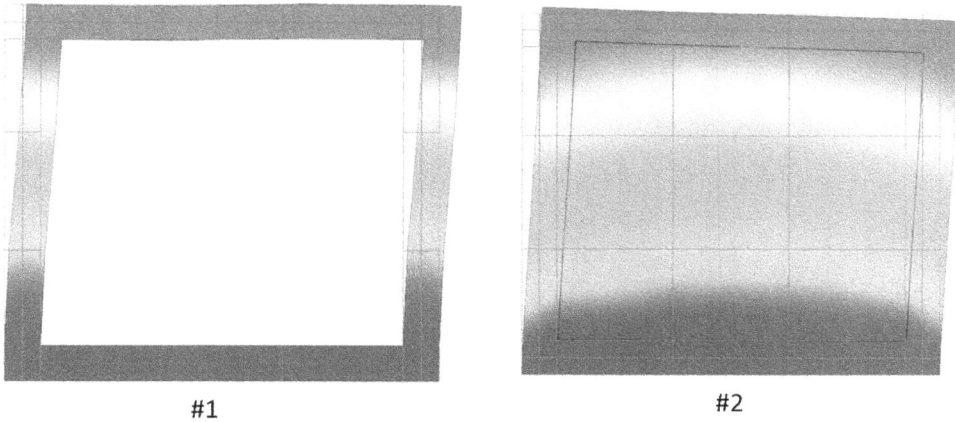

#1   #2

Figure 6.25 Deformation response for Mode 1. (a) Bare frame and (b) frame with infill wall.

Table 6.1 Computed frequency and period of single-frame structures

Model	Frequency	Period
#1	11.923	0.083
#2	57.345	0.017

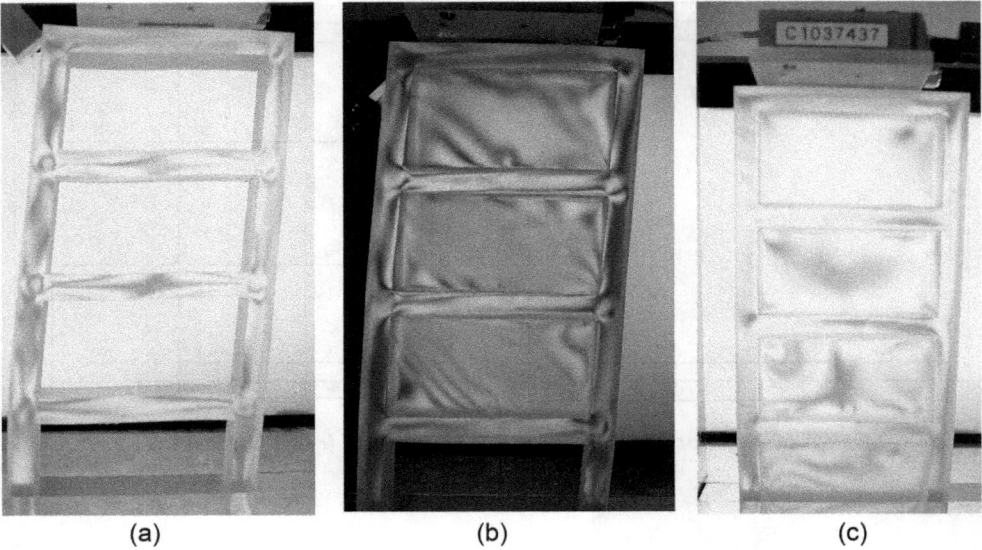

Figure 6.26 Four-story framed structure models subjected to free vibration. (a) Without any infill wall, (b) soft (weak) floor condition, and (c) with infill walls.

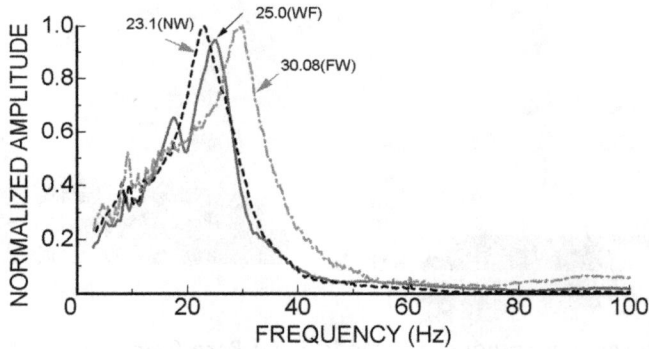

Figure 6.27 Fourier spectra of four-story framed structures subjected to free vibration.

using wireless accelerometers set to triggering mode. The four-story framed structure models were as follows (Figure 6.26):

a.   Framed structure without any infill wall
b.   Framed structure with soft (weak) floor (without any infill wall at the ground floor)
c.   Framed structure with infill walls

Figure 6.27 shows the Fourier spectra of four-story framed structures. As expected, the dominant natural frequency of the structure with soft (weak) floor condition is the

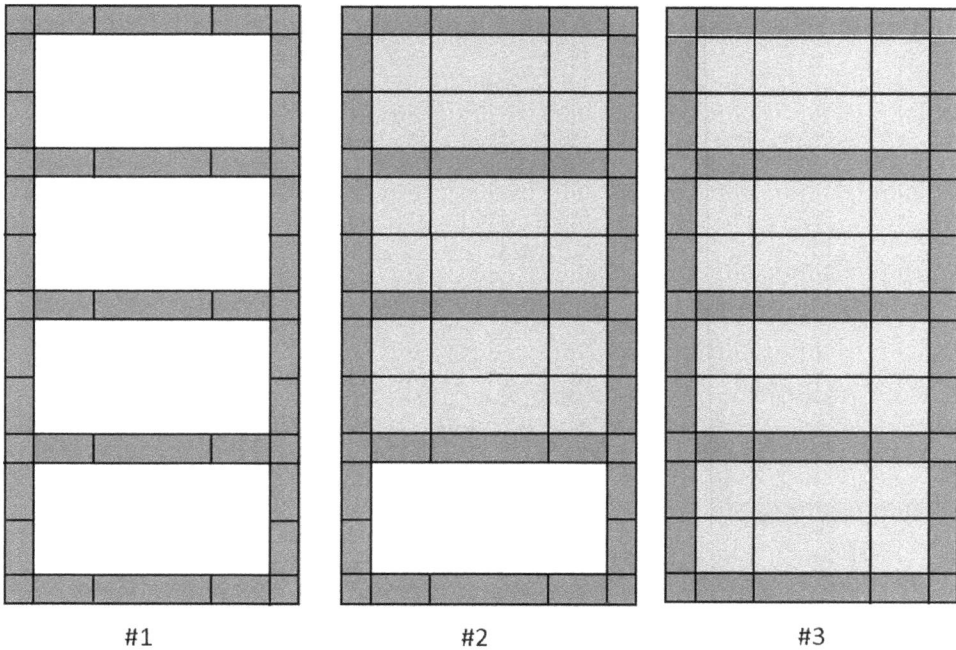

Figure 6.28 Finite element models for Eigen value analyses.

middle among all structures. The structure with infill walls has the greatest natural frequency. It is also interesting to note that the dominant natural frequency of the framed structure without infill walls is slightly lower than that for the structure with soft (weak) floor condition.

A series of Eigen value analyses using the finite element method were carried using the models shown in Figure 6.28. Figure 6.29 shows the deformation response for Mode 1 and computed Eigen values for the four-story frame structures are given in Table 6.2. The computed results are somewhat different from measured values and they are higher from the measured results. One of the main reasons might be mechanical properties and constraint conditions. Nevertheless, the computational results confirm that the frequency of the frames with infill wall would be greater than that of bare frame structures.

6.6.4 Beam models

The natural frequency of a beam, which is 170 cm long, 20 cm thick and 11.5 cm wide, reinforced with two steel bars, was determined using wireless accelerometers in drop-weight tests and a micro-tremor device with ambience-induced vibration. The beam was cracked during drop-weight tests. An SPC-51A micro-tremor device was used to determine the natural frequency of the cracked beam, as shown in Figure 6.30.

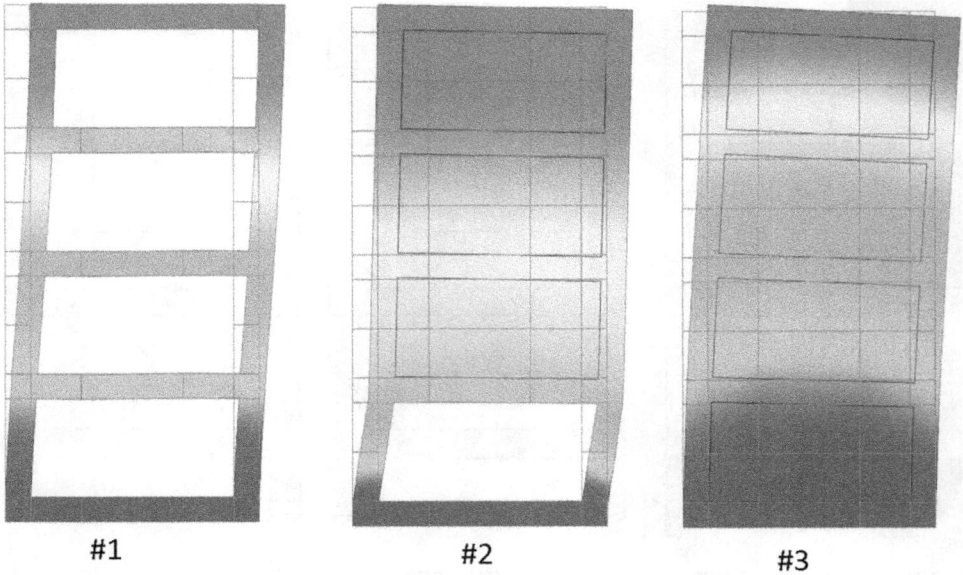

#1   #2   #3

Figure 6.29 Computed deformation response for Mode 1.

Table 6.2 Computed and measured frequencies and periods of frame structures

Model	Feature	Frequency (Hz)	Period (s)
#1	Bare	12.035	0.083
#2	Weak floor	15.882	0.063
#3	Infilled	29.054	0.034

Figure 6.30 Monitoring of a cracked beam using a SPC-51A micro-tremor device.

The natural frequencies of the un-cracked and cracked beams are determined and compared as shown in Figure 6.31. As noted from the figure, the natural frequency of the cracked beam decreases compared to that of the un-cracked beam. In the same figure, the H/V spectra obtained from the micro-tremor measurements is also plotted. It is interesting to note that the micro-tremor method distinctly yielded natural frequencies of the beam for the cracked and un-cracked states. However, the amplitude for the un-cracked state is slightly lower than that for the cracked state.

A pipe-like beam with an external diameter of 50 mm and length of 5,600 mm was subjected to free vibration. In the first experiment, the beam was subjected to distributed mass over its entire length (Figure 6.32). In the second experiment, an additional concentrated mass was added to the centre of the beam. Figure 6.33 compares the Fourier spectra of the acceleration records for both cases. As noted from the figure, the natural frequency of the beam with concentrated load decreases compared with that of the beam with distributed mass only. In view of theoretical considerations, the natural frequency should be affected by the total mass of the beam and the increase in the total mass would result in lower value natural frequencies.

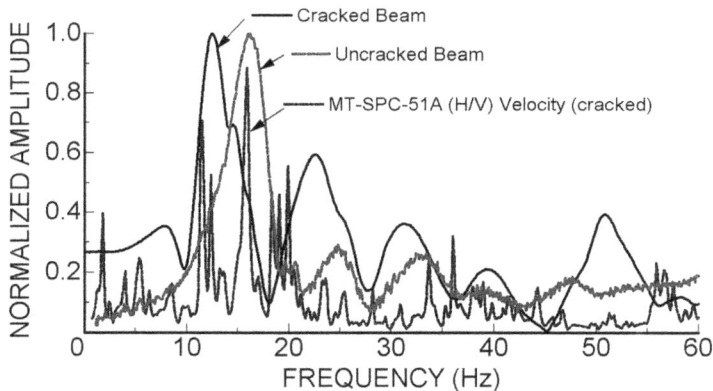

Figure 6.31 Comparison of Fourier spectra of un-cracked and cracked beams.

Figure 6.32 View of the measurement setup.

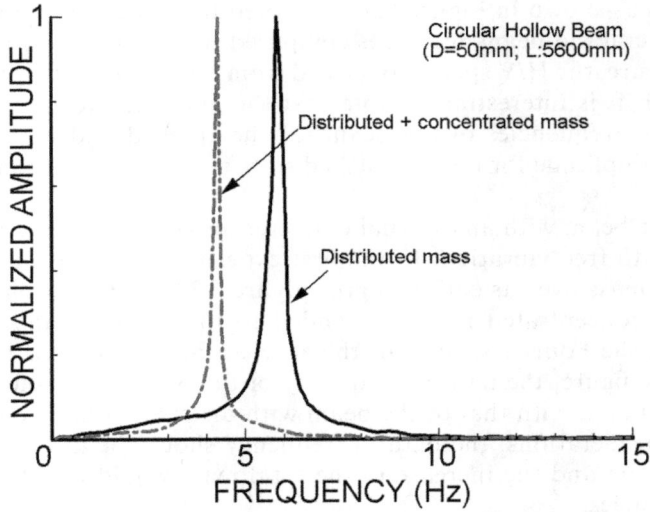

Figure 6.33 Comparison of Fourier spectra of beams with different masses.

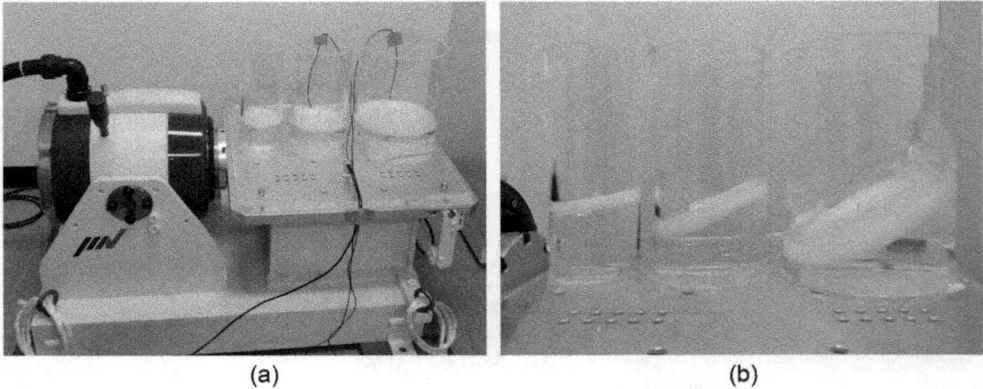

Figure 6.34 (a) View of the tank sloshing education setup and (b) the sloshing phenomenon during a test.

6.6.5 Tanks

Aydan (2019c) developed an education model test package (OA-EP-TSM71015) for studying sloshing phenomenon of liquids in tanks as a part of the natural disaster demonstration facility. The package consists of three acrylic tanks with a diameter of 70, 100 and 150 mm, respectively, and they are equipped with water pressure sensors. Tanks with floating lids that were fixed to an acrylic base were put on the shaking table as shown in Figure 6.34a. The tanks were subjected to sweeping-type shaking

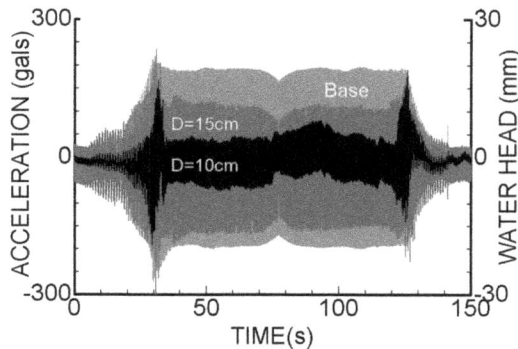

Figure 6.35 Water pressure variation in relation to base shaking.

(1–70–1 Hz) and the sloshing behaviour was investigated. Figure 6.34b shows a view of the test, and the sloshing behaviour could be easily seen from Figure 6.34b. Figure 6.35 shows the measured water pressure as a function of the base shaking. One can easily note the sloshing phenomenon in relation to the frequency content of shaking.

6.7   ACTUAL STRUCTURES

6.7.1   Bridge of the University of the Ryukyus

There is a reinforced concrete bridge spanning over the reservoir of an earth-fill dam in the Senbaru Campus of the University of the Ryukyus (Figure 6.36a). The bridge is 85 m long, 3 m wide (Figure 6.36b). This bridge suffered an alkali–silica reaction (ASR) problem and it was repaired recently. Natural frequencies of this bridge were measured using drop-weight test and micro-tremor method. The measurements were taken before and after the repair, and they indicated that there were almost no major differences in natural frequencies, as seen in Figure 6.37. Furthermore, the different techniques yielded almost the same results. Compared to the micro-tremor device, the drop-weight test is cheaper and quite effective in determining the vibration characteristics of structures.

6.7.2   Vibration of Yofuke Bridge due to passing trucks

As pointed out, passing vehicles may also induce vibration in structures. The vibrations of Yofuke Bridge due to passing trucks (about 10 tonf) were measured using wireless stand-alone accelerometers set to trigger mode (Figure 6.38). Figure 6.39 shows the Fourier spectra of the measured acceleration wave forms. Measurements were made in 2015 and 2017, and the results indicated that the dominant natural frequency of the bridge remained the same. Nevertheless, some additional frequencies were noted and these may imply that some degradation of the bridge has been taking place.

(a)

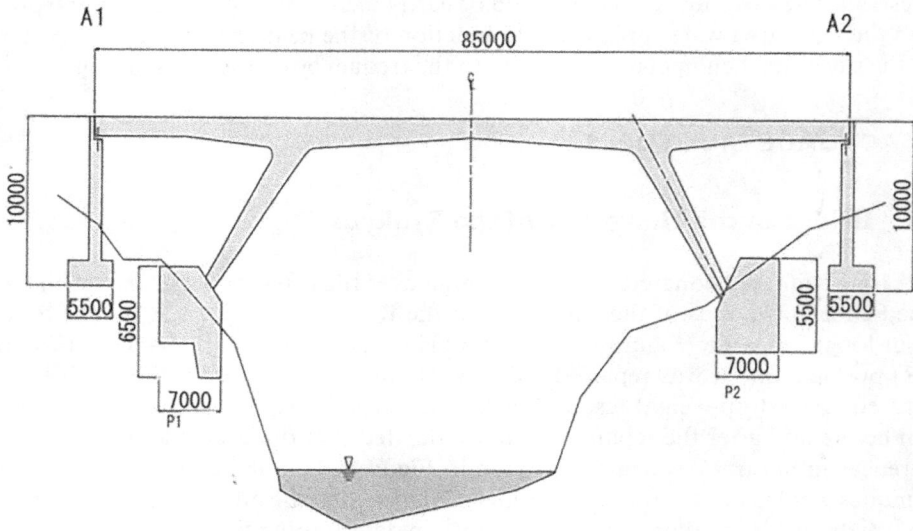

(b)

Figure 6.36 View of the bridge of the University of the Ryukyus. (a) Single frame without infill wall (b) single frame with infill wall.

6.7.3 Pole for hybrid wind and solar energy

A unique pole has a wind turbine and a solar panel to produce a combined renewable energy in the campus of the University of the Ryukyus. The rotation axis of the wind turbine coincides with the vertical axis of the pole. Figure 6.40 show a view of the pole and instrumentation. Two accelerometers are attached to the pole. One of the

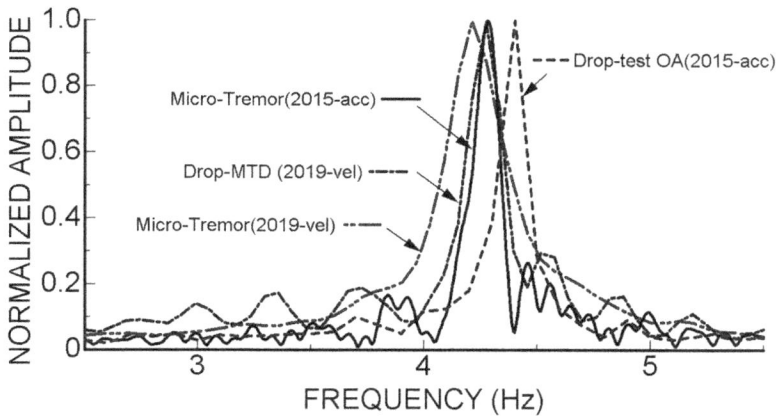

Figure 6.37 Comparison of Fourier spectra of the bridge of the University of the Ryukyus using different measurement techniques.

Figure 6.38 Views of Yofuke Bridge and instrumentation.

accelerometer was attached to the pole at about 130cm from the base plate and the other one was attached to the base plate. Micro-tremor sensors were directly placed on the base plate. Figure 6.41 compares the Fourier spectra of acceleration records obtained from the wireless accelerometer and the micro-tremor sensor. Despite slight

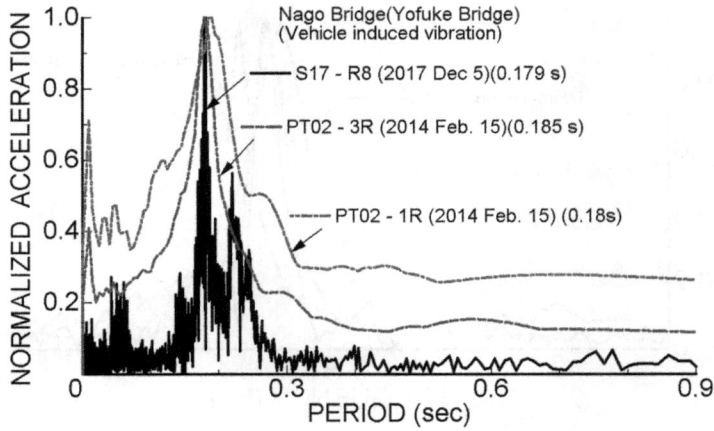

Figure 6.39 Comparison of Fourier spectra of truck-induced vibration measurements at the Yofuke Bridge in Nago City at different times.

Figure 6.40 View of the pole and instrumentation.

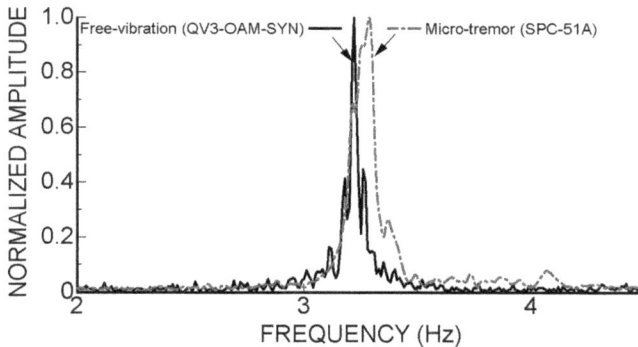

Figure 6.41 Comparison of the Fourier spectra of acceleration records obtained from the wireless accelerometer and micro-tremor sensor.

Figure 6.42 Views of (a) old and (b) new houses.

differences, the dominant natural frequency of the pole is almost the same. The noise-like components may be due to the interaction between the base plate and the base concrete foundation.

6.7.4 Wooden houses

The natural frequencies of two old and new wooden two-story buildings were measured using SPC-51A micro-tremor device (Figure 6.42). The new building was built over the site of the old building, which was about 36 years old at the time of demolition. Figure 6.43 compares the natural frequencies of the wooden house using H/V spectra technique. While the natural frequency of the new building is about 10.39 Hz (0.96 seconds), the old building has the dominant natural frequency at 6.13 Hz (0.163 seconds). This result implies that the old building underwent some degradation over the years.

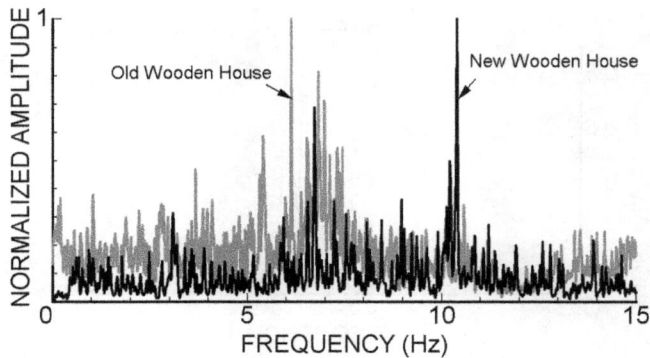

Figure 6.43 Comparison of the H/V Fourier spectra of acceleration records for old and new wooden two-story buildings.

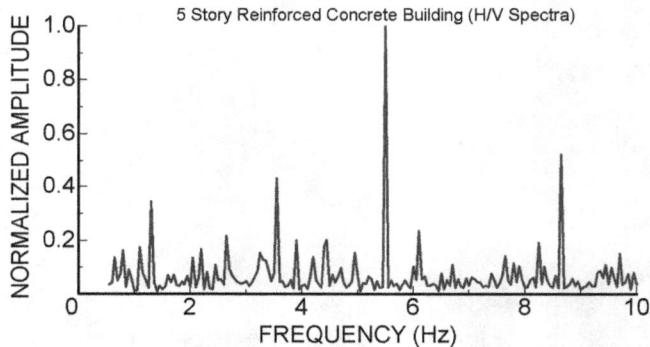

Figure 6.44 Comparison of the H/V Fourier spectra of acceleration records for the reinforced concrete five-story building.

6.7.5   Reinforced concrete building

A reinforced concrete five-story building was built about 35 years ago in Shimashi district of Ginowan City. The measurements were carried out on the third floor. The dominant natural frequency of this building was about 4.09 Hz (0.244 seconds) (Figure 6.44).

6.8   PAST STUDIES ON THE NATURAL FREQUENCY OF BUILDINGS

There are some past studies on the natural periods of buildings. These results were compiled by Aydan et al. (1999b, 2000b). The results are shown in Figure 6.45. In the same figure, some empirical estimations by Aydan et al. (2000b) and seismic codes of Turkey and the United States are plotted. Aydan et al. (2000b) concluded that most of the data implies that the natural frequencies of various types of buildings obey the following relation as a function of the number of floors:

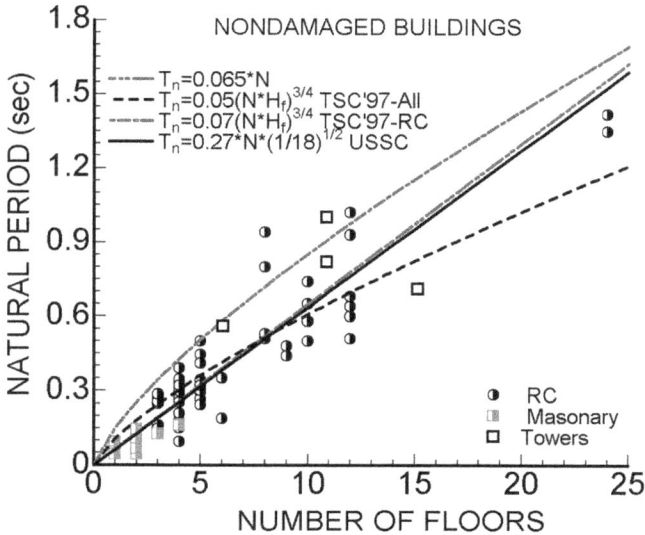

Figure 6.45 The natural frequencies of various types of buildings as a function of floor number (Aydan et al. 2000b).

$$T_n = 0.065N \qquad (6.62)$$

In addition, the natural periods of the structures may also be estimated from several simple theoretical formulas, which are given below:

Shear Model

$$T_o = \frac{2\pi}{\bar{V}_s} H = \frac{2\pi H_f}{\bar{V}_s} N \qquad (6.63)$$

Bending Model with concentrated mass at top

$$T_o = \frac{2\pi}{\bar{V}_s} H \frac{H}{t} = \frac{2\pi H_f^2}{\bar{V}_s t} N^2 \qquad (6.64)$$

Bending Model with distributed mass

$$T_o = \frac{2\pi}{\bar{V}_s} \frac{3}{2} H \frac{H}{t} = \frac{3\pi H_f^2}{\bar{V}_s t} N^2 \qquad (6.65)$$

where $\bar{V}_s$, H_f, H, t and N are shear wave velocity, floor height, width and floor number of a typical building. Nevertheless, it is strongly recommended to use the actual data whenever possible.

6.9   DAMS

There are different types of dams, and micro-tremor measurement at a rockfill dam is described as an example herein. The Haneji dam, located in Okinawa Island (Figure 6.46), was built in 1977 and is a rockfill dam with a clay core. It is 66.5 m high and the crest is 10 m wide and 198 m long. The inclination of the downstream side is 1:2.2, while the upstream side is 1:2.7. Micro-tremor measurements were carried out in the middle of the crest. The Fourier spectra of micro-tremor measurements are shown in Figure 6.47. The spectral analysis indicated that three components had a common peak at 6.68 Hz.

(a)   (b)

Figure 6.46 Views of the dam and micro-tremor measurements. (a) An aerial view and (b) micro-tremor monitoring.

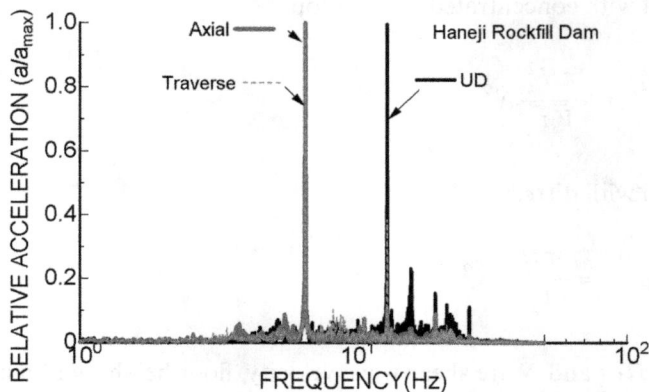

Figure 6.47 Fourier spectra of measured components.

6.10   WIND TURBINES

Wind turbines have become one of the most common renewable-energy-producing facilities. The measurements were implemented using portable accelerometers (QV3-OAM-SYN) as well as micro-tremor measurements (Figure 6.48). The Fourier spectra of vibration measurements were calculated using portable accelerometers, and are shown in Figure 6.49. The spectral values correspond to the total structure. The

(a)   (b)

Figure 6.48 Views of the measurement site. (a) Single frame without infill wall (b) single frame with infill wall.

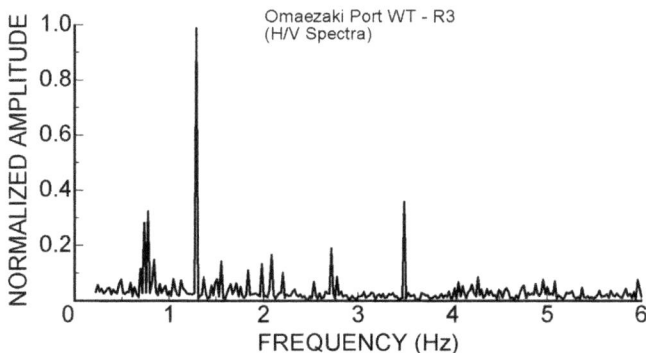

Figure 6.49 Fourier spectra of measured components.

tower is 80.4 m high and blade radius is 39 m. The dominant frequency was observed as 2.44 Hz. However, other frequencies were also noted as expected.

6.11   ABANDONED MINES

Aydan and Genis (2008a) carried out a series of elasto-plastic dynamic response analyses of an abandoned lignite mine in Mitake town in Japan. Figure 6.50 shows a three-dimensional view of the mine and the in situ stress state inferred from the fault striations using Aydan's method. The material properties used in the analyses are given in Table 6.3. The input ground motion is assumed to be sinusoidal with a chosen period. The responses for actual ground motions induced by nearby earthquakes can be found in another article (Aydan and Genis 2008b). Figure 6.51 shows the three-dimensional geological structure on the adopted numerical mesh. For these particular analyses, the FLAC3D developed by Itasca (2005) is used. The method is based on the finite difference technique utilizing silent boundary conditions and Rayleigh-type damping.

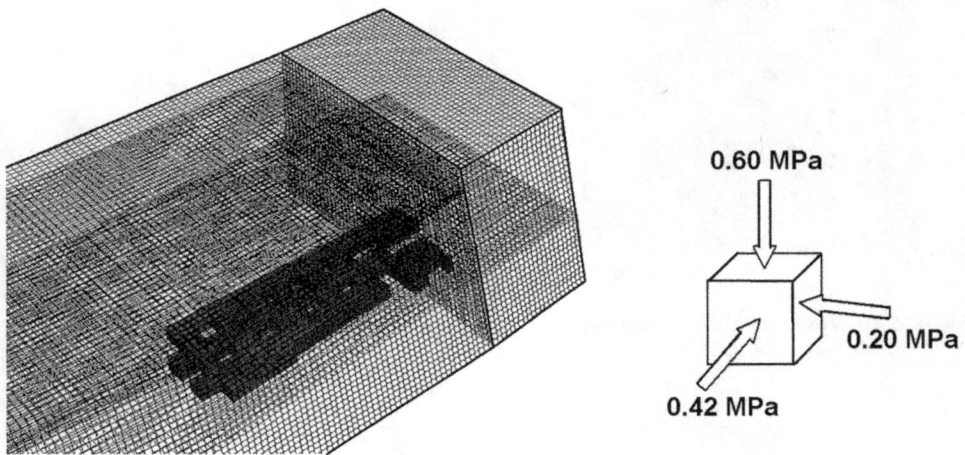

Figure 6.50 Finite difference mesh and assumed in situ stress state.

Table 6.3 Rock material properties used in numerical analyses

Unit	Uniaxial compressive strength (MPa)	Friction angle (°)	Elastic modulus (MPa)	Unit weight (kN/m³)	Tensile strength (MPa)	Cohesion (MPa)	Shear modulus (MPa)
Lignite	4	53	557	20	0.4	0.669	228.8
Sandstone	2.3	53	257	20	0.06	0.385	102.8
Mudstone	0.96	29	76	20	0.05	0.283	30.4

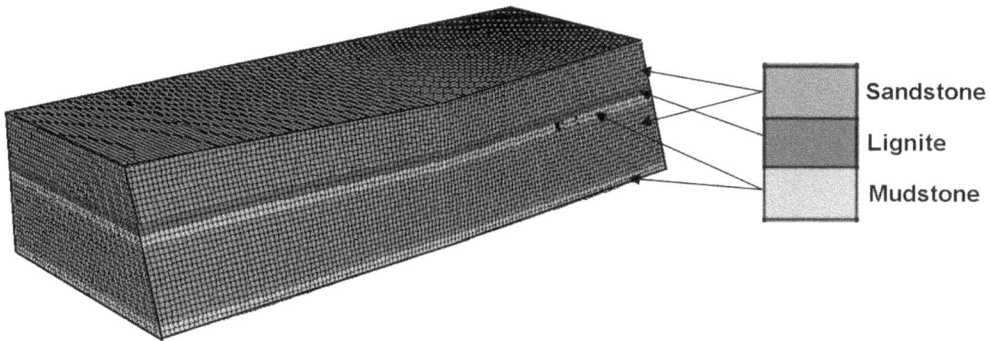

Figure 6.51 Three-dimensional view of the numerical model of the mine and geologic formations.

The maximum amplitude of the acceleration was selected as 20 and 80 gals with a frequency of 2 Hz. The acceleration is applied for a period of 1.7 seconds. Figure 6.52 shows the yield zone development for assumed ground motions. While the yielding occurred due to tensile stresses for the ground motion with an amplitude of 20 gals, the yielding took place due to tensile and shearing for the ground motion with an amplitude of 80 gals. The shear yielding occurred in pillars, as seen in Figure 6.52b.

Figure 6.53 shows the acceleration responses at some selected points. As noted from Figure 6.53, the maximum ground acceleration takes place at the crest of the slope. The ground acceleration at the pillar adjacent to the slope is three times higher than that at the base. This tendency is quite similar for ground motions with amplitudes of 20 and 80 gals.

Another computation was carried out for the case, in which the properties of rock mass was reduced to 1/8 of those of the intact rock, while the ground acceleration amplitude was assumed to be 80 gals. This situation corresponds to a state of degradation of rock mass properties with time. Figure 6.54 shows the yield zone development at the model surface and at EW and NS sections. This situation probably corresponds to a state of the total collapse of the abandoned mine. As seen in Figure 6.54, the yielding at ground surface extends to a large area and the yielding in EW (A-A') cross section resembles a slope failure containing the abandoned mine.

6.12 RESPONSE OF HORONOBE UNDERGROUND RESEARCH LABORATORY DURING THE 2018 JUNE 20 SOYA REGION EARTHQUAKE AND 2018 SEPTEMBER 6 IBURI EARTHQUAKE

6.12.1 Characteristics of the Soya region earthquake

The Soya region earthquake occurred on June 20, 2018, at 5:28 AM. The moment magnitude of the earthquake was 4.0 according to F-NET of NIED. The focal mechanism of the earthquake was estimated to be a thrust fault (Figure 15.55a), which is a

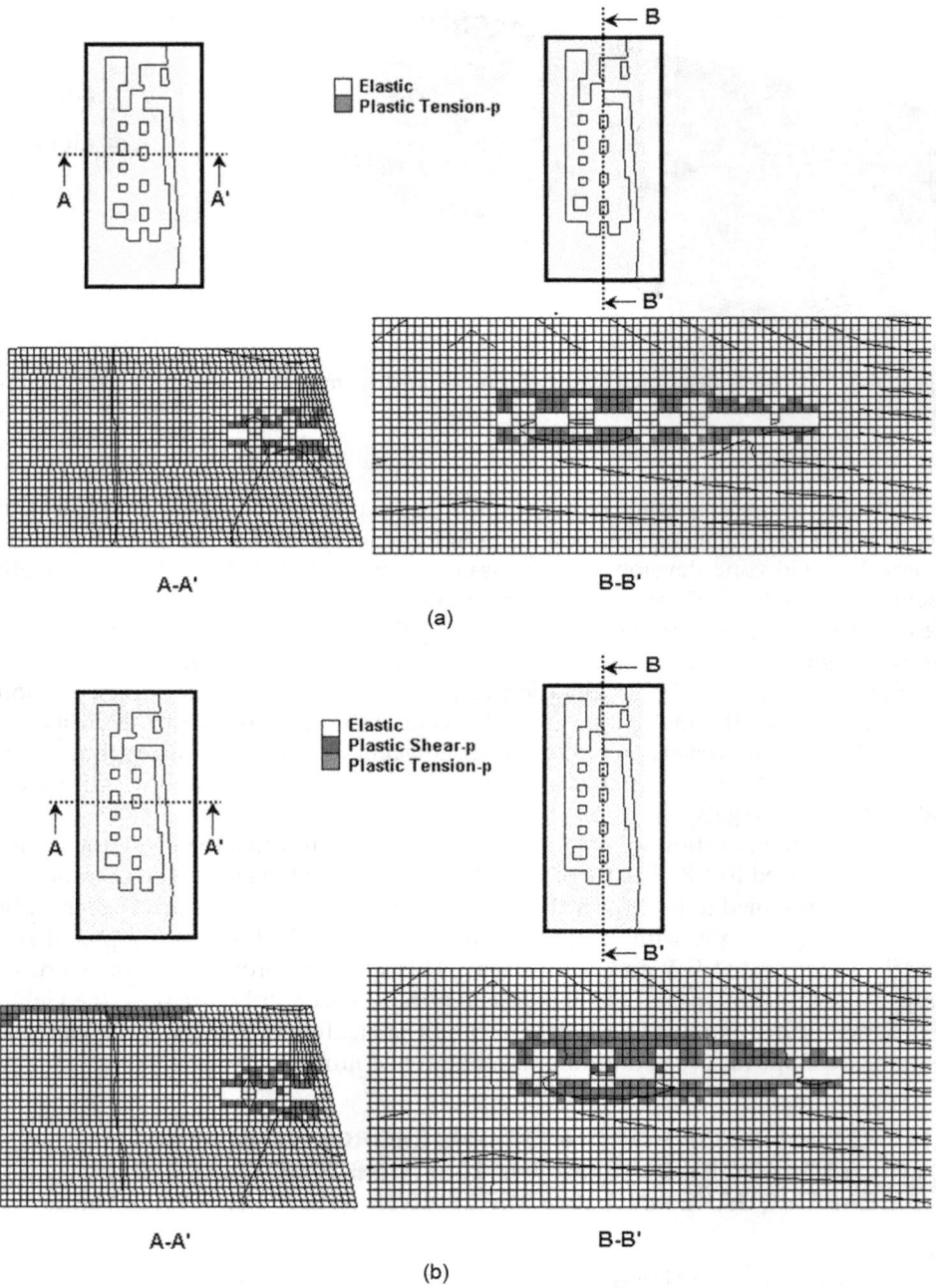

Figure 6.52 Yield zones around openings and pillars. (a) $A_{max} = 20$ gals and (b) $A_{max} = 80$ gals.

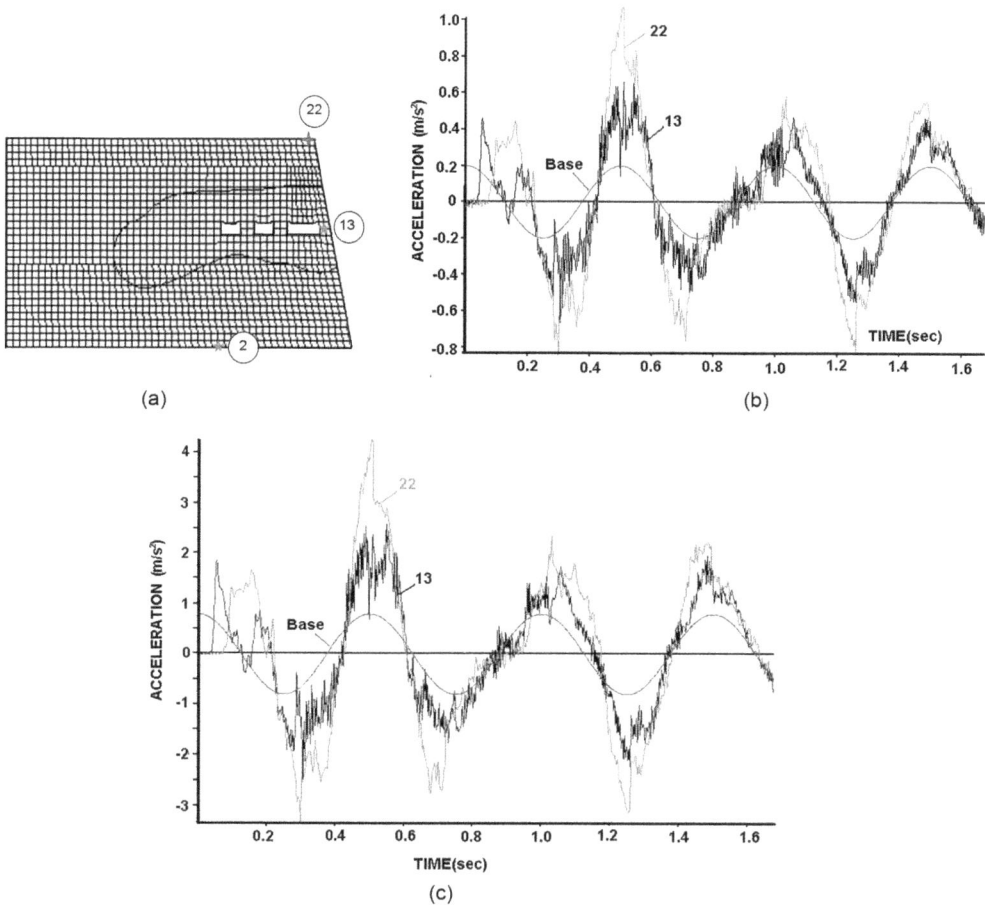

Figure 6.53 Acceleration responses of several selected points. (b) $A_{max} = 20$ gals and (c) $A_{max} = 80$ gals.

consistent mechanism in view of tectonics of the Soya region. Figure 6.55b shows the inferred stress state for the earthquake. The maximum horizontal stress acts almost in the EW direction.

6.12.2 Characteristics of Iburi earthquake

Another major earthquake with a moment magnitude of 6.6 (Mj 6.7) occurred on September 6, 2018, at 3:08 a.m. in Iburi Region of Hokkaido Island, which is about 260 km away from Horonobe. The focal mechanism of this earthquake was due to a blind, steeply dipping thrust fault. The earthquake was felt in Horonobe as recorded by the KiK-Net network. However, the maximum ground acceleration was 8.0 gals.

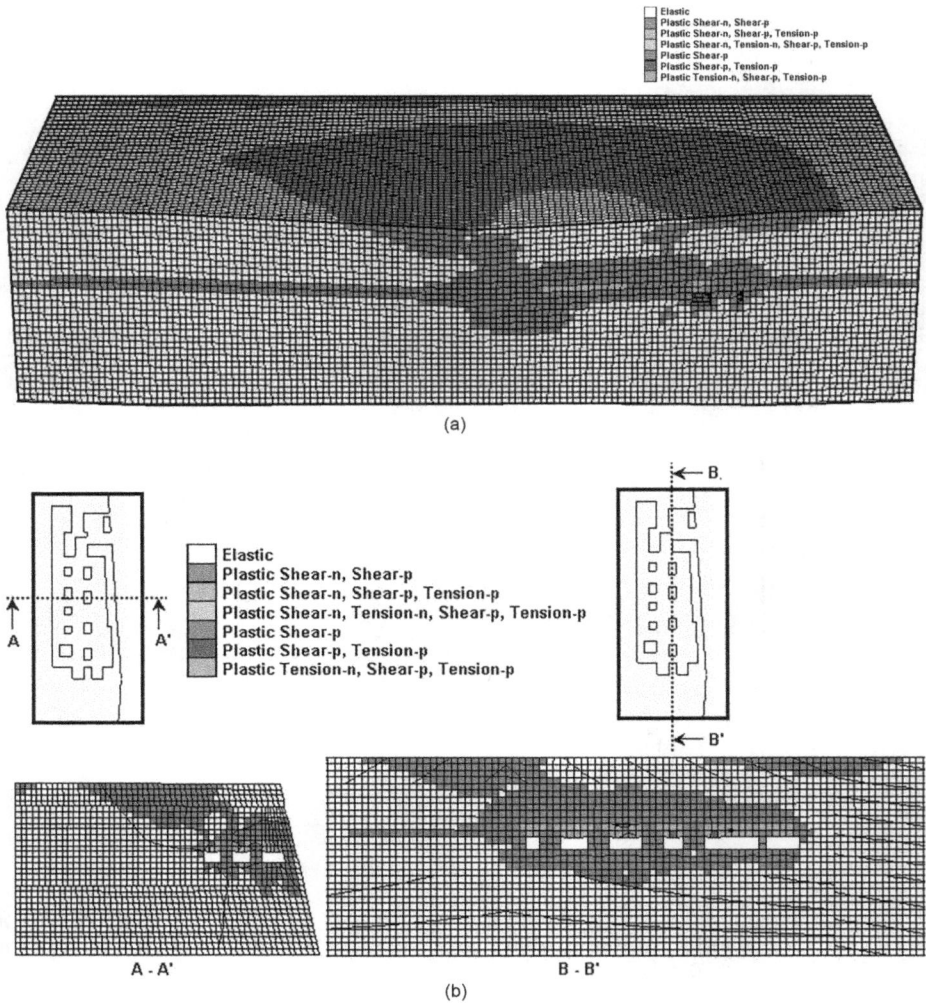

Figure 6.54 Yield zone around mine for ground with reduced strength properties. (a) Yield zone development at model surface and (b) yield zone development at EW and NS cross sections.

6.12.3 Acceleration records at Horonobe URL

Accelerometers are set at the ground surface, GL. −250 m and GL. −350 m galleries in Horonobe URL (Sato et al. 2019). Figure 6.56 shows the seismic records of the 2018 June 20 Soya region earthquake observed at GL. −350 m gallery and shaft bottoms (Figure 6.57). Table 6.4 compares the maximum acceleration at each strong motion station installed at various depths. The ground motions are amplified towards the ground surface. The data even in the same level is scattered, which may imply some

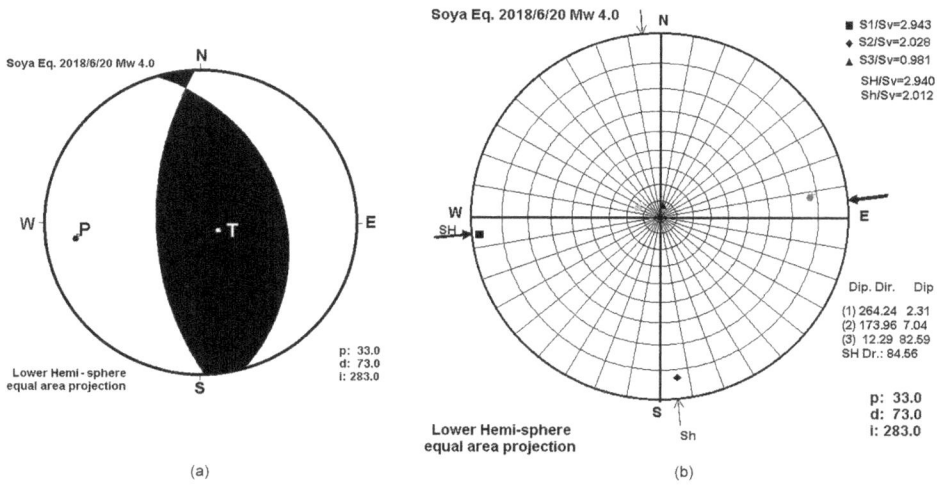

Figure 6.55 (a) Redrawn focal mechanism obtained by F-NET and (b) inferred stress state for the focal mechanism obtained by F-NET using Aydan's method (2000a).

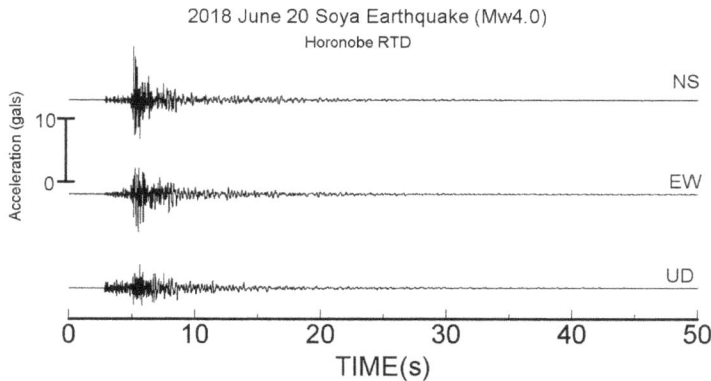

Figure 6.56 Seismic records of the 2018 June 20 Soya region earthquake (20 June 2018, $M = 4.8$) observed in Horonobe URL (maximum accelerations are 8.0, 5.8 and 2.0 gals for N-S, E-W and Z directions, respectively).

local factors, such as geological conditions and the geometry of the opening where the devices are installed.

The National Research Institute for Earth Science and Disaster Prevention of Japan (NIED) has been operating the KiK-Net and K-Net strong motion networks. There is a strong motion station of the KiK-Net in Horonobe town, and accelerations were recorded at the ground surface and at a depth of 100 m from the ground surface (-70 m); both records are shown in Figure 6.58. Table 6.5 gives the maximum ground accelerations and their amplifications. Theoretically, the amplification is expected to

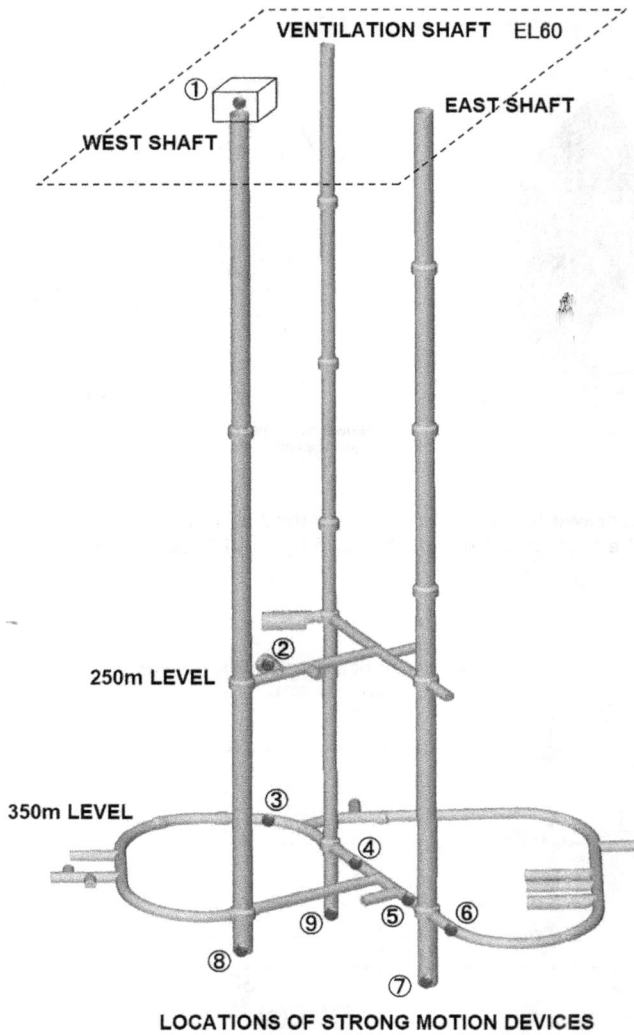

Figure 6.57 Locations of strong motion observation stations.

Table 6.4 Maximum acceleration and amplification

Locations	NS (gals)	EW (gals)	UD (gals)
Surface (+60 m)	8.0	5.8	2.0
250 m level	8.72	6.04	3.8
350 m level	10.4	8.76	3.45
West shaft (365 m)	7.79	5.46	2.74
Vent. shaft (380 m)	0.87	0.64	0.94

KiK-Net and K-Net data

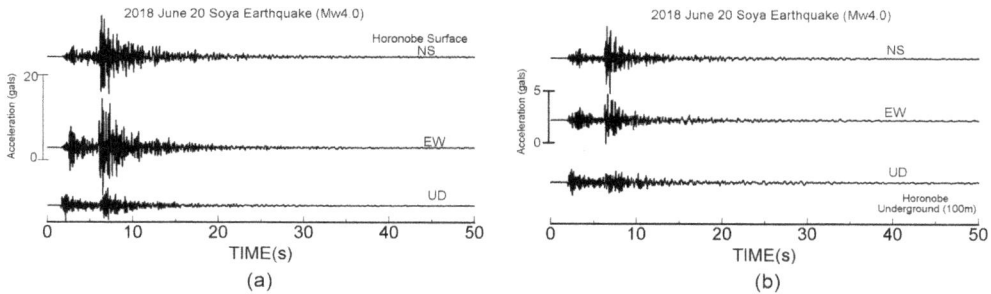

Figure 6.58 The acceleration records taken at the ground surface and at the base. (a) Ground surface (30 m) and (b) base (100 m below the ground surface).

Table 6.5 Maximum acceleration and amplification

	NS (gals)	EW (gals)	UD (gals)
Surface (+30 m)	3.4	2.5	1.3
Base (−70 m)	10.7	11.8	4.2
Amplification	3.15	4.72	3.23

be greater than 2 for an elastic ground (Nasu 1931). The comparison indicates that the amplification of acceleration is more than three times. Compared to data from the KiK-Net, the measurements at the Horonobe URL are somewhat scattered.

6.12.4 Fourier and acceleration response spectra analyses

Fourier and acceleration RSA have been carried out for each strong motion station. Some of the results are described herein.

6.12.4.1 Fourier spectra analyses

The Fourier spectra analyses of acceleration records measured by the KiK-Net (RMIH01) at the ground surface and base are shown in Figure 6.59. As noted from the figure, the dominant frequency ranges between 4 and 8 Hz, and the Fourier spectra characteristics do not change with depth, although the amplitude of the ground surface is at least three times that at the base.

Figure 6.60 shows the FFT of records taken at the ground surface and at a depth of 380 at the shaft bottom in Horonobe URL. The FFT amplitude of the shaft bottom records is almost the same as that of the ground surface. The frequency characteristics are also quite similar, and they resemble those of the KiK-Net records.

Figure 6.61 shows the FFT spectra of the record taken at No. 9 observation station during the 2018 Iburi Earthquake. Except the up-down (UD) component and amplitude, the other components are quite similar to those of the Soya region earthquake

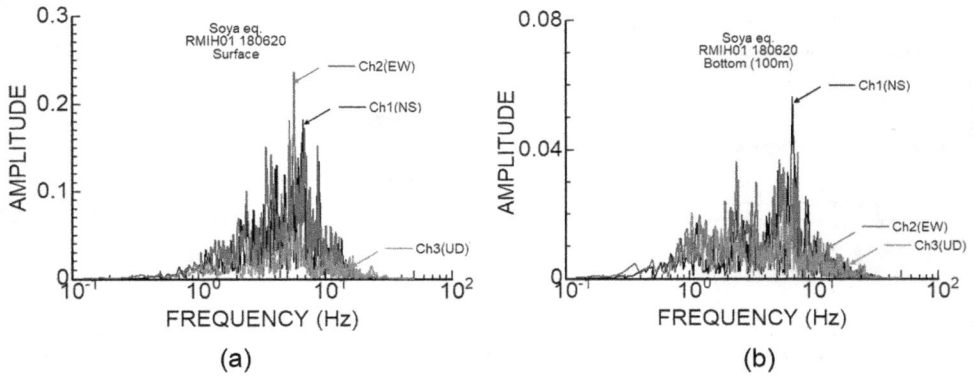

Figure 6.59 FFT of records at (a) the ground surface and (b) base. Taken at RMIH01 strong motion station.

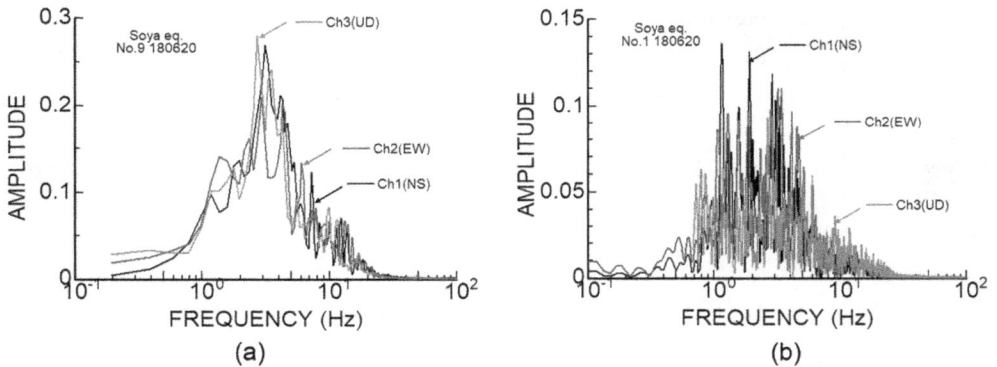

Figure 6.60 FFT of records at (a) the ground surface and (b) base. Taken at the shaft bottom of Horonobe URL.

shown in Figure 6.60. The normalized amplitude may be useful for comparison purposes.

6.12.4.2   Acceleration response spectra analyses

A series of ARA were carried out. Figure 6.62 shows the acceleration response spectra for the RHIM01 and Horonobe URL No. 1 strong ground motion stations. The amplitude and frequency characteristics are somewhat different. The ground conditions at the RHIM01 may be softer than those at the Horonobe URL site.

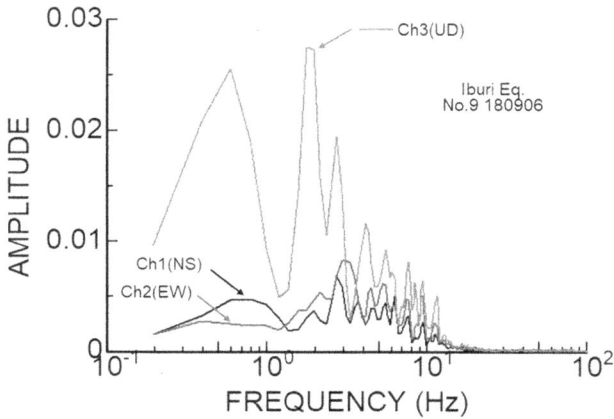

Figure 6.61 FFT of records at the ground surface for the records due to 2018 Iburi Earthquake of 2018 September 06.

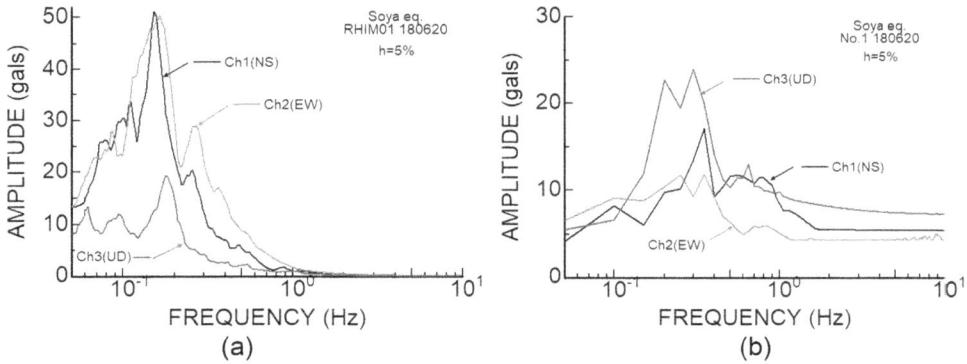

Figure 6.62 Comparison of acceleration response spectra for (a) RHIM01 and (b) Horonobe URL site.

6.13   SLOPES

6.13.1   Characteristics of shaking table

The shaking table used for model tests was produced by AKASHI. Its operation system was recently updated by IMV together with the possibility of applying actual acceleration wave forms from earthquakes. Table 6.6 gives the specifications of the shaking table and monitoring devices. The size of the shaking table is $1,000 \times 1,000\,mm^2$. The maximum acceleration is 600 gals for a model with a weight of 100 kgf. The displacement responses of models were monitored using laser displacement transducers, and the input acceleration of the shaking table and acceleration response of the retaining wall were measured using two accelerometers.

Two shaking test (ST) devices were used. The shaking table at Nagoya University (NU) was used for studying the response of models slopes for unbreakable material and the shaking table at the University of the Ryukyus (UR) was used to study the response of models slopes with breakable material (Aydan et al. 2019b). The main features of the shaking table apparatuses are given in Table 6.7. Figure 6.63 shows sketches of the devices together with the model and instrumentations. Slope models were two-dimensional and were mounted upon the table through metal frames. The metal frame at NU-ST was 1,200 mm long and 800 mm high, while it was 1,000 mm long and 750 mm high at UR-ST. The frame width was 100 mm wide in both shaking table experiments. Acceleration responses of the slope at several locations and input waves were measured using accelerometers.

Table 6.6 The specifications of monitoring sensors and shaking table

Shaking table and sensors	Specifications	
Shaking table – AKASHI	Frequency	1–50 Hz
	Stroke	100 mm
	Acceleration	600 gals
Accelerometers	Range	10 G
Laser displacement transducer		
OMRON	Range	0–300 mm
KEYENCE	Range	0–100 mm

Table 6.7 Specifications of shaking tables

Parameters	NU-shaking table	UR-shaking table
Vibration direction	Uniaxial	Uniaxial
Operation method	Electro-oil servo	Magnetic
Table size	$1,300 \times 1,300$	$1,000 \times 1,000$
Load	30 kN	6 kN
Stroke	150 mm	100 mm
Amplitude	5G	0.6G
Wave form	Harmonic, triangular, rectangular, random	Harmonic, triangular, rectangular, random

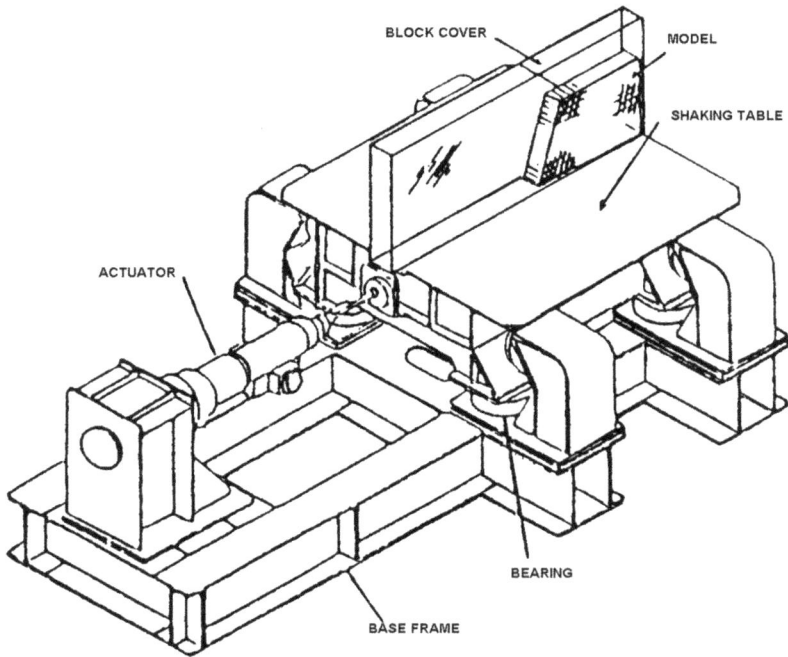

(a)

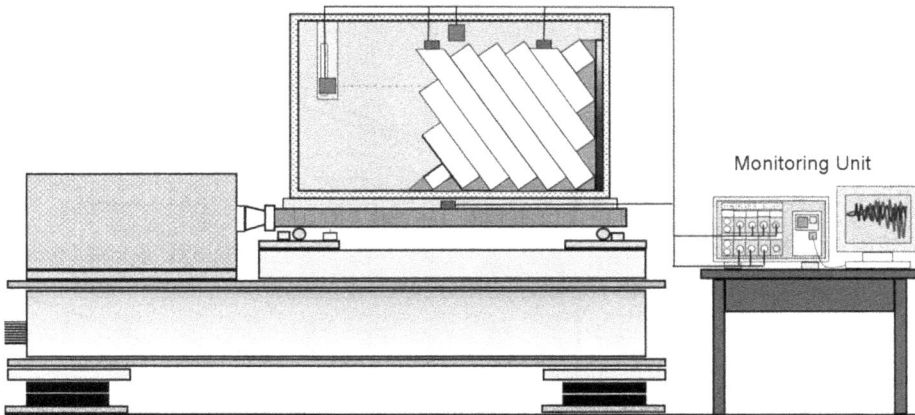

(b)

Figure 6.63 Illustration of shaking tables and instrumentation. (a) NU-shaking table and (b) UR-shaking table.

6.13.2 Applications to slopes and cliffs

6.13.2.1 Model materials

6.13.2.1.1 NON-BREAKABLE MATERIALS

Blocks with dimensions of $10 \times 10 \times 100$ and $10 \times 20 \times 100$ mm^3 were made of wood and used to simulate the discontinuity sets in rock masses. Direct shear tests were carried out on discontinuities between wood blocks and results together with shear strength envelopes are shown in Figure 6.64.

6.13.2.1.2 BREAKABLE MATERIALS

Breakable blocks are made of BaSO$_4$, ZnO and Vaseline oil, which is commonly used in base friction experiments (Aydan and Kawamoto 1992). Properties of materials of blocks and layers are described in detail by Aydan and Amini (2009) and Egger (1979). Figure 6.65 shows the variation of the strength of the model material with respect to compaction pressure. The material can be powderized and reused after each experiment. The friction angle of interfaces between blocks are tested and shown in Figure 6.66.

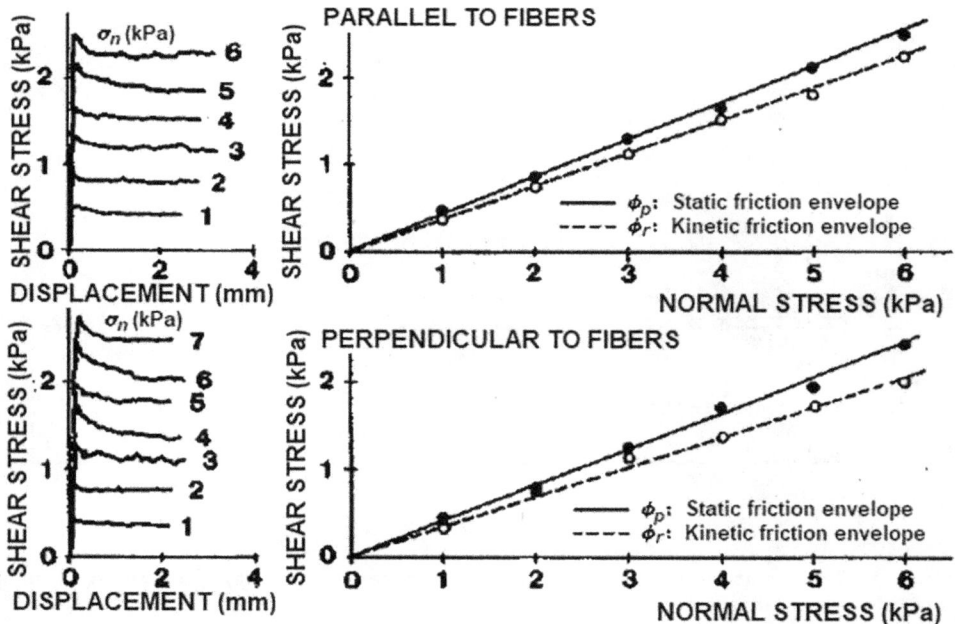

Figure 6.64 Shear tests on interfaces between wood blocks.

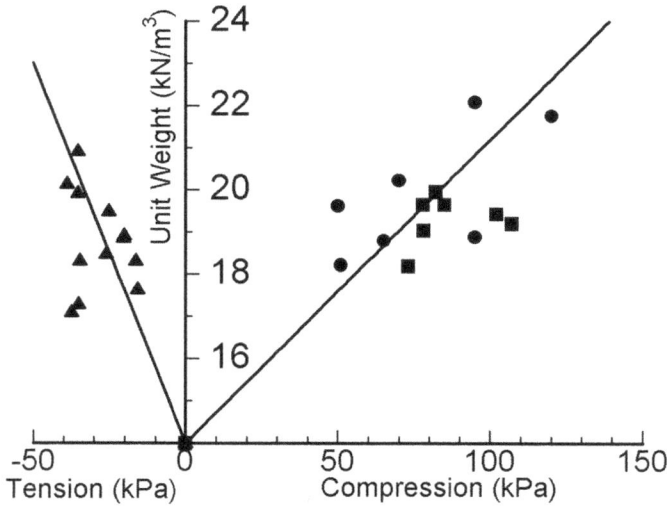

Figure 6.65 Variation in tensile strength of model materials.

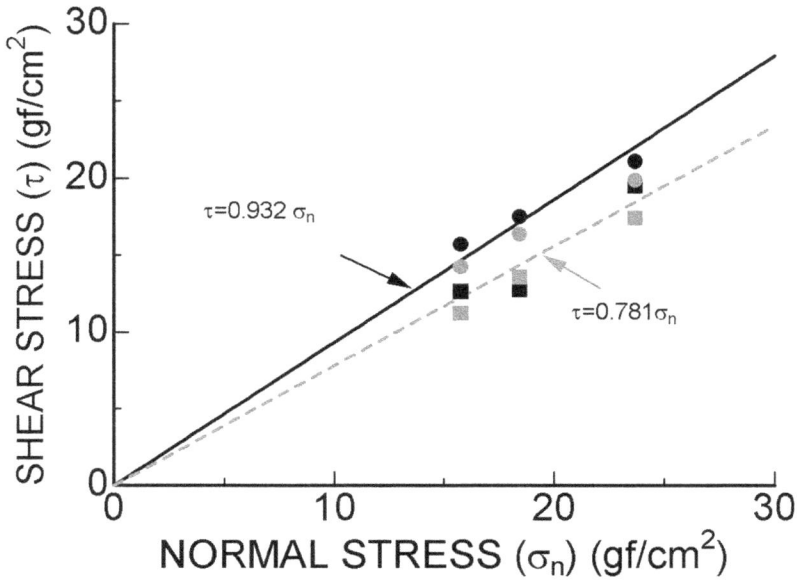

Figure 6.66 Shear tests on interfaces between breakable blocks.

6.13.2.2   Testing procedure

The metal frames have some special attachments to generate different discontinuity patterns. The models were subjected to some selected forms of acceleration waves through a shaking table. The acceleration responses of model slopes were measured using accelerometers installed at various points in the slope.

6.13.2.2.1   NON-BREAKABLE BLOCKS

Model slopes were prepared by arranging wood blocks in various patterns to generate discontinuity sets with different orientations in space. Slope angles were 45°, 63° and 90°, and the height and base width of model slopes were 800 and 1,200 mm, respectively. The intermittency angle ξ of cross joints were 0° and 45° (Aydan et al. 1989) and one discontinuity set was always continuous as such sets in actual rock masses always do exist.

The inclination of the thoroughgoing (continuous) set was varied from 0° to 180° by 15°. At some inclinations, model slopes were statically unstable and at such inclinations no tests were done. Besides varying the inclination of the continuous set, the following cases were investigated:

CASE 1: Frequency was varied from 2.5 to 50 Hz while the amplitude of the acceleration was kept at 50 or 100 gals.

CASE 2: The amplitude of the acceleration waves was varied until the failure of the slope occurred, while keeping the frequency of the wave at 2.5 Hz.

6.13.2.2.2   BREAKABLE BLOCKS

The inclinations for the thoroughgoing discontinuity set were selected as 0°, 45°, 60°, 90°, 120°, 135° and 180°. Before forcing the models to fail in each test, vibration responses of some observation points in the slope were measured with the purpose of investigating the natural frequency of slopes and amplification through sweep tests with a frequency range between 3 and 40 Hz. Also, deflection of the slope surface was monitored by laser displacement transducers and acoustic emission sensors.

6.13.3   Model experiments

Various parameters such as the effect of the frequency and the amplitude of input acceleration waves were investigated in relation to discontinuity patterns and their inclinations and the slope geometry for the model slopes with non-breakable and breakable models. The model slopes were finally forced to fail by increasing the amplitude of input acceleration waves, and the forms of instability were investigated.

6.13.3.1   Natural frequency of model slopes

6.13.3.1.1   MODEL SLOPES WITH NON-BREAKABLE BLOCKS

Figure 6.67 shows the amplification of waves measured at selected points in relation to the variation of input wave frequency. The inclination of the thoroughgoing set for

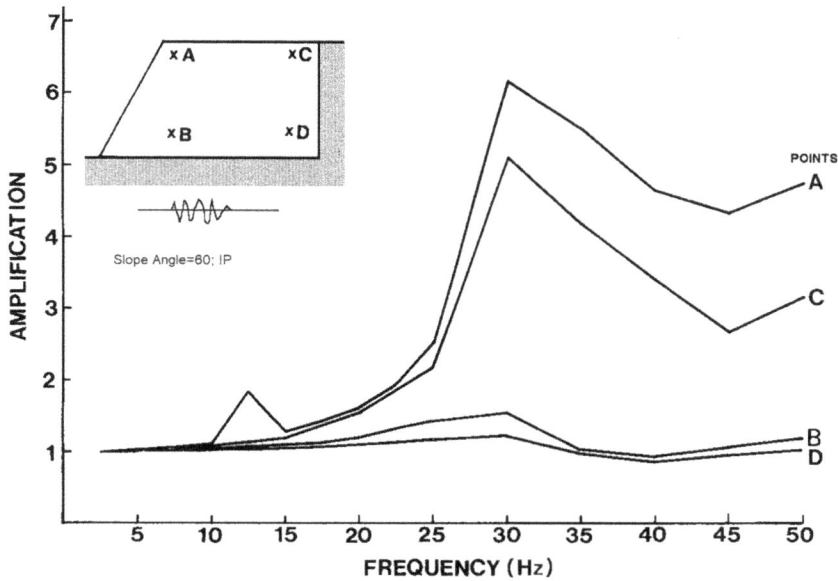

Figure 6.67 Variation in amplification with respect to the frequency of model slopes and measurement locations.

both discontinuity patterns was 75°. The letter on each curve indicates the selected points within the model slopes. It is noted that if the natural frequency of the slopes exists, it varies with the spatial distributions of the sets and the structure of the mass.

In the following, we discuss and compare the frequency responses for each respective inclination of the thoroughgoing discontinuity set for point A (see Figure 6.68 for location) as shown in Figure 6.68. The slope angle was 63° in the sweep tests shown in Figure 6.69. The results for each discontinuity set pattern are indicated in the figure: IP for intermittent pattern and CCP for cross-continuous pattern.

Inclination 0: The natural frequency of the slope is 10 Hz for the cross-continuous pattern and 20 Hz for the intermittent pattern, respectively. Therefore, the natural frequencies of the slopes for intermittent and cross-continuous patterns are different even when the slope geometry and intact material are same. This may be related to the resulting slender columnar structure of the mass in the case of the cross-continuous pattern.

Inclinations 15°, 30°, 45°: The slopes at these inclinations of the thoroughgoing set could not be tested as they were statically unstable at the slope angle of 60°.

Inclination 60°: From the figure, the natural frequencies for both patterns coincide and they have a value of 30 Hz. This may be attributed to the similarity of the structure of the mass for this inclination of the thoroughgoing discontinuity set.

Inclination 75°: Natural frequencies of the slopes for both patterns are almost the same, at 17.5 Hz. Similar reasoning as in the 60° inclination case can be stated for this case.

Inclination 90°: No tests for this inclination could be made.

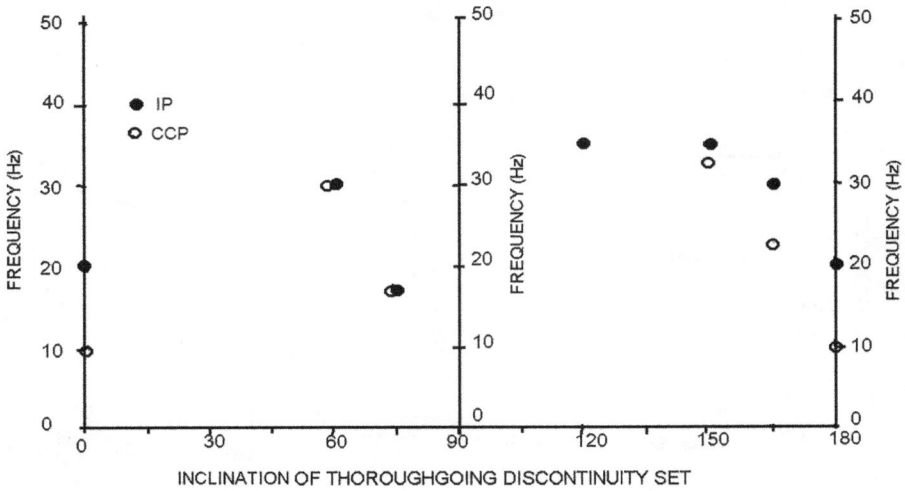

Figure 6.68 Variation in natural frequency of model slopes with respect to the inclination of the thoroughgoing discontinuity set.

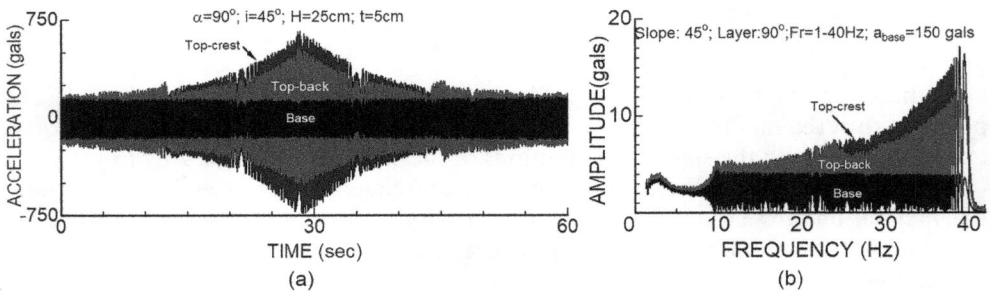

Figure 6.69 Acceleration responses of selected points on model slopes made with breakable blocks. (a) Input and measured acceleration responses in sweep tests and (b) Fourier spectra of acceleration responses in sweep tests.

Inclination 120°: Slopes having intermittent pattern were only tested as slopes having cross-continuous pattern could not be tested as they were statically unstable. For this inclination of the thoroughgoing set, the natural frequency of the slope has a value of 35 Hz.

Inclination 150°: Natural frequencies of the slopes for both patterns are almost the same, at 35 Hz.

Inclination 165°: The natural frequency of the slope is 22.5 Hz for the cross-continuous pattern and 30 Hz for the intermittent pattern, respectively. In addition, the natural frequencies of the slopes for intermittent and cross-continuous patterns are different.

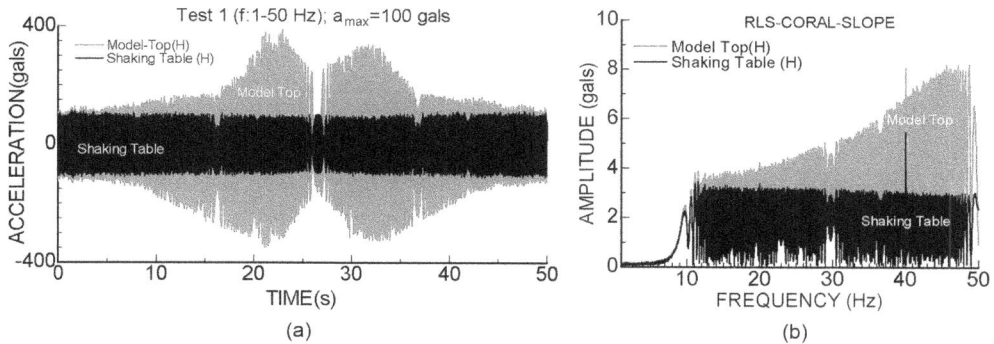

Figure 6.70 Horizontal acceleration records at the top of the model rock slope and applied base acceleration on shaking table, and their Fourier spectra. (a) Acceleration responses and (b) Fourier spectra.

6.13.3.1.2 MODEL SLOPES WITH BREAKABLE BLOCKS

Fundamentally, the vibration responses of model slopes are quite similar to those of model slopes made with unbreakable blocks. Figure 6.69a shows the input and measured wave forms at two selected points on the slope. The amplification of the vibration response is highest at the slope crest and the amplification at the top-back (ACC-TB) is a bit smaller, as seen in Figure 6.69b. From this figure, we can clearly state that amplification of the acceleration waves increases towards the slope (free) surfaces. In addition to this, the amplifications are larger at the top and have the maximum value at the crest of the slope (ACC-TC) as it was also noted in Figures 6.67 and 6.68 for model slopes made with non-breakable blocks.

6.13.3.1.3 UN-REINFORCED LAYERED ROCK SLOPES

A series of model tests on rock slopes using layered coral limestone were carried out. Before each failure test, a sweeping test was carried out to check the natural frequency characteristics of model rock slopes with a frequency ranging between 1 and 50 Hz at constant acceleration of 100 gals. Figure 6.70 shows horizontal acceleration records of the accelerometers fixed to the shaking table (Shaking table (H)) and the top of the model slope (model top (H)) as shown in Figure 6.63b. The results clearly indicate that model rock slopes has some natural frequency characteristics.

6.14 RETAINING WALLS

6.14.1 Model setup

An acrylic transparent box with 630 mm in length, 300 mm in height and 100 mm width was used as shown in Figure 6.71. The wall thickness was 10 mm so that the box was relatively rigid and the frictional resistance of sidewalls was quite low. The blocks used

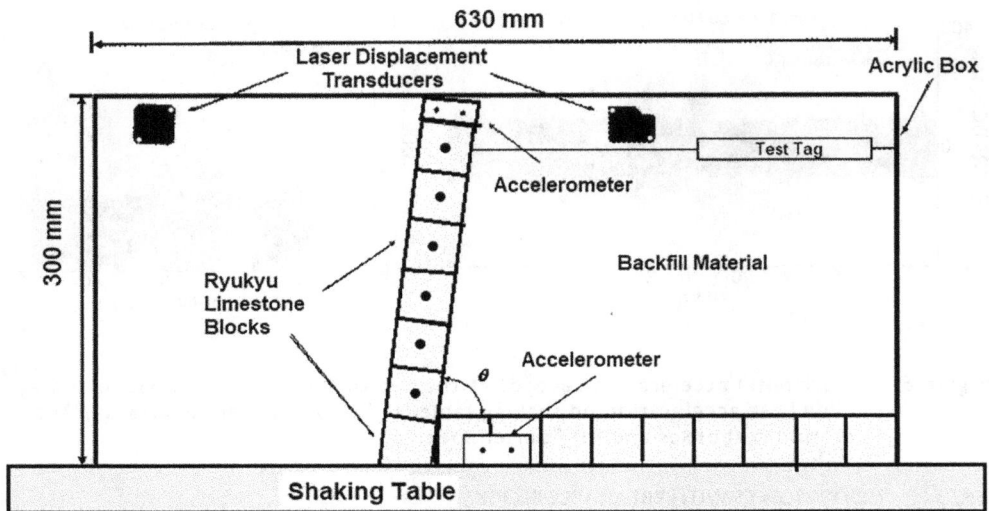

Figure 6.71 Illustration of model box.

were made of Ryukyu limestone with a size of $40 \times 40 \times 99.5\,mm^3$ with the consideration of materials used for the retaining walls of historical castles in the Ryukyu Archipelago. Furthermore, the base block was such that the overall wall inclination can be chosen as 70°, 83° and 90°. The base block was fixed with two-sided tapes to the base of the acrylic box. In addition, a Ryukyu limestone of the same size was laid over the base as seen in Figure 6.71. This was expected to provide a condition similar to the actual conditions observed in many historical castles in the Ryukyu Archipelago. The wall height was 240 mm and the ratio of the height to width was 1/6. At a retaining-wall inclination of 90° without backfill material, it is expected that the wall would start rocking at an acceleration level of 167 gals.

6.14.2   Backfill materials and their properties

Three different backfill materials were chosen (Figure 6.72). Glass beads were chosen to represent the lowest shear-resistant backfill material while angular fragments of Motobu limestone were selected as the highest shear-resistant backfill material. The third backfill material were rounded river gravels having a shear resistance between those of the other two materials.

A special shear testing setup was developed to obtain the shear strength characteristics of backfill materials under low normal stress levels, which are quite relevant to the model tests to be presented in this study. Figure 6.73 shows the shear strength envelopes for three backfill materials. As noted from the figure, shear strength of rounded river gravel is in-between the shear strength envelopes of glass beads and Motobu limestone gravel. The strength of backfill materials is frictional and the friction angle of the glass beads is about 21.68°.

Glass beads   Rounded river gravel   Motobu limestone gravel

Figure 6.72 Views of backfill materials.

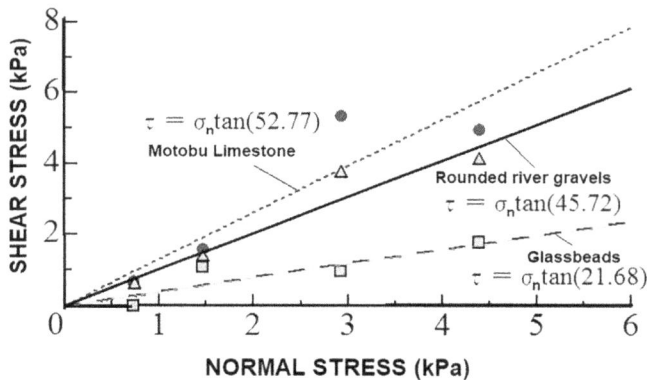

Figure 6.73 Shear strength envelopes for backfill materials.

Table 6.8 Friction angle between Ryukyu limestone and backfill materials

Parameter	Glass beads	Rounded river gravel	Motobu limestone fragments
Friction angle	12.5–16.8	25.0–27.5	25.9–27.8

Another important factor for the stability of the retaining walls of historical cas-tles as well as other similar structures is the frictional resistance between the backfill material and retaining-wall blocks. For this purpose, tilting experiments were carried out. The backfill material contained in a box put upon the Ryukyu limestone platens and tilted until it slides. This response of the backfill material contained in the box was measured using laser displacement transducers. The inferred friction angles are given in Table 6.8. The lowest friction angle was obtained for the glass beads as expected.

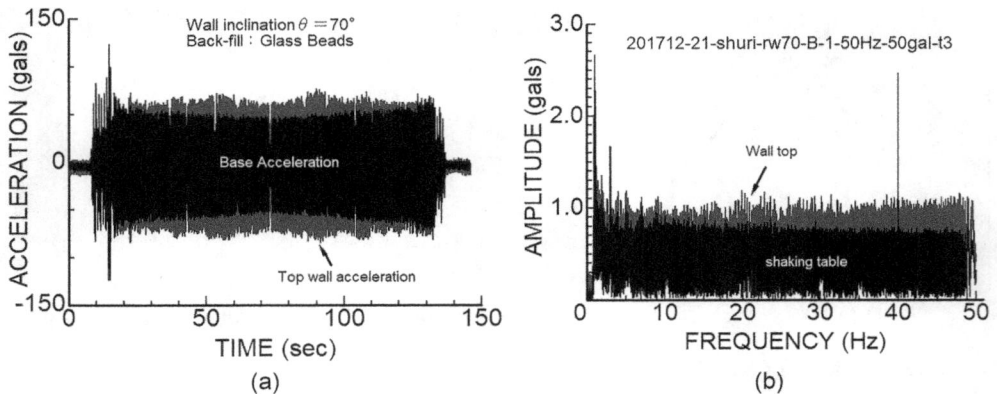

Figure 6.74 (a) Acceleration records of the shaking table and top of the retaining wall, and (b) Fourier spectra of acceleration records.

6.14.3   Shaking table tests on retaining walls with glass beads backfill

A series of sweep tests were carried out before the failure tests. Regarding the glass beads backfill material, the retaining walls were statically unstable at 90°, while they failed during the sweep test on the retaining walls at an inclination of 83°. Therefore, we could only show one example for retaining walls, at an inclination of 70° (Figure 6.74a). Its FSA is shown in Figure 6.74b. The results indicated no apparent natural frequency was dominant. The situation was quite similar in all experiments. Therefore, more emphasis will be given to the failure experiments.

Although the test on the retaining wall with an inclination of 83° was intended for a sweep test, it resulted in failure. Figure 6.75 shows the displacement and base acceleration during the test. Failure tests on the retaining walls with an inclination of 70° were carried out by applying sinusoidal waves with a frequency of 3 Hz. The amplitude waves were gradually increased until failure occurred. Figure 6.76 shows an example of failure. The yielding initiated at about 110 gals, and total failure occurred when the input acceleration reached 215 gals. Figure 6.77 shows the retaining wall before and after the failure test. The retaining wall failed due to toppling failure although some relative sliding occurred with the block at the toe of the model retaining wall.

6.14.4   Shaking table tests on retaining walls with river gravel backfill

A series of sweep tests were carried out before the failure tests, as explained in the previous section. Regarding the rounded river gravel backfill material, the retaining walls were statically unstable at 90°, while a sweep test on the retaining walls at inclinations of 83° and 70° could be carried out. We show one example for retaining walls for the inclination at 83° in Figure 6.78a and its FSA in Figure 6.78b. The results indicated there

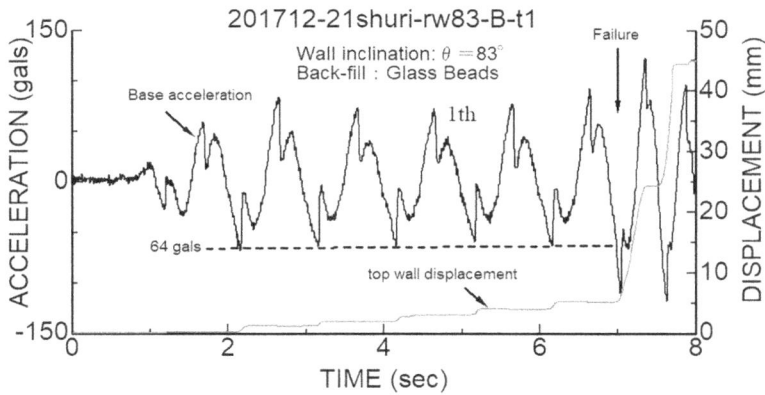

Figure 6.75 Acceleration and displacement responses on the retaining wall at an inclination of 83°.

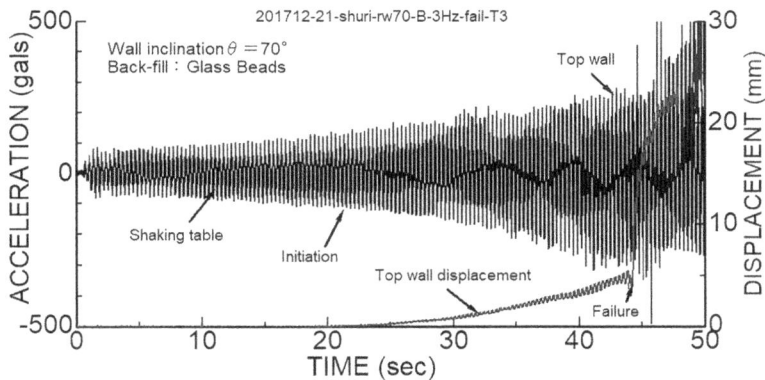

Figure 6.76 Acceleration and displacement responses on the retaining wall at an inclination of 70°.

Figure 6.77 Views of the model retaining wall at an inclination of 70° (a) before and (b) after the test.

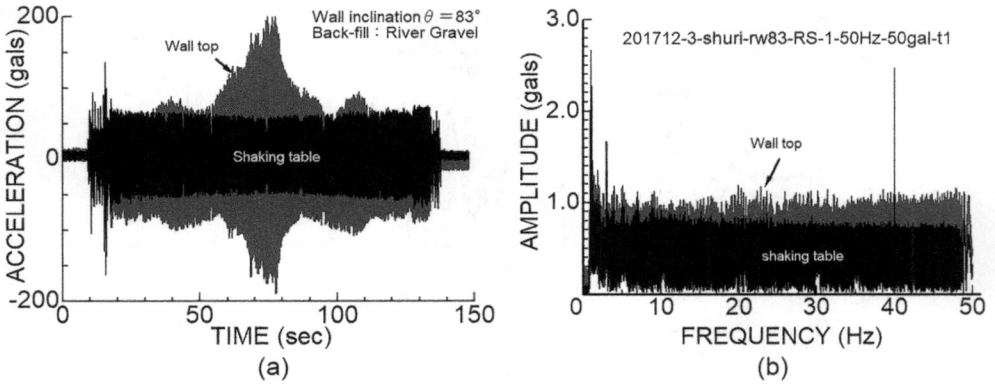

Figure 6.78 (a) Acceleration records of the shaking table and top of the 83° retaining wall with rounded river gravel backfill, and (b) Fourier spectra of acceleration records.

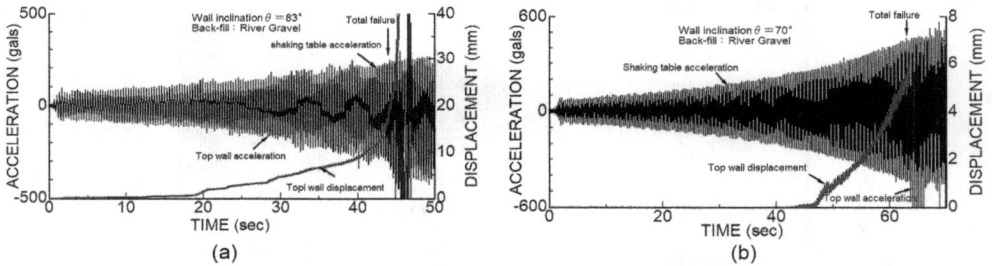

Figure 6.79 (a) Acceleration and displacement responses on the retaining wall at an incli- nation of 83°, and (b) acceleration and displacement responses on the retain- ing wall at an inclination of 70°.

was no dominant natural frequency for the given range of frequency. The situation was quite similar in all experiments for the 83° and 70° retaining-wall models.

Failure tests on the retaining walls with inclinations at 83° and 70° were carried out by applying sinusoidal waves with a frequency of 3 Hz. The amplitude waves were gradually increased until failure occurred. Figure 6.79a and b shows acceleration and displacement responses of retaining walls at inclinations of 83° and 70° as examples of failure tests. The yielding initiated at about 100 gals, and total failure occurred when the input acceleration reached 210 gals for 83° retaining walls. On the other hand, the yielding initiated at 220 gals and total failure occurred when the input acceleration was 430 gals for 70° retaining walls as seen in Figure 6.79. The retaining wall failed due to toppling failure although some relative sliding occurred with the block at the toe of the model retaining wall (Figure 6.80).

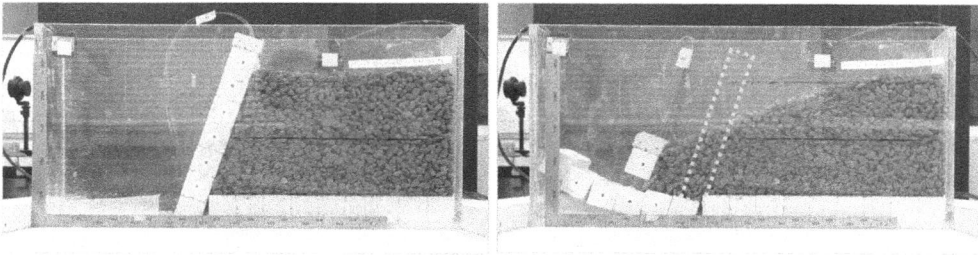

Figure 6.80 Views of the model retaining wall at an inclination of 70° (a) before and (b) after the test.

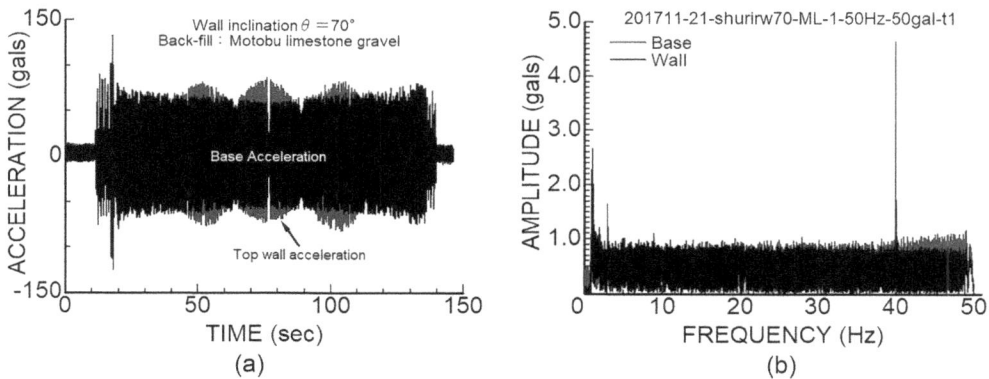

Figure 6.81 (a) Acceleration records of the shaking table and top for the 70° retaining wall with rounded river gravel backfill, and (b) Fourier spectra of acceleration records.

6.14.5 Shaking table tests on retaining walls with Motobu limestone gravel backfill

A series of sweep tests were carried out before failure tests as explained in the previous section. Regarding the angular Motobu limestone gravel backfill material, the retaining walls were statically unstable at 90° with a height of 240 mm. However, they were stable when the height was reduced to 160 mm. The sweep test on the retaining walls at inclinations of 90°, 83° and 70° was carried. We show one example for retaining walls at the inclination of 70° in Figure 6.81a and its FSA in Figure 6.81b. Again, the results indicated there was no dominant natural frequency for the given range of frequency. The situation was quite similar in all experiments for 90°, 83° and 70° retaining-wall models.

Failure tests on the retaining walls at inclinations of 90°, 83° and 70° were carried out by applying sinusoidal waves with a frequency of 3 Hz. The procedure was the same as that in previous experiments. Figures 6.82–6.84 show acceleration and displacement

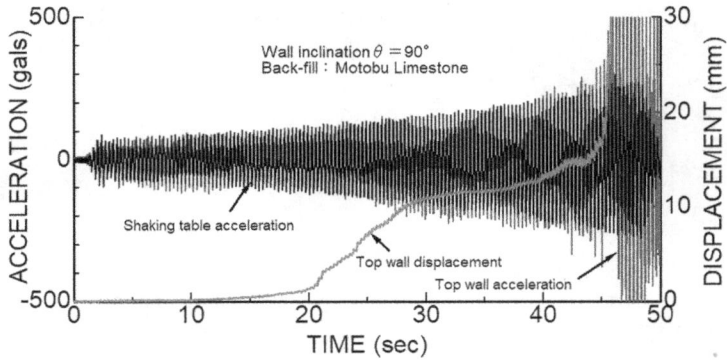

Figure 6.82 Acceleration and displacement responses on the retaining wall at an inclination of 90°.

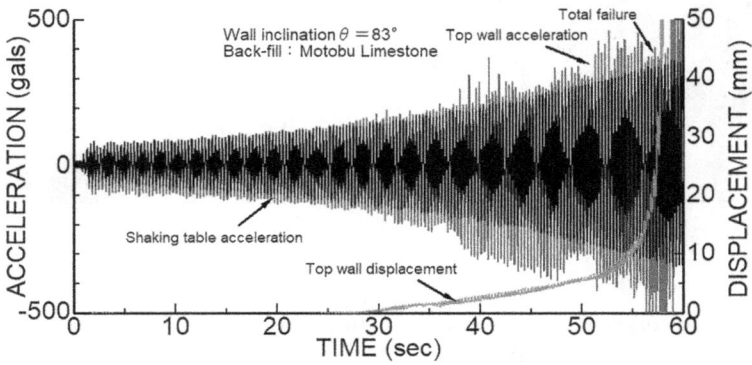

Figure 6.83 Acceleration and displacement responses on the retaining wall at an inclination of 83°.

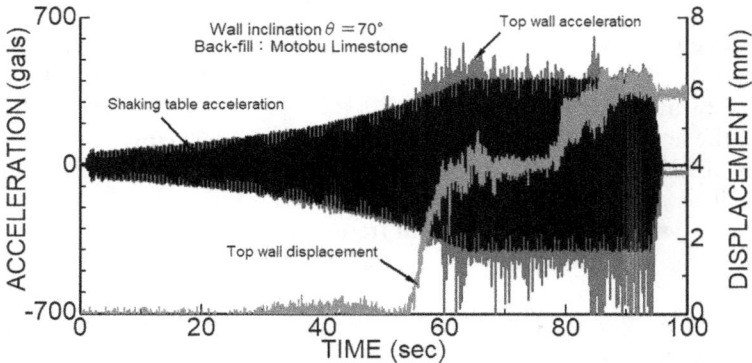

Figure 6.84 Acceleration and displacement responses on the retaining wall at an inclination of 70°.

Figure 6.85 Views of the model retaining wall at an inclination of 90° (a) before and (b) after the test.

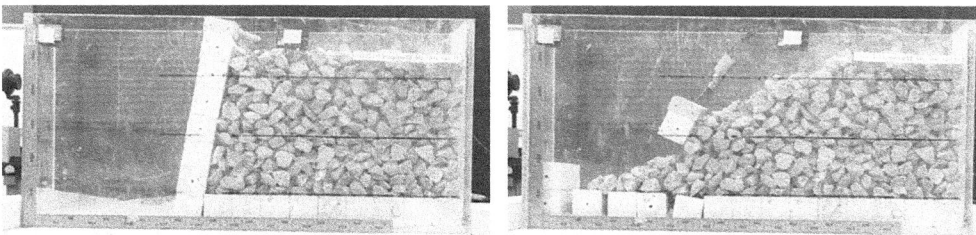

Figure 6.86 Views of the model retaining wall with an inclination of 83° (a) before and (b) after the test.

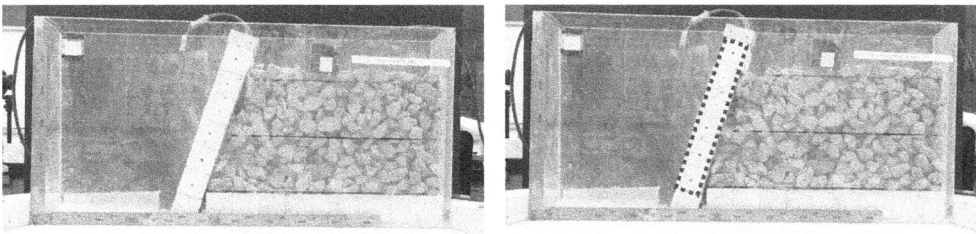

Figure 6.87 Views of the model retaining wall with an inclination of 70° (a) before and (b) after the test.

responses of retaining walls at inclinations of 90°, 83° and 70° as examples of failure tests. Failure states are also shown in Figures 6.85–6.87. The yielding initiated at about 110 gals and the total failure occurred when the input acceleration reached 260 gals for 90° retaining walls. On the other hand, the yielding initiated at 130 gals, and total failure occurred when the input acceleration was 300 gals for 83° retaining walls as seen in Figure 6.83. The retaining walls failed due to toppling (rotation) failure.

The retaining walls with 70° inclination and height of 240 mm did not fail during the entire test up to 400 gals as seen in Figures 6.84 and 6.87. Some relative sliding occurred with the blocks at the toe of the model retaining wall when the base acceleration reached 300 gals (Figure 6.87). However, some settlement of the backfill occurred and the retaining wall was pushed in passive sliding mode.

Chapter 7

Effects of earthquakes associated surface ruptures on engineering structures

The damaging effects of earthquakes on engineering structures may be due to ground shaking and permanent deformation resulting from surface ruptures due to faulting and ground deformation associated with lateral spreading or slope failures and rockfalls. In this chapter, some specific examples of the damaging effects of earthquakes on engineering structures are described in view of the author's reconnaissance of many earthquakes worldwide since the 1992 Erzincan earthquake in Turkey (Aydan 1996, 2003b, 2006a,b, 2007a,b, 2008a,b, 2009, 2012a, 2013; Aydan and Hamada 1992, 2006; Aydan and Kawamoto 2004; Aydan and Kumsar 1997a, 2002; Aydan and Tano 2012; Aydan and Ulusay 2002; Aydan et al. 1998, 1999a,b, 2000a,b, 2005a–c, 2006a, 2007a,b, 2008a,b, 2009a–f, 2010a–c, 2012a; Kawashima et al. 2010; Ulusay et al. 2001, 2002, 2003a,b; Ulusay and Aydan 2005, 2011).

7.1 EFFECTS OF GROUND SHAKING ON ENGINEERING STRUCTURES

7.1.1 Buildings

Buildings can be classified into three broad categories:

a. Reinforced concrete buildings
b. Masonry buildings
c. Timber buildings

Masonry buildings may also be divided into several categories:

 i. Adobe-type buildings
 ii. Burned brick buildings
iii. Random or cut stone buildings with mud, lime or cement mortar

DOI: 10.1201/9781003164371-7

7.1.1.1 Reinforced concrete buildings

The reinforced concrete frame buildings are generally constructed with unreinforced masonry infill walls which may be integrated or nonintegrated into the frame system. The infill walls are of various types, namely clay brick masonry in cement mortar, large block cut stone masonry in cement mortar, small block cut stone masonry in cement mortar, cement blocks in cement mortar and hollow cement blocks in cement mortar. Some causes of damage to reinforced concrete buildings are as follows in view of the 1999 Kocaeli earthquake (Figure 7.1):

a. Long duration of shaking
b. Concentration of damage to areas where the land was reclaimed
c. The number of stories of buildings is four or more
d. The top soil is sandy clay with sandy soil below
e. Collapsed buildings having isolated footings and the footings not connected to each other with lateral tie beams
f. Constructions not following the earthquake design code
g. Collapsed buildings having pilotis and no shear walls at the ground floor
h. Ground amplification due to geological and geotechnical conditions
i. Poor workmanship
j. Corrosion of steel bars due to the use of saline sand and gravel
k. Pounding and torsion

Figure 7.1 Damage to reinforced concrete buildings due to ground shaking.

7.1.1.2 Masonry buildings

As seen in the epicentral area of the 2009 L'Aquila earthquake and other earthquakes, older nonengineered dwellings made with load-bearing masonry walls supporting tiled roof or reinforced concrete slab roof. The different types of masonry are random rubble stone with mud/lime/cement mortar, small block cut stone in mud/lime/cement mortar, large block cut stone in mud/lime/cement and brick masonry in mud/lime/cement mortar. Most dwellings made of mud mortar generally either totally collapse or are heavily damaged during major earthquakes. Figure 7.2 shows several examples of the heavy damage to masonry structures with mud mortar. On the other hand, if masonry buildings are constructed using lime or cement mortar, they perform much better and the total collapses are much less as seen in the same figure. Most of the standing masonry buildings in the pictures are such structures. Especially, the use of cement mortar together with well-spaced concrete slabs improves the structural integrity of buildings. There are also hybrid-type buildings combining wooden beams and frames and masonry-type infill walls as seen in Figure 7.3.

7.1.1.3 Timber buildings

In many earthquakes, old wooden houses are mainly damaged as seen in Figure 7.4. The mode of damage is quite similar to those shown in Figure 7.4. The damage mechanism fundamentally involves the hinging of wooden columns at the base and also at the

Figure 7.2 Damage to masonry buildings in various earthquakes.

Figure 7.3 Damage to hybrid-type masonry buildings in various earthquakes.

Figure 7.4 Damage to wooden houses and schools.

Figure 7.5 Secondary structural damage due to collapse of ceilings.

connections between first and second floors as a result of large horizontal earthquake forces. As noted in one example, a very simple yet light wooden structure that failed during the 2004 Chuetsu earthquake illustrates this mechanism, as seen in Figure 7.4. Many wooden houses suffer cracking in the walls.

7.1.1.4 Secondary-type damage in buildings

Large halls for various social events at hotels, conference centres, indoor sport centres, restaurants, train stations and airports generally have suspended ceiling panels. The fall of ceiling panels injures peoples. For example, such an incident occurred at the second floor of a drive-in restaurant nearby Seki IC during the 2007 Kameyama earthquake despite its magnitude being less than 6. Similar incidents also occurred in indoor sports halls and swimming pools in Suzuka City in the 2007 Kameyama earthquake. Similar events were observed in the 2009 Suruga earthquake, the 2011 Great East Japan Earthquake (GEJE) and the 2016 Kumamoto earthquakes, as seen in Figure 7.5. Particularly the collapse of suspended ceilings of huge halls presents a major risk in many countries as observed in great earthquakes worldwide.

7.1.2 Dams

Dams can be structurally classified as earth or rockfill, gravity or arch dams. Such dam failures were seen in the past. For example, the Sheffield dam failed during the 1925 Santa Barbara earthquake while the San Fernando dam failed during the 1971 San Fernando earthquake, and Niteiko water reservoir in the 1995 Kobe earthquake.

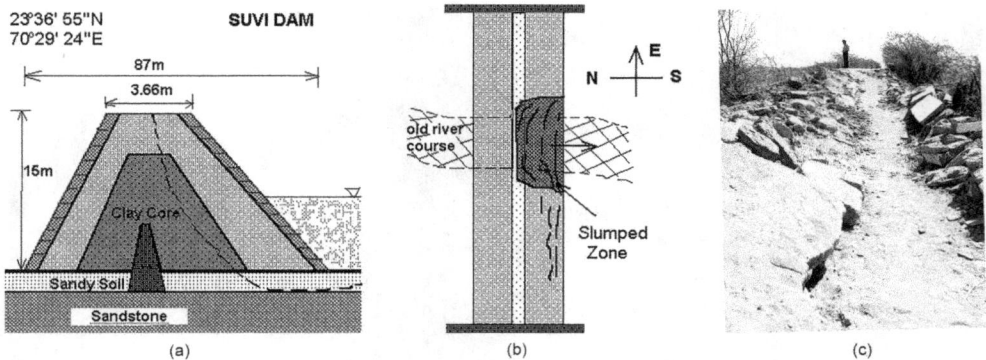

Figure 7.6 Illustration of damage at the Suvi dam and the settlement at the crest. Side and top sketches and settled crest.

During the 2001 Kutch earthquake in Gujarat Province of India, several earthfill dams were damaged by the earthquake. One of the damaged dams was the Suvi dam, which was built near Lilpar village. The crest of the dam is 3.66 m wide and the bottom is 87 m wide. The crest height is about 15 m. The upstream side of the dam had multiple longitudinal cracks indicating movement towards the reservoir (Figure 7.6). No visible cracks were observed on the downstream side as no water exists on this side. The longitudinal direction of the dam was EW. The direction of movements were towards the reservoir. A huge sliding occurred along a section, which was an old river course. It seemed that the soil below was probably liquefied. The foundation rock was sandstone and the soil thickness was about 1–2 m. Some lateral spreading of ground was observed. This was due to soil liquefaction as quartzite sand with a repose angle of 32° was observed along the cracks.

Damage to dams by the GEJE was described in detail by Matsumoto et al. (2011). Despite the large magnitude of the earthquake, many dams performed well in spite of ground acceleration up to 0.5 g. A small dam called Fujinuma Reservoir with a height of 18 m failed and the post-failure debris flow caused some casualties downstream. Although the exact reason is still unknown, the embankment probably failed due to breaching from overtopping of the reservoir water body.

Regarding the 2008 Wenchuan earthquake, there were around 400 dams in the epicentral area of the earthquake and four dams were classed as major dams with heights more than 100 m. The nearest dam to the epicentre was the Zipingpu dam built in 2006 with a height of 158 m, and it is a concrete-faced rockfill dam (Figure 7.7). The surveying indicated that the abutments of the dam were downward and the maximum displacement was 101.6 mm. The settlement of the crest of the dam was 734.6 mm with a 179.9 mm displacement towards the downstream side. The seepage of the dam changed from 10.38 to 15.07 L/s. The Zipingpu dam was designed for a base acceleration of 0.26 g but the acceleration at the crest during the earthquake was >2 g. The gates and power-generating units were damaged by the earthquake. Another major dam is the Shapai dam, which is a 32 m high arch roller compacted concrete dam. It was only located 12 km from the epicentre and no damage occurred. This dam was designed for a base acceleration of 0.138 g.

Figure 7.7 Views of the Zipingpu dam and structural damage.

There was no total failure of gravity and arch dams due to ground shaking although some cracking due to ground shaking or impacts of rock falls took place. However, it is well known that Pacoima arch dam had experienced large ground motions >1 g during the 1971 San Fernando earthquake. It is reported that the Sefidrud buttress dam had some horizontal cracking in the concrete body due to the 1990 Manjil earthquake.

7.1.3 Bridge and viaduct damage

Damage to bridges due to earthquake shaking generally occurs when the earthquake's magnitude is >6. It is generally difficult to identify the causes of ground shaking as the damage involves some ground failures at abutments. Damage to bridges and viaducts associated with ground motions caused by some selected earthquakes, the reconnaissance of which the author had directly participated in, are described herein.

The first selected earthquake is the 1995 Kobe (or Great Hanshin) earthquake, which caused extensive damage to bridges and viaducts. Extensive damage to road and railway bridges occurred in Kansai, Takarazuka, Itami, Amagasaki, Nishinomiya, Ashiya, and Kobe cities. Damaged road and railway lines were located along a narrow corridor, and included Routes 2 and 43 of the National Highway, Routes 3 and 5 of the Hanshin Expressway, Meishin and Chugoku Expressways, Sanyo-Shinkansen Train Line, Tokaido Train Line, Hankyu Railways and the Hanshin Electric Railway. Figures 7.8 and 7.9 show the damage to bridges and viaducts of expressways and railway lines. The estimated causes of damage were high ground shaking, permanent ground deformation due to faulting, and ground liquefaction. As seen in the figures, the girder falls, rupture of piers and the failure of reinforced concrete frame of viaducts of railways were observed. One of the most striking features is the rupture or failure of piers at the mid-level, which cannot be explained purely by high ground shaking despite

Figure 7.8 Views of damage to expressway and viaducts.

Figure 7.9 Views of damage to railway bridges and viaducts.

ground motions up to 0.8 g were recorded during the earthquake. As the expressways and elevated railways are linear structures, it was very likely affected by other motions.

The 2005 Kashmir earthquake with a moment magnitude of 7.6 occurred in the Azad Kashmir region of Pakistan. There are three bridge types in the epicentral area.

Figure 7.10 Damaged bridges at Balakot and in Jhelum valley.

Old bridges were either stone masonry arch bridges or truss bridges. New bridges were mainly cast-in-place concrete bridges. Heavily damaged bridges were observed in valleys running parallel to the strike of the causative fault. Furthermore, the longitudinal axis of the damaged bridges was parallel or sub-parallel to the slip direction of the fault. The largest damaged bridge is located in Balakot town (Figure 7.10). The girder of the bridge was displaced by about 1 m in the southward direction. While the west side of the bridge sits over the piers beneath, the eastern side of the bridge was offset from the pier by about 35–50 cm.

Several bridges along the Jhelum valley between Muzaffarabad and Chakoti were damaged. The collapse or permanent movements of masonry abutments mainly caused damage to bridges as seen in Figure 7.10. The top part of arch masonry superstructure collapsed partially as a result of earthquake shaking. The west abutment of the third bridge along the Jhelum valley was partially damaged by the earthquake. The bridges in Muzaffarabad city were nondamaged, although the bridges were about 2–3 km away from the causative fault and they are on the footwall side of the fault.

The 2008 Wenchuan earthquake with a moment magnitude of 7.9 occurred in the Shichuan Province of China. The earthquake involved two fault segments with slightly different faulting mechanism. Among all bridges affected by the earthquake, Miaziping (Miaotzuping) bridge receives great attention and this bridge passes over the Zipingpu dam reservoir in the epicentral area of the 2008 Wenchuan earthquake. The bridge is 1,436 m long with a height of 100 m (Figure 7.11). The main bridge is a long-span box girder bridge with 19 approaches on T-girders. The construction of the bridge was completed at the time of the earthquake. However, it was not open to traffic yet. The earthquake caused the shifting of bridge girders longitudinally and laterally. One of the T-type girder approaches collapsed as seen in Figure 7.11. The surveying indicated that the distance between piers was increased by more than 50 cm and there were all-around fractures, spalling and bending cracks in the pier just above its foundation. While there were stoppers for lateral movements, there were no stoppers longitudinally. Furthermore, the sliders (pedestals) were 40 by 40 cm. The fall of the girder was presumed to be due to a large relative displacement as the piers were very high, and there were no stoppers against the slippage at the pedestals.

7.1.4 Overturning or derailment of vehicles due to ground shaking

Derailment of trains has occurred in many earthquakes worldwide. At the time of the 1995 Kobe earthquake, trains were in their train yards and they were derailed under

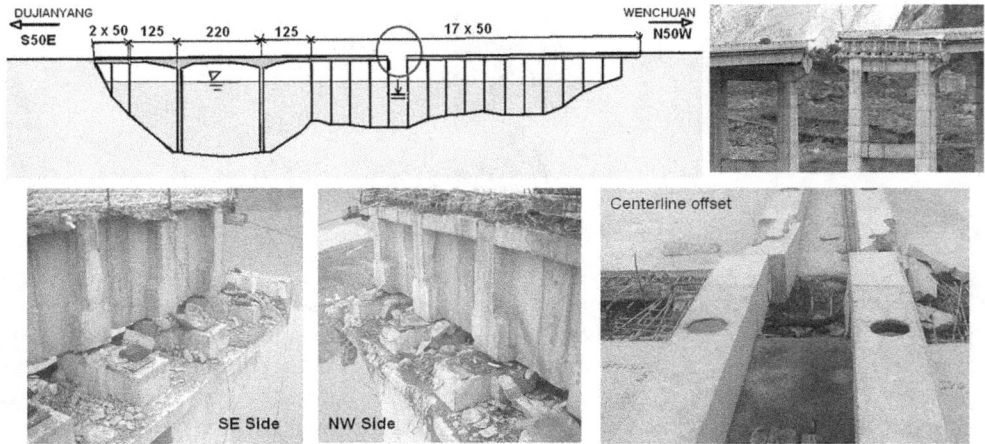

Figure 7.11 Sketch and views of the fallen section of the Miaoziping Bridge.

Figure 7.12 Derailment of the stationary trains due to the 1995 Kobe earthquake.

stationary conditions, as seen in Figure 7.12. This lucky condition probably resulted in no casualties from derailment considering the speed of the Shinkansen and local trains.

The overturning or derailment of vehicles in many countries has started to receive great attention following the derailment of the Shinkansen train TOK I-325 of the Joetsu Shinkansen line during the 2004 Niigata-ken Chuetsu earthquake (Aydan 2004c). The Shinkansen train was travelling at a speed of 200 kmph when the earthquake hit. The event took place when the train appeared from the northern portal of Takiya Tunnel and it stopped at a distance of about 1,500 m from the portal. The wagons at the ends of the train derailed with widespread damage to the railways (Figure 7.13). Following this event, it was explained that the train would derail if the waves >100 mm displacement amplitude with 1 Hz frequency acted on the train laterally. Some emphasis was put upon the height of piers of the viaducts as they become higher from the tunnel portal towards Nagaoka station. There is no doubt that the shaking period of the viaducts

Figure 7.13 Maximum ground accelerations along Joetsu Shinkansen and views of the derailed Shinkansen train. (Arranged from Aydan (2004c).)

would become larger as the height of structure becomes higher. Furthermore, the liquefaction of foundation soil should also cause the prolongation of acceleration waves. In addition, permanent ground deformations can cause offsets of the rails, which should result in further undesirable shaking and tilting effects on the train. One of the reasons for nondevastating train derailments was the introduction of precautions against such incidents during heavy snow fall in the region. In other words, such precautions may be necessary for railway lines in highly seismic areas.

Similar events also took place in other countries. For example, the AMTRACK train in the United States was also derailed during the Hector Mine earthquake in 1999. It was indicated that the strike-slip faulting caused some offsets of the rails at their rail joints. A similar event was also observed in the 1906 San Francisco earthquake and the 1978 Santa Barbara earthquake. The expected ground accelerations at the site estimated to be about 400 gals were inflicted by the 1999 Hector Mine earthquake. The overturning limit of the AMTRACK train for the static case is about 390 gals, and the train was travelling at a speed of 90 km/h at the time of earthquake. The permanent ground displacement due to faulting caused some offsets of the rails at their joints. This caused further undesirable effects from derailment.

Vehicles such as cars, trucks, buses may also be overturned during earthquakes. Figure 7.14 shows two examples of overturning of vehicles observed in the 1999 Düzce-Bolu earthquake and the 2004 Niigata-ken Chuetsu earthquake. The maximum ground acceleration measured at Düzce was more than 500 gals and the shaking was prolonged due to ground softening as a result of liquefaction and other causes. The overturning limit of the truck shown in Figure 7.14 is about 450 gals. The overturning of the car caused by the 2004 Niigata-ken Chuetsu earthquake was probably a combined effect of high ground shaking and failed roadway pavement.

7.1.5 Tanks

The first well-known earthquake, which caused some damage to oil and refinery facilities, is the 1933 Long Beach earthquake. The 1964 Niigata and Alaska earthquakes are

Figure 7.14 Overturned vehicles. (a) The 1999 Düzce-Bolu earthquake and (b) the 2004 Niigata-ken Chuetsu earthquake.

the next major earthquakes to be mentioned. In the California earthquakes, some storage tanks for oil, water and other liquids were also damaged, which may be relevant for assessing the seismic response of oil tanks. During the 1978 Miyagi-oki earthquake, tanks were ruptured by ground shaking, although they did not cause any fire. The most spectacular example of fire damage are the storage tanks of the Tüpraş oil refinery in Kocaeli prefecture of Turkey during the 1999 Kocaeli earthquake. The refinery was just 3–4km away from the rupture of the 1,500km long North Anadolu Fault (NAF). The most recent refinery damage was caused by the 2003 Tokachi-oki earthquake. Some crude oil and naphtha storage tanks of the Komakomai Refinery of Idemitsu Corporation, which is far from the epicentre, caught fire. In this section, damage caused to oil facilities or similar types of facilities by major earthquakes is described following a brief description of the fundamental damage types and loads involved.

7.1.5.1   Classifications of damage to oil tanks

Damage to storage tanks for oils or other liquids as a result of ground shaking may be classified as follows:

- Shell buckling (elephant foot buckling)
- Roof damage (fixed or floating)
- Anchorage failure (anchored tanks)
- Tank support/column system failure (pertains to elevated tanks)
- Foundation failure (largely a function of soil failures or fault break)
- Hydrodynamic pressure failure
- Connecting pipe failure
- Manhole failure
- Fire

These damages would entirely depend upon the shaking characteristics at the site of plants and ground conditions. The most characteristic type of liquid storage tank damage is a circumferential "elephant's foot" bulge that can form near the base of the tank due to excessive compressive loads in the tank wall (e.g. Nyman and Kennedy 1987; EERI 1986; Haroun 1983). Excessive sloshing of tank contents has often resulted in damage to floating and fixed roofs, and tank settling, sliding or rocking has caused breakage or pull-out at pipe connections. Differential settlements of the foundation have also led to tank failure. In the following sections, damage to tanks by the selected earthquakes is described.

7.1.5.2 Damage by the 1995 Kobe earthquake

The 1995 Great Hanshin earthquake ($M = 6.9$), or the Kobe earthquake, was one of the most devastating earthquakes to hit Japan; more than 5,500 people were killed and more than 26,000 injured. The economic loss has been estimated at about US $200 billion. Peak accelerations as large as 0.8 g were recorded in the near-fault region on alluvial sites in Kobe and Nishinomiya. Widespread ground failure was observed throughout the strongly shaken region along the margin of Osaka Bay. On the islands of Rokko and Portopia, which are reclaimed lands in Osaka Bay near Kobe, lique-faction caused subsidence in the range of 50–300 cm, and large volumes of silty soil was ejected. Lateral spreading of soils occurred along quay walls in many parts of the extensive port facilities in Kobe. Due to large-scale liquefaction of reclaimed land in coastal regions, many storage tanks for oil, petroleum products and high-pressure gas were greatly tilted. However no tanks collapsed. This is considered to be due to the relatively short time period over which earthquake motion continued. Despite strong tremors specific to in-land earthquakes, the Kobe earthquake had a short duration of earthquake motion.

There are numerous tank farms in the area most severely affected by the earth-quake (Figures 7.15 and 7.16). Some of the tank farms are associated with the port facilities, while others provide storage for refineries, fossil fuel power plants and chem-ical manufacturing facilities. Most of the tanks surveyed from the air appeared to have performed well, and there were no reports of widespread damage to tanks even within port areas disturbed by severe liquefaction effects. The generally good performance

Figure 7.15 Views of various types of damages to storage tanks in Kobe.

Figure 7.16 Damage to various types of pipes by the 1995 Kobe earthquake.

of tanks may be due to the apparent practice of placing tanks on pile foundations. However, there were a few exceptions to the generally good performance; these few instances of poor performance may have been due to use of mat foundations in areas where liquefaction occurred.

7.1.5.3   Damage by the 1999 Kocaeli earthquake

The most widely publicized and spectacular damage occurred at the massive Tüpraş Refinery in Yarımca caused by the 1999 Kocaeli earthquake. This refinery is the largest one accounting for about 1/3 of Turkey's oil, and is a major supplier to much of the industry in the area. The annual refined petroleum is 270,000 m^3 and it is the seventh largest plant in Europe. The plant owned by the state oil company was designed and constructed in 1961 by the US firm CALTEX (now defunct).

The refinery is located along the shore at Tütünçiftlik of the western Kocaeli province. The ground is firm, and no ground failure occurred except some liquefaction of reclaimed land during the earthquake. The refinery has three crude oil and three vacuum distillation units, three hydrodesulphurization (kero-diesel) units, one hydrocracker, two unifer/reformer units, two FCC units, one isomerization unit, one asphalt unit, one sulphur recovery unit, one isopentane unit, one naphtha sweetening unit and related utility units. The products are naphtha, gasoline, jet oil and kerosene. The 860,000 ton crude oil is stored in 14 large cylindrical tanks, and the 840,000 ton semi-products are stored in 86 middle and small-size cylindrical tanks. Figure 7.17 shows the locations of damaged and burned tanks.

7.1.5.3.1   FIRE IN TANKS WITH FLOATING ROOFS

Six cylindrical tanks having floating roofs burned due to the earthquake four middle-size tanks with a diameter of 20–25 m and 2 small-size tanks with a diameter

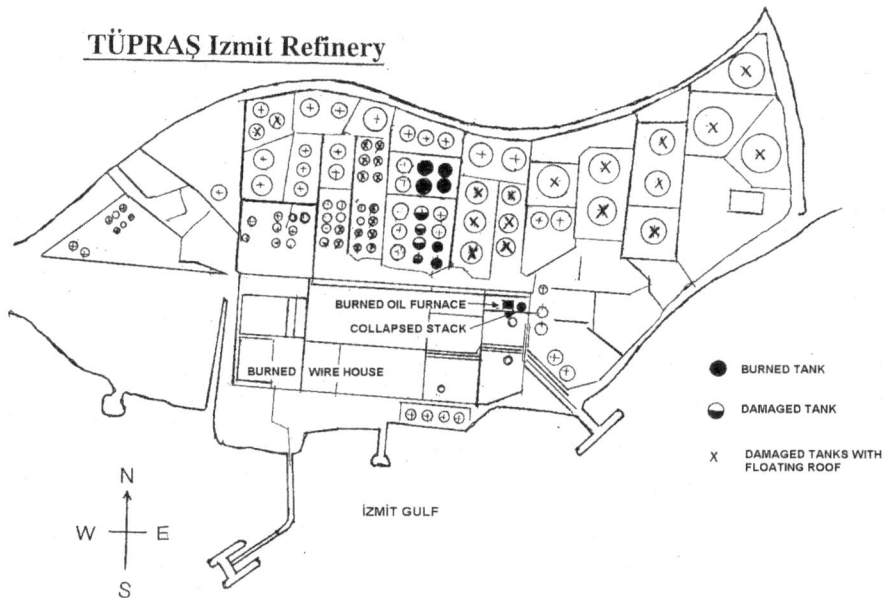

Figure 7.17 Locations of damaged and burned tanks.

of 10 m. Naphtha in the middle-size tanks was completely burned. Tanks were damaged as a result of thermal deformation. The estimated naphtha was about 36,000 ton. Figure 7.18 shows the deformed state of burned tanks. The fire in a naphtha tank farm was considered to be initiated by sparks due to bouncing of the floating roof in one of the tanks during the earthquake. The sparks ignited the naphtha. There were 46 tanks with floating roofs and all were damaged irrespective of the size of tanks.

Most of these tanks were built in 1961 according to the earthquake design code of California for a Level 4 earthquake shaking. In addition to cylindrical tanks, there are some spherical tanks in the plant. None of these was damaged. As seen in Figure 7.19, the piers and bracing of these spherical tanks show no visible damage. Therefore, they should have had enough resistance against the ground shaking during the earthquake. The sealing of the tanks with the floating roof built in 1961 was mainly metallic.

7.1.5.3.2   FIRE DUE TO COLLAPSED HEATER STACK

The collapse of one of the crude oil stacks with a height of 105 m caused the crude oil unit fire. The temperature of the collapsed part was more than 500°C. Figure 7.20 shows the burned crude oil unit and collapsed heater stack. The height heater stacks ranges from 90 to 115 m, and there are five heater stacks in the plant. The collapsed heater

Figure 7.18 Burned and damaged cylindrical tanks.

Figure 7.19 Undamaged spherical tank.

Figure 7.20 Damaged crude oil furnace, collapsed stack and damaged pipe rack hit by the collapsed stack.

stack was 105 m high. The heater stacks were built in 1981. The reason for collapse of the heater stack may be associated with material degradation as a result of corrosion due to alkali gases. It is difficult to consider that the main cause of the collapse of the stack is ground shaking alone. Both fires started soon after the earthquake. The fire of the crude oil unit extinguished at the night of August 17. While extinguishing the fire at the crude oil unit, the fire at the tank farm became stronger and unmanageable, which was completely extinguished only on August 20, 1999 (4 days after the earthquake). A part of the collapsed heater stack hit the pipe rack and caused extensive damage, as seen in Figure 7.20.

7.1.5.3.3   OTHER DAMAGES

A fire started at the tele-communication room. However, this fire was quickly extinguished. Several concrete pipe frames were broken. Pipes were nondamaged and the damaged concrete frames were re-supported by steel elements. The pipes having a diameter of 700 mm installed over a concrete embankment were fallen towards the seaside for a distance of 150 m. Nevertheless, the pipes were almost nondamaged. The total damage was about US $500 million.

7.1.5.4   The 2001 Kutch earthquake (India)

Kutch (Kachchh) region in Gujarat state of India was severely shaken by a powerful earthquake at 8:46 a.m. on January 26, 2001, India Standard Time, which has been the most damaging earthquake in the last five decades in India. This M7.9 earthquake is the first to hit metropolitan cities of India and its modern industrial constructions in recent times. Seismological data indicate that a severe earthquake measuring Ms 7.7 (NEIC) Mw 7.7 (NEIC) Mw 7.6 (HRV) occurred at 03:16 UTC (8:46 a.m. local time) in the southwestern province of Gujarat, near the Gulf of Kutch. The hypocenter was placed at a shallow depth of about 17 km below the surface and both the NEIC and Harvard (HRV) fault plane solutions indicated predominantly reverse faulting along a moderately dipping, nearly east-west trending fault plane with a slight sinistral sense. The entire Kutch region of Gujarat sustained highest damage with maximum intensity of shaking as high as X on the MSK intensity scale. The strong motion records obtained from the region at the Passport Office Building in Ahmedabad city (200 km away from epicentre) indicate a peak ground acceleration of about 0.11 g measured by Roorkee University. The peak acceleration exceeds 1 g nearby the earthquake fault: 600–800 gals estimated for the villages and towns of Bachau, Anjar, Chobari and Gandhidam, and 500–600 gals for Bhuj.

7.1.5.4.1   TANK YARD FRIENDS ASSOCIATION

This tank yard is for liquid-type chemicals and oils. There are 66 cylindrical tanks having a diameter ranging between 7 and 26 m. The pump houses were reinforced concrete type and suffered structural damage. No damage to pumps was reported. The tank yard is built on reclaimed land. The top soil was excavated to a depth of 2 m after constructing an all-around wall and drained. Then lateritic soil was filled to a height of 3.5 m after compaction and tanks were built over this soil basement, as illustrated in Figure 7.21.

Two tanks were heavily tilted and connections of valves to pipelines were damaged due to tilting and settlement of the tanks (Figure 7.22). It seems that the tilting and settlement were due to foundation failure associated with the base soil. Considering the bulging and motion of the ground, it seems that the ground moved in the direction of S40W–S50W. Besides the two heavily tilted tanks, some differential settlements, tilting and settlement of tanks were observed in the newly reclaimed tank yard. In the oil tank yard, which was built for more than 20 years ago, the settlement of the ground was not observed.

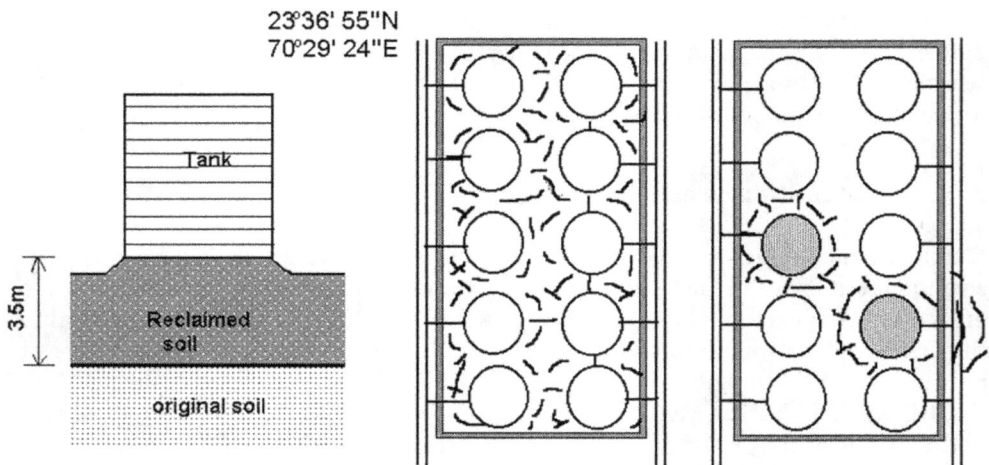

Figure 7.21 Foundation construction procedure and a sketch of damage to tank farms.

Figure 7.22 Tilted and damaged tanks at the Friends Tank Farm.

Figure 7.23 Damaged tanks in the Oil India tank farm.

7.1.5.4.2   TANK FARM OF OIL INDIA

Most of the tanks were nondamaged in the tank farm of Oil India. Nevertheless, some oil tanks were buckled or tilted. Some of damage to tanks was due to the liquefaction of the foundation soil (Figure 7.23).

7.1.5.4.3   TRISUNS CHEMICAL IND. LTD.

There were six steel tanks with a conical fixed roof. No damage to tanks was observed although it was very close to the epicentre of the earthquake. The tanks had 2 m thick concrete basement. The buildings in this complex, however, were heavily damaged.

7.1.5.4.4   PIPELINES

Product pipelines from Vadinar to Kandla and Kandla to Bhatinda were shutdown. The pipe racks were made of steel and were bent or buckled in the vicinity of the port.

Figure 7.24 Damage to (a) pipe stoppers and (b) ground failure nearby pipeline.

There was no breakage of pipelines at any location although they were displaced from their original locations. The steel strip-type fixtures were either broken or separated from their welding locations (Figure 7.24a). The pipes were displaced both laterally 30 cm and longitudinally 40 cm. The direction of movement was NE towards the epi-centre. The liquefaction and lateral movement of the ground were observed at the base of pipelines as shown in Figure 7.24b.

Crude oil pipelines from Salaya, feeding Koyali/Mathura/Panipat refineries re-sumed operations since January 27 after due checking. The oil pipelines and jetties at the Kandla port were severely damaged (Figure 7.25). The pipelines are supported by two-layer steel frames, and each frame supports three pipelines. Two of them were filled with oil during earthquake. The pipelines run towards the NS direction, and the south ends are curved to the west direction. Because the ground motion of the NS direction was dominant and the shaking modes were different for the three parallel pipelines which have different curvatures, the steel frames buckled and collapsed with the pipelines. Traces of strong shaking were observed on the pipelines.

7.1.5.5 The 2003 Tokachi-oki earthquake

A severe earthquake struck Hokkaido, Japan, at 4:50 a.m. on September 26, 2003. The epicentre of the earthquake was about 80 km east-southeast of Erimo Cape, offshore of Tokachi. The depth of the epicentre was 42 km, and its magnitude 8.0 (M_{JMA}). This magnitude of the earthquake is ten times as large as that of the Kobe earthquake.

While there was no damage in Tomakomai City because of the earthquake, at 4:52 a.m. on September 26, fire following the earthquake burned down a crude oil tank (32,778 kL) in the Hokkaido Refinery in Tomakomai (Idemitsu Kosan Co. Ltd.). Simultaneously, oil leaked from the piping in the refinery, and resulted in a fire, which was brought under control at 12:09 p.m. on September 26. The roof type of the tank that caused the fire is called a "floating roof". The fire may have initiated due to "slosh-ing" actions of the floating roof. The roof swung together with the oil in the tank that was shaken by the earthquake ("sloshing"), and then the oil spilled out onto the roof. The fire ignited the oil due to the friction between the roof and sidewall of the tank.

(a)   (b)

(c)

Figure 7.25 Damage to pipeline racks and jetty at Kandla port. (a) broken racks, (b) fallen pipes (c) jetty

A naphtha tank (26,000 kL) in the same refinery initiated a fire at 10:36 on September 28. The floating roof had been tilted due to sloshing and the oil had spilled on the roof. It is considered that the spilled oil on the roof had caught fire for some reason. In the refinery, there were about 90 oil tanks out of which oil spilled onto the floating roof due to sloshing (Figure 7.26). Maximum ground acceleration was 86 gals at the plant. Sloshing period of tanks was estimated to be around 3–8 seconds. Maximum ground velocity was more than 100 cm/s.

7.1.6   Sinkholes due to abandoned mines and natural caves

The seismic response and stability of settlement areas situated above the abandoned lignite mines in Japan due to urbanization in recent years are of great concern. The July 26, 2003, Miyagi-Hokubu earthquake (Mw 6.1) caused damage in the areas of abandoned lignite mines nearby Yamoto town, which was located just above the hypocenter of the earthquake (Figure 7.27a). It is also known that many collapses occurred

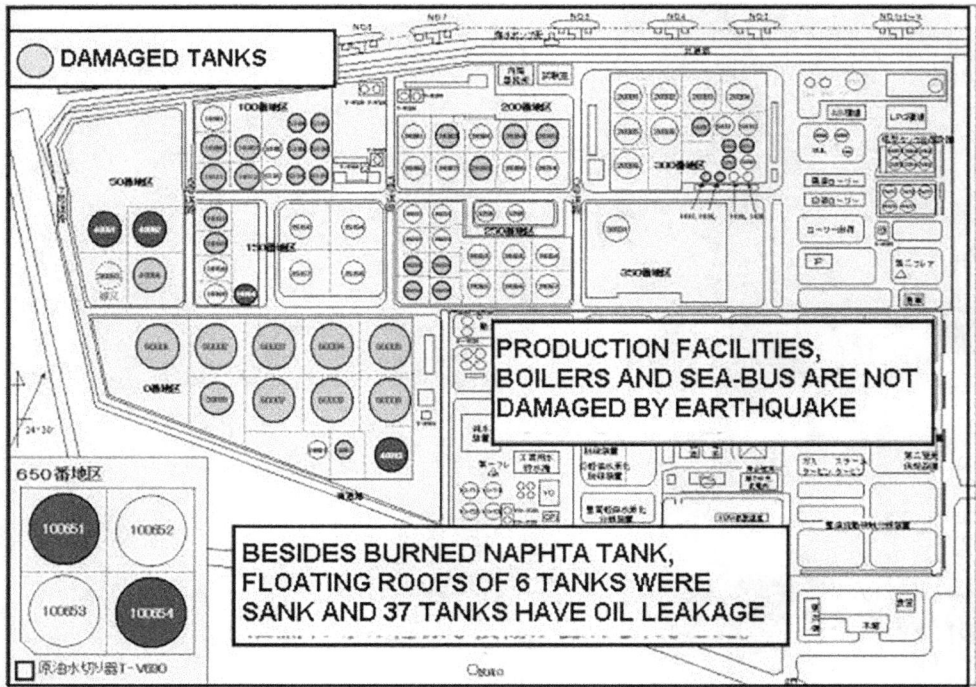

(a)

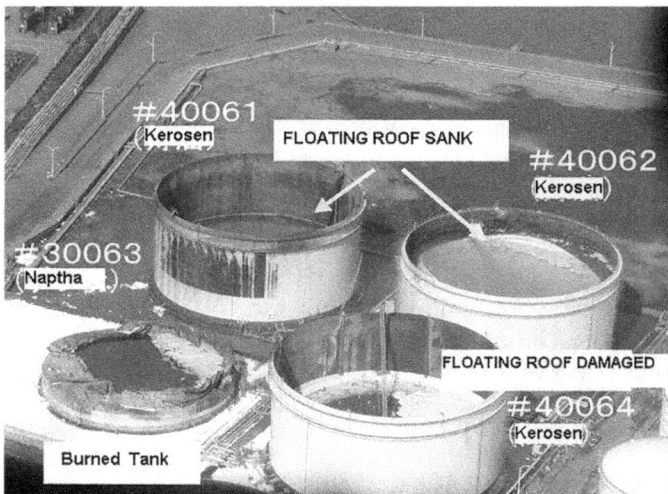

(b)

Figure 7.26 Location and view of burned or damaged tanks at Tomakomai Plant. (a) location
(b) view of tanks.

(a)   (b)

Figure 7.27 Sinkholes at (a) the Yamoto abandoned mine and (b) L'Aquila. (From Aydan and Kawamoto (2004) and Aydan et al. (2009a,b).)

in abandoned mines exploited by room and pillar technique in past earthquakes in Japan. The GEJE caused sinkholes at 5 locations in Iwaki City in Fukushima Prefecture, 11 locations in Kurihara City, 7 locations in Osaki, 11 locations in Higashi Matsushima and 3 locations in Kurogawa in Miyagi Prefecture, 11 locations in Ichinoseki City and 10 locations in Oshu City in Iwate Prefecture according to the first preliminary report (Aydan and Tano 2012). However, the press release on July 25, 2011, by the METI revealed that the number of events were more than 316 (Aydan and Tano 2012).

The 2009 L'Aquila earthquake caused two sinkholes in L'Aquila City (Figure 7.27b). One of the sinkholes was well publicized worldwide; its width was about 10 m while its depth is not known. A car had fallen into this sinkhole. Another sinkhole was observed in Castelnuovo with a length and width of 5 and 3 m and a depth of 5 m.

7.1.7 Damage to tunnels and underground shelter

7.1.7.1 Damage to tunnels

Earthquakes caused damage to tunnels from time to time. The damage to underground structures due to the 1992 Erzincan, 1999 Kocaeli, 2005 Kashmir and 2008 Wenchuan earthquakes may be classified as follows:

a. Shaking-induced damage (Figures 7.28)
b. Portal damage (Figure 7.29)
c. Damage induced by permanent ground deformations (see the next section).

Shaking-induced damage in tunnels is almost none unless the tunnels pass through geologically weak zones, as reported by Asakura and Sato (1998) and Aydan et al. (2010a,b). The damage shown in Figure 7.28 is the collapse of the Bolu tunnel face reaching the ground surface where the tunnel suffered a squeezing problem (Aydan et al. 2000c). Tunnel portals are generally prone to slope failures.

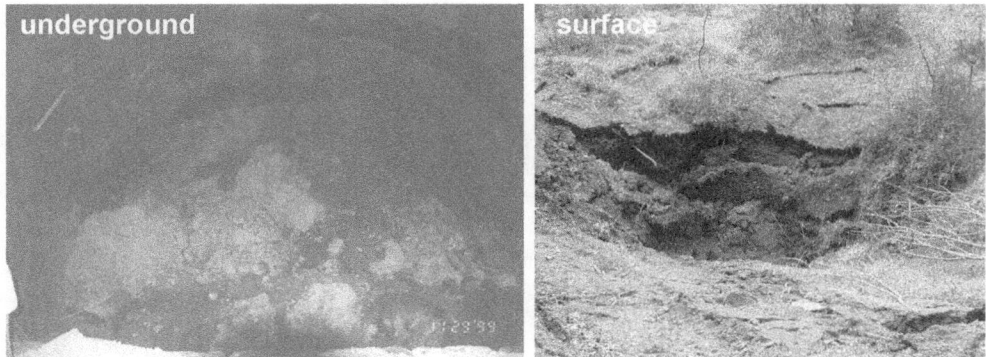

Figure 7.28 Views of the collapsed section of Bolu tunnel and its surface depression (Aydan et al. 2010a,b).

Figure 7.29 Examples of damaged portals of tunnels (Aydan et al. 2010a,b). (a) The 1992 Erzincan, (b) the 2005 Kashmir, and (c) the 2008 Wenchuan earthquakes.

7.1.7.2 Damage to the Bukittingi underground shelter

Japanese Imperial Army constructed an underground shelter along Sianok valley in Bukit Tinggi, which was hit by the 2007 Sinkarak earthquake. Sianok valley is created by relative dextral movements along the Sumatra Fault zone (Aydan et al. 2011). The ground consists of pyroclastic flow deposits. Following the immediate thin soil deposits, there is a pyroclastic flow deposit numbered Layer 1. This layer looks like a pumice and it is whitish. The second pyroclastic flow deposit numbered Layer 2 is slightly welded and it is more resistant. The underground shelter is mainly excavated in Layer 2. The access to the underground shelter is a 64 m long inclined shaft with 1,000 stairs. The layout of the underground shelter and a cross section is shown in Figure 7.30a and b. The underground shelter is about 28 m deep from ground surface and it is connected to ground surface by an inclined shaft entrance, and two adits are open to the Sianok valley. The configuration of adits changes from location to location and based on their functions. The inclined shaft has an arched roof and it is 2.4 m high and 3.0 m wide (Figure 7.30c). The main adit has a trapez shape with a height range of 1.8–2.1 m and a base width range of 2.4–2.6 m. The rooms between adits are larger and their width is

Open to valley

0   50 m

Open to valley

Entrance

Sianok
Valley

Pyroclastic I
(pumic-like)

Open to valley

Entrance

Pyroclastic II
(tuff-like)

0   50 m

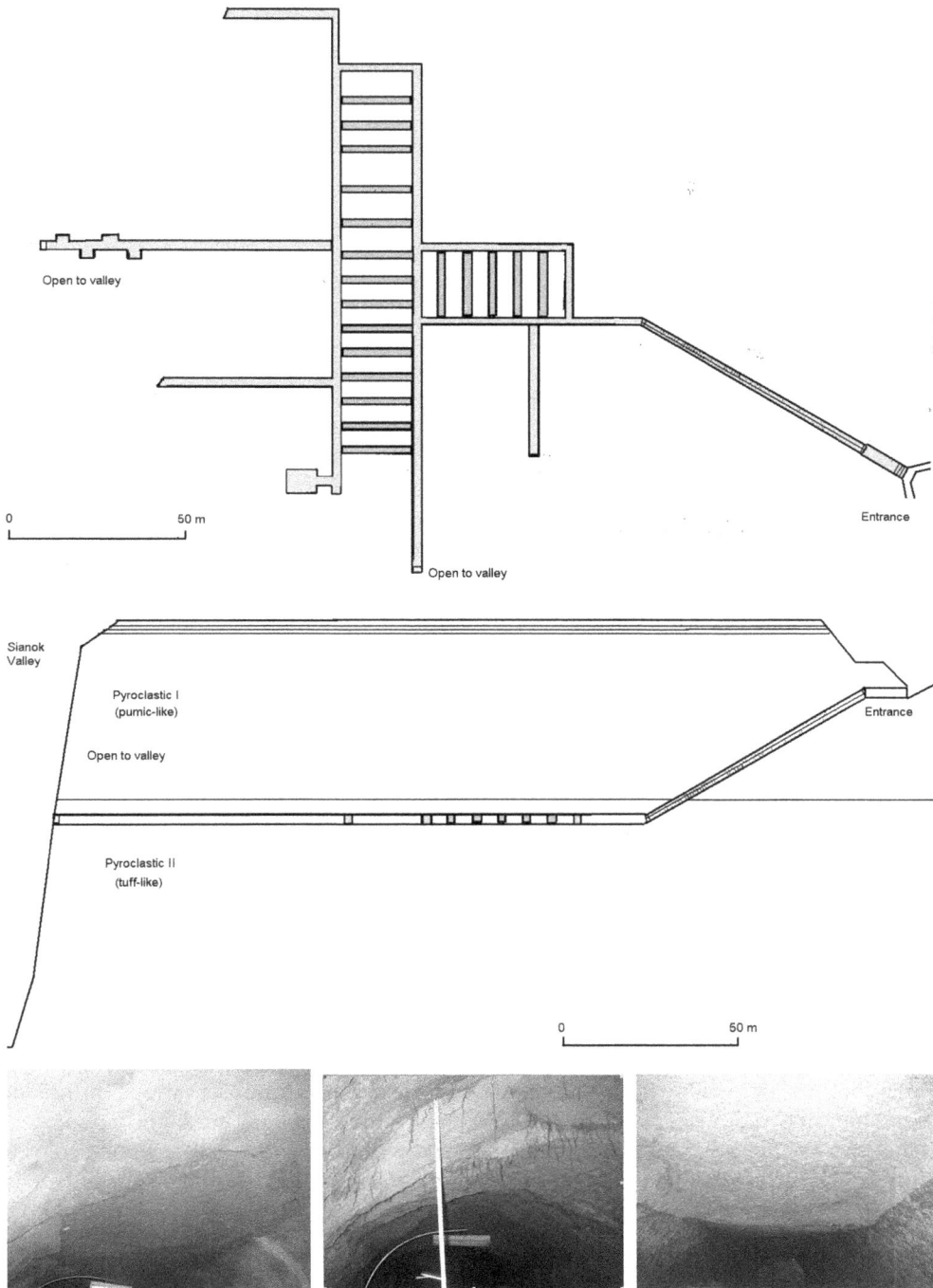

Figure 7.30 (a) Plan, (b) cross section and (c) views of spalling in several locations at the Bukittingi underground shelter.

about 4 m with a height of 2 m. The adits were probably supported by wooden supports at the time of the construction. However, the wooden supports rotted with time and they were taken away at the time of opening of the shelter to touristic visits. The earthquake caused some spalling from the roof of the adits as seen in Figure 7.30c.

7.1.8   Slope failure

7.1.8.1   The 1999 Chi-Chi earthquake

This earthquake occurred near Chi-Chi in the centre of Taiwan, about 160 km SSW of the capital city of Taipei, at 01:47 a.m. local time. It was a shallow-thrust-type earthquake, caused by the collision between the Philippine Sea and Eurasian plates. The ground motions were high particularly on the hanging-wall side of the earthquake fault and the faulting also induced permanent ground deformations. Numerous rock slope failures were induced by the Chi-Chi earthquake of September 21, 1999 (Aydan 2013, 2015b). Slope failures mostly occurred on the hanging-wall block of the Chelungpu earthquake fault (Aydan et al. 1999b, 2000d). Among these slope failures, the Tsaoling (also called Chaoling) and Chiufengershan (also called Mt. Jio-Fun-Ell-Shan or Nankang) slope failures are of great interest in view of their geometric dimensions (Figure 7.31).

7.1.8.1.1   TSAOLING SLOPE FAILURE

The rock mass at this location consists of intercalated sandstone and shale. The overall inclination of the bedding plane was about 14°. A 50–300 m thick, 1,800 m wide and 2,000 m long mass of sandstone and shale was involved in sliding. The slide surface was somewhat stepped. The estimated volume of the sliding body is about 120 million cubic metres. It is also reported that the earthquakes in 1862 and 1941 (Ishihara 1985) caused partial slope failures at this location. Slides in 1942 and 1979 were associated with heavy rainfalls.

   Figure 7.31a shows the stereo projections of principal geological features with a friction angle of 30°. The kinematics analysis implies that the slope should be stable under static conditions. However, a simple analysis with the use of the seismic coefficient ($\eta = a/g$) method would reduce the effective frictional resistance as illustrated in the same figure. In view of the maximum accelerations measured nearby these sites, it is quite possible that the earthquake could induce slope failures along the bedding planes. The sliding mass should have moved almost as monolithic bodies until they struck the valley bottom in view of both theoretical considerations and observations on model rock slopes (Aydan et al. 1989, 1999b). Reports of nondamaged or slightly damaged structures on the sliding bodies are in agreement with the previous observations of the author. The distance of the slope failure was 40 km from the earthquake's epicentre.

7.1.8.1.2   CHIUFENGERSHAN SLOPE FAILURE

The rock mass at this location also consists of intercalated sandstone and shale. The overall inclination of the bedding plane is about 20°–27°. A 30–50 m thick, 1,000 m

(a)

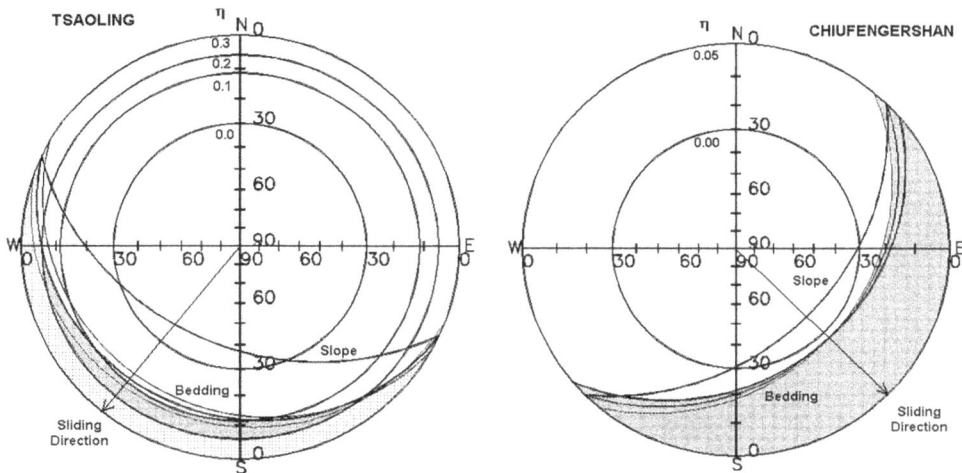

(b)

Figure 7.31 Views of large slope failures caused by the Chi-Chi earthquake and their stereo-net projections) ($\eta = a/g$ is seismic coefficient). (a) Tsaoling and (b) Chiufengershan.

wide and 1,000 m long mass of sandstone and shale was involved in sliding. The sliding surface was steep compared to that of the Tsaoling slope failure (Figure 7.31b). The estimated volume of the sliding body was about 90 million cubic metres, and there is no report about any previous sliding at this site. The distance of the slope failure was about 9 km from the earthquake's epicentre. Figure 7.31b shows the stereo projections of principal geological features with a friction angle of 30°. The kinematics analysis implies that the slope should be stable under a dry-static condition when the cohesion is assumed to be nil. However, if the seismic coefficient method is used, the slope must fail at a horizontal seismic coefficient ($\eta = 0.05$) for non-cohesive dry condition, where ϕ and α are friction angle and the inclination of sliding plane, respectively.

Figure 7.32 A view of the Terano slide.

7.1.8.2   The 2004 Chuetsu earthquake

The Niigata-ken Chuetsu earthquake occurred on October 23, 2004, at 17:56 on JST and it had a magnitude (Mj) of 6.8 on the magnitude scale of the Japan Meteorological Agency. One of the most striking characteristics of this earthquake is extensive slope failures that caused extensive damage on roadways and expressways, as well as destroying homes. The most extensive slope failures were observed in Yamakoshi. The area mainly consists of Neogenic mudstone. This mudstone near ground surface is highly weathered and it had become clayey soil. Figure 7.32 shows an example large slope failure at Terano. The slope failure in this area is associated with bedding planes which dip NE with an inclination ranging between 10° and 25°. Before the earthquake, Typhoon No. 23 passed over the earthquake epicentral area and resulted in very heavy rainfall. Therefore, rock mass and soil layers are expected to be fully saturated at the time of earthquake. The most spectacular slope failures occurred at Takezawa, Terano and Shiraiwa. The slope failures were categorized as curved deep-seated slope failures, shallow seated slope failures, planar sliding and rock falls.

7.1.8.3   The 2005 Kashmir earthquake

The 2005 Kashmir earthquake was quite devastating and the earthquake occurred near Muzaffarabad on October 8, 2005. The hypocenter depth of the earthquake was estimated to be about 10 km and it had a magnitude of 7.6 (Aydan and Hamada 2006; Aydan et al. 2009f). The earthquake killed more than 82,000 people in Kashmir, Pakistan. The earthquake resulted from the subduction of the Indian plate beneath the Eurasian plate, and the faulting mechanism solutions indicated that the earthquake

Figure 7.33 Views of major slope failures caused by the 2005 Kashmir earthquake. (a) Muzaffarabad and (b) Hattian.

was due to thrust faulting. The surface expression of the causative fault follows the valley between Bagh to Balakot through Muzaffarabad. The city of Muzaffarabad and Balakot town were the nearest settlements to the epicentre and the fault, and they were the most heavily damaged settlements. Valleys were filled with weakly cemented conglomeratic deposits. Fast-flowing rivers cut through these deposits, resulting in very steep slopes. Many cities, towns and villages were located on these types of terraces. The earthquake caused extensive damage to housing and structures found on these steep soil slopes. Furthermore, extensive rock slope failures occurred along Neelum and Jhelum Valleys, which obstructed both river flow and roadways.

There were numerous slope failures particularly on the hanging-wall side of the earthquake fault compared to those on the footwall side of the earthquake fault. Furthermore, the areal extension of the slope failures is much larger on the hanging-wall side than that on the footwall side. They were generally associated with the whitish 100 m thick dolomite layer, which was a highly deformed and fractured rock unit (Figure 7.33a). The wedge-like sliding failure at Hattian was quite large in scale (Figure 7.33b). This slope failure caused a 150 m high slope failure dam with a base length of 1,200 m. This slope failure dam created two lakes. The sliding area was about 1.5 km long and 1.0 km wide. The rock mass consisted of shale and sandstone and it constituted a syncline. The estimated wedge angle was about 100° and it was asymmetric. The friction angle of the shale via a tilting test was estimated at more than 35° with an average of 40°.

7.1.8.4 The 2008 Wenchuan earthquake

The 2008 Great Wenchuan (Sichuan) Earthquake with a moment magnitude of 7.9 occurred in Wenchuan County of Sichuan Province of China and caused extensive damage to buildings, infrastructures and it caused the failure of natural slopes as well as cut-slopes. More than 85,000 people lost their lives, besides heavy structural damage. The earthquake occurred at the well-known and well-studied Longmen Shan Fault Zone via thrust faulting with a dextral component. Preliminary analyses indicated

Figure 7.34 Views of some large slope failures caused by the 2008 Wenchuan earthquake.

that the rupture process activated a 300 km long fault section. Unfortunately, the strong motion records are not available to the scientific community. However, the preliminary estimations indicate that the ground motions were high in the epicentral area. The maximum ground acceleration and velocity were expected to exceed 1 g and 100 kines, respectively (Aydan et al. 2009c–e). One of the most distinct characteristics of 2008 Wenchuan earthquake was the widespread slope failures all over the epicentral area (Figure 7.34).

Slope failures, which are fundamentally similar to those observed during the 2005 Kashmir earthquake, may be classified into three categories as soil slope failures, surficial slides of weathered rock slopes and rock slope failures. Rock slope failures are further subdivided into curved or combined sliding and shearing failures, planar sliding, wedge sliding failures and flexural or columnar toppling. In addition to the active forms of failures, some passive forms of sliding and toppling failures were observed. To initiate passive modes of sliding and toppling failures, very high ground accelerations are necessary. Furthermore, the relative displacement of the block on the sliding plane at the toe of the slope should be greater than its half-length for the collapse of the slope in a passive sliding mode. When such relative movements take place, the initial sliding movement changes its characteristic to a rotational mode. As a result, the unstable part rolls down the slope. As for the passive toppling mode, the blocks in the upper part of the slope rotate and roll down. The many rock falls observed during this earthquake were related to these two phenomena.

The satellite images and in situ investigations indicated that there were very-large-scale slope failures along the earthquake's fault zone. These large-scale events took place in Xuankou (southwest end of the fault), Beichuan (central part, three large-scale slope failures including the Tangjiashan slope failure) and Donhekou village (northeast end of the fault). The characteristics of these large-scale slope failures are briefly explained in this section. The Xuankou slope failure (Figure 7.34) involved mainly limestone, which dips towards the valley side with an inclination of 20°–25°. Furthermore, there is a fault dipping parallel to the failure surface within the rock mass. The angles of the lower and upper parts of the failed slope are 60° and 30°, respectively. In addition to this new slope failure, one can easily notice the existence of a paleo-slope failure in the foreground. The failure plane of this slope failure is bi-planar and it involved the fault and bedding plane. The fault plane probably assisted the separation of the failed body from the rest of the mountain. The in situ tilting tests on the limestone joint surfaces yielded the friction angle as 38°–42°. If the resistance is purely frictional the expected horizontal seismic coefficient should be >0.36 g. As this earthquake heavily damaged Yingxiu town, it is more likely that such high ground acceleration did act on this slope.

There were several large slope failures in Beichuan county and its close vicinity including the Tangjashan Slope failure (Figure 7.34). In association with the motion of the earthquake fault, NW- or SE-facing slopes failed during this earthquake. There were two large-scale slope failures in Qushan town (Beichuan county), which destroyed numerous buildings and facilities. The NW-facing slope failure (Jingjiashan) involved mainly limestone while the SE-facing slope failure (Wangjiaya) involved sedimentary rocks and Quaternary deposits (Figure 7.34). Limestone layers dip towards the valley side with an inclination of about 30°. Furthermore, there are several faults dipping parallel to the failure surface within the rock mass. The angles of the lower and upper parts of the failed slope are 60° and 30°–35°, respectively. The existence of several faults dipping parallel to the slope with an inclination of about 60°–65° creates a stepped failure surface. If the resistance is assumed to be purely frictional, the expected horizontal seismic coefficient should be >0.18 g.

The SE-facing slope (Wangjiaya slope) failure (Figure 7.34) may have involved a slippage along the steeply dipping bedding plane and shearing through the layered rock mass or sliding along a secondary joint set. In other words, it may be classified as a combined sliding and shearing failure(Aydan et al. 2009c–e). The angles of the lower and upper parts of the failed slope are 40°–45° and 30°–35°, respectively. The layers dip at an angle of 40° towards the valley, and the shearing plane is inclined at an angle of 20°. Since material properties are unknown for this site yet, no estimation on possible ground acceleration has been done. Nevertheless, it is more likely that such a high ground acceleration did act on this slope, as Qushan town was heavily damaged by this earthquake.

Tangjiashan slope failure (Figure 7.34), which obstructed Jian River, is located about 2 km NW of Qushan town. The slope failure faces NW. The rock units involved in this slope failure were limestone (top part) and intercalated sandstone and shale (lower part). Rock layers dip towards the valley at an inclination of 40°–50°. Although the upper part of the mountain has a slope angle of 40°–45°, the lower part is more steeply inclined up to 70° due to the toe erosion by Jian River. The earthquake induced almost a planar sliding of a 600 m wide and 450 m long body of rock mass. This slope failure blocked Jian River for a length of 800 m and created a slope failure dam with a height of 124 m.

The Donghekou slope failure (Figure 7.34) occurred at Donghekou village of Hongguang Township, Qingchuan County, and it is located at the NE tip of the earthquake fault. The upper part of the mountain was covered by limestone, while the lower part consisted of phyllite, shale and sandstone. This phyllite formation is considerably thick and it is inclined at an angle of 30°–45° to NE and it is slightly discordant with upper limestone formation. A paleo-slope failure exists to the east of the new slope failure source area. The in situ tilting tests on the phyllite schistosity (bedding) plane surfaces yielded a friction angle of 22°–24°, which is considerably small. In other words, slopes facing NE cannot be stable even under static conditions if the resistance is only frictional. The initial movement of the unstable part of the mountain should have been in the NE direction and then its direction of motion rotated towards the SW. Further investigations and laboratory tests on samples from this site should provide essential parameters for investigating the conditions for the failure of the slope at this site.

7.1.8.5 The 2008 Iwate-Miyagi intraplate earthquake

2008 Iwate-Miyagi intraplate earthquake occurred on June 14, 2008, at 8:43 a.m. and had a moment magnitude (M_w) of 6.9 (M_{JMA} is 7.2). While the damage to residential buildings was quite limited, it caused extensive slope failures on the hanging-wall side of the earthquake fault (Figure 7.35). The earthquake occurred in a region of upper-plate contraction zone within the complicated tectonics of the Ou Backbone Range, known to have hosted several large earthquakes in historic times. The earthquake was caused by a thrust fault, which was not designated as an active fault in the active fault map of Japan. The observed maximum horizontal accelerations exceeded

Figure 7.35 Views of the Aratozawa deep-seated slope failure caused by the 2008 Iwate-Miyagi intraplate earthquake. (From Aydan (2015a).)

l g at several sites and 0.5 g at about 20 sites. The unusual maximum ground acceleration was measured at the IWTH25 station of KIK-NET (2011). However, the nonsymmetric nature of the acceleration records at ground surface compared to that in the base rock implies some peculiar amplification of ground motions.

A very large slope failure occurred on the northern side of the reservoir of the Aratozawa dam (Aydan 2013, 2015a). The total displacement was more than 300 m. The rock mass consisted of intercalated loosely cemented sandstone and mudstone (marl-like) and/or siltstone. The inclination of layers was about 10° to the east. Furthermore, fault traces were observed on the approach road at the right abutment of the dam with about 40 cm upward offset. The high ground acceleration, which was about 1 g at the base of the dam, and high water content of loosely cemented sandstone were probably the major players in the slope failure on a gently bedding surface. The failed rock mass body in Aratozawa dam slope failure did not slide into the reservoir, indicating that the incident was not similar to that of the Vaiont dam.

7.1.9 Embankment failure

The failure of embankments is quite common in many earthquakes. Some embankment failures observed during the 2007 Noto-Hanto earthquake deserves mentioning (Aydan et al. 2007a). The Noto Peninsula (Noto-Hanto) earthquake occurred at 9:42 AM JST on March 25, 2007, and it had a magnitude (Mj) of 6.9 on the magnitude scale of the Japan Meteorological Agency. Strong ground motions were quite high in the epicentral area with high-frequency components. The earthquake induced many slope failures along the steep shores and mountainous terrain. Embankment failures and settlements were widespread particularly along Noto Tollway, rivers and creeks. The embankments with a height ranging between 20 and 30 m and inclination of 26°–33° were failed along a section of the tollway between Noto Airport and Tokuda-Otsu Interchanges (Figure 7.36). The embankments facing east mainly failed due to strong

(a)   (b)

Figure 7.36 Failure of an embankment of Noto Tollway.

ground shaking. The embankment material is a soil of volcanic origin (Aydan et al. 2007a). Embankment materials along the tollway originate from the weathered volcanic rocks and it comes from the slope cuts for the tollway construction. This embankment shown is about 35 km from the epicentre. The height of the embankment is about 18 m, and its inclination (1:1.5) is about 33.7°. The inclination of the initial failure plane is about 25°, which gradually changes its configuration.

7.1.10 Retaining-wall failure

There are different types of retaining walls. Old retaining walls are generally of dry masonry type. Stones used are either rounded or cut to rectangular prism shape and the size changes depending upon their height, location and purpose. Some recent retaining walls were constructed using pre-cast concrete blocks. The inclination of a retaining wall generally ranges between 70° and 80°. The failures were generally induced by the displacement of a stone or several stones in the wall, which serves as a point of singularity to induce the failure of subsequent falling of stones. Several examples of retaining-wall failures observed in recent earthquake are shown in Figure 7.37. The retaining walls have generally some backfill material, and depending upon the characteristics and geometry of the backfill material, failure may be initiated.

Figure 7.37 Some examples of the failures of retaining walls in recent earthquakes.

Retaining-wall failures were observed at the remains of the Sunpu castle due to the 2009 Suruga Bay earthquake, and the most publicized retaining-wall failure is shown in Figure 7.37. Retaining walls of about 8–9 m high failed at three locations in the NS direction, indicating the directivity effect of the strong ground motions. Two failures occurred at the south and north side of the outer moat (soto-bori) and one failure occurred in the south side of the inner moat (uchi-bori). The size of corner stones is generally large (the longest side length is about 200 cm), and they are placed in an intermittent pattern with an average inclination of 70°. The sizes of blocks in other parts are highly variable. Nevertheless the longest side of the stones cut into pyramid shape is generally more than 80 cm and it is placed into the inside of the wall. Old stones are generally made of andesite and porphyrite, while the newly restored parts consist of basalt blocks. It is well known that the retaining walls of the Sunpu castle collapsed during the 1854 Ansei earthquake. Some part of the restored retaining wall failed again in the 1935 Shizuoka earthquake and the same location failed in the 2009 Suruga earthquake.

A retaining-wall failure took place in Elmalı village due to the 2007 Çameli earthquake. Figure 7.37 shows a view of this failure. As noted from the figure, the failure plane of the backfill is inclined at an angle of about 62°–65° from the horizontal.

The 2007 Kameyama earthquake caused some retaining-wall failure at the Kameyama castle, which was built in 1590 on a 10 m high hill (Aydan et al. 2008a). The inclination of south, east and west walls was about 50° while the inclination of north wall is much steeper and it is about 70°–75°. As seen in Figure 7.37, the earthquake caused the collapse of NW corner of the 5 m high northern wall, where a stone masonry stair is located. The collapsed section was 2 m wide and this section was actually damaged by a typhoon in 1972 and it was repaired (Aydan et al. 2008a). The stones used at the NW corner were typically 30 cm wide, 30 cm high and 55 cm long, and the rock itself is either andesite or diorite.

The 2010 Okinawa earthquake caused the collapse of the north wall of the Katsuren castle, which was about 5 m high with an inclination of about 70° with block sizes ranging between 50 and 60 cm as seen in Figure 7.37. The block size of older parts of the castle walls is more than 100 cm and their inclination is about 50°. Furthermore, the other walls of the castle are more than 10 m high.

The collapse of the Kumamoto Castle walls caused by the 2016 Kumamoto earthquakes clearly show the effect of backfill material, as seen in Figure 7.37. The backfill material consisted of rounded river boulders in order to increase the permeability of the retaining walls so that water pressure would not exist. However, the overall shear resistance of this backfill material was quite low and exerted high lateral forces on the retaining walls of the castle, which led them to collapse.

7.2 EFFECTS OF SURFACE RUPTURES INDUCED BY EARTHQUAKES ON ENGINEERING STRUCTURES

In this section, typical examples of damage to various structures induced by the fault breaks observed in recent large earthquakes since 1995 are presented, and details can be found in the quoted references (e.g. Aydan et al. 2011; Aydan 2003b, 2004c, 2012a; Ohta et al. 2014; Ulusay et al. 2001).

Figure 7.38 (a) The collapse of the overpass and (b) collapse of Pefong Bridge.

7.2.1 Bridges and viaducts

Along the damaged section of the TEM motorway mentioned above, there were several overpass bridges. Among them, a four-span overpass bridge at Arifiye junction collapsed as a result of faulting (Figure 7.38a). The fault rupture passed between the northern abutment and the adjacent pier. The overpass was designed as a simply supported structure according to the modified AASHTO standards, and its girders had elastomeric bearings. However, the girders were connected to each other with prestressed cables. The angle between the motorway and the strike of the earthquake fault was ~15° while the angle between the axis of the overpass bridge and the strike of the fault was 65°. The measurements of the relative displacement in the vicinity of the fault range between 330 and 450 cm. Therefore an average value of 390 cm could be assumed for the relative displacement between the pier and the abutment of the bridge. The Pefong Bridge collapsed due to thrust faulting in the 1999 Chi-Chi earthquake, which passed between the piers near its southern abutment as seen in Figure 7.38b.

The heaviest damage occurred at the four-span, reinforced Hsiaoyudong arch bridge in Longmenshan town, from which the Longmenshan fault zone was named. The arch sections of the bridge were sheared, as shown in Figure 7.39. Besides high ground motions in the vicinity of the bridge, the shearing of the arched section in the longitudinal axis of the bridge implies permanent ground movements, which may result from thrust faulting and ground liquefaction beneath the riverbed.

7.2.2 Dams

The Shihkang dam, which is a concrete gravity dam with a height of 25 m, was ruptured by a thrust-type faulting during the 1999 Chi-Chi earthquake (Figure 7.40). The relative displacement between the uplifted parts of the dam was more than 980 cm. The Liyutan rockfill dam with a height of 90 m and a crest width of 210 m, which was on the overhanging block of Chelongpu fault was not damaged even though the acceleration records at this dam showed that the acceleration was amplified 4.5 times of that at the base of the dam (105 gals). The deformation zone of faulting during the 2008 Wenchuan earthquake caused some damage in the Zipingpu dam with concrete facing (see Figure 7.7).

Figure 7.39 Collapsed Hsiaoyudong bridge in Longmenshan town.

Figure 7.40 Failure of Shihkang dam due to thrust faulting.

Another example of dam damage by faulting occurred at Ohkirihata Dam during the 2016 Kumamoto earthquakes (Aydan et al. 2018a,b). Ohkirihata Dam, which is an earthfill dam with a height of 23 m for irrigation purpose, was damaged by earthquake faulting and leakage of reservoir water occurred. The earthquake fault passed through the dam embankment near the right abutment and the relative displacement of the fault was about 140 cm in the vicinity of the dam (Figure 7.41).

7.2.3 Tunnels and subways

Past experience on the performance of tunnels through active fault zones during earthquakes indicates that the damage is restricted to certain locations. As reported by

Figure 7.41 Views of damage and faulting at the Ohkirihata Dam.

Aydan et al. (2010a,b), portals and the locations where a tunnel crosses a fault may be damaged, as occurred in the 2004 Chuetsu, 2005 Kashmir and 2008 Wenchuan earthquakes (7.29). A section nearby the Elmalık portal of Bolu tunnel collapsed (Figure 7.28). The collapsed section of the tunnel was excavated under very heavy squeezing conditions. Well-known examples of damage to tunnels at locations where the fault rupture crossed the tunnel are mainly observed in Japan. The Tanna fault ruptured during the 1930 Kita-Izu earthquake caused damage to a railway tunnel and the relative displacement was about 100 cm. The 1978 Izu-Oshima Kinkai earthquake induced damage to the Inatori railway tunnel. Similar types of damage with small amounts of relative displacements due to motions of the Rokko, Egeyama and Koyo faults to the tunnels of Shinkansen and the subway lines through Rokko mountains were also observed. During the 1999 Chi-Chi earthquake, the portal of the water in-take tunnels was ruptured for a distance of 10 m as a result of thrust faulting. Except this section, the tunnel was undamaged for its entire length.

The Jiujiaya tunnel is a 2282 m long double lane tunnel and it is 226.6 km away from the earthquake epicentre and about 3–5 km away from the earthquake fault of the 2008 Wenchuan earthquake. The tunnel face was 983 m from the south portal at the time of the earthquake. The concrete lining follows the tunnel face at a distance of ~30 m. Thirty workers were working at the tunnel face and one worker was killed by

Figure 7.42 Earthquake damage at Jiujiaya tunnel due to permanent deformations.

the flying pieces of rock bolts, shotcrete and bearing plates caused by intense deformation of the tunnel face during the earthquake. The concrete lining ruptured and had fallen down at several sections (Figure 7.42). However, the effect of the unreinforced lining rupturing was quite large and intense in the vicinity of the tunnel face. The rupturing of the concrete lining generally occurred at the crown sections although there was rupturing along the shoulders of the tunnel at several places. Furthermore, the invert was uplifted due to buckling in the middle sections.

The Kumamoto earthquake on April 16, 2016, caused heavy damage to several tunnels in the vicinity of Tateno and Minami-Aso villages. Damages to Tawarayama Roadway Tunnel and Aso Railway Tunnel and Minami-Aso Tunnel were publicized. The damage to Tawarayama tunnel occurred at two locations (Figure 7.43). The first damage occurred ~50 to 60m from the west portal of the tunnel and the concrete lining was displaced by about 30cm almost perpendicular to the tunnel axis. The heaviest damage occurred for a length of 10m about 1,600m away from the west portal and about 460m from the east portal. The angle between the relative movement and tunnel axis was about 20°–30°. At this location, the non-reinforced concrete lining collapsed for a length of about 5m. Although the tunnel is located about 2km away from the main fault, it was damaged by secondary faults associated with the trans-tension nature of the earthquake fault.

Figure 7.43 Views of damages and their locations at Tawarayama tunnel.

The behaviour of subways in active fault zones are basically quite similar to that of tunnels. The Daikai station of the subway line in Kobe was caused by the lateral strike-slip movement of the Egeyama fault just beneath this station although some tried to associate the collapse of the station with the intensity of shaking (Figure 7.44). The investigation of the collapse of this station by the author showed that the collapse was not simply due to shaking as the central columns of the station were subjected to torsional failure resulting from permanent ground displacement, which was consistent with the lateral strike-slip movement of the Egeyama fault (Aydan 1996).

7.2.4 Slope failures and rockfalls

The recent large earthquakes caused mega scale slope failures and rockfalls particularly along the surface ruptures on the hanging-wall side of the fault (Aydan 2015b, 2017). The slope failure induced in Beichuan town during the 2008 Wenchuan earthquake is of great interest. In association with the sliding motion of the earthquake fault, NW- or SE-facing slopes failed during this earthquake. There were two large-scale slope failures (landslides) in Qushan town, which destroyed numerous buildings and facilities. The NW-facing landslide (Jingjiashan) involved mainly limestone while

Figure 7.44 The collapse of the Daikai station.

Figure 7.45 Views of landslides in Beichuan.

the SE-facing landslide (Wangjiaya) involved phyllite (mudstone according to some) rock units (Figure 7.45). Limestone layers dip towards the valley side at an inclination of about 30°. Furthermore, there are several faults dipping parallel to the failure surface within the rock mass. The angles of the lower and upper parts of the failed slope are 60° and 30°–35°, respectively. The existence of several faults dipping parallel to the slope with an inclination of about 60°–65° creates a stepped failure surface.

The SE-facing slope (Wangjiaya landslide) may involve a slippage along the steeply dipping bedding plane (fault plane?) and shearing through the layered rock mass. In other words, it may be classified as a combined sliding and shearing failure (Aydan et al. 1991, 1992a). The angles of the lower and upper parts of the failed slope are 40°–45° and 30°–35°, respectively. The layers dip at an angle of 40° towards the valley and the shearing plane is inclined at an angle of 20°.

Figure 7.46 Damage to pylons due to faulting. (a) 1999 Kocaeli earthquake, (b) 1995 Kobe earthquake, and (c) 1999 Chi-Chi earthquake.

7.2.5 Pylons

Power transmission lines generally consist of pylons and power transmission cables. The design of pylons and cables are generally based on the wind loads resulting from typhoons or hurricanes. Additionally, the possibility of slope failure of the pylon foundations is considered in the design of pylons. The cables do not fail during earthquakes unless the pylons are toppled due to faulting, shaking or slope failure (e.g. Aydan 2012a; Aydan et al. 2018b). During the 1999 Kocaeli earthquake, only one pylon was damaged nearby Ford-Otosan automobile factory at Kavaklı district of Gölcük town. At this site a normal fault, which is a secondary fault to the main lateral strike-slip faulting event, crossed through the foundations of the pylon and its vertical throw was about 240 cm. One of the foundations of the pylon was pulled out of the ground and was exposed as seen in Figure 7.46a. Some of its truss elements were slightly buckled. Nevertheless, the damage to the pylon was quite limited and this damage could not hinder its function. Similar types of damages to pylons straddling the Nojima fault break in the 1995 Kobe earthquake (Figure 7.46b) and the Chelungpu fault during the 1999 Chi-Chi earthquake were observed as shown in Figure 7.46c.

7.2.6 Linear and tubular structures

Tubular structures may be specifically designated as petrol and gas pipelines, water pipes and sewage systems. They can also be classified as line-like structures. These structures may fail either by buckling or separation during a faulting event (e.g. Aydan et al. 2011, 2018a). Five such incidents were observed during the 1999 Kocaeli earthquake (Figure 7.47). One of the incidents involved the separation of a ductile iron pipe as a result of faulting near the collapsed overpass bridge. The second incident took place at the pumping facility of the Seka papermill plant at Sapanca Lake.

Figure 7.47 Damage to pipes in the 1999 Kocaeli and 1999 Düzce earthquakes.

The third incident occurred near Tepetarla village where the railways were buckled. The fourth and fifth incidents took place at Arifiye and nearby Başiskele. The fifth incident was quite important since the fault caused a heavy damage to the main water pipe having a diameter of 2 m. Similar types of failures took place in the sewage pipe networks whenever faulting breaks were observed. The natural gas pipelines crossing the İzmit Gulf between Yalova and Pendik were undamaged. Some brittle asbestos water pipes were also damaged in Kaynaşlı and Fındıklı due the fault rupture of the 1999 Düzce earthquake. Although it is difficult to prevent damage to tubular structures, the use of flexible joints may be effective in such active fault zones when they are embedded.

The 2002 Denali earthquake in Alaska showed the importance of preparedness for the relative displacements of the pipelines. The Trans-Alaska oil pipeline was designed to accommodate a 6 m lateral and 1.5 m vertical relative displacement over a 600 m wide zone of the Denali fault in the 1970s. This design concept proved to be appropriate during the 2000 Denali earthquake. The lateral and vertical relative displacement caused by the 2002 Denali earthquake was 420 and 75 cm, respectively. If the location and seismicity of the active faults are well defined, these types of counter-measures could be good examples for seismic design against permanent displacements caused by faults.

Figure 7.48 (a) Behaviour of a single-story reinforced concrete building at Fındıklı village and (b) collapse of apartment blocks at Kullar village.

7.2.7   Buildings

Many buildings along the earthquake faults of the 1999 Kocaeli, Düzce and Chi-Chi earthquakes behaved differently. During the site investigation of the 1999 Kocaeli earthquake, one could see either totally collapsed, severely damaged or intact buildings just on or next to the traces of the fault breaks (e.g. Aydan 2003b, 2012a; Aydan et al. 1999b, 2011, 2018a,b). The examples are many and it is quite difficult to quote all of them here. Two typical examples are given and briefly discussed. Nevertheless, this topic deserves more thorough investigations. The first example is a single-story reinforced concrete house with a raft foundation in Fındıklı village. The fault passed just beneath the building (Figure 7.48a). The relative displacement of the fault break was about 200 cm with a 100 cm downthrow. The building was tilted but no damage was observed. A very peculiar behaviour of an apartment complex consisting of 8 five-story apartment blocks was observed in Kullar village as shown in Figure 7.48b. Seven apartment blocks failed in a pancake mode while one apartment block remained self-standing. One of the failed apartment blocks just crossed by the fault break, which had a relative horizontal displacement of 240 cm and 20–25 cm vertical throw (north side down). The ground surface was sloping toward the north. One of two apartment blocks on the southern side was damaged while the other one collapsed towards east in a pancake mode in accordance with the movement of its foundation. The five blocks on the northern side completely collapsed in a pancake mode towards the west in accordance with the direction of shaking. Except the apartment block over the fault break, the failure of five blocks on the northern side of the fault break may be considered to be purely due to shaking. Although the intensity of shaking on the southern side of the fault break should be the same, the damaged self-standing apartment block deserves some special consideration. Whatever the reason is, it is of great interest that the most vulnerable buildings may also survive within a distance of 5–6 m to the fault break during in-land earthquakes.

Buildings having a frame structure together with shear walls performed very well during the 1999 Chi-Chi earthquake; they behaved differently despite the vertical

Figure 7.49 Post-earthquake view of Bailu school's reinforced concrete building.

uplift relative displacement being up to 8 m. The Bailu secondary school, a reinforced concrete building, survived the 2008 Wenchuan earthquake even though the surface rupture of the thrust fault passed underneath the building (Figure 7.49). Although the intensity of shaking on the southern side of the fault break should be the same, the damaged self-standing apartment block deserves some special consideration. Whatever the reason is, it is of great interest that the most vulnerable buildings may also survive even within a distance of 5–6 m to the fault break during in-land earthquakes.

7.3   DAMAGE BY GROUND LIQUEFACTION AND LATERAL SPREADING

Liquefaction is generally associated with noncemented granular sedimentary deposits (Figure 7.50). It can take place in any granular soil as long as the pore-pressure due to seismic shaking is sufficient to cause liquefaction. The liquefaction was initially known as quick-sand phenomenon, it was renamed to "soil liquefaction or simply liquefaction" following the 1964 Niigata earthquake (Seed 1979; Seed and Idriss 1971, 1981; Seed and DeAlba 1986; Seed et al. 1985). Liquefied soil appears on the ground surface as sand volcanoes, and Figure 7.50 shows an eruption of soil in a baseball ground during the 1983 Nihonkai-Chubu earthquake. Liquefaction depends entirely on shear resistance and permeability of granular soil. Although sand deposits much more easily liquefy than other types of soils, liquefaction could both theoretically and experimentally take place in gravelly or clayey soils, as also observed in the Hyogo-ken Nanbu earthquake of January 15, 1995 (see Figure 7.50, Kobe - Port Island) (Atak et al. 2004; Aydan 1995a, 2006a, 2007b, 2008b, 2012a; Aydan and Kumsar 1997b; Bardet et al. 1999; Dobry and Baziar 1992; Hamada and Wakamatsu 1998; Hamada et al. 1986; Ishihara 1993; Kanibir et al. 2006; Rathje et al. 2004; Shamoto et al. 1998; Towhata and

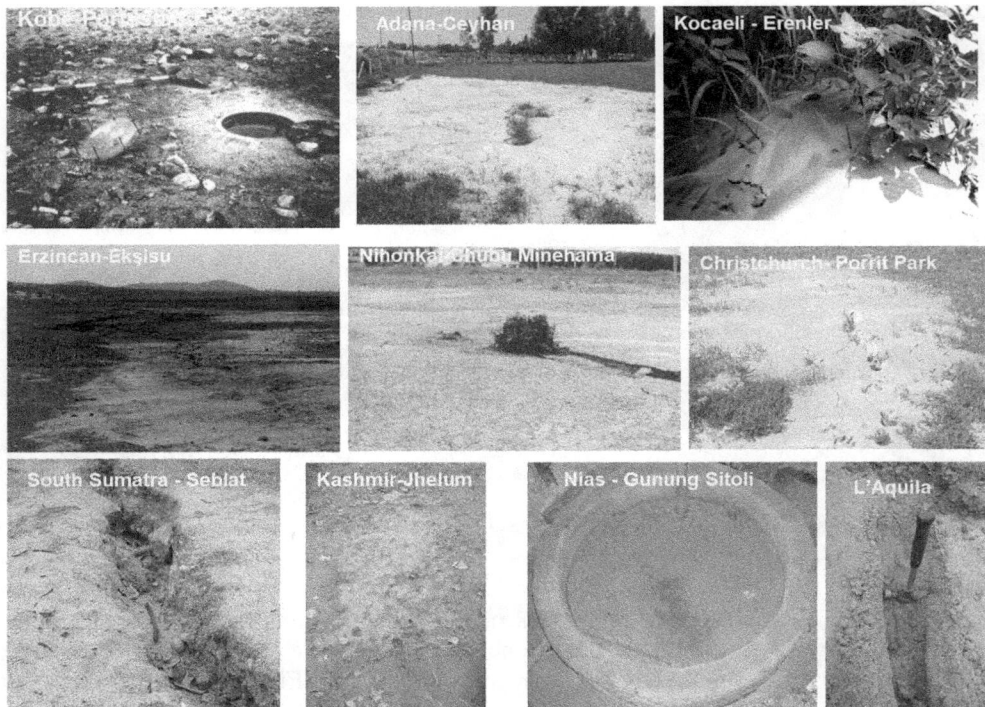

Figure 7.50 Some examples of liquefied ground in various earthquakes worldwide.

Matsumoto 1992; Ulusay and Aydan 2005, 2011; Ulusay et al. 2002, 2003a,b; Waka-matsu 1991, 1993). The existence of a nonpermeable top layer results in the prolongation of the liquefaction state of ground. The liquefaction of the ground is achieved when the effective stress acting on the ground becomes zero and the ground behaves like liquid. As a result, the ground would lose its bearing capacity and the light and heavy structures would behave according to the Archimedes principle. Specifically, the light structures would be uplifted while the heavy structures would sink or settle into the ground. When ground surface is inclined or the boundary between liquefied and nonliquefied ground is inclined, the ground would flow laterally. This phenomenon is known as lateral spreading. When there is nonliquefied soil at the top, some extension-type fractures would occur.

The author has been involved in the reconnaissance of many large earthquakes since the 1992 Erzincan earthquake, as mentioned previously. Among all these earthquakes, the 1992 Erzincan, 1998 Adana-Ceyhan, 1999 Kocaeli and Düzce and 2003 Bingöl earthquakes in Turkey; the 1995 Kobe, 2004 Chuetsu, 2009 Suruga, 2011 Great East Japan earthquakes in Japan; the 2005 Nias, 2007 Bengkulu (South Sumatra), 2009 Padang-Pariaman (West Sumatra) earthquakes in Indonesia; and the 2010 Darfield and 2011 Christchurch earthquakes in New Zealand provided many important case history data on ground liquefaction phenomenon and associated damage to structures and built environment, as shown in Figures 7.51–7.55. One of the most important

Figure 7.51 Views of lateral spreading due to ground liquefaction.

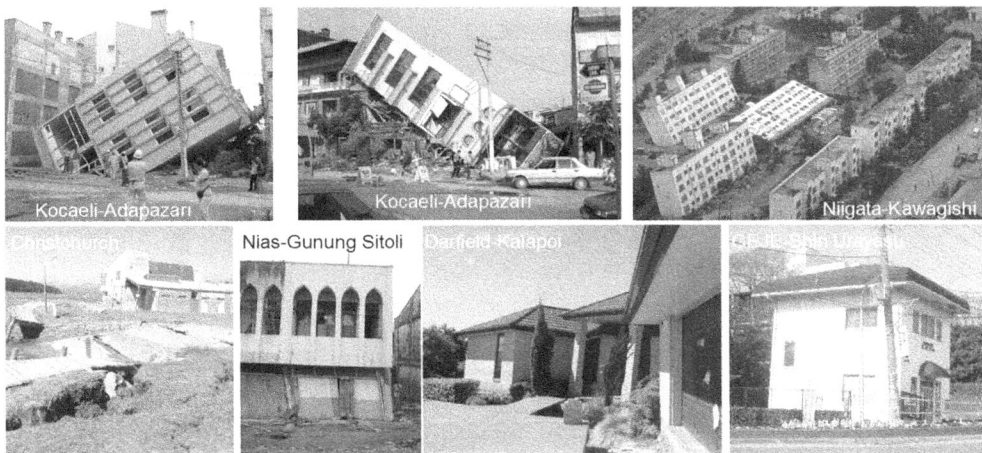

Figure 7.52 Effect of ground liquefaction on buildings (tilting, settlement, rotation).

Figure 7.53 Uplifted structures due to ground liquefaction.

Figure 7.54 Settled or sunken structures due to ground liquefaction.

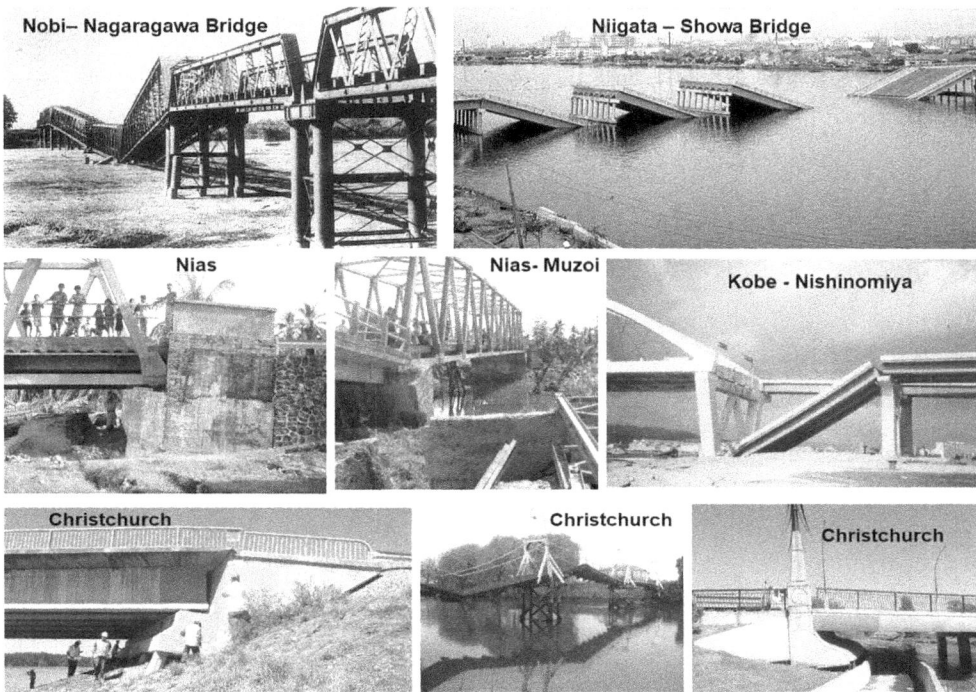

Figure 7.55 Effect of ground liquefaction on bridges.

geotechnical issues were the permanent ground deformations due to widespread ground liquefaction on sites founded at delta fans (Figure 7.51). The lateral spreading of ground nearby bridge abutments were almost entirely associated with liquefaction of sandy soil layers (Figure 7.55). The lateral spreading of ground was particularly amplified on the convex side of riverbanks as the ground can freely move towards the river. Similar situations were also observed on reinforced concrete bridges.

7.4   EFFECT OF ROCKFALLS ON BUILT ENVIRONMENT

Although rockfall is a special form of slope failure, the effects of rockfalls may be tremendous when the size of the falling rock boulder is >3 m in diameter and/or has a great impact velocity. Recent earthquakes showed that there may be disastrous effects on built environments as seen in Figure 7.56. It is observed that rockfalls cause rupture or destruction of structures such as pylons, buildings, roadways, concrete dams, power houses and penstocks. For example, rockfalls on roadways and expressways in mountainous regions become a serious issue as observed in the 2008 Wenchuan earthquake, the 2009 Pariaman-Padang (West Sumatra) earthquake, the 2015 Nepal-Gorkha earthquake and the 2016 Kumamoto earthquakes. The T-type pier of the Toshita Bridge was broken and fallen to the bottom of the valley as seen Figure 7.56. The fallen rock boulders at the portals of the tunnels may cause some severe accidents. For example,

Figure 7.56 Illustration of damage to built environments by rockfalls.

the train with oil tankers hit the fallen boulders just outside of a tunnel, causing a huge fire in the Wenchuan earthquake. Similarly, rockfalls can be quite disastrous for concrete dams as well as for their spillways, power houses and penstocks. Furthermore, rock boulders may destroy or inflict heavy damage to buildings as seen in the 2008 Wenchuan, 2009 L'Aquila and 2011 Christchurch earthquake. For example, a huge boulder passed through a timber house at Raika district in the 2011 Christchurch earthquake.

Chapter 8

Seismic design of structures

8.1 FUNDAMENTAL APPROACHES

The seismic design and foundation of any structure must be based on the solution of the equation of motion for a given dynamic loading condition together with appropriate mechanical models of materials constituting the structure and design specifications similar to the approaches described in Chapter 6. Figure 8.1 illustrates some common engineering structures, which may be classified as surface, semi-underground or underground structures. The constitutive laws in the analyses should be generally elasto-visco-plastic together with yield conditions appropriate for the materials constituting the structure (Figure 8.2). When structures are situated on/in rock mass, the characteristics of rock masses must be considered, which may require discontinuum-type modelling and considerations of geologic discontinuities (Figure 8.3). While simple cases may be analysed using analytical solutions, the utilization of numerical techniques such as finite difference, finite element and boundary element methods would be essential for general situations.

Although any structure is a part of the earthquake-affected area, it is quite common to consider a certain region in the close vicinity of the structure, which may be classified as ground (soil) – structure interaction analyses (Figure 8.4). As it is quite difficult to model structures with their tiny details, the fundamental approach is to separate the domain into two parts, namely ground and structures. The unbounded ground is modelled as dynamic subsystem and the common interface between the ground and structure is modelled through the described stiffness and damping characteristics to obtain interface force and displacement representing the interaction between ground and the structure for a given seismic motion. The boundary of ground is represented by the silent (absorbing) boundary, which is also known as Lysmer boundary, in order to represent transient seismic motion. This boundary is represented by viscous elements such that imposed seismic motion does not remain within the domain of ground analysed. The structure itself is represented with dashpot elements at the interface. The dynamic response of the structure is analysed by subjecting the system to an interface displacement field. However, there is also a recent trend to represent the structure and ground together with silent boundaries and/or utilization of infinite elements in view of advances in computational techniques and hardware.

The simplest approach would be the one described in Section 6.2, by isolating the structure from its foundation and assuming a linear (elastic or visco-elastic) constitutive law. The foundation would be treated as seismic loading boundary, which would

DOI: 10.1201/9781003164371-8

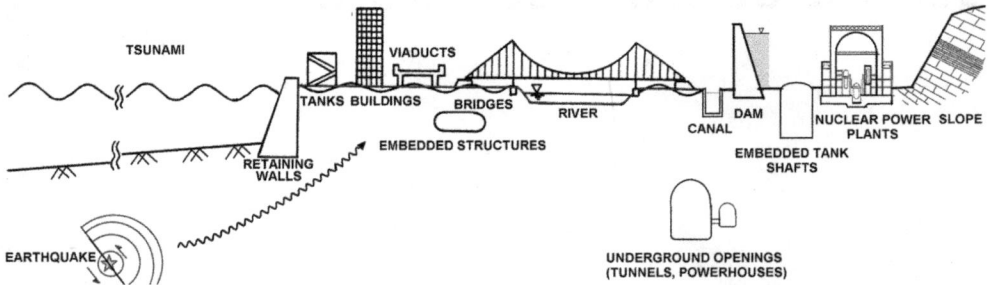

Figure 8.1 Illustration of structures.

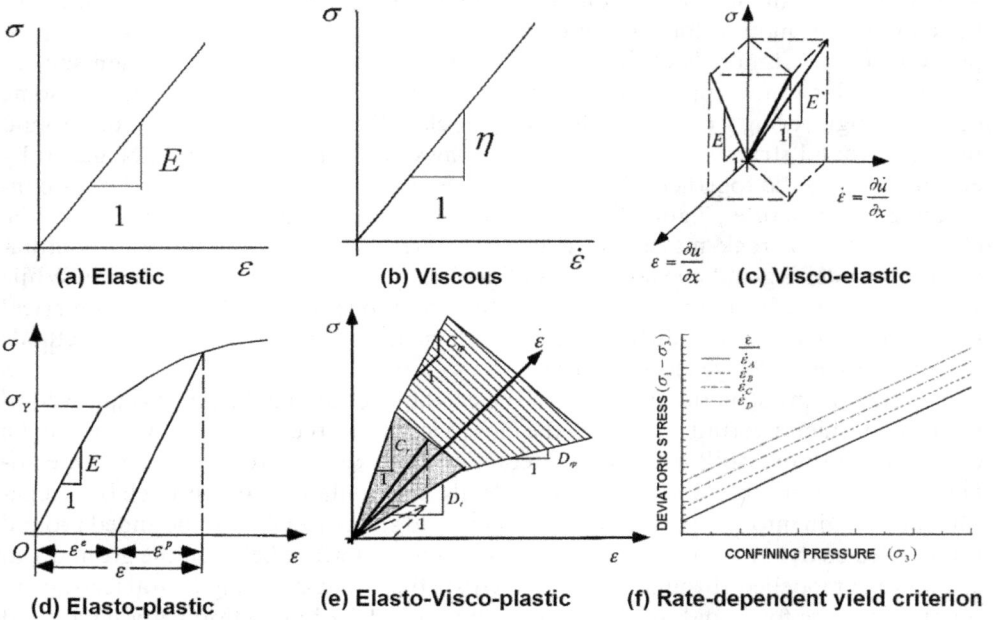

(a) Elastic

(b) Viscous

(c) Visco-elastic

(d) Elasto-plastic

(e) Elasto-Visco-plastic

(f) Rate-dependent yield criterion

Figure 8.2 Illustrations of some constitutive models (Aydan, 2021).

be generally represented as a prescribed displacement boundary. The constitutive law governing the behaviour of materials may generally be nonlinear depending upon the stress and strain induced within the structure under the given seismic loading conditions. Therefore, some design procedures may allow a certain amount of nonlinear behaviour, although the general design is likely to assume a linear behaviour.

As the loads acting on structures are transient, the use of a hyperbolic-type solution is necessary with the consideration of inertia force as well as damping characteristics. Nevertheless, the solution procedure may be quite simplified and the solution may be pseudo-dynamic, in which the transient load is applied like a static load in

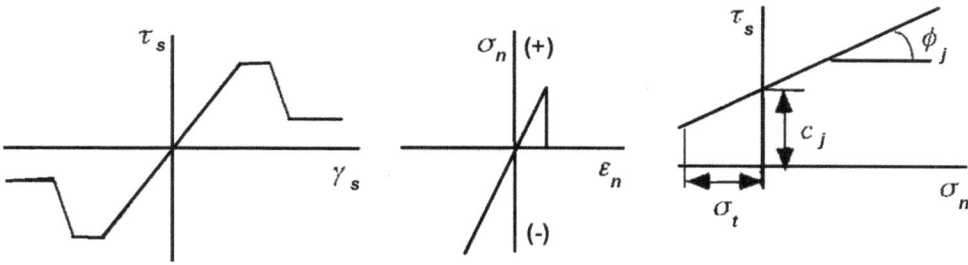

Figure 8.3 Elasto-plastic constitutive models for geological discontinuities (Kawamoto and Aydan, 1999).

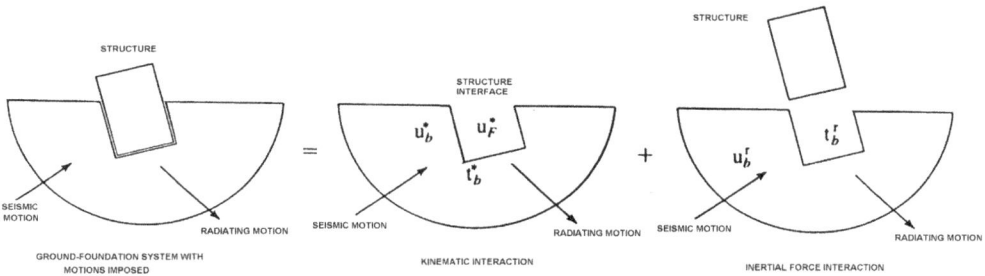

Figure 8.4 Illustration of ground–structure interaction analysis.

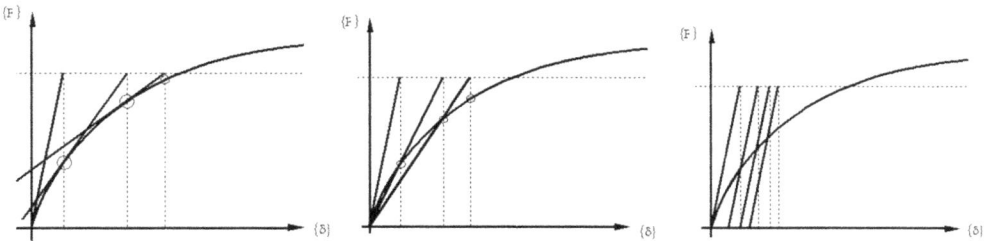

Figure 8.5 Methods for dealing with nonlinear behaviour. (a) Tangential stiffness, (b) secant stiffness and (c) initial stiffness.

incremental manner and inertia force is omitted. One of such analyses is called pushover analyses while allowing the nonlinear behaviour and yielding of materials and/or components of the structure. In such analyses, there are different techniques to deal with nonlinear responses (Owen and Hinton 1980). These techniques are tangential stiffness, secant stiffness and initial stiffness methods as illustrated in Figure 8.5. If the overall response shows a hardening response, tangential stiffness may be used for quick solutions. However, if the system starts to show softening behaviour, the use of

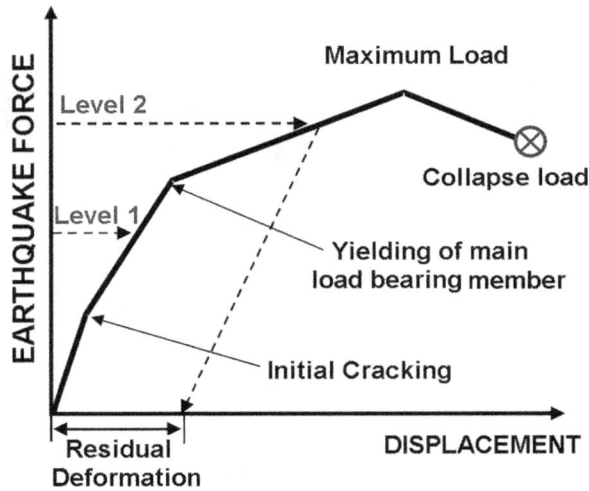

Figure 8.6 Two-level seismic design concept of structures (JSCE-EEC 2000).

secant stiffness and initial stiffness methods is necessary. Particularly, the initial stiffness method would yield absolute convergence.

The recent tendency in design has been moving from an elastic limit to allow a certain amount of nonlinear behaviour. The Japan Society of Civil Engineers (JSCE-EEC 2000) revised the design codes and adopted two design levels of load that structures can accommodate. The principle is illustrated in Figure 8.6. Although the seismic design of a viaduct was chosen as an illustrative case, the same concept is adopted for foundations and other civil engineering structures. Up to Level 1, the structure is presumed to behave almost elastically with little residual deformation. The design capacity up to Level 2 presumes that some yielding of the main members of the structural system would take place and some permanent deformation would occur. Nevertheless, the structure would continue to resist seismic loads without total collapse.

The basic philosophy of seismic design requires that there is no total collapse of the structure. As there are different types of civil engineering structures, the adoption of this philosophical concept in their seismic design may require certain considerations in relation to the characteristics of the structures. Particularly slopes; embankments; embedded underground structures, such as conduits, immersed tunnels, shafts, semi-underground tanks; and underground structures such as tunnels, underground power houses and storage caverns would require the definition of the principal structural member with quantified load-bearing capacity and allowable displacement. In some structures, this may be a critical issue. For example, the main structural member in tunnels would not be the concrete liners or primary support systems consisting of rock bolts, shotcrete and steel sets unless the surrounding rock yields and collapses, falling into the opening. During the 2004 Chuetsu earthquake, the surrounding rock, which is the main load-bearing member, was stable, while the concrete liners of the Wanazu roadway tunnel and Uruma railway tunnels collapsed, and during the 2008 Wenchuan earthquake, the Longqi (Longxi) and Jiujiaya roadway tunnels collapsed

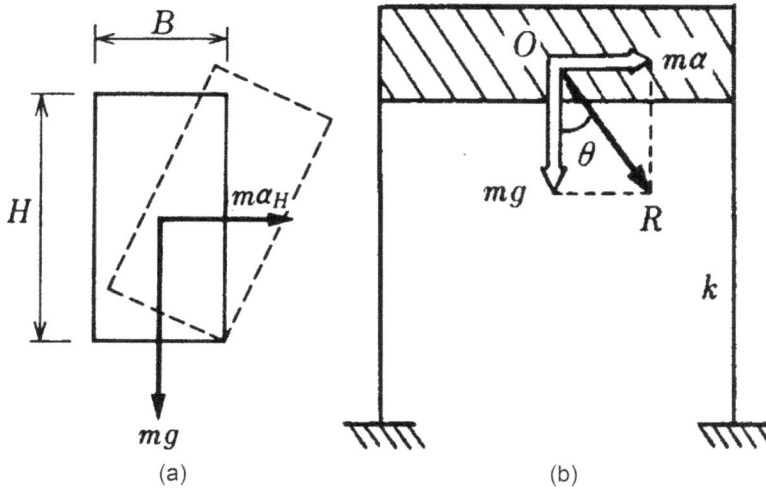

Figure 8.7 Illustration of seismic coefficient concept. (a) Block model and (b) structure model. (Modified from Okamoto (1973).)

(Aydan et al. 2009d,e, 2010a,b). Similar situations may be observed in large caverns for storage of oil, gas and accommodating power generation units and in rock slopes next to roadways, railways, dams and nuclear power plants. Furthermore, the yielding of discontinuities of rock masses on/in which structures are constructed may not be critical. In other words, certain considerations on the definition of yielding in rock mass are necessary and the concepts utilized for concrete, reinforced concrete or steel may not be relevant at all.

One of the simplest techniques is to utilize the seismic coefficient method. The seismic load acting on the structure is assumed to be proportional to the mass of the structure as illustrated in Figure 8.7. As discussed in Chapter 5, the toppled or slid simple structures such as stone blocks, columns and walls could be used to infer the possible ground motions in an earthquake-affected area. As strong motion devices do not always exist in very close vicinity of epicentres, or countries may be poorly instrumented for economic and other reasons, they are still used to infer strong motions from those simple toppled or displaced structures. Figure 8.7a illustrates a possible lateral seismic coefficient value for the toppling of a block, while Figure 8.7b represents how the lateral seismic force is considered for structural seismic analyses. This concept basically replaces the transient seismic force by an equivalent static force proportional to its mass. It is always discussed how to assign the value of seismic coefficient. The simple approach would be to assign the value equivalent to that starting its toppling or sliding over the base. Toppling would require just the height and width of the block while a sliding condition requires the friction coefficient of the block. When the ratio of width over height is <0.5, it is very likely that the toppling mode would be prevailing. The common values used for lateral seismic coefficient are given in Table 8.1. One of the major issues of this approach is that the lateral seismic force act statically and continuously while the dynamic forces are transient. However, there is still a strong

Table 8.1 Common values used for lateral seismic coefficient

Structure	Seismic coefficient value
Ordinary buildings/bridges	0.2
Oil and petrochemical facilities	0.3–0.6
Dams	0.15
Port and harbour facilities	0.15–0.2
Nuclear facilities	0.2–0.6

Table 8.2 Values of C_Z

Structure	Coefficient value
Region 1	1.0
Region 2	0.9
Region 3	0.8
Region 4 (Okinawa)	0.7

Table 8.3 Values of C_G

Structure	$T_G(s)$	Coefficient value
Ground I(stiff)-GI	$T_G < 0.2$	0.8
Ground II(soft)-GII	$0.2 \leq T_G < 0.6$	1.0
Ground III(very soft)-GIII	$0.6 < T_G$	1.2

T_G: Dominant period of ground.

interest among researchers on how to assign the lateral seismic coefficient as it is the simplest approach.

As shown in Chapter 6, the seismic force is transient and its value changes with respect to natural period (frequency) and damping characteristics of structures. It is now quite common that the damping coefficient is 5% on the basis of measurements. By considering the importance of the structure, ground conditions and regional seismicity, the seismic coefficient is given by the following formula (e.g. Usami 1980):

$$k_h = C_Z C_G C_I C_T k_{h0} \tag{8.1}$$

where

k_{h0} is the horizontal seismic coefficient. Although it is generally assumed to be 0.2, its value may be different as given in Table 8.1

C_Z is the regional modification coefficient

C_G is the ground condition coefficient

C_I is the structural importance coefficient

C_T is the spectral modification coefficient

For example, the values of coefficients of the formula above are assigned as given in Tables 8.2–8.5 for Japan (e.g. Okamoto 1973, Usami 1980).

Table 8.4 Values of C_I

Structure	Coefficient value	
Class 1	1.0	Bridges for expressway, national routes and main transportation routes
Class 2	0.8	Roadway bridges of towns and provinces

Table 8.5 Values of C_T

Structure	Natural period of structure		
Ground I	$T < 0.1$ $C_T = 2.69T^{1/3}$, provided $C_T \geq 1.0$	$0.1 \leq T < 1.1$ $C_T = 1.25$	$1.1 < T$ $C_T = 1.33T^{-2/3}$
Ground II	$T < 0.2$ $C_T = 2.15T^{1/3}$, provided $C_T \geq 1.0$	$0.2 \leq T < 1.3$ $C_T = 1.25$	$1.3 < T$ $C_T = 1.49T^{-2/3}$
Ground III	$T < 0.34$ $C_T = 1.8T^{1/3}$, provided $C_T \geq 1.0$	$0.34 \leq T < 1.5$ $C_T = 1.25$	$1.5 < T$ $C_T = 1.64T^{-2/3}$

Table 8.6 Values of parameters in Eq. (8.2)

Structure	T_G (s)	$a_{max}(g)$	C_A	T_A	T_B	T_C
GI	$T_G < 0.2$	1.2	2.5	0.2	1.0	4.0
GII	$0.2 \leq T_G < 0.6$	1.0	2.5	0.5	1.5	4.0
GIII	$0.6 < T_G$	0.8	2.5	0.8	2.0	4.0

Following the 1999 Kobe earthquake, the seismic design code of Japan has been modified. After each modification, the amplitude of spectral acceleration has been changed and the most recent modifications are given in Table 8.6 and spectral acceleration values are obtained from the following equations (Kawashima 2018). As noted from Table 8.6, the maximum ground acceleration (a_{max}) for computing the spectral acceleration value has become twice that of the 1990 seismic design code of Japan.

$$S_A(T) = \begin{cases} \left\{ 1 + (C_A - 1)\dfrac{T}{T_A} \right\} a_{max} & 0 \leq T < T_A \\ C_A a_{max} & T_A \leq T < T_B \\ C_A a_{max} \dfrac{T_B}{T} & T_B \leq T < T_C \\ C_A a_{max} \dfrac{T_B T_C}{T^2} & T >\leq T_C \end{cases} \qquad (8.2)$$

The 2011 Great East Japan earthquake was an interplate earthquake with a moment magnitude of 9.0 (KiK-Net, 2011). Figure 8.8 compares the spectral acceleration of several strong motion stations in the earthquake-affected area for three different ground

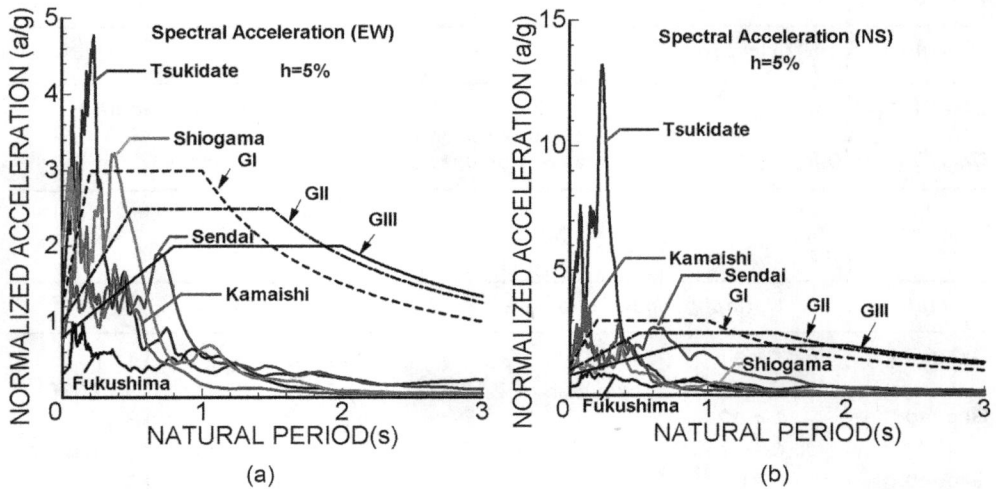

Figure 8.8 Comparison of spectral acceleration with measurements for the 2011 Great East Japan earthquake. (EW direction, NS direction).

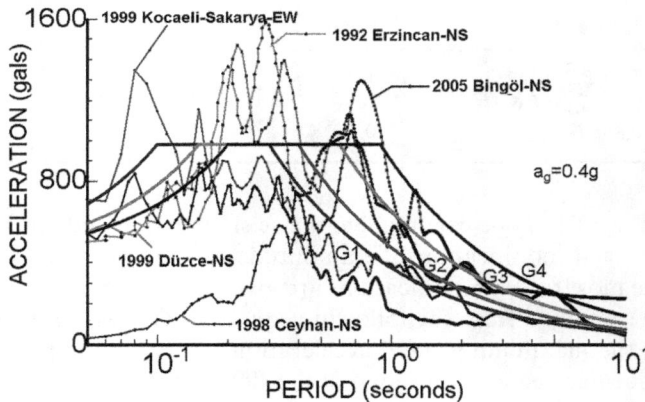

Figure 8.9 Comparison of spectral acceleration designations with measurements for major earthquakes since the 1992 Erzincan earthquake.

conditions. As noted from the figure, the spectral acceleration values exceed several times the peak values. Particularly, the spectral acceleration of the Tsukidate station, which is located on GI ground, is 4.33 times the seismic design values.

Figure 8.9 compares the spectral acceleration values of major Turkish earthquakes with seismic design spectral accelerations for four different ground conditions (G1, G2, G3 and G4) since the 1992 Erzincan earthquake. The maximum acceleration in the 2007 Turkish Seismic Code (AFAD 2007) is 0.4 g. The maximum spectral acceleration is 2.5 times the base acceleration. As also noted from the figure, the measured spectral accelerations exceed the seismic design spectral acceleration values.

In the following sections, the common design procedures for various structures are presented separately as each structure has their own characteristics.

8.2 SEISMIC DESIGN OF BUILDINGS

8.2.1 Framed structures (timber, steel and reinforced concrete structures)

Timber, steel and reinforced concrete buildings are generally modelled as frame structures or moment-resisting structures. The column and beam are modelled by beam elements together with axial stiffness. Fundamentally, they are one-dimensional elements in their local coordinate system while they become three-dimensional structures in the global coordinate system. The utilization of the finite element method would be the most appropriate procedure for solving the equation system given in Chapter 6. Although the details would not be given herein, the stiffness matrix for a two-dimensional frame structure in local coordinate system can be derived as given below (see Figure 8.10) as (e.g. Reddy 1993):

$$[K]_{ab}^{l} = \begin{bmatrix} K_{11}^{a} & 0 & 0 & K_{12}^{a} & 0 & 0 \\ 0 & K_{11}^{b} & K_{12}^{b} & 0 & K_{13}^{b} & K_{14}^{b} \\ 0 & K_{21}^{b} & K_{22}^{b} & 0 & K_{23}^{b} & K_{24}^{b} \\ K_{21}^{a} & 0 & 0 & K_{22}^{a} & 0 & 0 \\ 0 & K_{31}^{b} & K_{32}^{b} & 0 & K_{33}^{b} & K_{34}^{b} \\ 0 & K_{41}^{b} & K_{42}^{b} & 0 & K_{34}^{b} & K_{44}^{b} \end{bmatrix} \tag{8.3}$$

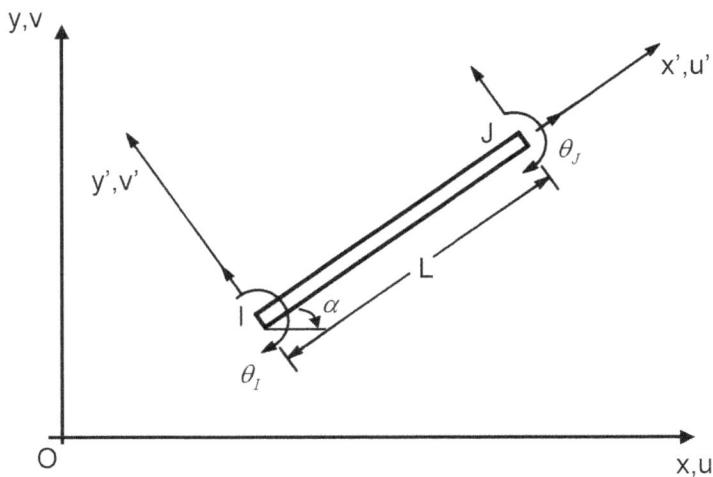

Figure 8.10 Illustration of a beam element in two-dimensional space.

where

$$[K']_b = \frac{EI}{L^3} \begin{bmatrix} 12 & 6L & -12 & 6L \\ 6L & 4L^2 & -6L & 2L^2 \\ -12 & -6L & 12 & -6L \\ 6L & 2L^2 & -6L & 4L^2 \end{bmatrix} \quad \text{or} \quad [K']_b = \begin{bmatrix} K_{11}^b & K_{12}^b & K_{13}^b & K_{14}^b \\ & K_{22}^b & K_{23}^b & K_{24}^b \\ & & K_{33}^b & K_{34}^b \\ \text{sym} & & & K_{44}^b \end{bmatrix}$$

$$[K']_a = \frac{EA}{L} \begin{bmatrix} 1 & -1 \\ -1 & 1 \end{bmatrix} \quad \text{or} \quad [K']_a = \begin{bmatrix} K_{11}^a & K_{12}^a \\ \text{sym} & K_{22}^a \end{bmatrix}$$

E, A, I and L are elastic modulus, area and inertia modulus of the member and element length.

The stiffness matrix given in the local coordinate system is transformed to that in the global coordinate system through the following transformation law:

$$[K]_{ab}^g = [T^*]^T [K]_{ab}^l [T^*] \tag{8.4}$$

where

$$[T^*] = \begin{bmatrix} c & s & 0 & 0 & 0 & 0 \\ -s & c & 0 & 0 & 0 & 0 \\ 0 & 0 & 1 & 0 & 0 & 0 \\ 0 & 0 & 0 & c & s & 0 \\ 0 & 0 & 0 & -s & c & 0 \\ 0 & 0 & 0 & 0 & 0 & 1 \end{bmatrix}; \quad \alpha = \tan^{-1}\left(\frac{y_J - y_I}{x_J - x_I}\right); \quad c = \cos\alpha; \quad s = \sin\alpha$$

Ohta (2011) reported some analyses on the response of framed building structures subjected to faulting movements at the foundation level. Figure 8.11 shows simple examples of representation of a finite element frame modelling of a reinforced concrete structure subjected to thrust and normal faulting conditions. Table 8.7 gives the material properties utilized in the analyses. The net fault movement was 238 cm for thrust and normal faulting models.

The utilization of a beam element concept corresponds to the replacement of a two-dimensional or three-dimensional structure by an equivalent one-dimensional member. However, the utilization of solid elements, including reinforcement elements, would yield almost the same results. The results of analyses shown in Figure 6.29 correspond to this type analyses.

Rodgers et al. (2004) reported a case study on the seismic response of a 13-story steel-moment-framed building in Alhambra, California, USA. They instrumented the

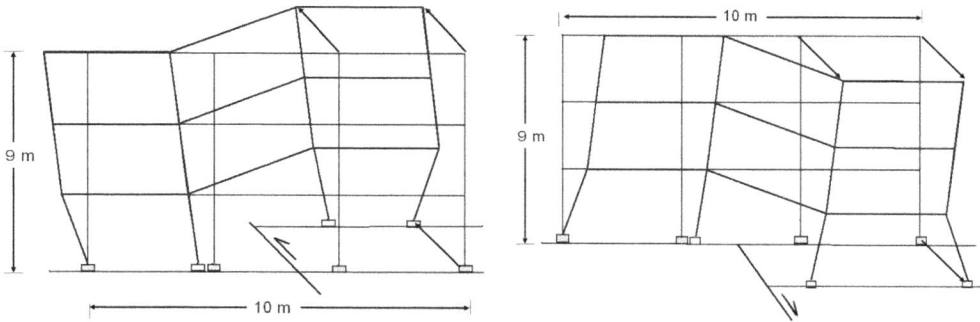

Figure 8.11 Deformation response of a moment-resisting building subjected to (a) thrust and (b) normal faulting action. (Arranged from Ohta (2011).)

Table 8.7 Material and geometrical properties of the building

Material	Elastic modulus (GPa)	Thickness (mm)	Width (mm)
Column/beam	21	300	300

283 cm relative normal or thrust faulting.

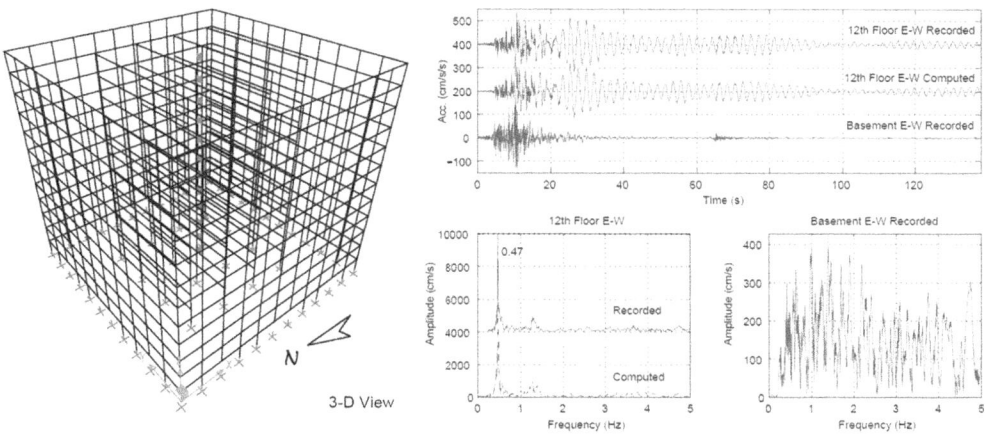

Figure 8.12 (a) 3D FEM model and (b) measured and computed responses. (Arranged from Rodgers et al. (2004).)

building and performed response and three-dimensional (3D) finite element modal analyses. The actual acceleration waveforms recorded at different major earthquakes were utilized to compute seismic responses and were compared with actual monitoring results. Figure 8.12a shows the 3D finite element mesh and Figure 8.12b compares the measured and computed responses for the record of the 1994 Northridge earthquake, which slightly damaged the building instrumented since 1971. They concluded that the 3D analyses could explain the responses monitored during actual earthquakes as well as the state of the building after each earthquake.

8.2.2   Masonry buildings

As mentioned previously, the main components of buildings may be stones or bricks together with/without mortar. Therefore, the actual representation should consider this blocky structure together with contact conditions. As failure involves sliding, rotational or combined sliding and rotational modes, simple methods based on frictional resistance or toppling mode may yield certain estimations of critical acceleration at the time of failure (Aydan 2002, 2017; see also Section 5.3.1). Modelling these types of structures using continuum-type finite element analyses may not be appropriate to estimate the seismic capacity of masonry buildings although some approximate evaluation of the stress state within the structure may be possible.

Discrete finite element method (DFEM) is capable of modelling of blocks and contacts within the structure. Several applications of the DFEM to fully dynamic and pseudo-dynamic analyses of masonry arch structures are described in this section (Aydan 1998; Aydan et al. 1996d,e, 2011; Tokashiki et al. 1997). The acceleration waves shown in Figure 8.13 and the material properties for blocks and interfaces given in Tables 8.8 and 8.9 are used in the analyses presented in this section.

8.2.2.1   Masonry tower or wall (out-of-plane)

Figure 8.14 shows initial and deformed configurations of a dry masonry tower or wall (out-of-plane) at time steps 22 and 50 (4.4 and 10 seconds) and responses of a nodal

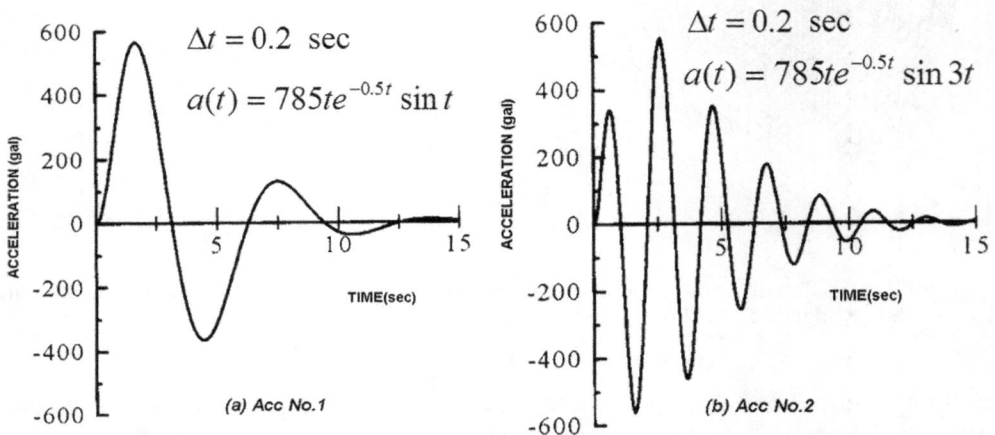

Figure 8.13 Imposed horizontal acceleration waves on foundations. (From Aydan (1998a) and Mamaghani et al. (1994, 1999).)

Table 8.8 Material properties of blocks

Unit weight (kN/m³)	λ (MPa)	μ (MPa)	λ^* (MPa·s)	μ^* (MPa·s)
25	30	30	30	30

Table 8.9 Material properties of contacts

λ (MPa)	μ (MPa)	λ^* (MPa·s)	μ^* (MPa·s)	h (mm)	c (kPa)	σ_t (kPa)	$\tan\phi$
5	2.5	5	2.5	5	0	0	0.7

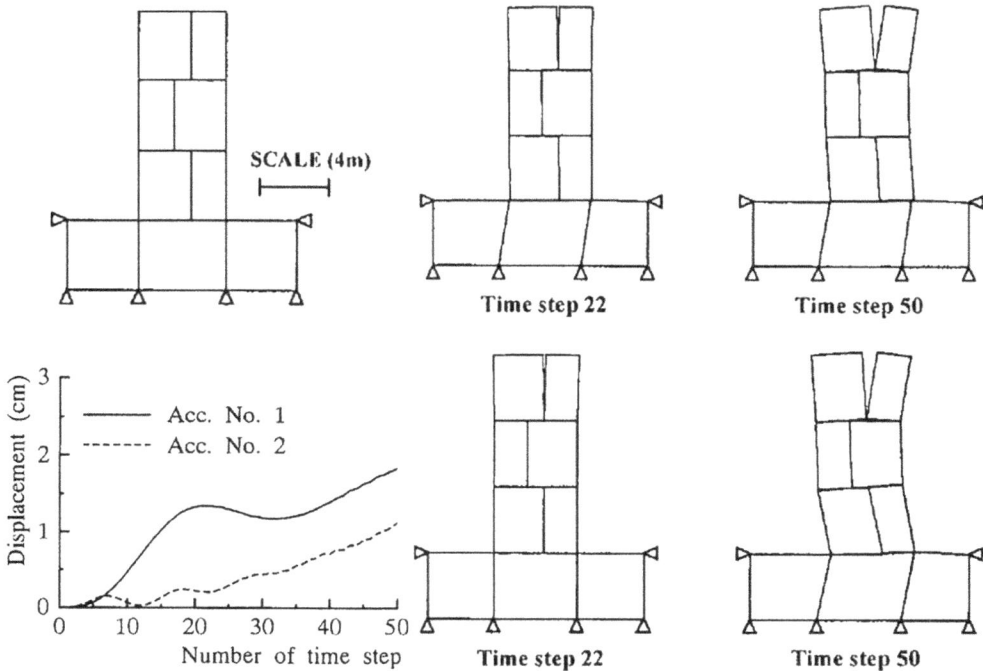

Figure 8.14 Initial and deformed configurations displacement response of a masonary tower. (From Aydan (1998a) and Mamaghani et al. (1994, 1999).)

point at the top of the structure to time. As seen from the figure, there is a relative sliding at the base of the wall and separation and rotation of blocks within the wall occur. Furthermore, the wall does not return to its original position at the end of shaking.

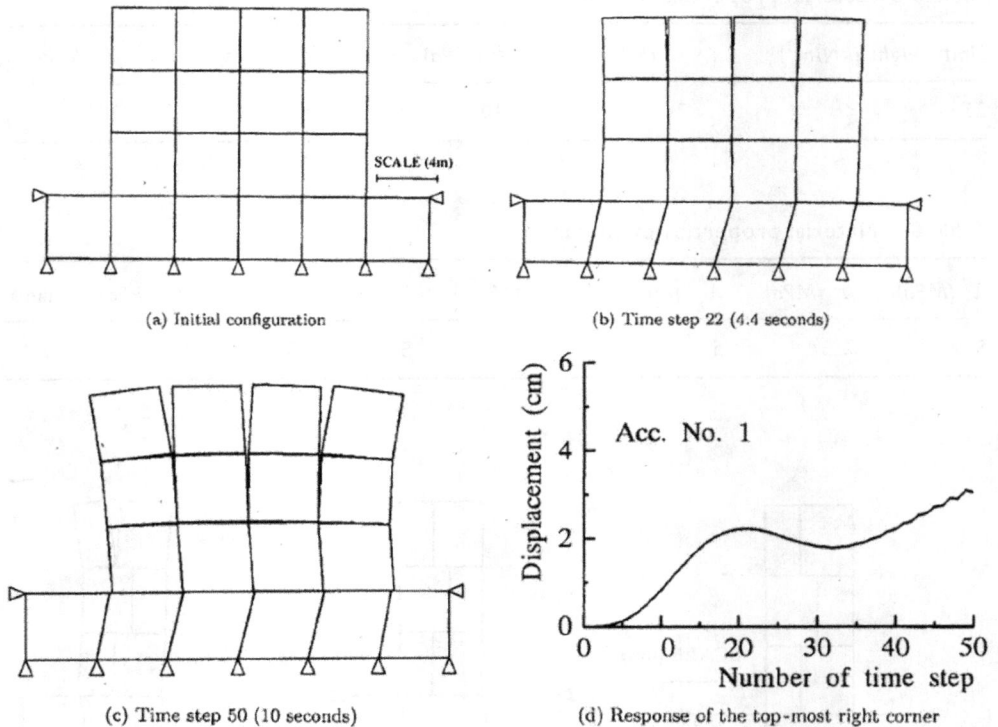

(a) Initial configuration

(b) Time step 22 (4.4 seconds)

SCALE (4m)

(c) Time step 50 (10 seconds)

(d) Response of the top-most right corner

Acc. No. 1

Figure 8.15 Initial and deformed configurations displacement response of a masonary wall. (From Aydan (1998a) and Mamaghani et al. (1994, 1999).)

8.2.2.2   Wall (in-plane)

In the analysis, the foundation of the structures was subjected to the horizontal acceleration waveform in plane as shown in Figure 8.15, which shows the initial and deformed configurations at time steps 22 and 50. The displacement response of the top of a masonry wall is also shown in the same figure. As noted from the figure, there is a relative sliding at the base of the wall and separation and rotation of blocks occur within the wall. Particularly the block separation and movement are quite amplified at the top of the wall. Furthermore, the wall does not return to its original position at the end of shaking, and some permanent relative displacements occur between the wall base and its foundation.

8.2.3   Seismic design of bridges and viaducts

The bridge types are various and they may range from a simple beam bridge, which consists of a beam made of wood, stone, reinforced concrete or steel, to complex suspension or cable-stayed bridge with single to multiple spans. The superstructure of bridges may be modelled as beams together with trusses, cables and/or arches

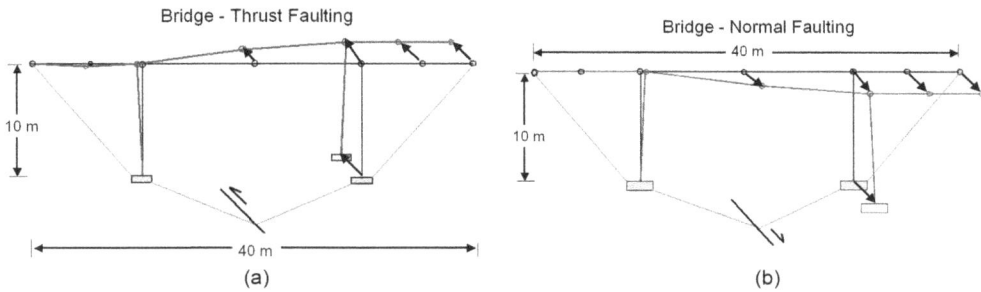

Figure 8.16 Deformed configuration of beam-type bridges under thrust and normal faulting movements. (a) thrust faulting (b) normal faulting.

Table 8.10 Material and geometrical properties of bridge

Material	Elastic modulus (GPa)	Area (m^2)	Inertia (m^4)
Girder	21	4(4 × 1)	1.333 (4 × 1^3/12)
Piers	21	4(2 × 2)	0.333 (2 × 2^3/12)

283 cm relative normal or thrust faulting.

while the sub-structure would be abutments founded on ground and piers/viaducts
fixed to ground with/without by piles. The superstructure of bridges having different
load-bearing components may be modelled as a truss structure, a frame structure or a
mixed truss and beam structure. Shell elements were also used for decks in 3D analyses. Regarding cable-stayed bridges and suspension bridges, pylons (towers) may also
be modelled by solid elements. On the other hand, the sub-structure may be modelled
as supports or piers.

The responses of beam-type bridges subjected to thrust and faulting movements
are shown in Figure 8.16 together with material properties given in Table 8.10. These
analyses were performed with the purpose of illustrating how bridges deform during
permanent ground movements due to faulting and where the highest member forces
develop.

Nakano and Ohta (2008) reported a dynamic finite element analyses of a bridge
subjected to transient displacement response on one side (specifically at support A and
P1) as shown in Figure 8.17. The transient displacement records shown in Figure 8.18
were obtained from acceleration records on the 1999 Chi-Chi earthquake from the
TCU068 strong motion station of Taiwan using the EPS method of Aydan and Ohta
(2010). Figure 8.19 shows the computed displacement responses at points B and C,
while Figure 8.20 shows the computed acceleration at the centre of the bridge (Point B).
Figure 8.21 shows the deformed configuration of the bridge along its longitudinal axis
when the motions ceased.

The mechanics of suspension bridges was first treated by Melan (1906) and later
by Steinman (1922) utilizing the bending theory of beams with the consideration of

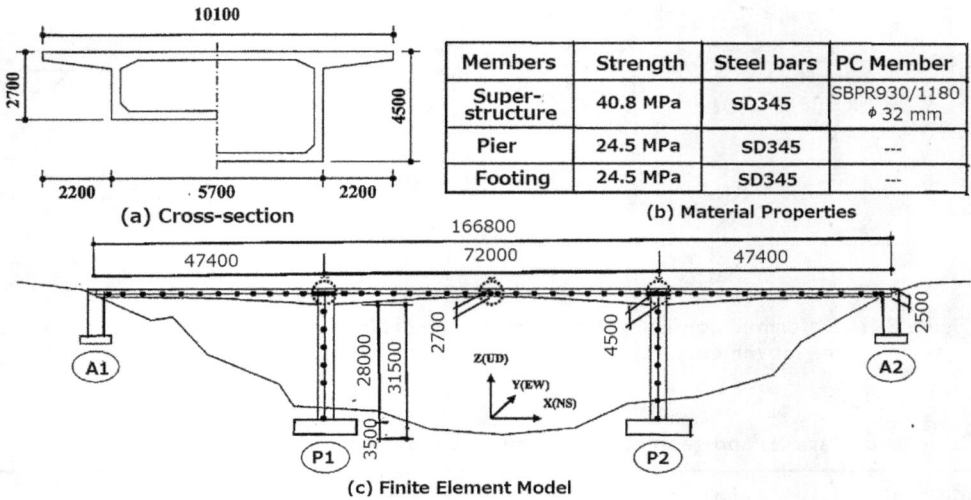

Figure 8.17 Details of geometry and FEM model of a bridge and material properties.

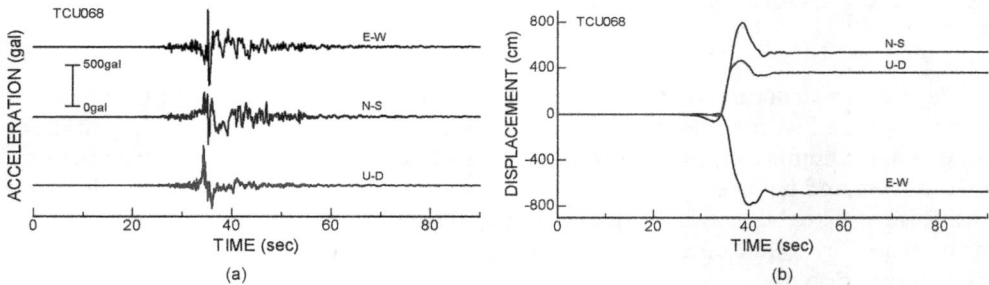

Figure 8.18 Utilized acceleration record at TCU68 and integrated displacement record by the EPS method. (From Ohta (2011).) (a) acceleration records, (b) integrated displacement.

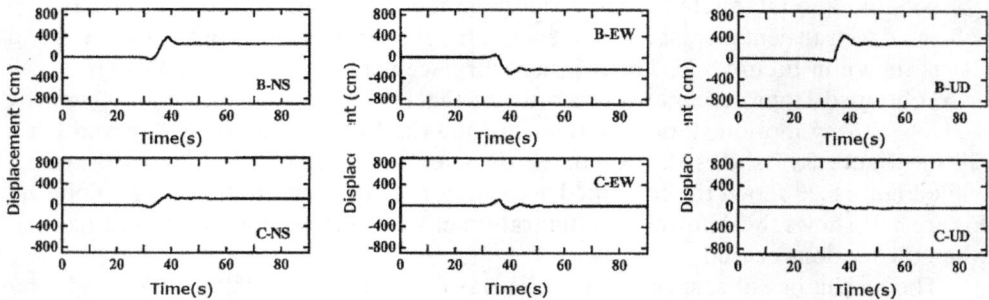

Figure 8.19 Computed displacement responses at points B and C.

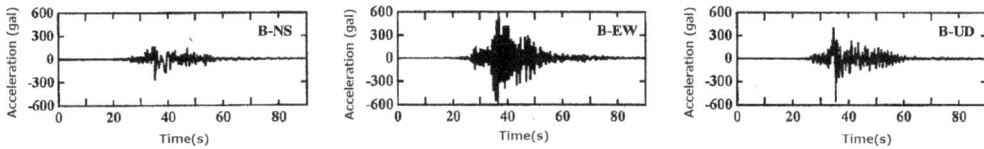

Figure 8.20 Computed acceleration at the centre of the bridge.

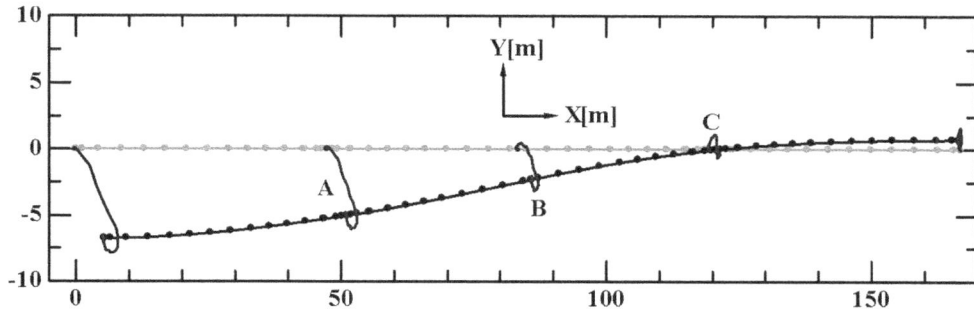

Figure 8.21 Final displacement configuration along the longitudinal axis of the bridge.

cables. Some analytical expressions for natural frequencies of suspension bridges for certain assumed conditions were developed. However, the numerical methods provide much easier evaluations of natural frequencies of suspension bridges. Konishi and Yamada (1960) studied the vibrations of suspension bridges by using a finite difference technique. As the finite element method is a much more flexible approach to consider different boundary, material property and loading conditions, it has now become the conventional design tool that is part of numerical techniques.

With these facts in mind, various studies on suspension bridges along the transportation project between Honshu and Shikoku Islands of Japan were initiated in 1955. As a specific example, the modelling of the Minami-Bisan-Seto Bridge is considered briefly herein. The bridge is a double-deck suspension bridge and it was planned in 1955 and it was constructed during 1978–1988. The centre span is 1,100 m, and the total length is 1,723 m. At the time of design, computer technology was not as advanced as today. Figure 8.22 shows the computational model for the suspension bridge. The superstructure was modelled as a multi-degree of freedom (MDOF) structure as illustrated in Figure 8.22a. Longitudinally the bridge was discretized into 59 elements while the pylons (towers) were discretized into 17 elements. Foundations (pier and anchorage) are modelled as rigid blocks supported by elastic springs, as shown in Figure 8.22c.

Piers of expressways can be modelled using solid elements in the finite element method together with the consideration of reinforcement. Furthermore, the longitudinal and traverse responses of piers would be different. Piers may have either circular or rectangular cross section. While reinforced concrete piers would be entirely solid, steel

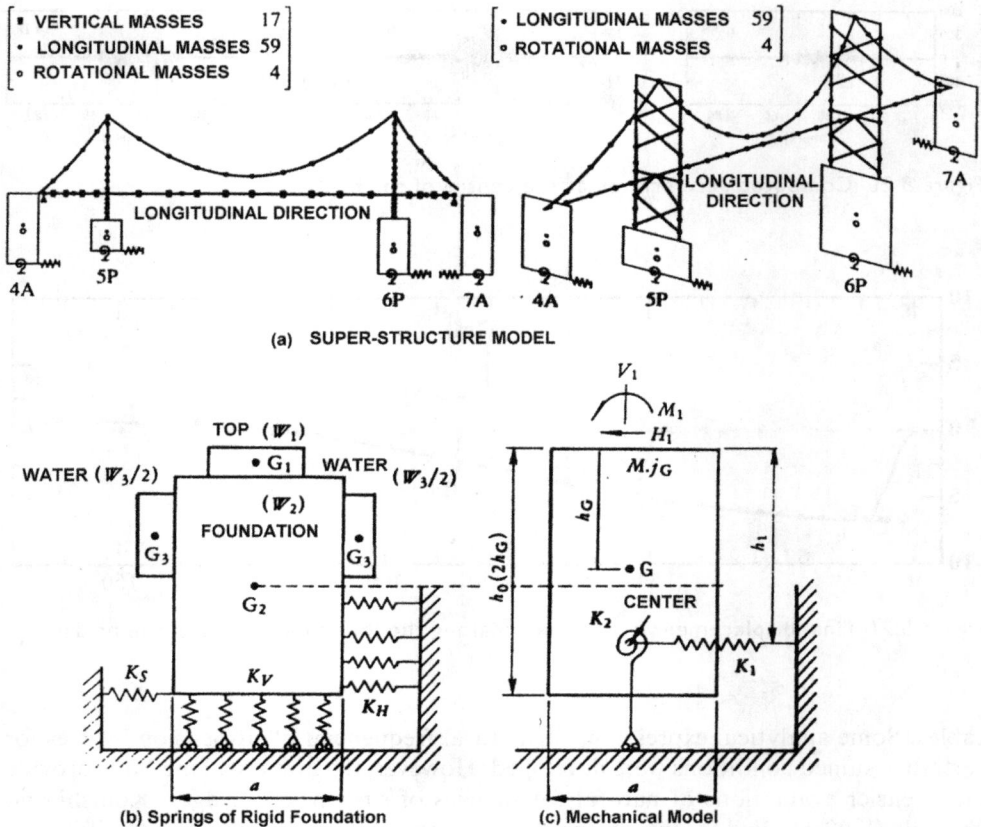

Figure 8.22 A numerical model of the Minami-Bisan-Seto Bridge used for dynamic analyses.

piers may be hollow. The simplest model of piers would be an inverted pendulum subjected to bending due to lateral seismic forces with a fixed or flexible base as illustrated in Figure 8.23. In many practical applications, piers with girders on top connected through supports with/without isolators would be represented a single or multiple beam elements. Their responses would be evaluated under given boundary, initial and boundary conditions. During earthquakes, the traverse response of piers would be most critical unless there are high shear forces resulting from earthquake faulting along the longitudinal direction. Figure 8.24 illustrates possible bending deformation responses of piers in lateral and longitudinal directions. During shaking, the behaviour of piers may become plastic at the ground level and such a situation must also be considered in computations. The piers may be fixed to the ground through piles. Figure 8.25 shows some examples of representation of foundations of piers, which generally incorporate the Pasternak-type beam elements for surrounding medium as well as the pile itself.

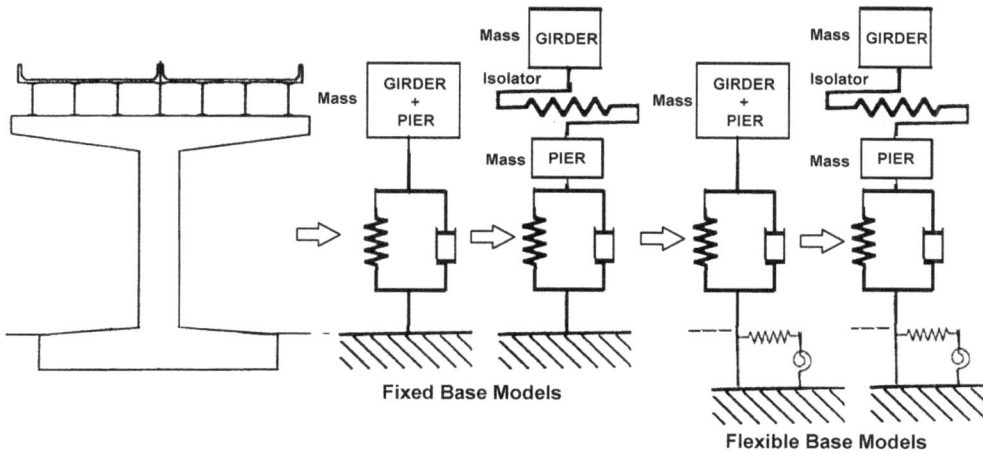

Figure 8.23 Some simple models for the seismic design and response analyses of a pier.

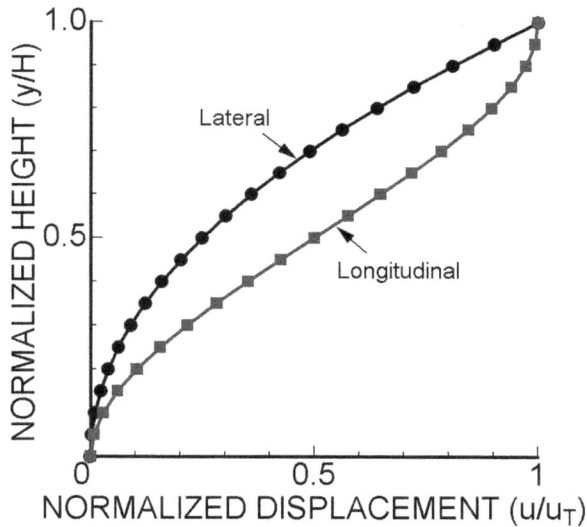

Figure 8.24 Bending deformation responses of a pier for lateral and longitudinal directions.

8.2.4   Pylons and truss structures

Pylons utilized for power lines and transmission towers are generally constructed using truss-type members. In structural mechanics, the member forces of such structures can be easily evaluated while calculating their displacement responses can be quite cumbersome. Nevertheless, it has nowadays become a very easy task to carry out such analyses by virtue of the development of the finite element method and computers. A truss is considered to be a member having only axial resistance and no moment

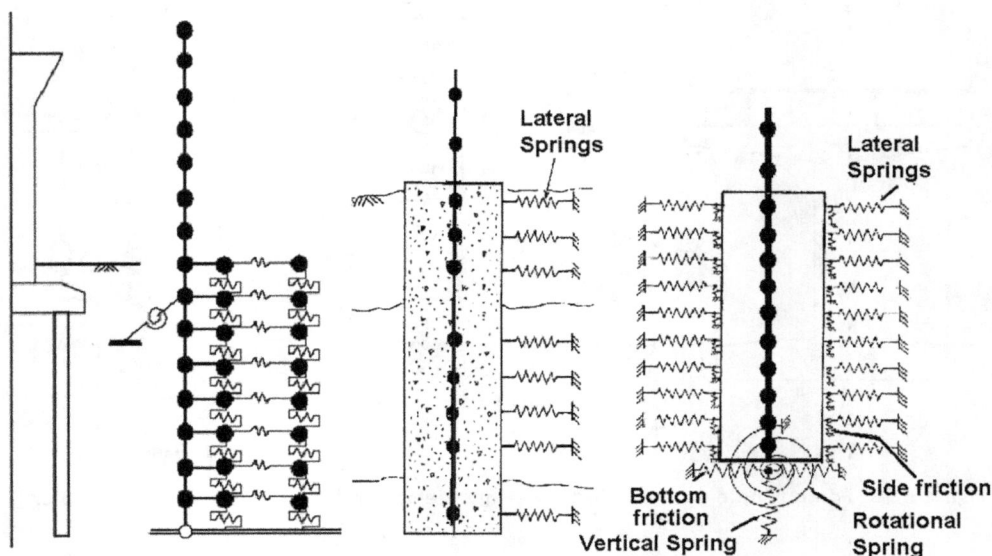

Figure 8.25 Examples of various representations of pier foundations.

resistance at hinges. With this assumption, the stiffness of a truss element in the local coordinate system can be shown to be as given below (e.g. Reddy 1993; Aydan 1989):

$$K' = \begin{bmatrix} K'_a & 0 & -K'_a & 0 \\ 0 & 0 & 0 & 0 \\ -K'_a & 0 & K_a & 0 \\ 0 & 0 & 0 & 0 \end{bmatrix} \tag{8.5}$$

where $K'_a = \dfrac{EA}{L}$, E, A and L are the elastic modulus and area of the truss and element length. It is transformed to the global representation through the following relation:

$$[K]_g = [T]^T [K'][T] \tag{8.6}$$

where

$$T = \begin{bmatrix} c & s & 0 & 0 \\ -s & c & 0 & 0 \\ 0 & 0 & c & s \\ 0 & 0 & -s & c \end{bmatrix}; \quad \alpha = \tan^{-1}\left(\frac{y_J - y_I}{x_J - x_I}\right); \quad c = \cos\alpha; \quad s = \sin\alpha$$

The responses of truss-type pylons and bridges subjected to thrust and faulting movements are shown in Figures 8.26 and 8.27 together with material properties given in Table 8.11, respectively. These analyses are just to illustrate how bridges deform during permanent ground movements due to faulting and where the highest member forces develop.

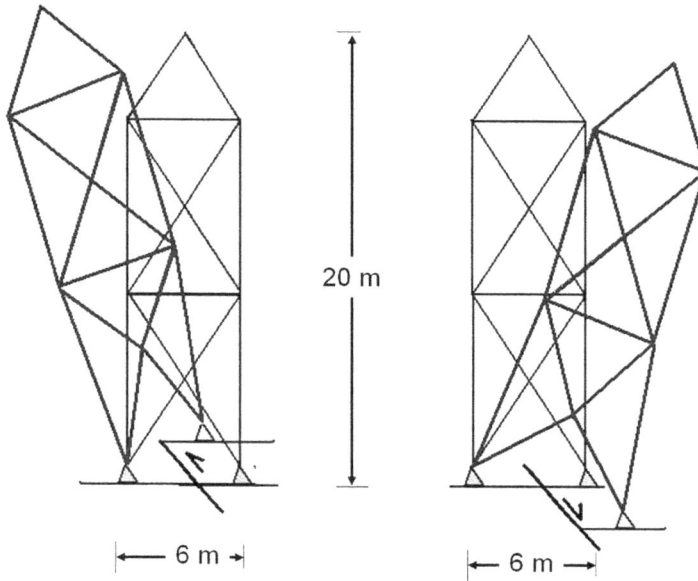

Figure 8.26 Deformed configuration of truss-type pylons under (a) thrust and (a) normal faulting movements.

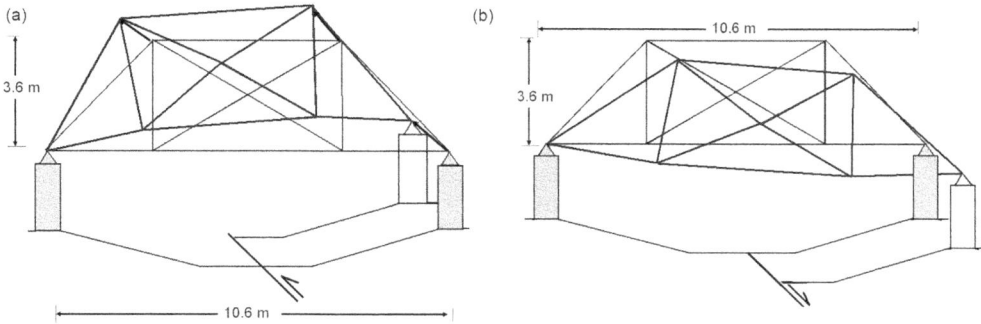

Figure 8.27 Deformed configuration of truss-type bridges under (a) thrust and (b) normal faulting movements.

Table 8.11 Material and geometrical properties of truss structures

Material	Elastic modulus (GPa)	Area (m²)
Bridge	210	0.025
Pylon	210	0.025

283 cm relative normal or thrust faulting.

Table 8.12 Materials properties used in discrete finite element method simulations

Material	λ (MPa)	μ (MPa)	γ (kN/m^3)	c (MPa)	ϕ (°)	σ_t (MPa)
Ground	2000	2000	26	-	-	-
Fault	50	50	-	0.0	40	0.0

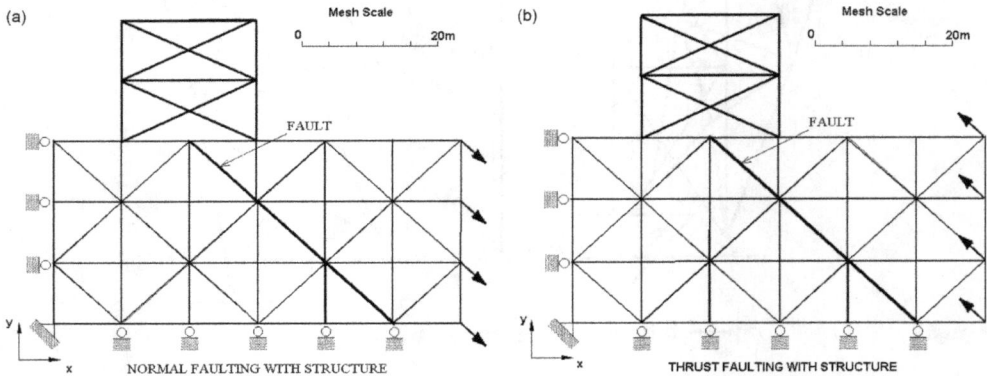

Figure 8.28 Finite element meshes and boundary conditions for fault-structure interaction simulations. (a) normal faulting (b) thrust faulting.

The next examples are concerned with the simulations of the response of a truss-type structure subjected to normal and thrust faulting using a pseudo-dynamic version of the DFEM as reported by Aydan (2003b). Table 8.12 gives the materials properties used in DFEM simulations. The thickness of the fault plane was selected as 10 mm. In the simulations, the fault plane was modelled through contact elements. The fault plane behaves elastically when the normal and shear stresses are below its yield strength. However, if yielding takes place, its behaviour is simulated as an elastic-perfectly plastic behaviour. In normal and thrust faulting the displacements having an amplitude of 10 cm are imposed at the boundary nodes as indicated in Figure 8.28 both in *x* and *y* directions.

The most important aspect in earthquake engineering is the interaction between structures and fault breaks. For this purpose, a truss structure straddling over the projected fault trace on the ground surface was considered, and normal faulting and thrust faulting conditions were imposed through prescribed displacement at selected points as in the previous computations. The material properties for ground and fault are the same as those given in Table 8.12. The elastic modulus and the cross section area of a typical truss were chosen as 90 GPa and 0.1 m^2, and their behaviour was assumed to be elastic. Figure 8.28 shows the finite element meshes and boundary conditions used in simulations. Figure 8.29 shows the deformed configurations at computation steps 1 and 10 for normal and thrust-type faulting modes. In both cases, the truss structure tilts. While the thrust-type faulting causes the contraction of trusses, the normal faulting condition results in the extension of trusses and separation of the supporting members fixed to the ground.

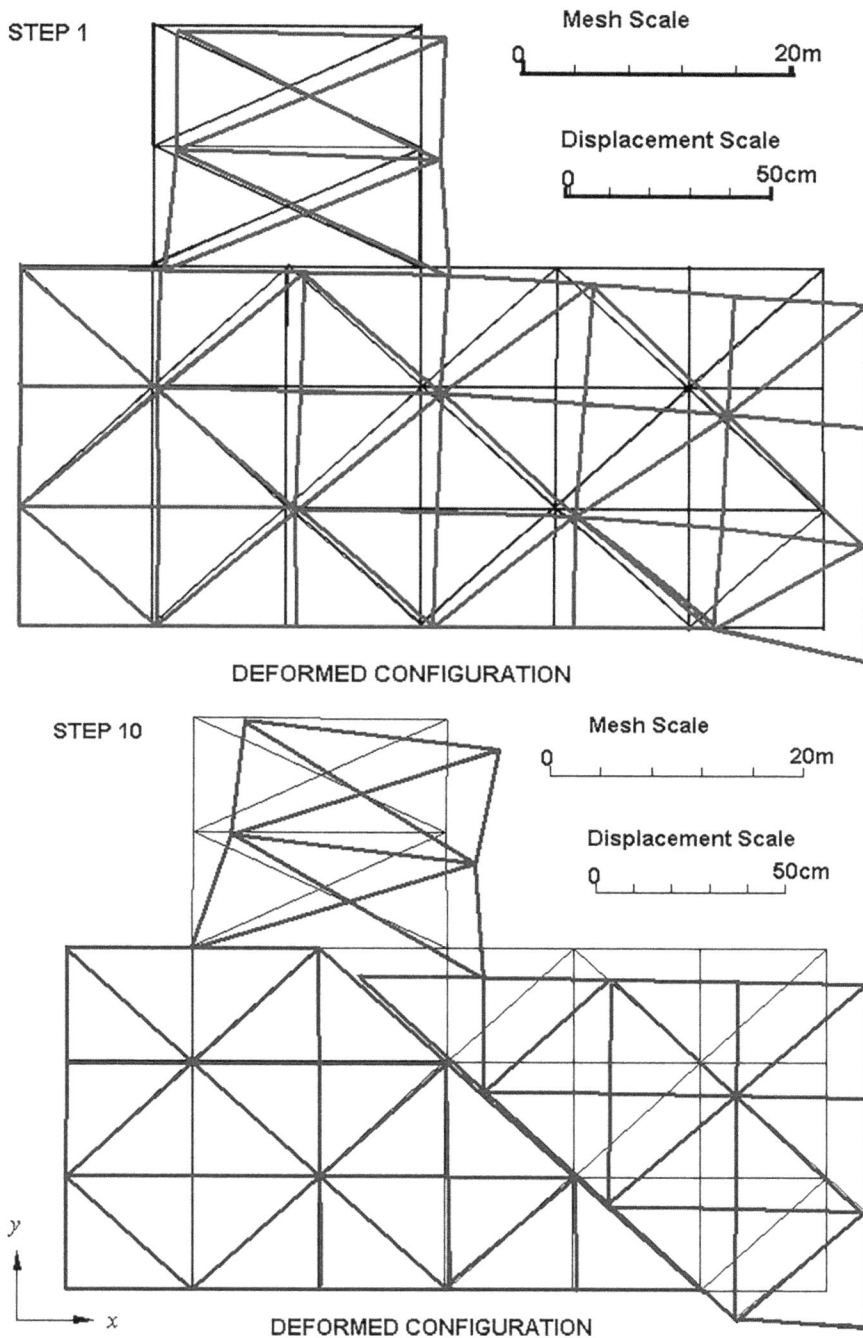

STEP 1

Mesh Scale

0 |_____| 20m

Displacement Scale

0 |_____| 50cm

DEFORMED CONFIGURATION

STEP 10

Mesh Scale

0 |_____| 20m

Displacement Scale

0 |_____| 50cm

y

x

DEFORMED CONFIGURATION

Figure 8.29 Simulations of the fault-structure interaction for (a) normal and (b) thrust faulting conditions. (Arranged from Aydan (2003b).)

(Continued)

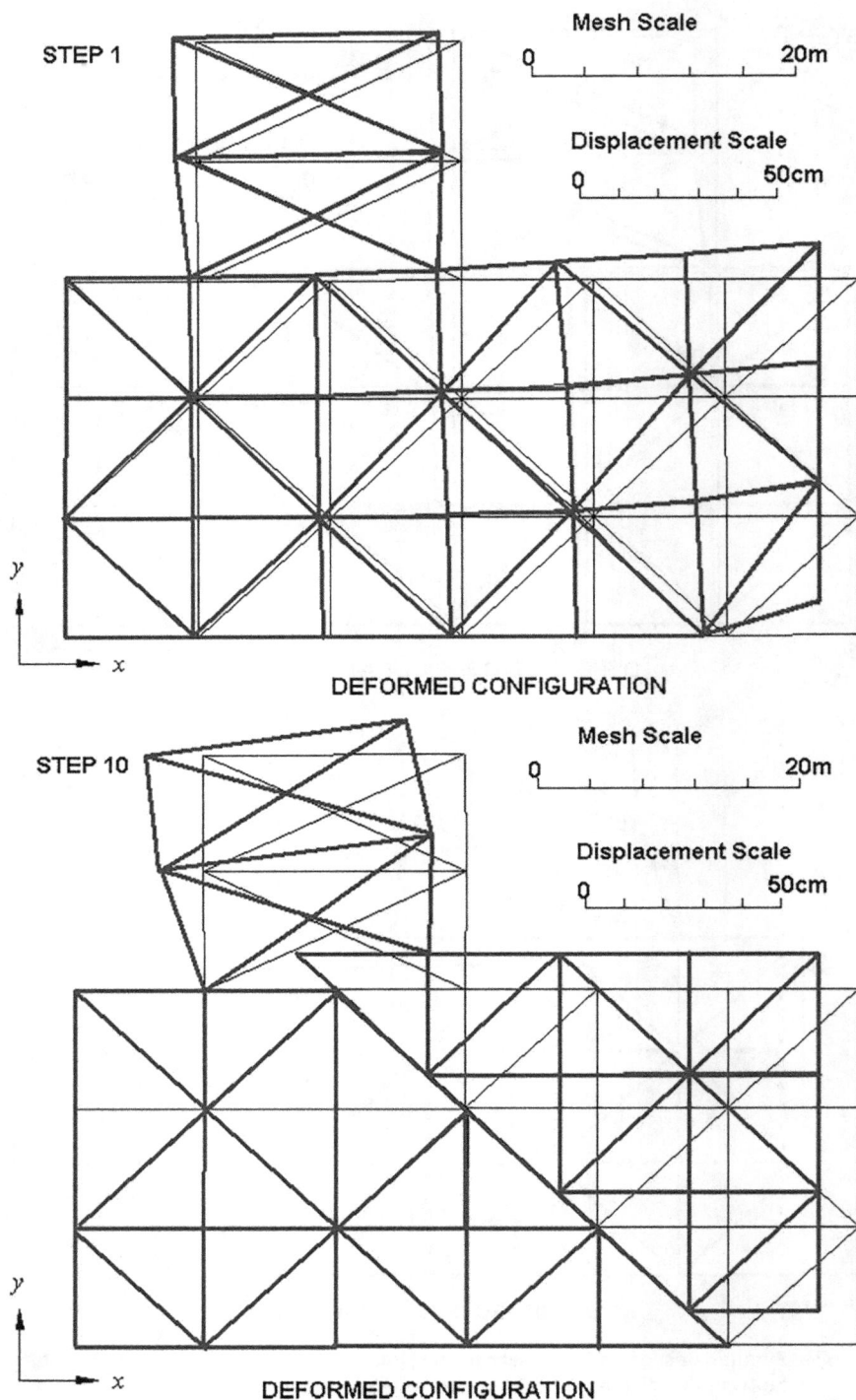

STEP 1

Mesh Scale

0 — 20m

Displacement Scale

0 — 50cm

DEFORMED CONFIGURATION

STEP 10

Mesh Scale

0 — 20m

Displacement Scale

0 — 50cm

DEFORMED CONFIGURATION

Figure 8.29 (Continued) Simulations of the fault-structure interaction for (a) normal and (b) thrust faulting conditions. (Arranged from Aydan (2003b).)

8.2.5 Liquid tanks on ground and elevated tanks

8.2.5.1 Liquid tanks on ground

There are different types of tanks for storage of liquids such as water, oil and other chemical products. It is well known that earthquake shaking may cause severe damages to tanks and there is report that 6,000 oil tanks were damaged in the 1923 Kanto earthquake. Particularly, damage due to the sloshing phenomenon is well known, and it is the most critical issue for seismic design of tanks. There are some theoretical and experimental studies since 1934 as reported by Jacobsen and Ayre (1951) and the work of Lamb (1932) on the modes of fluid in circular tanks, as shown in Figure 8.30.

The most practical method for designing tanks to withstand seismic motions was developed by Housner (1957, 1963). On the basis of experimental observations, Housner (1957, 1963) suggested that a larger volume of fluid remains stationary (assumed to be rigid) during shaking while a certain top part of fluid (convective mass) oscillates. The fluid surface remains inclined proportional to basal acceleration initially and gradually resembles Mode 1 shown in Figure 8.30. Figure 8.31 illustrates Housner's modelling of a liquid storage tank subjected to shaking. The formulation of Housner (1957, 1963) is a quite practical yet the most outstanding method to evaluate the response of fluids in tanks subjected to ground shaking and their seismic design, which is commonly used worldwide. In Housner's 1957 paper, various conditions were considered, including the flexibility of tank wall. In his paper, dynamic fluid loads acting on tank wall and tank base were also evaluated for the seismic design of tanks. Since then, there have been many studies on the seismic responses of tanks, which incorporate the foundation conditions, compressibility of fluid and the existence of floating lids. Table 8.13 summarizes major parameters of seismic design of circular and rectangular-shaped tanks proposed by Housner (1957, 1963).

Figure 8.32 compares the natural frequencies of circular and rectangular tanks in relation to the level of fluid as a function of tank diameter or length. As noted from the

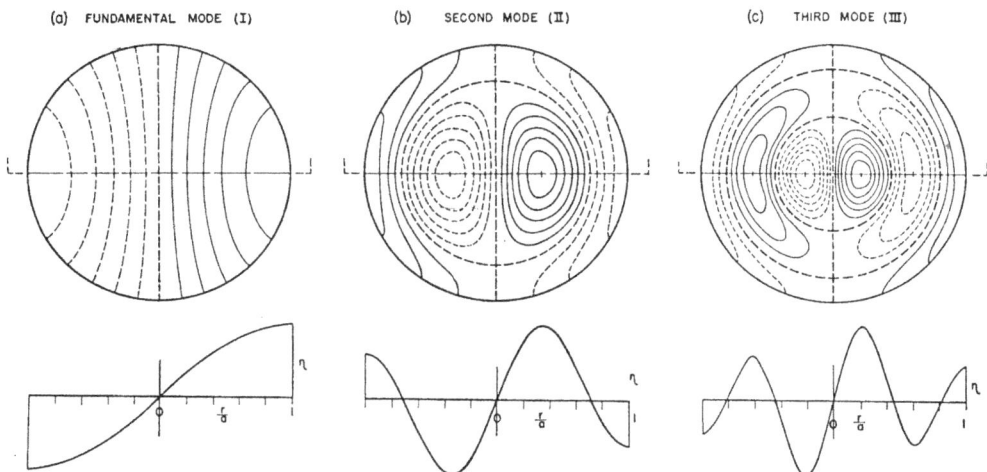

Figure 8.30 Modes of fluid surface during shaking. (From Jacobsen and Ayre (1951).)

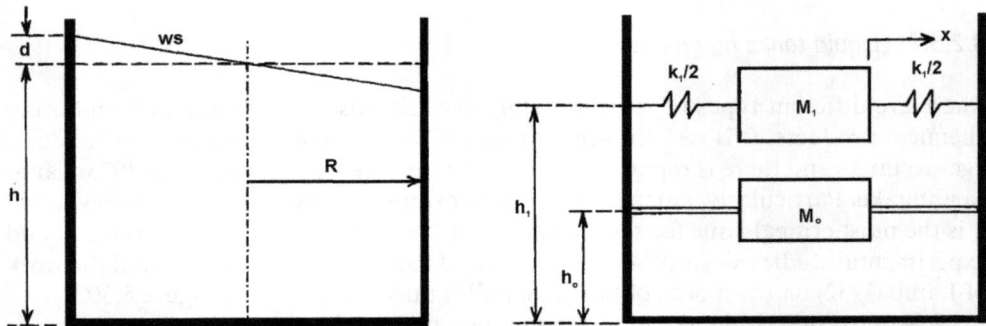

Figure 8.31 Housner's model (redrawn from the original figure).

Table 8.13 Major parameters of seismic design of tanks

Parameter		Circular tank	Rectangular tank
Natural period	T_1	$2\pi\sqrt{\dfrac{R}{g}1.84\coth\left(1.84\dfrac{h}{R}\right)}$	$2\pi\sqrt{\dfrac{\pi}{2}\dfrac{L}{g}\coth\left(\dfrac{\pi}{2}\dfrac{h}{L}\right)}$
Stiffness	k_1	$\dfrac{g}{R}1.84\tanh\left(1.84\dfrac{h}{R}\right)M_1$	$\dfrac{g}{L}1.58\tanh\left(1.58\dfrac{h}{R}\right)M_1$
Convective mass	M_1	$0.318\dfrac{R}{h}\tanh\left(1.84\dfrac{h}{R}\right)M$	$0.527\dfrac{L}{h}\tanh\left(1.58\dfrac{h}{L}\right)M$
Impulsive mass	M_0	$0.575\dfrac{h}{R}\tanh\left(1.74\dfrac{R}{h}\right)M$	$0.577\dfrac{L}{L}\tanh\left(1.73\dfrac{L}{h}\right)M$
Convective mass height	h_1	$h\left[1-0.545\dfrac{R}{h}\coth\left(1.84\dfrac{h}{R}\right)\right]$	$h\left[1-0.632\dfrac{L}{h}\coth\left(1.58\dfrac{h}{L}\right)\right]$
Impulsive mass height	h_0	$0.375h$	$0.375h$
Total mass	M	$\rho h\pi R^2$	$2\rho hLw$

figure, the natural period of circular and rectangular tanks are quite similar to each other. The difference between circular shape and rectangular shape tanks is limited to 0.5 seconds under full condition. Fundamentally, the motion of fluid in the tanks resembles that in u-tubes. If an equivalent length of the fluid-filled tank can be defined, the motion of the fluid can even be approximated by the solution of the differential equation of oscillating fluids in u-tubes.

8.2.5.2 Elevated tanks

Elevated liquid tanks are fundamentally treated by combining the models used for tanks on ground and the support system. The modelling of support system is based

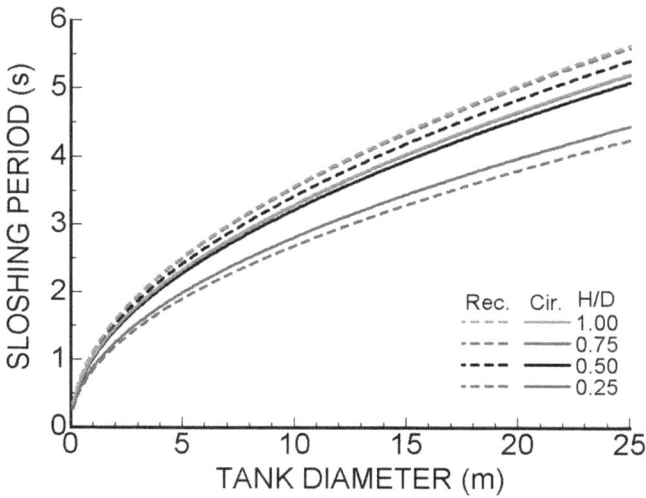

Figure 8.32 Comparison of natural periods of circular and rectangular tanks.

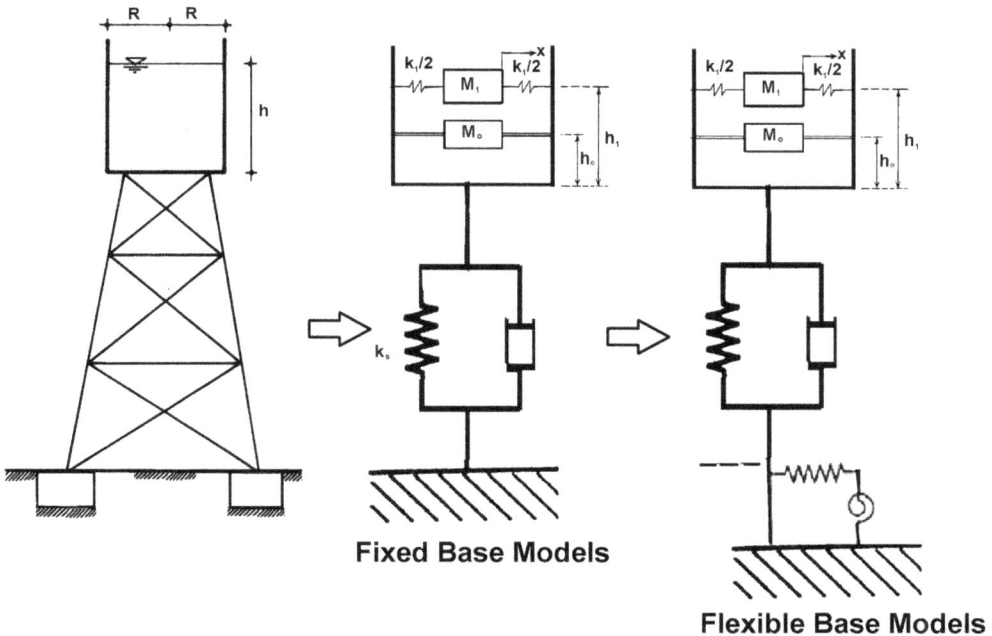

Figure 8.33 Simple mechanical models for elevated liquid tanks.

on the models used for piers as illustrated in Figure 8.33. Including Housner's model, one can find some studies reported by several researchers (e.g. Chandrasekaran and Krishna 1954; Livaoğlu and Dogangün 2006).

8.3   GEOTECHNICAL STRUCTURES

8.3.1   Seismic design of embankments

Embankments are simple geotechnical structures used along river banks, roadways and railway lines as illustrated in Figure 8.34, and they are constructed by banking soil. The compaction of the embankment during construction differs depending upon the importance of the structure and there may be some special treatment of the interface between the existing ground and embankment body. If there is no special treatment of the interface, it may be quite critical during earthquakes where high excess pore pressure may develop and it may lead to the failure of embankment. There are methods for analysing the stability of embankments, such as the pseudo-dynamic method based on the seismic coefficient concept, pure dynamic limiting equilibrium method (DLEM) and numerical techniques. In this section, the available methods and their principles are outlined.

8.3.1.1   Pseudo-dynamic method

The Swedish or circular sliding method is probably one of the earliest applications. The interslice force concept was later introduced to handle the variations of ground properties, partial surface loading, ground water conditions and stabilization forces (i.e. anchors). The most simple form of interslice force approach is that of Bishop

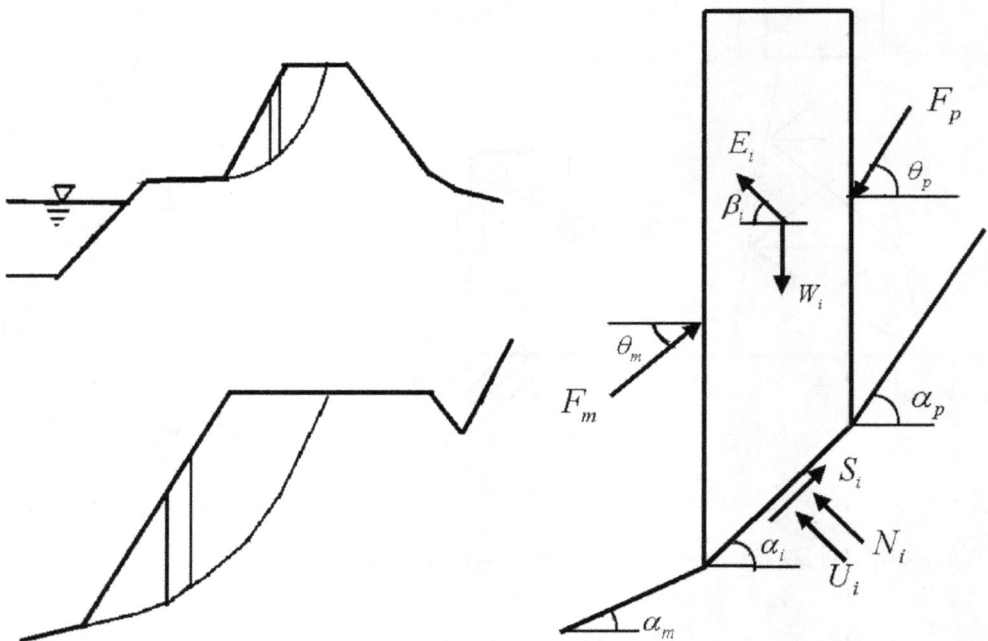

Figure 8.34 Embankment types and force condition on a typical slice *i*.

(1955), that is the normal forces at the artificial slice boundaries are transmitted while tangential (shear) forces are assumed to be nil. Spencer (1967) replaced this approach by introducing a coefficient in order to incorporate the effect of tangential interslice force on stability computations. This coefficient is generally assumed to be equal to the friction angle of the sliding body, which may be interpreted as internal frictional yielding. Morgenstern-Price (1965) introduced an initially unknown varying interslice force coefficient and the stability problem was solved through an iterative approach. Aydan suggested an excess force concept, in analogy to the theory of computational plasticity, to determine the interslice force coefficient. This approach implicitly shows that there is no internal yielding within the sliding body. However, it can be easily modified to handle the internal yielding within the body subjected to sliding. Nevertheless, the internal yielding may appear following the body that starts to move.

Let us consider a typical slice within a potentially unstable body and also let us assume the forces acting on the typical slice can be visualized as shown in Figure 8.34. The force equilibrium equations of the slice numbered k for s and n directions may be written as follows:

$$\sum F_s = W_k \sin\alpha_k + E_k \cos(\alpha_k + \beta) + F_p \cos(\alpha_p - \alpha_k)$$
$$- F_m \cos(\alpha_k - \alpha_m) - S_k = 0 \tag{8.7a}$$

$$\sum F_n = W_k \cos\alpha_k - E_k \sin(\alpha_k + \beta) + F_p \sin(\alpha_p - \alpha_k)$$
$$+ F_m \sin(\alpha_k - \alpha_m) - U_k - N_k = 0 \tag{8.7b}$$

where
W_k is the weight of slice k
E_k is the earthquake force acting on slice k
F_p is the force transmitted from slice $(k+1)$
F_m is the force transmitted from slice $(k-1)$
U_k is the uplift force due pore water pressure at the base of slice k
S_k is the shear force acting at the base of slice k
N_k is the normal force acting at the base of slice k
α_k is the basal inclination of slice k
α_{k+1} is the basal inclination of slice $k+1$
α_{k-1} is the basal inclination of slice $k-1$

The earthquake force E_k can be related to the weight of the slice through a seismic coefficient η and it is assumed to be proportional to the weight of the slice as given below:

$$E_k = \mu W_k \tag{8.8}$$

The uplift force due to pore pressure may be taken into account in the following forms:

a. If water table is known:

$$U_k = \gamma_k \frac{h_p + h_m}{2} l_k \tag{8.9}$$

where γ_k, h_p, h_m, l_k are unit weight of water, water heads and base length
   b. If water table is unknown:

$$U_k = r_w W_k \quad \text{with} \quad 0 \le r_w \le \frac{\gamma_w}{\gamma_s}. \tag{8.10}$$

where r_w is uplift water pressure coefficient. The shear resistance of geotechnical materials is generally given in the following form:

$$T_k = cl_k + N_k \tan\phi \tag{8.11}$$

where c and ϕ are the cohesion and friction angle, respectively.

The safety factor is commonly used in this type of formulations, which may be defined as follows:

$$SF = \frac{T_k}{S_k} \tag{8.12}$$

The interslice force F_m may be easily obtained in the following form through some manipulations of Eqs. (8.7)–(8.12):

$$F_m = \frac{P_1 + P_2 + P_3 - U_k \tan\phi - cl_k}{\sin(\alpha_k - \alpha_m) \cdot \tan\phi + \cos(\alpha_k - \alpha_m) \cdot SF} \tag{8.13}$$

where
$$P_1 = W_k \left[\sin\alpha_k - \cos\alpha_k \tan\phi \right]$$
$$P_2 = \eta W_k \left[\cos(\alpha_k + \beta) \cdot SF + \sin(\cos\alpha_k + \beta)\tan\phi \right]$$
$$P_3 = F_p \left[\cos(\alpha_p - \alpha_k) \cdot SF - \sin(\alpha_p - \alpha_k)\tan\phi \right]$$

For a given value of safety factor (SF), the above equation can be solved by a step-by-step method for a boundary condition of $F_{n+1} = 0$ and the stability of the slope having n slices can be assessed through the utilization of the following criteria:

$$\begin{aligned}
F_1 &> 0 \quad \text{slope is unstable} \\
F_1 &= 0 \quad \text{slope is at limiting equilibrium state} \\
F_1 &< 0 \quad \text{slope is unstable}
\end{aligned} \tag{8.14}$$

The SF for the slip surface analysis may be written as

$$SF = \frac{\sum \left[W_k(\cos\alpha_k - \eta\sin(\alpha_k + \beta)) - U_k \tan\phi - cl_k \right]}{\sum \left[W_k(\sin\alpha_k + \cos(\alpha_k + \beta)) \right]} \tag{8.15}$$

8.3.1.2   Dynamic limiting equilibrium method

8.3.1.2.1   MODIFIED NEWMARK METHOD: SINGLE BLOCK MODEL

The combined effect of seismic loads and the apparent changes in shear strength result in an overall decrease in the stability of an affected slope. Aydan and Ulusay (2002) suggested a method to compute displacement and velocity response of embankments subjected to static and dynamic loading during failure. The basic concept of the method is quite similar to that presented by Newmark (1965) and Aydan et al. (1989, 1996a,b). The model presented herein basically deals with the mode of planar sliding failure.

A body situated on a plane at an inclination of α subjected to gravitational loading and earthquake loading is considered. The dynamic force equilibrium condition for this body may be written in the following form for directions s and n (Figure 8.35):

$$\sum F_s = W\sin\alpha + E_H\cos\alpha - E_V\sin\alpha - S = \frac{W}{g}\ddot{s} \tag{8.16a}$$

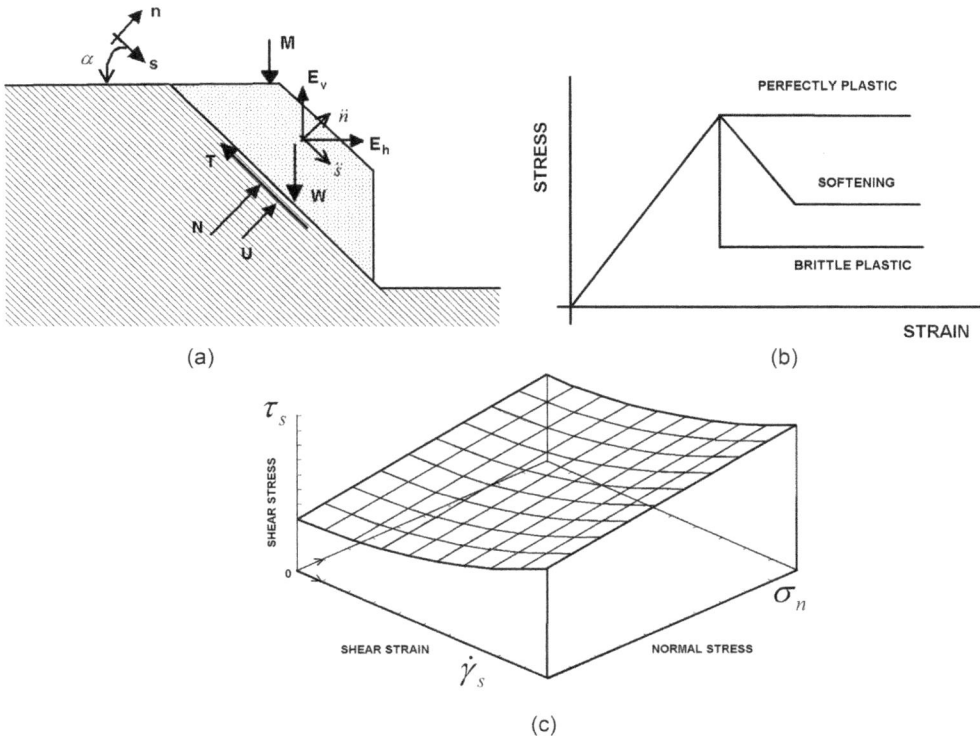

Figure 8.35 Modelling the embankment and yielding characteristics. (a) Mechanical model, (b) constitutive model and (c) yield criterion.

$$\sum F_n = -W\cos\alpha + E_H\sin\alpha + E_V\cos\alpha + U_w + N = \frac{W}{g}\ddot{n} \qquad (8.16b)$$

The uplift force U_w of water can be assumed to be a sum of static and dynamic components. U_w^s, U_w^d, which are specifically given by

$$U_w = U_w^s + U_w^d \qquad (8.17)$$

The static component of the uplift force of water can be easily obtained from the geometry of water table and failure surface. It can also be given as a fraction of the weight of the sliding body given by Bishop (1955) and Aydan et al. (1996a,b).

$$\beta_s = \frac{U_w}{W} \qquad (8.18)$$

On the other hand, the dynamic component of the uplift force due to the dynamic excitation of water is extremely difficult to determine. Nevertheless one may assume that it is proportional to the inertia force induced by the earthquake. It is generally assumed that the geo-materials obey the Mohr–Coulomb yield criterion given for a uniform distribution of stress over the total failure surface as

$$T = cA + N\tan\phi \qquad (8.19)$$

where c, ϕ and A are cohesion, internal friction angle and the area of the failure plane, respectively. During sliding, the Mohr–Coulomb parameters may also vary. For example, it is easy to implement the failure condition of brittle materials, which implies the abrupt reduction of cohesion and friction of the material (Figure 8.35b). It is also known that the shear strength of geo-materials and interfaces increases as a function of the loading or deformation (strain) rate (i.e. Aydan and Nawrocki, 1998). The simplest form of the yield criterion of Bingham type with the consideration of the loading and deformation rate may be written as (Aydan et al. 1996a,b)

$$T = cA + N\tan\phi + \eta\frac{(W\sin\alpha + E_H\cos\alpha - E_V\sin\alpha)}{g}\dot{s} \qquad (8.20)$$

where $\dot{s}$ is the velocity of the sliding body and η is the viscosity resistance coefficient. Both horizontal and vertical earthquake loads may be written in terms of horizontal and vertical ground accelerations and the weight of sliding block as

$$E_H = \frac{a_H(t)}{g}W, \quad E_V = \frac{a_V(t)}{g}W. \qquad (8.21)$$

If the dynamic component of uplift force of water is assumed to be proportional to the inertia force induced on the water contained in the sliding body, the following relation may be written.

$$U_w^d = U_w^s\left(\frac{a_H(t)}{g}\sin\alpha + \frac{a_V(t)}{g}\cos\alpha\right) \qquad (8.22)$$

Normalizing the above relation by the weight of the sliding body, one may write the following:

$$\beta_d = \frac{U_w^d}{W} = \frac{U_w^s}{W}\left(\frac{a_H}{g}\sin\alpha + \frac{a_V}{g}\cos\alpha\right). \tag{8.23a}$$

or

$$\beta_d = \beta_s\left(\frac{a_H}{g}\sin\alpha + \frac{a_V}{g}\cos\alpha\right) \tag{8.23b}$$

Neglecting the acceleration component in the direction n, one may easily obtain the following relation from Eqs. (8.16), (8.19), (8.20), (8.21) and (8.23), provided that the dynamic equilibrium condition is satisfied at any time during sliding:

$$\ddot{s} + A(t)\dot{s} - B(t) = 0 \tag{8.24}$$

where

$$A(t) = \eta\left(\sin\alpha + \frac{a_H(t)}{g}\cos\alpha - \frac{a_V(t)}{g}\sin\alpha\right) \tag{8.25a}$$

$$B(t) = B_g + B_H - B_V - B_c + B_w \tag{8.25b}$$

$$B_g = g(\sin\alpha - \cos\alpha\tan\phi) \tag{8.25c}$$

$$B_H = a_H(\cos\alpha + \sin\alpha\tan\phi) \tag{8.25d}$$

$$B_V = a_V(\sin\alpha - \cos\alpha\tan\phi) \tag{8.25e}$$

$$B_c = \frac{cA}{W}g \tag{8.25f}$$

$$B_w = \beta_s(g + a_H\sin\alpha + a_V\cos\alpha)\tan\phi \tag{8.25g}$$

The solution of Eq. (8.24) yields both the velocity and the displacement of the sliding body for rigid body motions. During sliding, the velocity of the body should remain positive. Otherwise, the body should not be moving relative to the sliding plane. It should also be noted that the above equation does not involve the elastic response of the sliding body. Although it is quite easy to solve the above equation for simple waveforms, it would be rather difficult to do so when actual earthquake waves are considered. Therefore, instead of this, some finite difference techniques for numerical solution of the above equation may be employed. The linear acceleration technique is chosen herein. As Eq. (8.24) holds for any time, it may be written for time step $n+1$ as

$$\ddot{s}_{n+1} + A_{n+1}\dot{s}_{n+1} - B_{n+1} = 0 \tag{8.26}$$

The Taylor expansion of acceleration $\ddot{s}$, velocity $\dot{s}$ and displacement s for a time step $n+1$ are written as given below:

$$\dddot{s}_{n+1} = \dddot{s}_n + \frac{\ddddot{s}_n}{1!}\Delta t + 0^2 \quad \text{or} \quad \ddddot{s}_n = \frac{\dddot{s}_{n+1} - \dddot{s}_n}{\Delta t}. \tag{8.27}$$

$$\dot{s}_{n+1} = \dot{s}_n + \frac{\ddot{s}_n}{1!}\Delta t + \frac{\dddot{s}_n}{2!}\Delta t^2 + 0^3 \tag{8.28}$$

$$s_{n+1} = s_n + \frac{\dot{s}_n}{1!}\Delta t + \frac{\ddot{s}_n}{2!}\Delta t^2 + \frac{\dddot{s}_n}{3!}\Delta t^3 + 0^4 \tag{8.29}$$

Inserting Eq. (8.27) into Eq. (8.28) and (8.29) yields the following equations for velocity and displacement of the sliding body at time $n+1$ as

$$\dot{s}_{n+1} = \dot{s}_n + \frac{\ddot{s}_n}{2}\Delta t + \frac{\ddot{s}_{n+1}}{2}\Delta t \tag{8.30}$$

$$s_{n+1} = s_n + \frac{\dot{s}_n}{1}\Delta t + \frac{\ddot{s}_n}{3}\Delta t^2 + \frac{\ddot{s}_{n+1}}{6}\Delta t^2 \tag{8.31}$$

Using Eq. (8.30) in Eq. (8.26) results in the following equation for acceleration $\ddot{s}_{n+1}$ at time step $n+1$:

$$\ddot{s}_{n+1} = \frac{1}{1 + A_{n+1}\dfrac{\Delta t}{2}}\left(B_{n+1} - A_{n+1}\left(\dot{s}_n + \frac{\ddot{s}_n}{2}\Delta t \right) \right) \tag{8.32}$$

The equation above is integrated step by step with the following initial conditions:

$$\dot{s}_o = 0; \ s_o = 0$$

Once the acceleration of the sliding body at time step $n+1$ is obtained, then its velocity and the displacement can be easily computed from Eqs. (8.30) and (8.31). It should be noted that the implementation of the brittle plastic behaviour or softening behaviour of the failure surface can be easily implemented. However, some additional information is necessary for the softening behaviour. Furthermore, the variation in inclination of the failure surface can also be considered during the numerical solution procedure, provided that the energy dissipation of the sliding body is negligible during the change in inclination of the failure surface.

8.3.1.2.2   MULTIPLE SLICE DYNAMIC LIMITING EQUILIBRIUM METHOD

Aydan (2016a) modified the single-body model and extended it to a multi-slice body to estimate the motions of embankments and similar types of geotechnical problems.

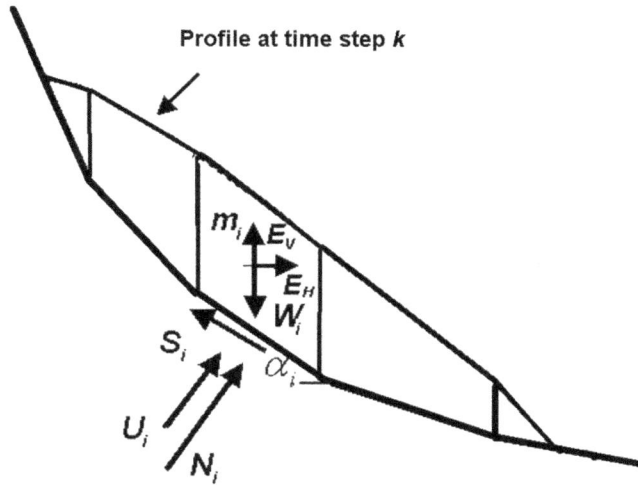

Profile at time step *k*

Figure 8.36 Mechanical model for estimating post-failure motions.

The method used for estimating post-failure motions of a failed body is based on earlier proposals by Aydan et al. (2006a, 2008b), Aydan and Ulusay (2002) and Tokashiki and Aydan (2010).

Let us consider an unstable body consisting of N number of blocks sliding on a slip surface as shown in Figure 8.36. If interslice forces are assumed to be nil as assumed in the simple sliding model (Fellenius-type), one may write the following equation of motion for the sliding body:

$$\sum_{i=1}^{n}(S_i - T_i) = \bar{m}\frac{d^2 s}{dt^2} \tag{8.33}$$

where $\bar{m}$, s, t, n, S_i and T_i are total mass, travel distance, time, number of slices, shear force and shear resistance, respectively. Shear force and shear resistance may be given in the following forms together with the Bingham-type yield criterion:

$$S_i = W_i\left(1 + \frac{a_H}{g}\right)\sin\alpha_i; \quad T_i = c_i A_i + (N_i - U_i)\tan\phi_i + \eta W_i\left(\frac{ds_i}{dt}\right)^b.$$

$$T_i = c_i A_i + (N_i - U_i)\tan\phi_i + \eta W_i\left(\frac{ds_i}{dt}\right)^b \tag{8.34}$$

where W_i, A_i, N_i, U_i, α_i, a_V, a_H, c_i, ϕ_i, η and b are weight, basal area, normal force, uplift pore water force, basal inclination, vertical and horizontal earthquake acceleration, cohesion, friction angle of slice i, Bingham-type viscosity and empirical coefficient, respectively.

If normal force and pore water uplift force is related to the weight of each block as given below:

$$N_i = W_i \left(1 + \frac{a_V}{g} \right) \cos \alpha_i, \quad U_i = r_u W_i. \tag{8.35}$$

one can easily derive the following equation with the use of Eqs. (8.33)–(8.35):

$$\frac{d^2 s}{dt^2} + \eta \left(\frac{ds}{dt} \right)^b - B(t) = 0 \tag{8.36}$$

where

$$B(t) = \frac{g}{\bar{m}} \left(\sum_{i=1}^{n} m_i \left[\sin \alpha_i \left(1 + \frac{a_H}{g} \right) - \left(\cos \alpha_i \left(1 + \frac{a_V}{g} \right) - r_u \right) \tan \phi_i \right] + \frac{c_i A_i}{g} \right) \tag{8.37}$$

In the derivation of Eq. (8.36), the viscous resistance of the shear plane of each block is related to the overall viscous resistance in the following form:

$$\eta \bar{m} g \left(\frac{ds}{dt} \right)^b = \sum_{i=1}^{n} \eta m_i g \left(\frac{ds_i}{dt} \right)^b \tag{8.38}$$

Equation (8.36) can be solved for the following initial conditions together with the definition of the geometry of the basal slip plane.

$$\text{At time } t = t_o, \quad s = s_o, \quad v = v_o \tag{8.39}$$

There may be different forms of constitutive laws for the slip surface (i.e. Aydan et al. 2006a, 2008b; Aydan and Ulusay, 2002). The simplest model to implement would be elastic-brittle plastic. If this model is adopted, the cohesion exists at the start of motion and it disappears thereafter. Therefore, the cohesion component in Eq. (8.37) may be taken to be nil as soon as the motion starts. Thereafter, the shear resistance consists of mainly frictional components together with some viscous resistance in post-failure regime.

8.3.1.2.3   APPLICATIONS TO ACTUAL EMBANKMENT FAILURES

Two examples of applications of the DLEM are explained herein. The first example is related to the embankment failure of the TOMEI Expressway near Makinohara caused by the 2009 Suruga Bay earthquake. Figure 8.37 shows aerial and side views of the embankment failure location. The failed embankment contains very large rock boulders as seen in Figure 8.37b.

K-NET Makinohara's strong motion acceleration record was used for evaluating the response of the failed portion of the embankment. First a series of pseudo-dynamic analyses were carried out and material properties were inferred. Then the motion of

Figure 8.37 Views of the Makinohara embankment failure. (a) Aerial view and (b) side view.

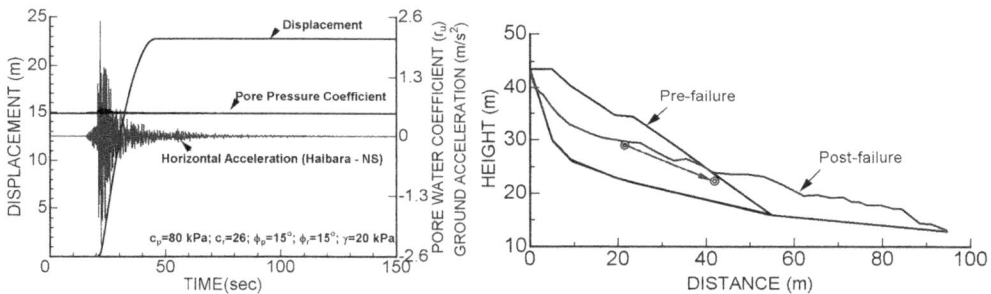

Figure 8.38 Simulation results for the Makinohara embankment failure.

the embankment during failure induced by the 2009 Suruga earthquake was simulated and results are shown in Figure 8.38.

The 2008 Noto Peninsula earthquake caused numerous embankment failures along the Noto Tollway. This model is utilized for estimating the response and post-failure motions of an embankment failure. One example of the embankment failures analysed by Aydan et al. (2007a) is reported herein. The strong motion records (EW and UD components) of K-NET Anamizu station are used and the peak cohesion and friction angle were taken as 300 kPa and 20°, which were determined for Noto Airport embankments. As the embankment fill material for Noto Airport was similar to that of the embankments of Noto Tollway, the use of such material properties would be appropriate. However, the post-failure properties are unknown. As the lower value of friction angle was 15°, it was taken as residual friction angle. The residual cohesion was taken as 6.5 kPa. In computations, the soil was assumed to be fully saturated. Figure 8.39a shows the displacement response of the mass centre during shaking while Figure 8.39b shows the trace of mass centre of the failed material during motion. The results are quite close to actual configurations and post-failure movements.

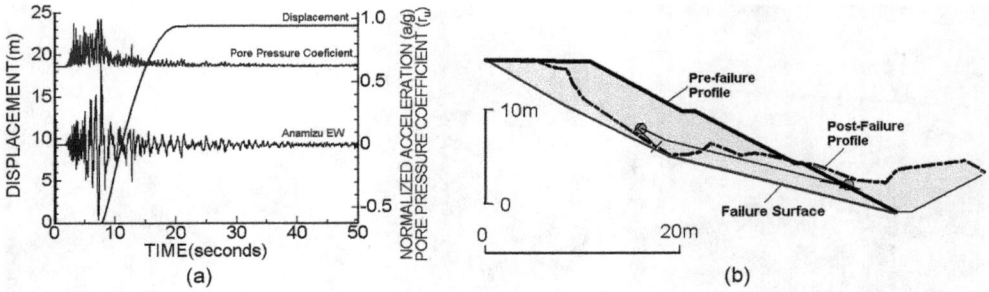

Figure 8.39 Simulation of an embankment failure on the Noto Peninsula Tollway. (a) Displacement response and (b) motion of the mass centre.

8.3.2 Retaining walls

8.3.2.1 Pseudo-dynamic method

There are four different situations for the failure of retaining walls: (1) stable, (2) sliding, (3) toppling and (4) toppling and sliding. If the seismic coefficient (pseudo-dynamic) method is employed, the initiation of sliding failure can be obtained in the following forms (Figure 8.40) (Aydan, 2017; Aydan et al. 2011; Aydan and Tokashiki, 2012):

a. Transition from stable to sliding mode

$$
\eta_s = \frac{a}{g} = \frac{\left(\sin\theta + \cos\theta\tan\phi_{ws}\right) - K\dfrac{\gamma_s}{\gamma_w}\dfrac{h}{t}\dfrac{\cos^2\theta}{2}\left(\cos\theta - \sin\theta\tan\phi_{ws}\right)}{\left(\cos\theta - \sin\theta\tan\phi_{ws}\right) + K\dfrac{\gamma_s}{\gamma_w}\dfrac{h}{t}\dfrac{\cos^2\theta}{2}\left(\cos\theta - \sin\theta\tan\phi_{ws}\right)} \tag{8.40}
$$

b. Transition from stable to toppling mode

$$
\eta_t = \frac{a}{g} = \frac{\left(h\sin\theta + t\cos\theta\right) - K\dfrac{\gamma_s}{\gamma_w}\dfrac{h}{t}\cos^2\theta\left(\dfrac{h}{3}\cos\theta - t\sin\theta\right)}{\left(h\cos\theta - t\sin\theta\right) - K\dfrac{\gamma_s}{\gamma_w}\dfrac{h}{t}\cos^2\theta\left(\dfrac{h}{3}\cos\theta - t\sin\theta\right)} \tag{8.41}
$$

where a is the maximum horizontal acceleration; g is the gravitational acceleration; θ is the wall base inclination; ϕ_{ws} is the friction angle between wall and backfill soil; γ_s is the unit weight of backfill soil; γ_w is the unit weight of wall; h is the length (height) of wall; t is the width of wall; K is the lateral force coefficient resulting from backfill

c. Transition from stable mode to combined sliding and toppling failure mode requires that both Eqs. (8.40) and (8.41) be simultaneously satisfied.

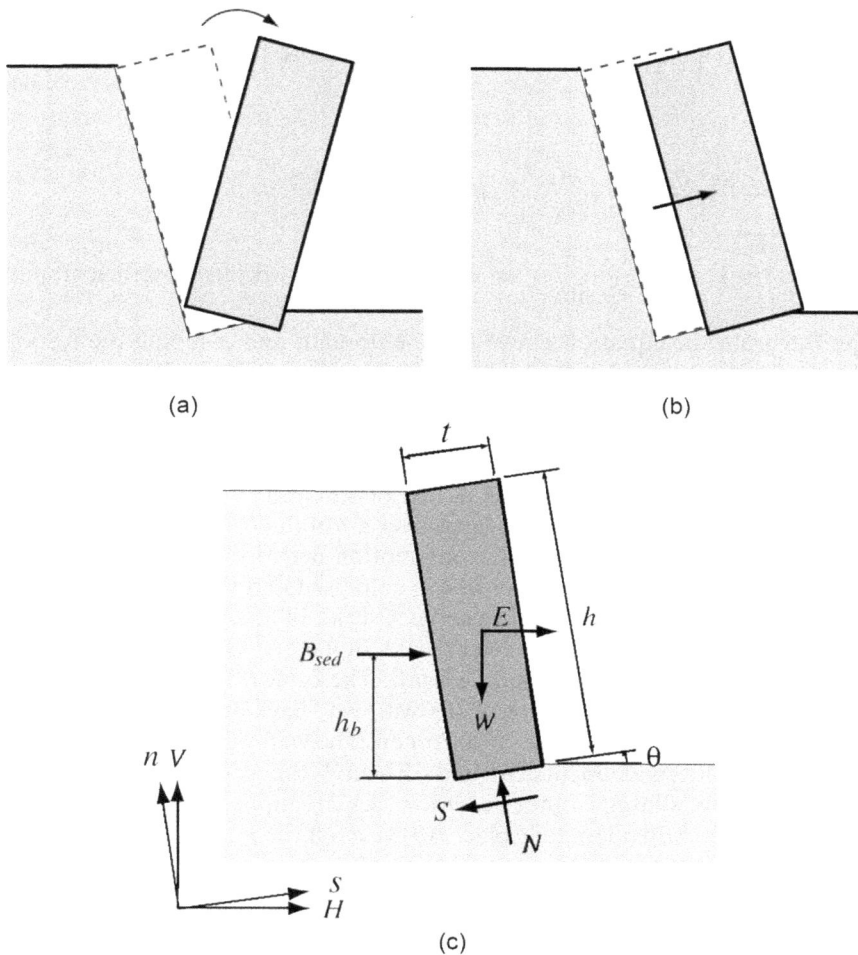

(a)

(b)

(c)

Figure 8.40 Failure modes and a mechanical model for retaining wall under ground shaking. (a) Toppling mode, (b) sliding mode and (c) mechanical model.

8.3.2.2 Dynamic limiting equilibrium method

It is also possible to model the rigid body translation and rotation of walls with the consideration of inertia term. In this type of formulation, the earthquake force is assumed to be proportional to the mass of the retaining wall. The differential equations can be obtained for sliding mode and toppling mode as follows (Tokashiki 2011, Aydan et al. 2002b, 2011):

Sliding mode

$$\frac{d^2 s}{dt^2} = g\left[A(t)\left(\cos\theta - \sin\theta \tan\phi_{ws}\right) - \left(\sin\theta + \cos\theta \tan\phi_{ws}\right)\right]. \tag{8.42}$$

Toppling mode

$$\frac{d^2\alpha}{dt^2} = \frac{3g}{R}\left[B(t)\left(\cos\theta\frac{h}{3t} - \sin\theta \right) + \frac{a(t)}{g}\left(\cos\theta\frac{h}{t} - \sin\theta \right) - \left(\sin\theta\frac{h}{t} + \cos\theta \right) \right] \quad (8.43)$$

where

$$B(t) = \frac{\gamma_s}{\gamma_w}\frac{h}{t}\frac{\cos^2\theta}{2} K_o\left(1 + \frac{a(t)}{g} \right); \quad A(t) = \left(B(t) + \frac{a(t)}{g} \right)$$

$$R = \sqrt{\left(\frac{h}{t}\right)^2 + 1}; \quad K_o = \frac{1 - \sin\phi_s}{1 + \sin\phi_s} \quad \text{or} \quad K_o = 1 - \sin\phi_s \quad \text{(Jacky load coefficient)}$$

One can integrate the equation above in time domain and compute the response of retaining walls as explained in the previous section (Eqs. 8.26–8.32).

Aydan et al. (2002b) have carried out numerous experiments on retaining walls with different configurations and backfilling materials. Figure 8.41 compares the measured response of retaining walls with those obtained from DLEM for retaining walls.

This method is also applied to the failure of Katsuren Castle caused by the 2010 Okinawa earthquake. The castle is located over a hill in Uruma City and the nearest strong motion station of the K-NET strong motion network is Gushikawa. The NW corner of the castle wall with a height of 4 m collapsed and there were numerous dislocations and rotation of blocks in the castle. A series of analyses were carried out using the DLEM and the acceleration records at Chinen and Gushikawa strong motion stations of the K-NET strong motion network. The typical size of the blocks ranges between 50 and 60 cm. The collapse of the wall was back-analysed using the strong motion records taken at Gushikawa and Chinen. The wall was stable against toppling mode for strong motions at Gushikawa and Chinen. If the records taken at Gushikawa are used, the relative sliding cannot be >10 cm, which implies that the wall should be stable although some slip might take place. However, if the records taken at Chinen are used the relative sliding could be >60 cm for $\theta = 5°$, which exceeds the half size of the

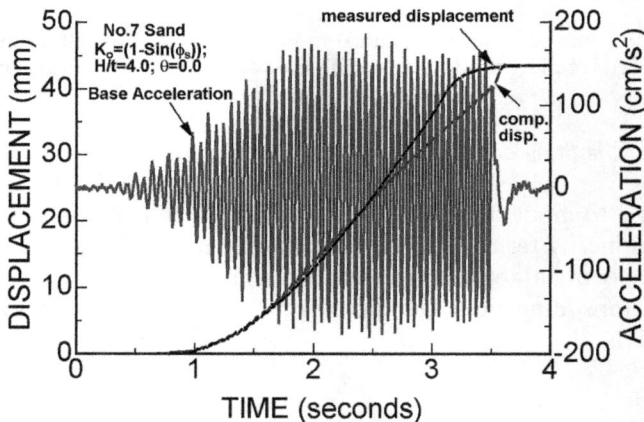

Figure 8.41 Comparison of measured responses with computed responses.

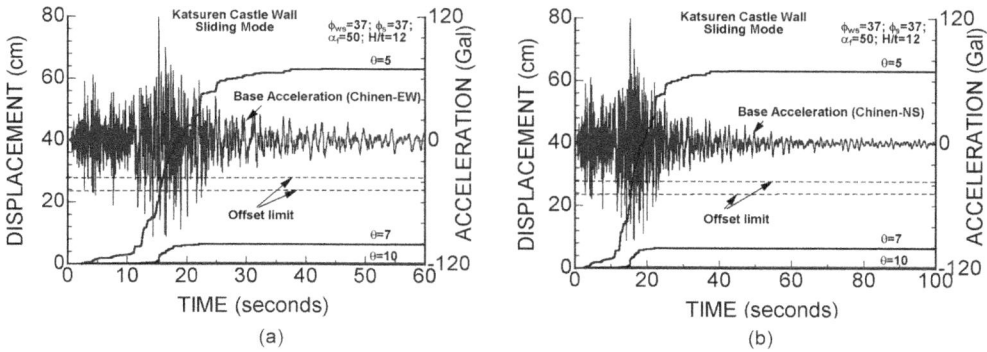

Figure 8.42 Sliding responses of the castle wall for Chinen record. (a) Computation for EW component (b) computation for NS component.

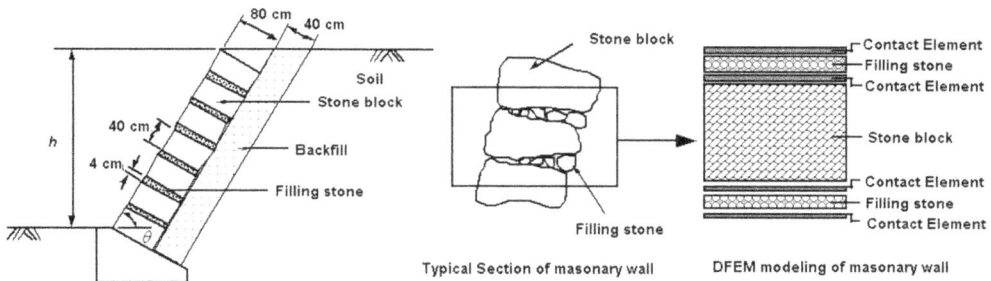

Figure 8.43 A typical masonry retaining wall and its DFEM representation.

block and this implies that the wall should collapse (Figure 8.42). The bulging of the wall and inclination of the foundation rock strongly support that this condition would be prevailing at the location of the collapse. As the castle is located on the top of the hill, it is likely that ground motions might have been also amplified.

Tokashiki et al. (2001, 2005) analysed the bulging phenomenon of retaining walls using the DFEM. The details of the analyses can be found in the respective references. The stability analysis of this masonry wall was first carried out under gravitational loading. Figure 8.43 shows how the retaining wall is modelled in the numerical analyses. The computed configuration is barrel-like and it is quite close to the actual observations (Figure 8.44a). A pseudo-dynamic simulation of the same retaining wall was carried out. Figure 8.44b shows the deformed configuration of the retaining wall for a horizontal seismic coefficient of 0.2. As expected, the horizontal displacement of the wall is larger than that for gravitational loading, and the effect of wall height on its deformation under gravitational and seismic loading. As expected the deformation of wall increases as a result of seismic loading and some relative displacements among blocks occur.

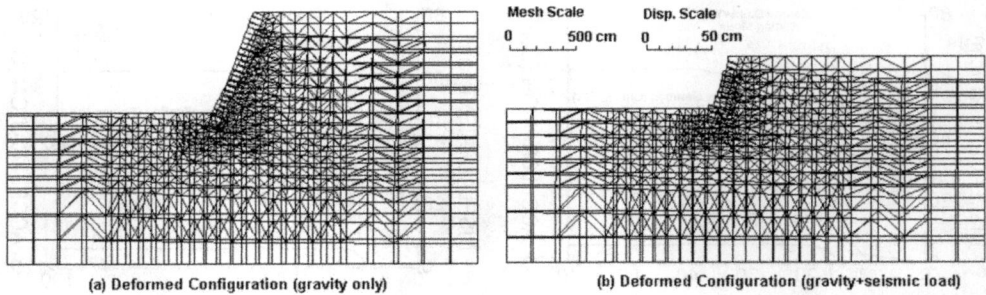

Figure 8.44 Comparison of deformed configurations for gravity only and gravity and seismic loading.

8.3.3   Seismic design of slopes

Slopes can be classified into two groups: soil slopes and rock slopes. Soil slopes can be treated using the methods described in Section 8.2.6 for embankments. Both pseudo-dynamic and DLEMs can be used. In addition, finite element method may also be used as an alternative.

Compared to the scale of soil slope failures, the scale and the impact of rock slope failures are very large, and the type of failure differs depending upon the geological structures of rock mass of slopes (Figure 8.45) (Aydan 1989; Aydan et al. 1989, 1992a,b, 1996a,b, 1997b, 2011; Horiuchi et al. 2018; Kumsar et al. 2000; Aydan and Kumsar 2010; Aydan and Kawamoto 1992). Furthermore, rock slope failures may involve both active and passive modes. However, passive modes are generally observed when ground shaking is quite large. In this section, the procedures and techniques for assessing the seismic stability of rock slopes are introduced.

8.3.3.1   Cliffs with toe erosion (bending failure)

Toe erosion of rock cliffs results in overhanging rock blocks. If the overhanging part of the cliff is continuously connected to the rest of rock mass, these rock blocks may be modelled as cantilever beams. However, depending upon the erosion type, their configuration may change from a rectangular prism to a triangular prism. If the bending theory is employed, one can easily derive the following set of equations by assuming that cliffs are subjected to gravitational and seismic loads as illustrated in Figure 8.46 for a unit thickness (Aydan 2019b; Horiuchi et al. 2018).

Equivalent beam thickness

$$h = h_b\left(1 - (1-\alpha)\frac{x}{L}\right) \tag{8.44}$$

Shear force

$$Q = V_o - (1 + k_v)\gamma h_b x\left(1 - (1-\alpha)\frac{x}{2L}\right) \tag{8.45}$$

FAILURES INVOLVING ONLY INTACT ROCK

SHEAR FAILURE   BENDING FAILURE

FAILURES INVOLVING DISCONTINUITIES AND INTACT ROCK

COMBINED SHEARING &   BUCKLING   FLEXURAL TOPPLING FAILURE
SLIDING

FAILURES INVOLVING ONLY DISCONTINUITIES

PLANE   SLIDING   WEDGE SLIDING

TOPPLING

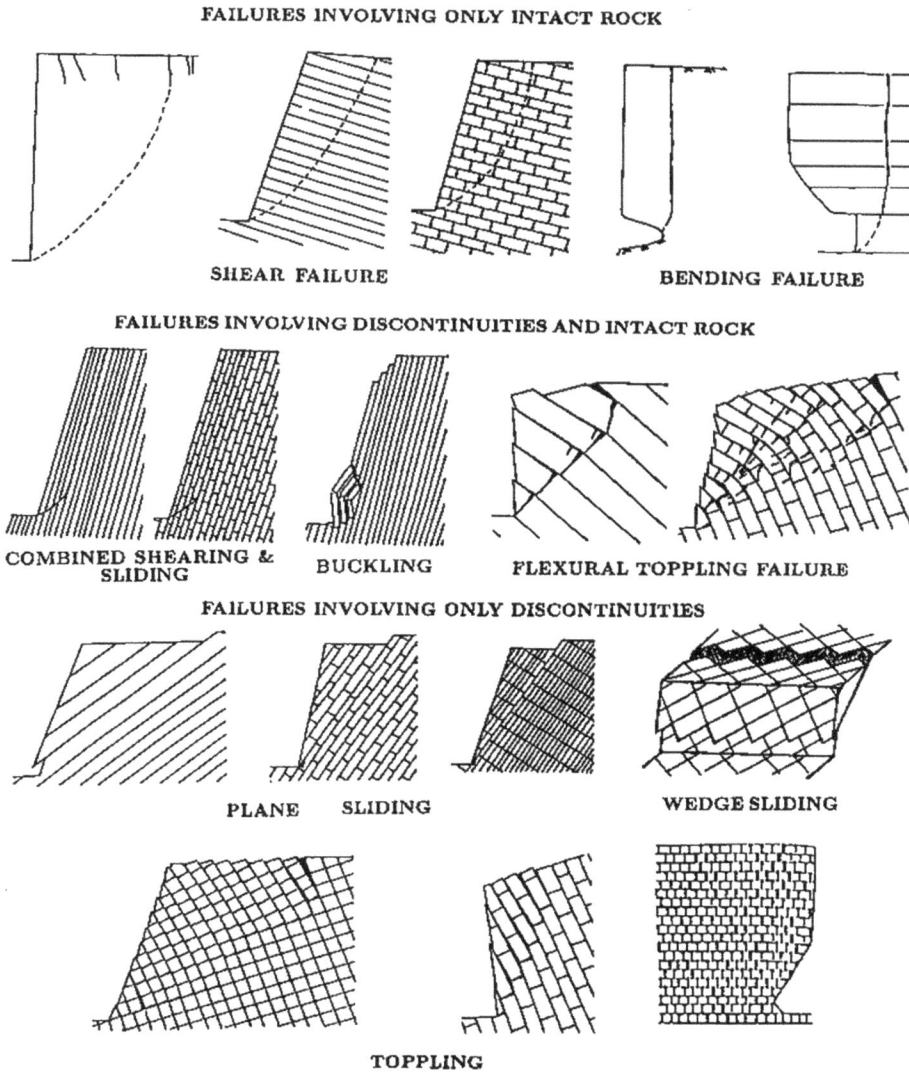

Figure 8.45 Active failure modes of rock slopes. (From Aydan (1989).)

Bending moment

$$M = M_o + V_o x - (1 + k_v)\gamma h_b x^2 \left(\frac{1}{2} - (1 - \alpha)\frac{x}{6L} \right) \tag{8.46}$$

Bending stress at the outer fibre

$$\sigma = k_h \gamma h_b L \left(\frac{1 + \alpha}{2} \right) + 6\frac{M}{h^2} \tag{8.47}$$

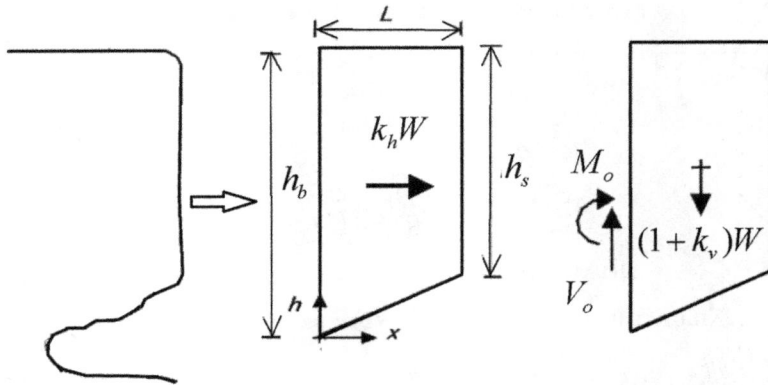

Figure 8.46 Modelling of overhanging cliffs. (a) Actual, (b) idealized and (c) mechanical model.

where

$$\alpha = \frac{h_s}{h_b}; \ \ V_o = (1+k_v)\gamma h_b L\left(1 - \frac{(1-\alpha)}{2}\right)$$ (8.48a)

$$M_o = -(1+k_v)\gamma h_b L^2\left(\frac{1}{2} - \frac{(1-\alpha)}{3}\right)$$ (8.48b)

h_b, h_s, γ and L are the beam height at the base and at the far end, unit weight of rock mass and erosion depth, respectively. k_h and k_v are the horizontal and vertical seismic coefficients.

Figure 8.47 shows the bending stress distributions along the outermost fibre of the beam for different geometrical configurations. The severest condition occurs when the beam has a rectangular shape and the value of the bending stress is much higher for the rectangular configuration. Tensile stress is also the largest at the base of the canti-lever beam. As discussed by Aydan and Kawamoto (1992), the cantilevers fail immedi-ately once the tensile stress exceeds the tensile strength of rock mass. Furthermore, the seismic loads in addition to gravitational load would make the cliffs more vulnerable to failure during earthquakes.

While the consideration of seismic loads is based on the seismic coefficient method in this section, one may assess any amplification from a response analysis if the fre-quency content of earthquake waves is of great importance for a given earthquake record using the following formula for the natural frequency (first mode) of cantilever beams as a first approximation:

$$f_1 = \frac{1.875^2}{2\pi}\sqrt{\frac{EI}{mL^4}}$$ (8.49)

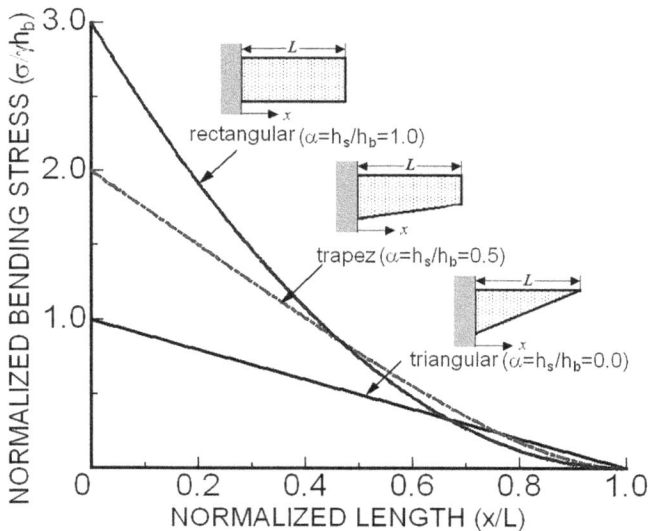

Figure 8.47 Comparison of distribution of bending stress at the outermost fibre of beam with different configurations.

where L is the erosion depth; E is the elastic modulus; m is the mass per unit length; I is the inertia moment of area.

The toppling and/or sliding of a block bounded by discontinuities at the cliff under earthquake shaking can also be analysed and its stability can be assessed using the dynamic limiting equilibrium method described in Section 8.2.7.2. It is also possible to introduce cohesion or tensile strength of discontinuities.

8.3.3.2 Shear and planar failure

Shear failure and planar failure under earthquake loading can be assessed using the methods described in Section 8.2.6. The most important aspect is the definition of failure plane. Planar failure is almost straightforward as the failure would be governed by a thoroughgoing discontinuity plane. Aydan (2019b) also proposed a method for reinforcement or rock slopes against planar sliding under dynamic conditions. If the plane is curved, it would be wise to carry out analyses using the pseudo-dynamic procedure. Once the failure plane is determined, then DLEM can be utilized. Of course, the numerical techniques based on finite element method, DFEM or other methods can also be used. The applications of the methods developed previously are given herein.

The first example is associated with the Chiufengershan large-scale slope failure (Aydan 2015b). The acceleration records at nearby stations denoted TCU089 (Central Weather Bureau, CWB, 1999) were used in computations. The computed response of displacement and velocity of the mass centre and its path are shown in Figure 8.48. The material and geometrical properties used in computation are also given in Figure 8.48. As noted from the figure, the path of the mass centre during motion is well estimated by

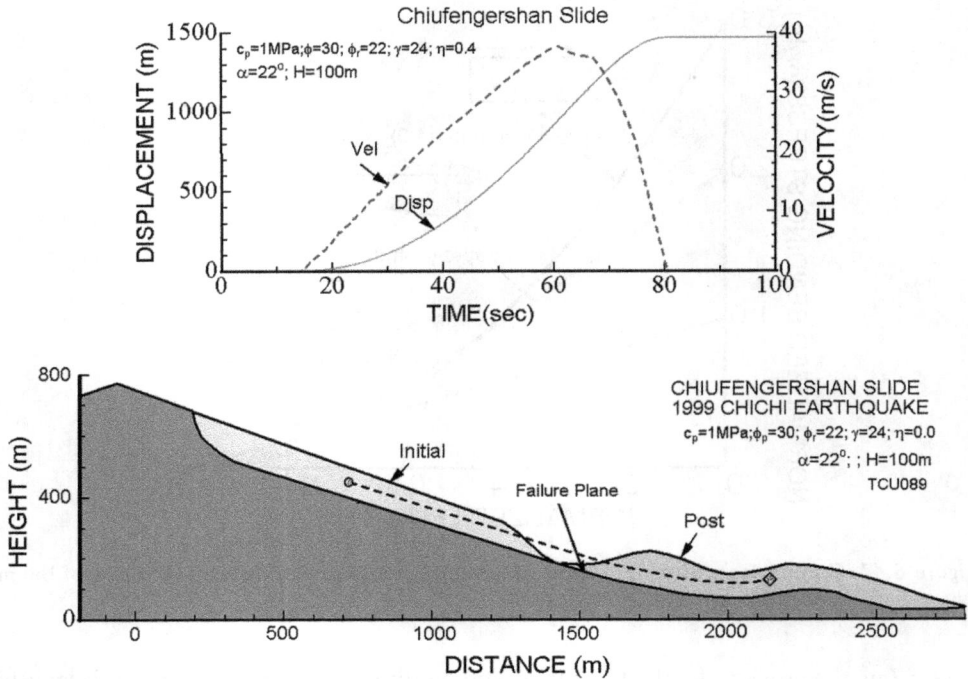

Figure 8.48 Comparisons of computed results for the Chiufengershan slope failure.

the mathematical model for the chosen material properties. Another example was associated with the Takezawa slope failure located in the Uji River valley caused by the 2004 Chuetsu earthquake. The rock mass here was intercalated weakly cemented sandstone and mudstone, and the failure took place along the contact between the mudstone and sandstone layers in the form of planar sliding. The acceleration records at Ojiya were used in computations. Figure 8.49 shows the absolute displacement and the horizontal and vertical position of the mass centre in time domain together with the EW component of the Ojiya strong motion record released by K-NET (2004). The shear strength properties of the failure surface are also shown in the respective figures together with the assumption of water head equivalent to 0.59 times the height of the failed body. As noted from the figure, the body slid for about 40 seconds and the first 10 seconds was associated with shaking induced by the earthquake. The total movement of the failed body was about 100 m.

8.3.3.3 Wedge failure

Kumsar et al. (2000) have advanced the method of stability assessment proposed by Kovari and Fritz (1975) for wedge failure of rock slopes under different loading conditions based on the seismic coefficient method and confirmed its validity through experiments. Aydan and Kumsar (2010) extended the method to evaluate sliding responses of rock wedges under dynamic loading conditions under submerged conditions with

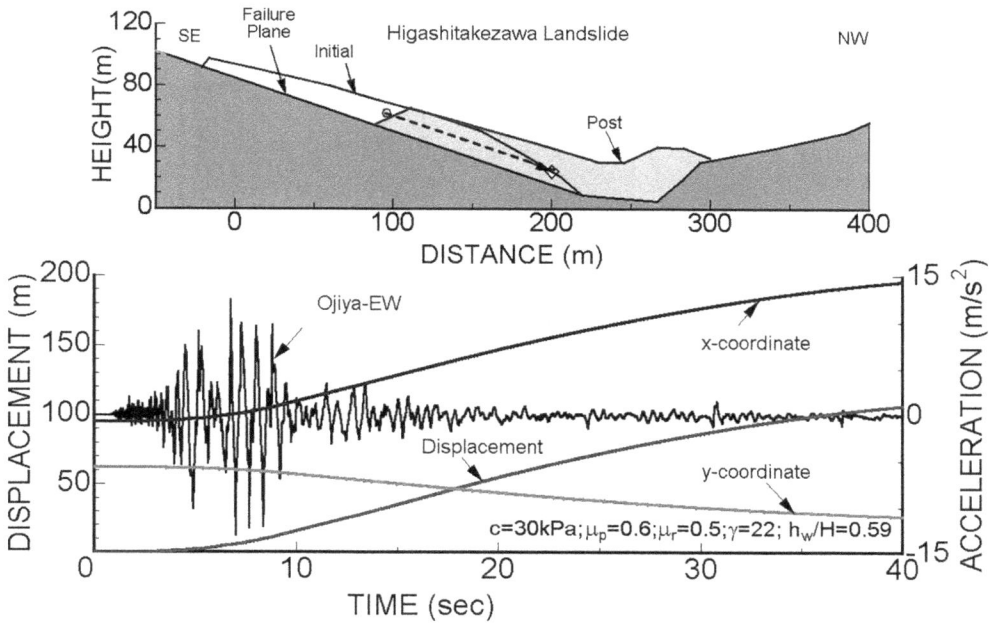

Figure 8.49 Comparisons of computed results for the Takezawa slope failure.

viscos resistance. A wedge subjected to dynamic and water loading is considered, as shown in Figure 8.50. One can easily write the following dynamic equilibrium conditions for the wedge during sliding motion on two basal planes in the coordinate system *Osnp* shown in Figure 8.50.

$$\sum F_s = (W + U_{td} - E_v)\sin i_a + (E_i - U_{sd})\cos i_a - S = m\frac{d^2 s}{dt^2} \qquad (8.50)$$

$$\sum F_n = (W + U_{td} - E_v)\cos i_a - (E_i - U_{sd})\sin i_a - N = m\frac{d^2 n}{dt^2} \qquad (8.51)$$

$$\sum F_p = -(N_1 + U_{b1d})\cos\omega_1 + (N_2 + U_{b2d})\cos\omega_2 + E_p = m\frac{d^2 p}{dt^2} \qquad (8.52)$$

where $N = (N_1 + U_{b1d})\sin\omega_1 + (N_2 + U_{b2d})\sin\omega_2$; W is the weight of wedge; E_v is the dynamic vertical load; E_i is the dynamic force in the direction of the intersection line; and E_p is the dynamic load perpendicular to the intersection line. The other parameters are shown in Figure 8.51.

Although the dynamic vectorial equilibrium equation is written in terms of its components, they correspond to a very general form for wedges sliding along the intersection line while being in contact with two basal planes. Furthermore, the earthquake force is decomposed to its corresponding components in the chosen coordinate system. It should be also noted that the introduction of water forces acting on the top

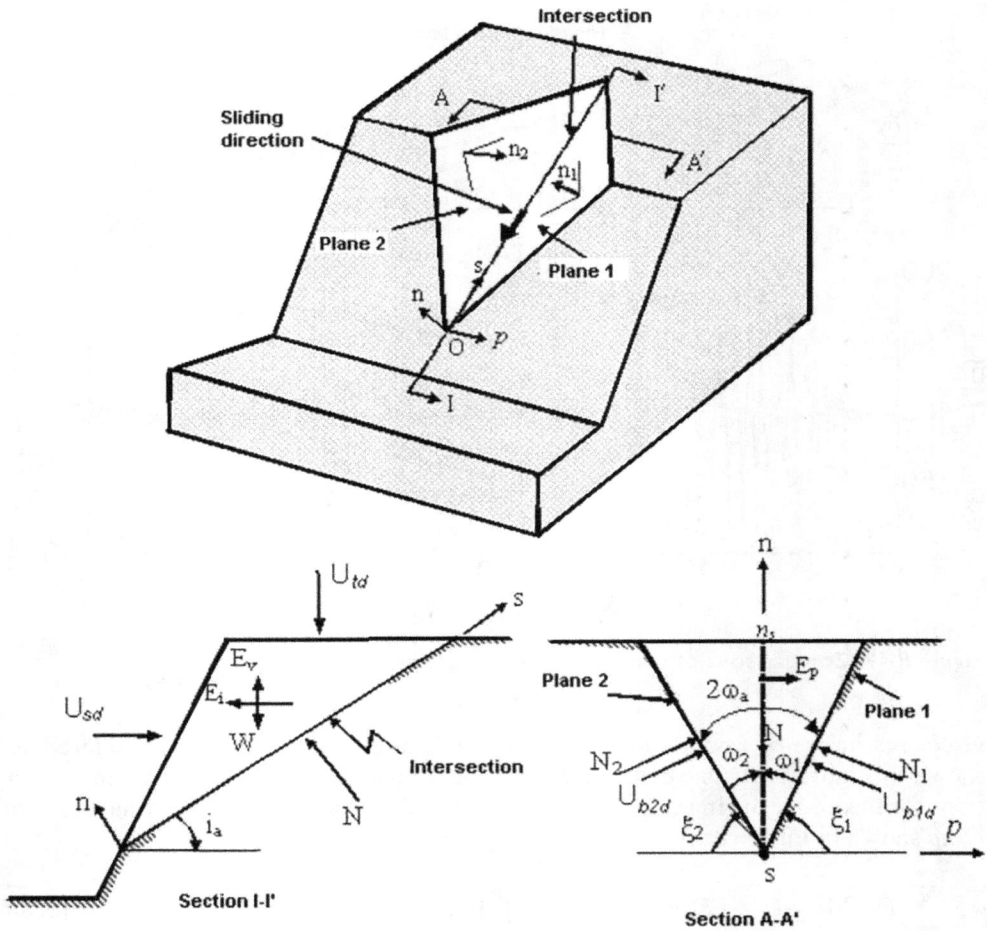

Figure 8.50 Mechanical model for wedge failure.

and side surfaces of wedges would enable one to consider various conditions such as fully or partially submerged, fully or partially saturated or dry condition. In addition to static water forces, water forces are assumed to be influenced by ground shaking with an additional dynamic component and they are given in the following forms:

$$U_{td} = U_{ts}(1+\beta); \ \ U_{sd} = U_{ss}(1+\beta); \ \ U_{b1d} = U_{b1}(1+\beta); \ \ U_{b2d} = U_{b2}(1+\beta). \tag{8.53}$$

The dynamic factor β in Eq. (8.53) is assumed to be of the following form on the basis of previous studies by Aydan and Ulusay (2002) and Aydan et al. (2008b):

$$\beta = |a_i \sin i_a + a_v \cos i_a| \tag{8.54}$$

Figure 8.51 View of the Hattian rock slope failure.

One can easily obtain the following identity from Eq. (8.52) by assuming that there are no motions upward and perpendicular to the intersection line:

$$N_1 + N_2 = \left[(W + U_{td} - E_v)\cos i_a - (E_i - U_{sd})\sin i_a\right]\lambda_i - E_p\lambda_p - U_{bd} \qquad (8.55)$$

where

$$\lambda_i = \frac{\cos\omega_1 + \cos\omega_2}{\sin(\omega_1 + \omega_2)}; \quad \lambda_p = \frac{\sin\omega_1 - \sin\omega_2}{\sin(\omega_1 + \omega_2)}; \quad U_{bd} = U_{b1d} + U_{b2d}. \qquad (8.56)$$

If the resistance is assumed to obey Mohr–Coulomb criterion together with a Bingham-type viscous resistance (Aydan and Ulusay 2002; Aydan et al. 2008b), one may write the following:

$$\begin{aligned} T = (N_1 + N_1)\mu_r + c_r(A_1 + A_2) \\ + \varsigma\frac{\left[(W + U_{td} - E_v)\sin i_a + (E_i - U_{sd})\cos i_a\right]}{g}\frac{ds}{dt} \end{aligned} \qquad (8.57a)$$

$$\mu_r = \tan\phi_r \qquad (8.57b)$$

where c_r and ϕ_r are residual cohesion and friction angle. A_2 and A_1 are the area of plane 1 and plane 2, ς is viscous resistance and g is gravitational acceleration, respectively. Under frictional condition, it should be noted that normal force $(N_1 + N_2)$ cannot be negative (tensile). If such a situation arises, normal force $(N_1 + N_2)$ should be set to 0 during computations. Let us introduce the following parameters:

$$\eta_v = \frac{E_v}{W} = \frac{a_v}{g}; \quad \eta_i = \frac{E_i}{W} = \frac{a_i}{g}; \quad \eta_p = \frac{E_p}{W} = \frac{a_p}{g}$$
$$\beta_{td} = \frac{U_{td}}{W}; \quad \beta_{sd} = \frac{U_{td}}{W}; \quad \beta_{bd} = \frac{U_{bd}}{W} \tag{8.58}$$

where $a_v; a_i; a_p$ are the acceleration components resulting from dynamic loading. $\beta_{td}; \beta_{sd}; \beta_{b1d}; \beta_{b2d}$ are the dynamic water force coefficients. The following dynamic equilibrium equation must be satisfied during the sliding motion of the wedge:

$$S = T \tag{8.59}$$

If the relations given by Eqs. (8.50), (8.56), (8.57) and (8.58) are inserted in Eq. (8.59), one can easily obtain the following differential equation:

$$\frac{d^2 s}{dt^2} - D\frac{ds}{dt} - gB - \frac{c_r}{mg}(A_1 + A_2) = 0 \tag{8.60}$$

where

$$B = (1 + \beta_{td} - \eta_v)(\sin i_a - \cos i_a \mu_r \lambda_i) + (\eta_i - \beta_{sd})(\cos i_a + \sin i_a \mu_r \lambda_i) - \mu_r(\eta_p \lambda_p + \beta_{bd})$$

$$D = \varsigma\left[(1 + \beta_{td} - \eta_i)\sin i_a + (\eta_i - \beta_{sd})\cos i_a\right]$$

As dynamic loads are very complex in time domain, the solution of Eq. (8.60) is only possible through numerical integration methods. The time-domain problems in mechanics are generally solved by finite difference techniques. For this purpose, there are different finite difference schemes. The solution of Eq. (8.60) is based on linear acceleration finite difference technique (i.e. Aydan and Ulusay 2002, Aydan et al. 2008b) and as described in Section 8.2.6.2.

As the resulting dynamic shear force exceeds the shear resistance of the wedge at time $(t = t_i = i\Delta t)$, one can easily incorporate the variation of shear strength of discontinuities from peak state to residual state through the following condition:

$$\ddot{s}_i = \left(\frac{d^2 s}{dt^2}\right)_{t=t_i} = gB_p + \frac{c_p}{mg}(A_1 + A_2); \quad \dot{s}_i = 0; \quad s_i = 0. \tag{8.61}$$

where

$$B_p = (1 + \beta_{td} - \eta_v)(\sin i_a - \cos i_a \mu_p \lambda_i) + (\eta_i - \beta_{sd})(\cos i_a + \sin i_a \mu_p \lambda_i) - \mu_r(\eta_p \lambda_p + \beta_{bd})$$

c_p and ϕ_p are the peak cohesion and friction angles

The method has been applied to model tests utilizing shaking tables as well as actual case histories in nature. Here one example of this method has been briefly described (e.g. Misawa et al. 1993; Aydan and Kumsar 2010). On October 8, 2005 at 8:50 (3:50UTC), a large devastating earthquake occurred in Kashmir region of Pakistan. The depth of the earthquake was estimated to be about 10 km and it had a magnitude of 7.6 (Aydan 2006a,b). The earthquake resulted from the subduction of the Indian plate beneath the Eurasian plate, and the earthquake was due to thrust faulting. Although there was no surface rupture as a result of the faulting, extensive slope failures were observed along the expected surface trace of the causative fault. The satellite images indicated that a large-scale slope failure occurred nearby Hattian in the vicinity of the SE tip of the causative fault (Figure 8.51). The slope failure was at Hattian (Dana Hill), and it was an asymmetric wedge sliding (Figure 8.52).

The wedge sliding failure at Hattian was quite large in scale. The sliding area was 1.5 km long and 1.0 km wide. Rock mass consisted of shale and sandstone, and it constituted a syncline. The estimated wedge angle was about 100° and it was asymmetric.

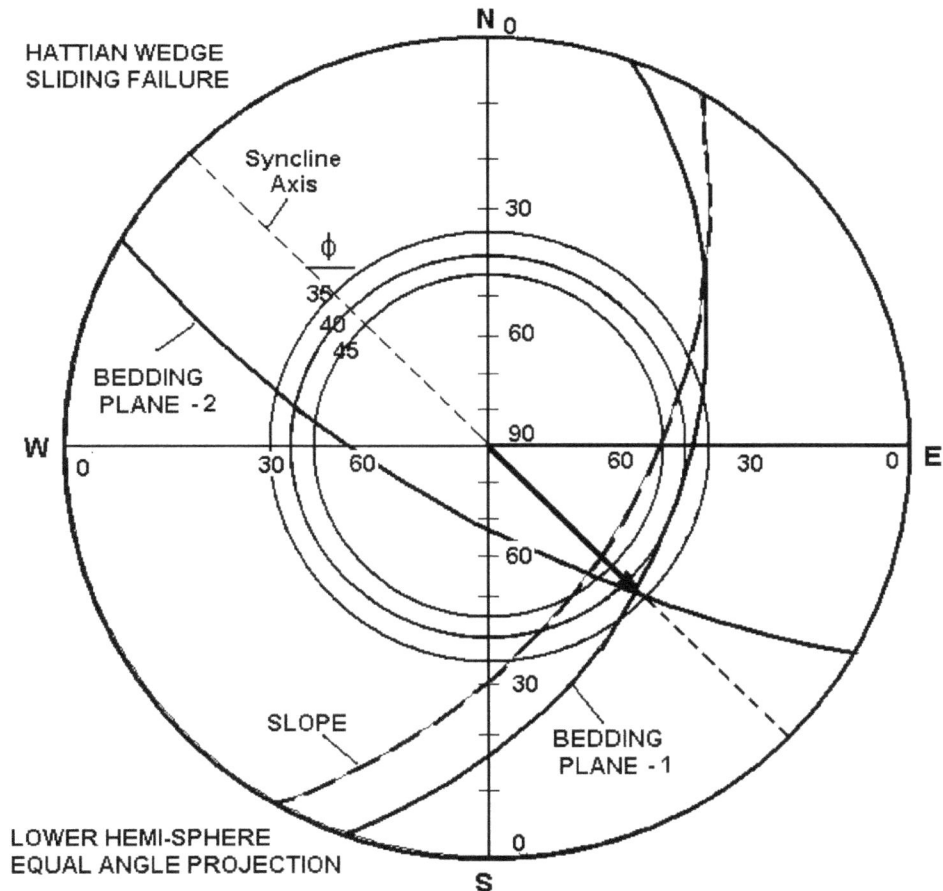

Figure 8.52 Kinematic analysis of the Hattian rock slope failure.

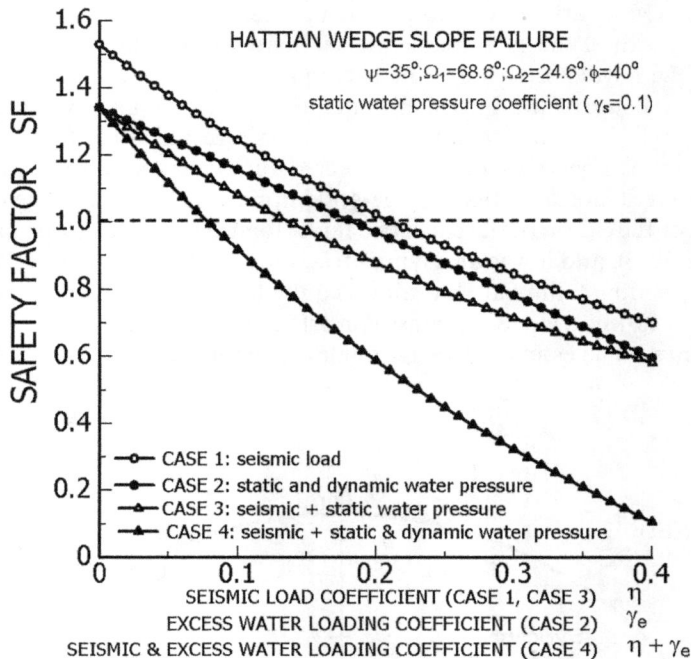

Figure 8.53 Results of parametric analyses.

The friction angle of shale from the tilting test was more than 35° with an average of 40°. Figure 8.52 shows the kinematic analysis of the Hattian slope through the projection of structural planes on equal angle stereo-net together with friction cones.

The limiting equilibrium analysis for wedge sliding failure (Kumsar et al. 2000; Aydan and Kumsar 2002) indicated that the SF of the slope would be 1.55 under dry static conditions (Aydan et al. 2009f). However, the mountain wedge becomes unstable when ground acceleration is equivalent to the horizontal seismic coefficient of 0.3 and the SF reduces to 0.9 under such a condition. Some parametric studies for the same wedge sliding at Hattian were carried out, and the results are shown in Figure 8.53. The result indicated that the seismic loading was the most critical parameter governing the wedge sliding.

Using the acceleration record of Abbotabad (Okawa 2005) and multiplying the record by an amplification factor of 1.27 to get a seismic coefficient value of 0.3, a dynamic simulation of the wedge failure was carried out. The residual friction angle was reduced to 28.5° from the peak friction after yielding. The results are shown in Figure 8.54. The slope becomes unstable by the earthquake-induced ground shaking, and the motion of the failed body increases with time. Unless the geometrical profile of the sliding surface changes, the sliding motion would continue with a constant velocity. Since the actual profile of the sliding surface is still not available, it is difficult to do so.

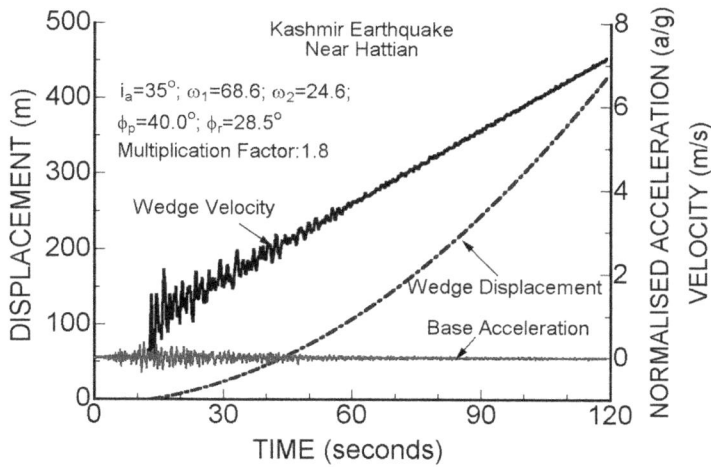

Figure 8.54 Results of dynamic analyses.

8.3.3.4 Combined shearing and sliding failure

Aydan et al. (1991, 1992a) was first to point out this failure form and formulated the stability analysis method. This method considers the effect of earthquakes on the slopes using the seismic coefficient method. Figure 8.55a illustrates how the force system acts on a typical layer of rock mass.

By considering the failure plane denoted as α_{min}, the force equilibrium for each direction is written as follows:

$$\sum F_{H_i} = E_i + (N_i + U_i)\sin\alpha_{min} + (Ns_{i-1} + Us_{i-1} - Ns_i - Us_i)\sin\theta_d$$
$$- S_i \cos\alpha_{min} + (Ss_{i-1} - Ss_i)\cos\theta_d = 0 \tag{8.62}$$

$$\sum F_{V_i} = W_i + (N_i + U_i)\cos\alpha_{min} + (Ns_{i-1} + Us_{i-1} - Ns_i - Us_i)\cos\theta_d$$
$$- S_i \sin\alpha_{min} + (Ss_{i-1} - Ss_i)\sin\theta_d = 0 \tag{8.63}$$

where W_i, U_i, E_i, S_i, N_i are the force components. Assuming that the shearing strength of the layer on the failure plane obeys the Mohr–Coulomb yield criterion together with a given SF, the following can be written:

$$S_i / SF = C \cdot l_i + N_i \tan\phi \tag{8.64}$$

In addition, the interlayer shear strength also obeys the Mohr–Coulomb criterion as given below:

$$Ss_i = C_d \cdot ls_i + Ns_i \tan\phi_d \tag{8.65}$$

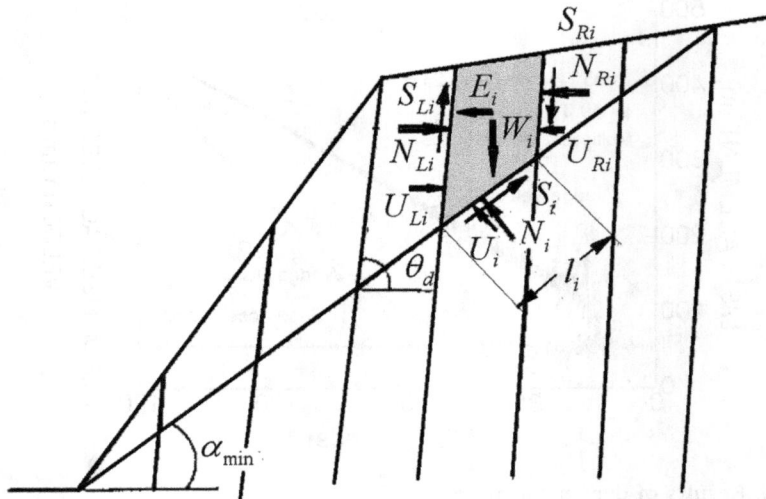

Figure 8.55 Visualization of force system acting on a typical layer.

In the last layer, the boundary condition is given by

$$N_L = Ns_n - Ns_{end} \tag{8.66}$$

The stability of the slope is assessed by introducing the following criteria:

If $N_L > 0$, slope is unstable
If $N_L = 0$, slope is at the limiting state   (8.67)
If $N_L < 0$, slope is stable

8.3.3.5   Flexural toppling failure

Flexural toppling failure was first formulated by Aydan and Kawamoto (1992) and the stability of slopes against flexural toppling failure was checked through model tests of rock slopes subjected to gravity and shaking through. The SF of a typical layer against flexural toppling is given by the following relation (Figure 8.56):

$$SF = \frac{\sigma_t}{\sigma_x} \tag{8.68}$$

The force equilibrium equations for shear (s) and normal (n) directions and the moment equilibrium of the typical layer numbered (i), can be written as given below:

$$\Sigma F_{Si} = S_i - W_i \sin(\theta_d - 90°) - E_i \cos(\theta_d - 90°) + Ns_i - Ns_{i-1} + Us_i - Us_{i-1} = 0 \tag{8.69}$$

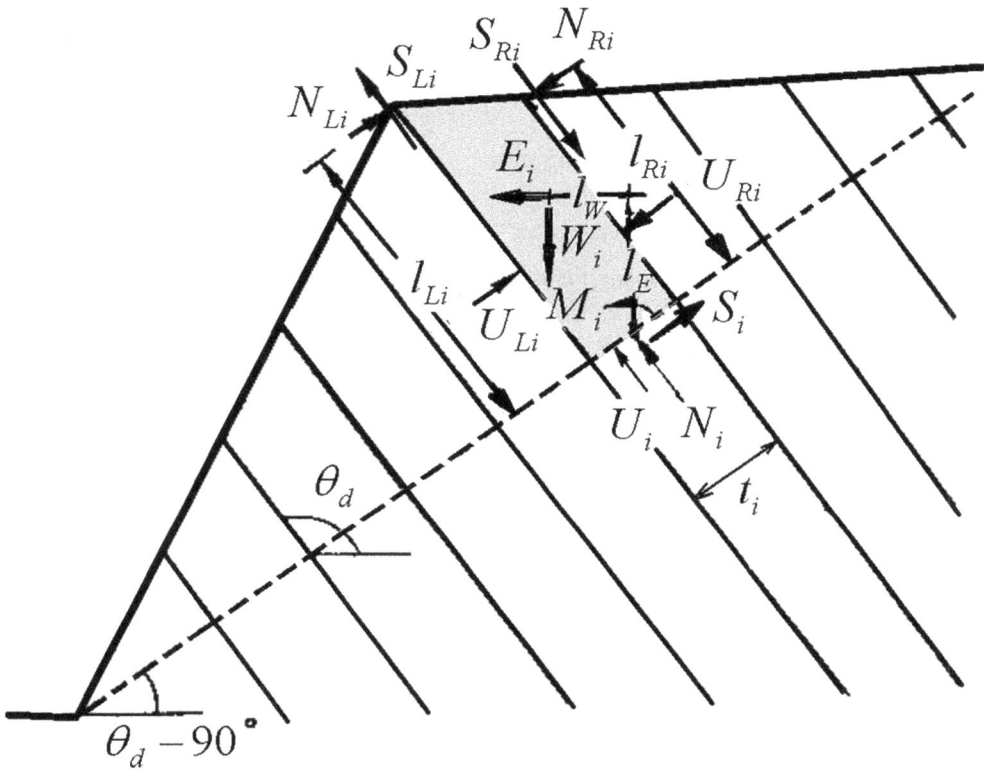

Figure 8.56 Visualization of force system acting on a typical layer.

$$\Sigma F_{Ni} = N_i - W_i \cos(\theta_d - 90°) + E_i \sin(\theta_d - 90°) + Ss_{i1} - Ss_{i-1} + U_i = 0 \qquad (8.70)$$

$$\Sigma M_{O_i} = M_i - ls_i \cdot Ns_i + ls_{i-1} \cdot Ns_{i-1} - \frac{1}{2} t \cdot Ss_i - \frac{1}{2} t \cdot Ss_{i-1}$$
$$+ l_{Ei} \cdot E_i + l_{Wi} \cdot W_i - \frac{1}{3} ls_i Us_i + \frac{1}{3} ls_{i-1} Us_{i-1} = 0 \qquad (8.71)$$

The earthquake force in this formulation is considered utilizing the seismic coefficient method. If the interlayer shear strength obeys the Mohr–Coulomb criterion, one can write the following:

$$Ss_i = C_d \cdot ls_i + Ns_i \tan \phi_d \qquad (8.72)$$

The tensile stress at the base of the typical layer on the anticipated failure plane takes the following form:

$$\sigma_{xi} = \frac{N_i}{A_i} + \frac{M_i}{I_i} \frac{t}{2} \qquad (8.73)$$

The unknowns are S_i, N_i, Ns_i, Ss_i, σ_{xi}. Starting from the uppermost layer, which is unstable under the given force condition, Ns_i, Ss_i are obtained for each layer using the step-by-step method. The following condition is introduced at the lowermost layer numbered (n) and its stability is assessed against the flexural toppling failure:

If $N_{Sn} > 0$, slope is unstable
If $N_{Sn} = 0$, slope is at the limiting state   (8.74)
If $N_{Sn} < 0$, slope is stable

8.3.3.6   Blocky columnar toppling failure

Toppling failure of a column consisting of several rectangular blocks of the same width can occur, as theoretically shown by Aydan et al. (1989). Therefore, Aydan et al. (1989) proposed a theoretical formulation for this particular situation and introduced the seismic coefficient method to consider the effects of earthquakes (Aydan et al. 1991). The force condition on a typical blocky column may be given in the following form:

$$\Sigma F_{Si} = S_i + U_{Si} - U_{Si-1} + N_{Si} - N_{Si-1} - W_i \sin\theta_{d2} - E_i \cos\theta_{d2} = 0 \qquad (8.75)$$

$$\Sigma F_{Ni} = N_i + Ss_i - Ss_{i-1} + U_i - W_i \cos\theta_{d2} + E_i \cos\theta_{d2} = 0 \qquad (8.76)$$

The disturbing and resisting moments acting may also be given in the following forms:
Disturbing moment

$$M_+ = ls_{i-1} \cdot Ns_{i-1} + l_{Ei} \cdot E_i + l_{Wi} \cdot W_i + \frac{1}{3} ls_i \cdot Us_{i-1} + l_i \cdot U_i \qquad (8.77)$$

Resisting moment

$$M_- = ls_i \cdot Ns_i + t_i \cdot Ss_{i-1} + \frac{1}{3} ls_i \cdot Us_i \qquad (8.78)$$

The SF against toppling may be defined as follows (Figure 8.57):

$$SF = \frac{M_-}{M_+} \qquad (8.79)$$

If the inter-column, shear strength obeys the Mohr–Coulomb criterion, the following can be written:

$$Ss_i = Cd_1 \cdot ls_i + Ns_i \tan\phi_{d1} \qquad (8.80)$$

One can obtain from Eqs. (8.77)–(8.80) N_i, S_i, Ns_i, Ss_i by using the step-by-step method. At the lowermost column, the following condition is obtained:

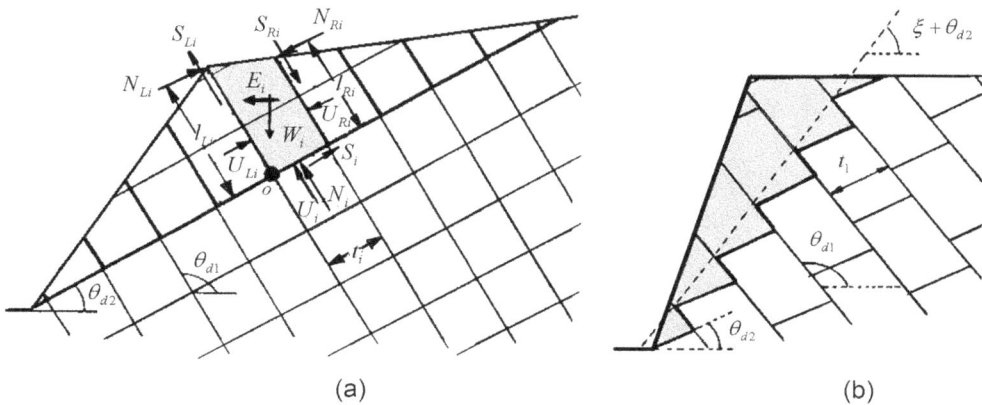

Figure 8.57 Visualization of force system acting on a typical layer and stepped failure plane. (a) Force system and (b) stepped failure plane.

$$N_L = Ns_n - Ns_{\text{end}} \tag{8.81}$$

Using SF and the value of N_L and introducing the following criterion, the stability of the slope is assessed against block columnar failure:

If $N_L > 0$, slope is unstable
If $N_L = 0$, slope is at the limiting state   (8.82)
If $N_L < 0$, slope is stable

When the failure is stepped, as happens in intermittent blocky rock slopes, the same procedure can be utilized by replacing the inclination of the failure plane by

$$\theta_{d2}^* = \xi + \theta_{d2} \tag{8.83}$$

where ξ is called the intermittency angle (Aydan et al. 1989).

8.3.3.7 Empirical relations between earthquake magnitude and limiting distance for slope failures

Keefer (1984) studied slope failures induced by earthquakes in the United States and other countries, and he proposed some empirical bounds for slope failures, which are classified as disrupted or coherent. The empirical bounds of Keefer (1984) is not specifically given as formula. The author compiled slope failures caused by recent worldwide earthquakes according to his classifications and plotted them as shown in Figure 8.58. Modifying the previous equation of Aydan et al. (2009a,b,f), the following empirical equation was proposed by Aydan et al. (2012a) for the maximum hypocentral distance of disrupted and coherent slope failures as a function of earthquake magnitude and fault orientation.

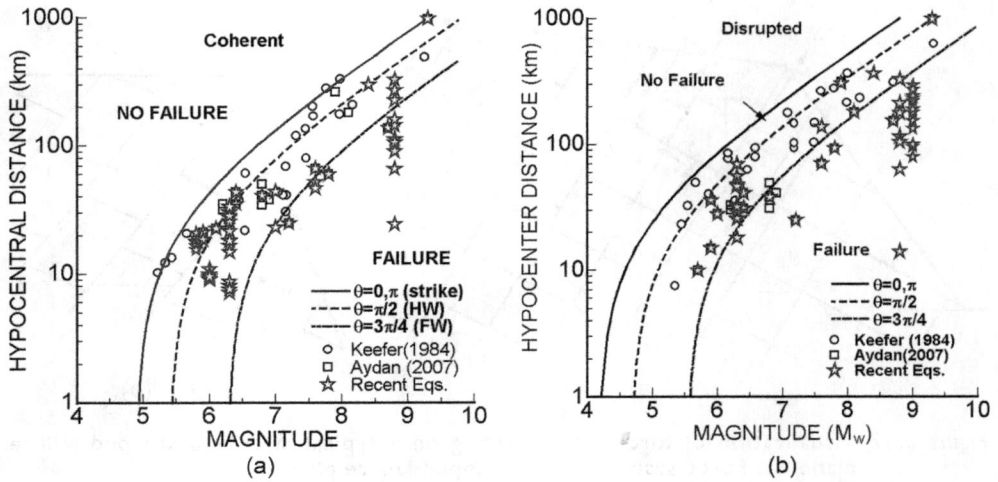

Figure 8.58 Comparison of empirical relations with observations. (a) coherent (b) disrupted.

Table 8.14 Parameters in Eq. (8.84) for disrupted and coherent Slope failures

Condition	A	B	C
Disrupted	0.3	0.8	25
Coherent	0.2	0.8	30

$$R_o = A * (3 + 0.5\sin\theta - 1.5\sin^2\theta) * e^{B \cdot M_w} - C \tag{8.84}$$

Constants A, B and C of Eq. (8.84) for disrupted and coherent landslides are given in Table 8.14. As ground accelerations differ according to the location with respect to fault geometry, the empirical bounds proposed herein can also provide some guidelines for the scattering range of observations.

8.3.3.8 Relation between thoroughgoing discontinuity inclination and slope angle

Aydan and co-workers (Aydan 2006a,b; Aydan et al. 1989, 1991) proposed a method based on the limiting equilibrium approach to determine the limiting stable slope angle under given seismic, geometrical and physical conditions. Figure 8.59 shows a plot of the slope angle of various rock slopes versus the inclination of the thoroughgoing discontinuity set whose strike is parallel or nearly parallel to the axis of the slope. Stable slopes are denoted by S and failed slopes by F. The plotted data include the data

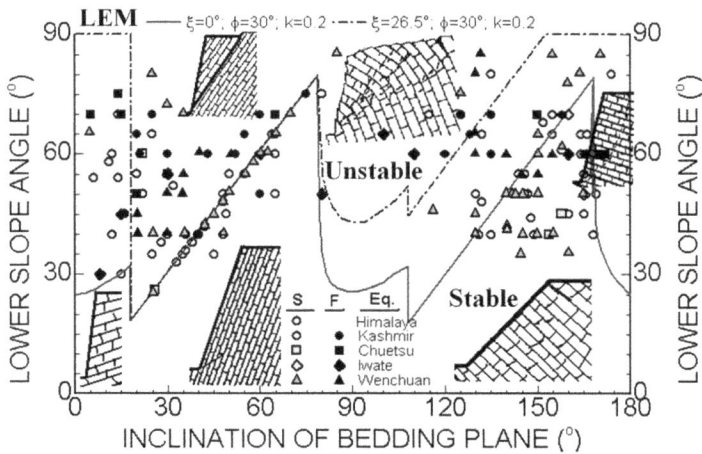

Figure 8.59 The relation between slope angle and bedding plane angle for stable (S) and failed (F) case histories in recent earthquakes. (From Aydan (2013).)

on presently stable natural rock slopes and rock slopes the failed due to earthquakes. Most of the data are compiled by the author. In the plots, the stability charts of a slope with a ratio of *t/H:1/75* for cross continuous and intermittent patterns ($\xi = 26.5°$) and $\eta = 0.0$ are also included to have a qualitative insight rather than a quantitative comparison (Aydan et al. 1989). The chosen value of *t/H* is arbitrary and may not correspond to the ratios of slopes plotted in the figure. Nevertheless, the plotted cases confirm the qualitative tendency described in Figure 8.59. It is also interesting to note there is almost no failed slopes when the slope angle is <25°–30° and most of the failed slopes have a slope angle >25°–30°. This is in accordance with the conclusion of Keefer (1984). Nevertheless, it is also noted that there are a great number of stable slopes with a slope angle >25°–30°. This implies that the angle and the height of slopes cannot be the only parameters determining the overall stability of natural rock slopes. Therefore, the orientations of discontinuity sets, their geometrical orientations with respect to slope geometry and their mechanical properties and loading conditions must also play a great role in determining the stable angles of natural rock slopes. The results shown in Figure 8.59 may serve as guidelines for a quick assessment of the stability of natural rock slopes and how to select the slope-cutting angle in actual restoration of failed slopes.

8.4   SEISMIC DESIGN OF UNDERGROUND STRUCTURES

Seismic design of underground structures can be divided into soil and rock underground structures. Accordingly, the design concepts would be different as soil cannot support the open space without any support while rock mass can be self-sustaining stress acting in the ground due to gravitational, seepage and seismic forces.

8.4.1　Tunnels

8.4.1.1　Shallow soil tunnels and conduits

Tunnels in soils generally have a shallow overburden. The tunnels are of either shield type or composite box type consisting of reinforced concrete and steel. Shield-type tunnels generally have a circular shape and are excavated through an earth-pressure balanced (EPB) tunnel boring machine (TBM), while precast-concrete-box-type tunnels are immersed tunnels and boxes are placed into specially excavated trenches beneath river or sea through specially navigated cranes. The shield tunnels in soil have an overburden <40–50 m and the stress state greatly depends upon the gravity, and the lateral stress coefficient value is <1.0. On the other hand, immersed tunnels have a backfill material over the structure as an overburden material mainly to protect precast boxes of 90–150 m long from ship anchors. Therefore, the seismic design concept greatly differs, and an overview is provided by Kawashima (1999) and Kiyomiya (1995). Precast boxes are generally modelled through beam elements with axial stiffness supported by springs to represent surrounding ground, as illustrated in Figure 8.60. The consideration of joints between tunnel segments is also of great importance and it requires different constitutive laws under compression and extension as well as shear behaviour.

One of the selected input seismic forms for representing earthquakes is a predescribed ground displacement function despite the complex three-dimensional geometry of the tunnel as well as ground conditions. The selection of the input displacement form is one of the most critical aspects in utilizing this procedure and the method itself is attributed to Newmark (1968). It is commonly utilized in the seismic design of immersed tunnels. Nevertheless, there are some attempts to represent the structural elements and ground conditions though three-dimensional finite element or finite difference methods employing different constitutive models. Figure 8.61 shows an application of the classical method and some three-dimensional representations

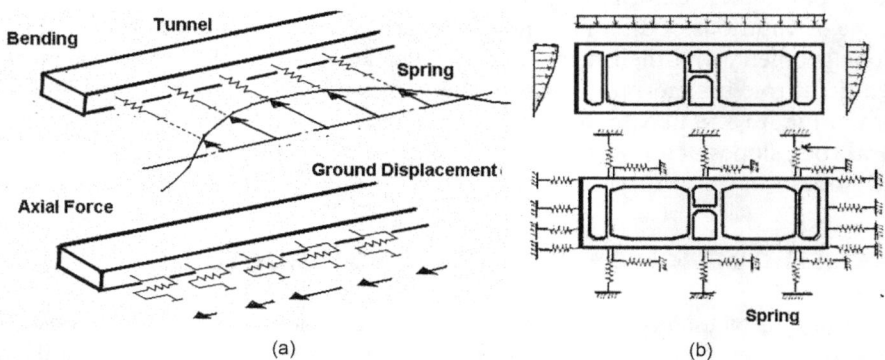

Figure 8.60 Simplified seismic design modelling of immersed tunnels. (a) Longitudinal representation and (b) traverse representation. (From Kiyomiya (1995) and Kawashima (1999).)

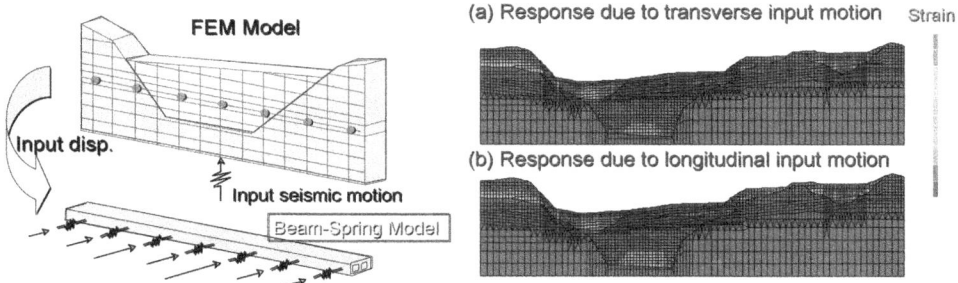

Figure 8.61 Modelling of the Marmaray tunnel. (From Yamamoto et al. (2014).) (a) model (b) analyses results.

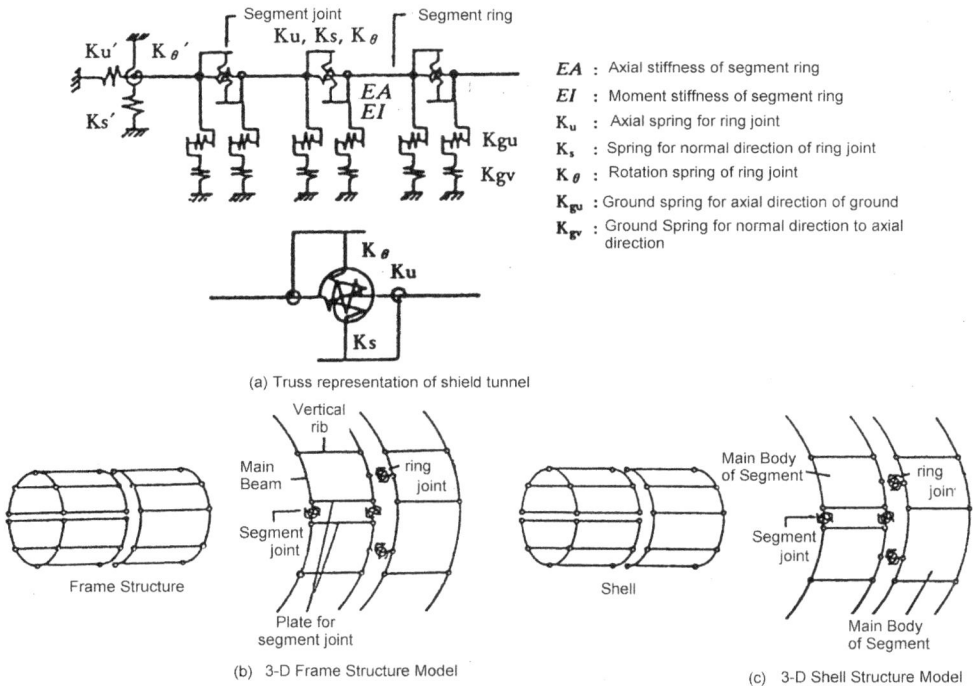

EA : Axial stiffness of segment ring
EI : Moment stiffness of segment ring
K_u : Axial spring for ring joint
K_s : Spring for normal direction of ring joint
K_θ : Rotation spring of ring joint
K_{gu} : Ground spring for axial direction of ground
K_{gv} : Ground Spring for normal direction to axial direction

(a) Truss representation of shield tunnel

(b) 3-D Frame Structure Model

(c) 3-D Shell Structure Model

Figure 8.62 Some models for shield tunnels (Anonymous).

of immersed tunnels together with bored tunnels in the Marmaray Project in Turkey (Yamamoto et al. 2014).

Shield tunnels are tunnels bored into natural ground. Nevertheless, the concepts developed for immersed tunnels or other shallow overburden soil tunnels are also utilized for shield tunnels. Figure 8.62 shows some examples. In some evaluations, the main emphasis is given to the shield segments and their interaction with the surrounding ground. The representation of shield segments is modelled through three-dimensional shell elements and the solution procedure is based on the finite element method. However, two-dimensional or three-dimensional numerical methods have been utilized for the seismic design and to check the design assumptions with the advance of three-dimensional modelling as well as computer technology.

8.4.1.2   Shallow underground openings in discontinuous rock mass

It is well known that shallow underground openings are more vulnerable to stability problems as compared to deep underground openings. The shallow underground openings may completely fail during earthquakes. There are also many reports of such case histories in literature (i.e. JSCE 1923; Kanai and Tanaka 1951; Wang et al. 2001; Asakura and Sato 1998; Asakura et al. 1996; Aydan 1986; Aydan and Kawamoto, 2004; Aydan et al. 1994, 2006b, 2009c,d, 2010a,b; Kawakami 1984; Komada and Hayashi 1980; Kuno 1935; Nasu 1931; Prentice and Ponti 1997; Rozen 1976; Sakurai 1999; Sharma and Judd 1991; TEC-JSCE 2005; Tsuneishi et al. 1978; Yashiro et al. 2007; Dowding and Rozen 1978; Hashimoto et al. 1999; Ueta et al. 2001). The author and his group have been studying the stability of shallow underground openings under both static and dynamic conditions (Aydan et al. 1994; Genis and Aydan 2002). The model tests (see also Figure 8.63) revealed that there are two or three regions potentially unstable (denoted regions I, II and III) in the close vicinity of shallow underground openings as illustrated in Figure 8.63. Aydan (Aydan et al. 1994) derived the following condition for the horizontal seismic coefficient (α_H) to initiate the sliding of a region along a discontinuity set emanating from the opening for a shallow underground opening as illustrated in Figure 8.63 in view of his experimental studies.

$$\alpha_H > \frac{\sin(\theta_B + \phi_B)\sin 2\phi_A - \dfrac{W_{\text{II}}}{W_{\text{I}}}\sin(\theta_A - \phi_A)\sin\theta*}{\sin(\theta_B + \phi_B - \beta)\sin 2\phi_A + \dfrac{W_{\text{II}}}{W_{\text{I}}}\sin(\theta_A + \beta - \phi_A)\sin\theta*} \tag{8.85}$$

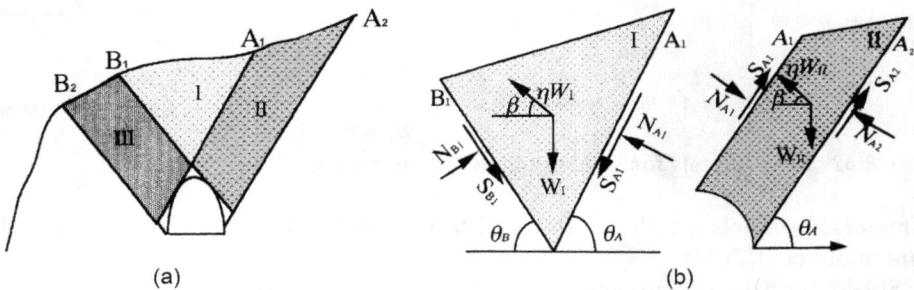

(a)   (b)

Figure 8.63 Illustration of the mechanical model for stability analysis of shallow underground openings. (From Aydan et al. (1994, 2011).) (a) physical situation, (b) mechanical model.

where $\theta^* = \theta_A + \theta_B + \phi_A + \phi_B$. $W_I, W_{II}, \theta_A, \theta_B, \phi_A, \phi_B$ and β are the weights of regions I and II, inclinations and friction angles of discontinuity sets A and B and the angle of the seismic force with respect to the horizontal, respectively. The same method can be used for the stability of region II. Aydan et al. (1994, 2011) have checked experimentally the validity of the limit equilibrium conditions for layered media and media with two joint sets and confirmed the appropriateness of this method.

8.4.1.3 Tunnels in rock mass

A series of parametric numerical analyses on the shape of underground openings under different high in situ stress regimes and the directions and amplitudes of earthquake-induced acceleration waves was carried out. The details of these numerical analyses can be found in publications by Genis (2002) and Genis and Gercek (2003). Figure 8.64 compares yield zone formations around circular and horseshoe-like tunnels subjected to in situ hydrostatic stress conditions ($P_o = 20\,\text{MPa}$) under static and

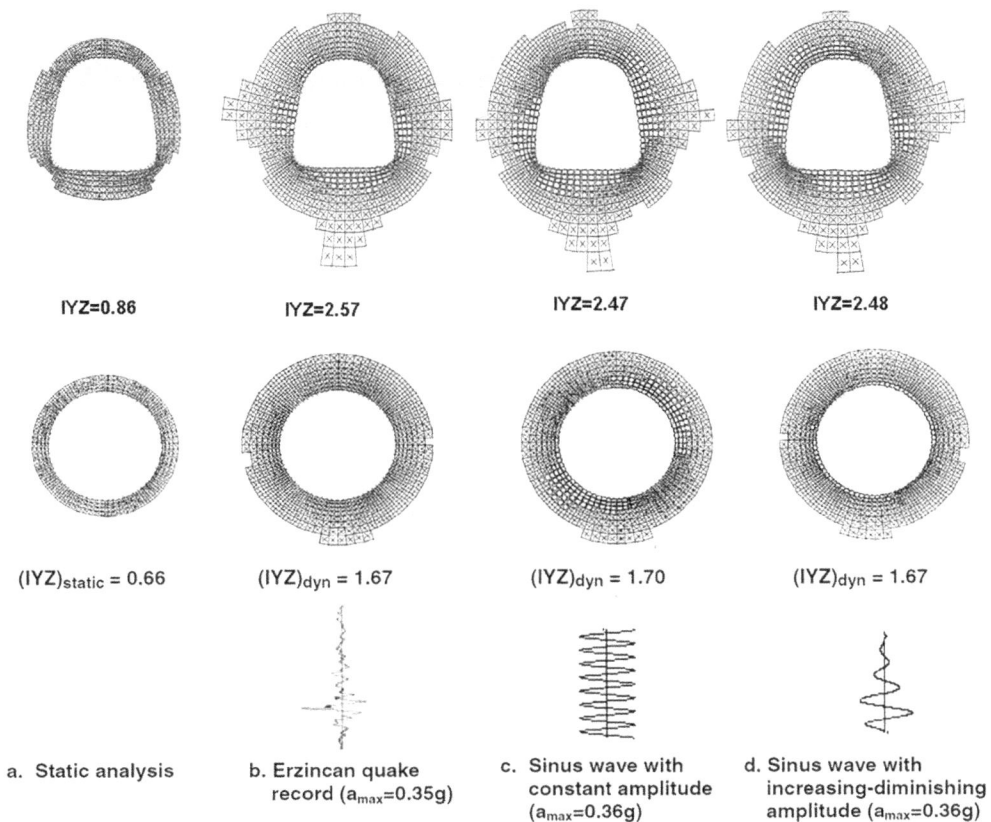

IYZ=0.86   IYZ=2.57   IYZ=2.47   IYZ=2.48

$(IYZ)_{static} = 0.66$ $(IYZ)_{dyn} = 1.67$ $(IYZ)_{dyn} = 1.70$ $(IYZ)_{dyn} = 1.67$

a. Static analysis   b. Erzincan quake record ($a_{max}=0.35g$)   c. Sinus wave with constant amplitude ($a_{max}=0.36g$)   d. Sinus wave with increasing-diminishing amplitude ($a_{max}=0.36g$)

Figure 8.64 Yield zone formations around a deep circular opening under different waveforms. (Partly from Genis and Gercek (2003).)

dynamic conditions. The reason for such an approach is to eliminate the effects of the in situ stress field on the geometry of the failure zone. In order to compare the yield zones determined in either static or dynamic analyses, a simple quantitative measure, i.e. "index of yield zone," or IYZ, was used. IYZ, simply, is the ratio of the total area of the yielded elements to the cross-sectional area of the opening. In the analyses, the rock mass behaviour is assumed to be elastic-brittle plastic. The properties of the elastic and yielded zones were assumed to obey the Hoek–Brown failure criterion with $\gamma = 25$ kN/m^3, $E_m = 27.4$ GPa, $v = 0.25$, $\sigma_{cm} = 14.1$ MPa and $m = 3.43$ for the elastic zone, and $\sigma_{cr} = 2.84$ MPa and $m_r = 2.23$ for the yielded zone. Three different acceleration records were used in these particular analyses. If the maximum amplitudes and dominant frequency characteristics of the earthquake records are almost the same, the yield zones that formed under dynamic conditions are same. They are almost circular for the circular tunnels while elliptical for horseshoe-like tunnels, although the acceleration record was uni-directionally applied.

8.4.2 Rock caverns

Genis and Aydan (2007) carried out a series of numerical studies for the static and dynamic stability assessments of a large underground opening for a hydroelectric powerhouse. The cavern is in granite under high initial stress condition and ~550 m below the ground surface. The area experienced the Nobi-Beya earthquake in 1891, which was the largest inland earthquake in Japan. In the numerical analyses, the amplitude, frequency content and propagation direction of waves were varied (Figure 8.65). The numerical analyses indicated that the yield zone formation is frequency and amplitude dependent. Furthermore, the direction of wave propagation also has a huge influence on yield zone formation around the cavern. When maximum ground acceleration exceeds 0.6–0.7 g, it results in the increase of plastic zones around the opening. Thus, there will be no additional yield zone around the cavern if the maximum ground acceleration is less than these threshold values.

8.4.3 Underground shelters

The elasto-plastic dynamic response of an underground shelter in Bukittinggi in Sumatra Island(Indonesia), which was excavated in a very soft rock in 1942 and experienced the M6.8 Singkarak earthquake in 2007, was chosen as an actual example, and its dynamic response and stability were analysed (Aydan and Genis 2008b). There was no acceleration record in the vicinity of the underground shelter. However, the inferred maximum ground acceleration was about 0.3 g as shown by Aydan (2007a). Therefore sinusoidal waves with frequencies of 0.3, 1 and 3 Hz and amplitude of 0.3 g were applied horizontally to the base of the model. Figure 8.66 shows a three-dimensional perspective view of the underground shelter, acceleration responses at selected positions where some seismic damage were actually observed, and displacement and yield zone distributions after the ground shaking disappeared. Although the analyses are limited to simple waveforms, the results can explain possible causes of the damage and its variation in the underground shelter.

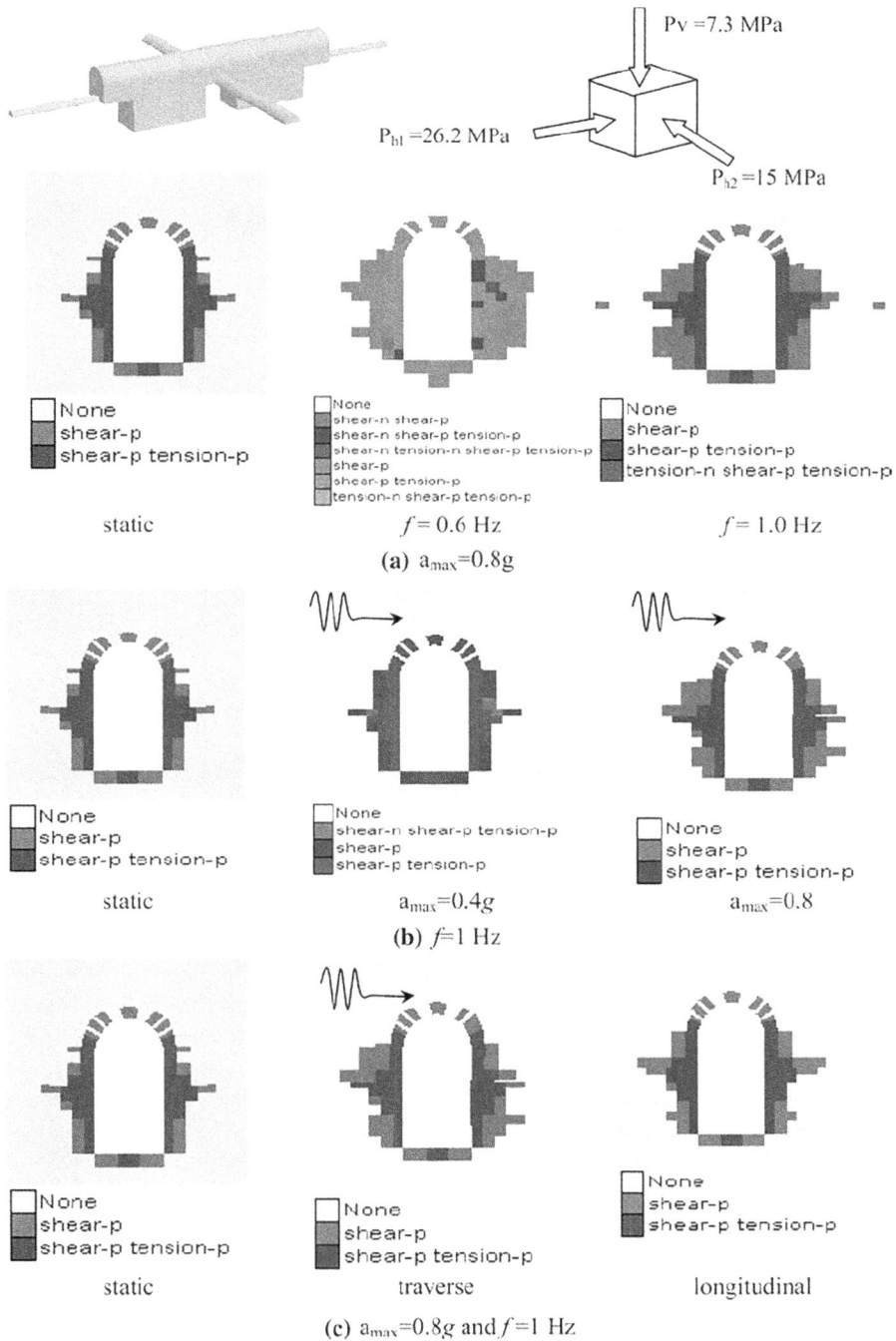

$P_v = 7.3$ MPa

$P_{h1} = 26.2$ MPa

$P_{h2} = 15$ MPa

None
shear-p
shear-p tension-p

static

None
shear-n shear-p
shear-n shear-p tension-p
shear-n tension-n shear-p tension-p
shear-p
shear-p tension-p
tension-n shear-p tension-p

$f = 0.6$ Hz

None
shear-p
shear-p tension-p
tension-n shear-p tension-p

$f = 1.0$ Hz

(a) $a_{max} = 0.8g$

None
shear-p
shear-p tension-p

static

None
shear-n shear-p tension-p
shear-p
shear-p tension-p

$a_{max} = 0.4g$

None
shear-p
shear-p tension-p

$a_{max} = 0.8$

(b) $f = 1$ Hz

None
shear-p
shear-p tension-p

static

None
shear-p
shear-p tension-p

traverse

None
shear-p
shear-p tension-p

longitudinal

(c) $a_{max} = 0.8g$ and $f = 1$ Hz

Figure 8.65 Yield zone formation of underground power house for different cases of input ground motions (Genis and Aydan 2007; Aydan et al 2010a).

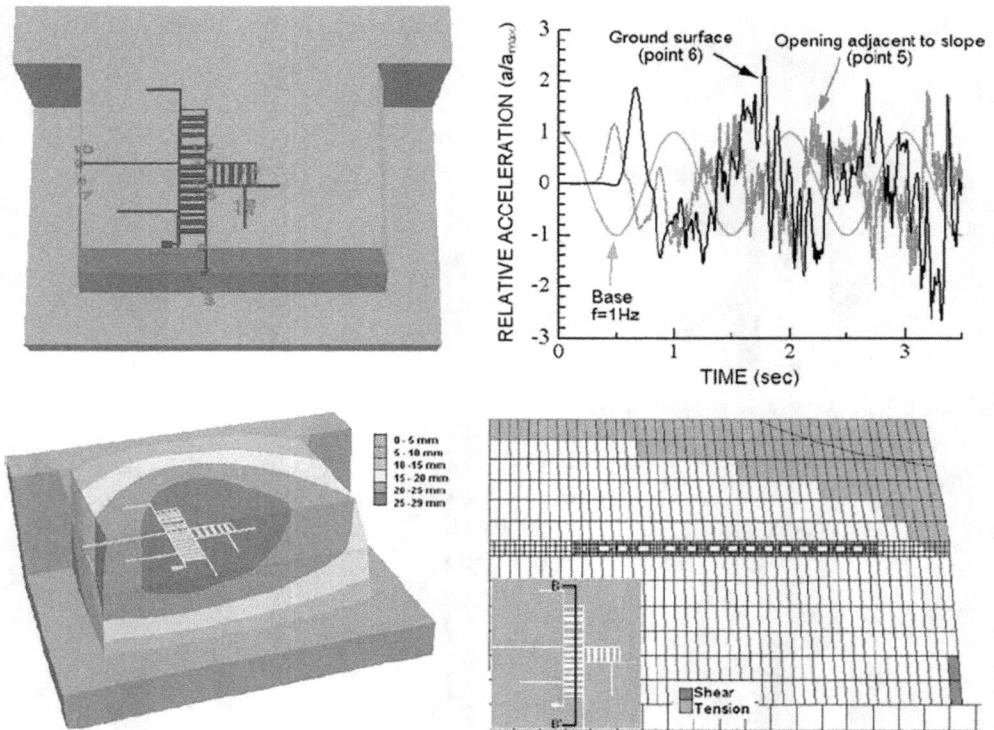

Figure 8.66 Displacement response and yield zone around the Bukittinggi underground shelter. (Arranged from Aydan and Genis (2008b).)

8.4.4 Tunnels below abandoned mines

Aydan and Genis (2014) have performed three-dimensional numerical studies on the response and stability assessment of abandoned mines above a new underground excavation. Figure 8.67 shows a 3D view of the layers, abandoned mine and circular tunnel. Table 8.15 gives the material properties used in numerical analyses. Under the static condition, no yielding occurred.

The response of an abandoned mine was investigated using the ground motion due to the anticipated M9 class Nankai, Tonankai and Tokai earthquakes at the site under consideration as shown in Figure 8.68, and it is based on the method proposed by Sugito et al. (2000). Figure 8.69 shows the yield zone formation in the computational model. As noted from the figure, some yielding occurred at pillars in the deeper sections of the model. The results indicated that under dynamic conditions, further yielding may occur in abandoned mines.

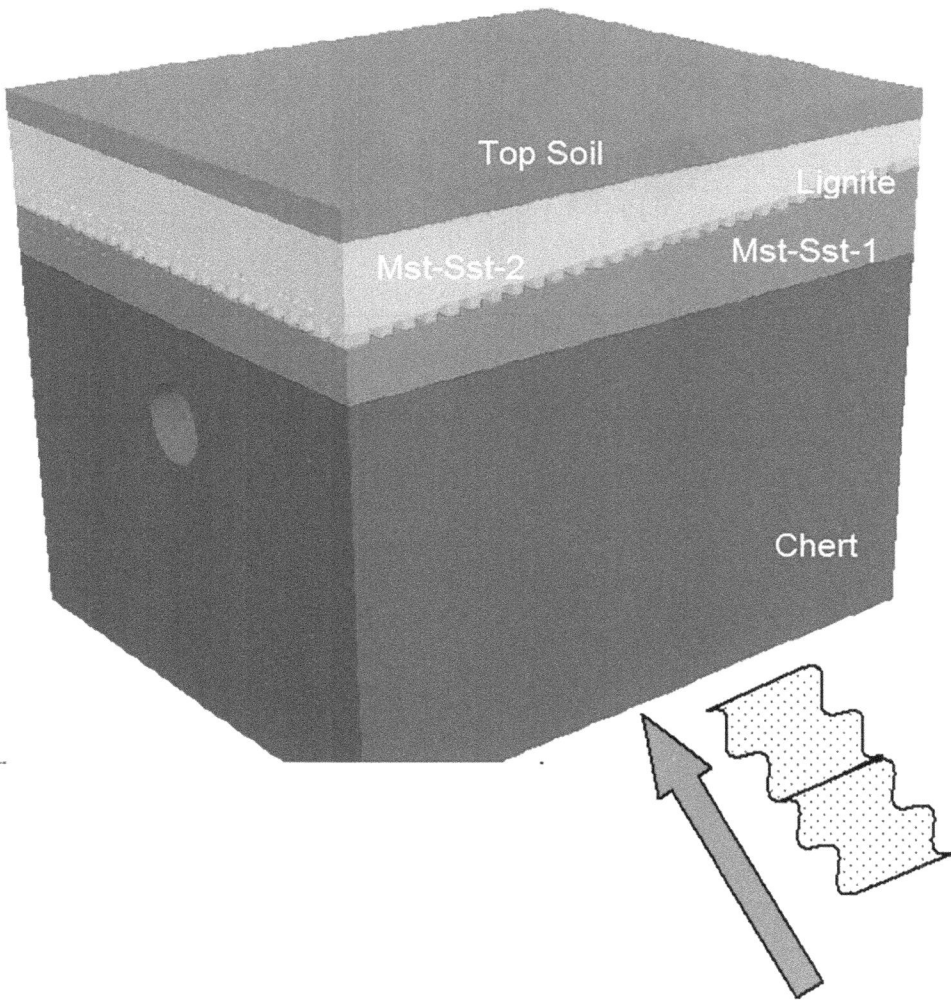

Figure 8.67 Layers of the model and direction of dynamic loading.

Table 8.15 Material properties used in numerical analyses

Layer	γ (kN/m^3)	C (kPa)	($\phi°$)	σ_t (kPa)	E (MPa)	υ
Soil	19	50	38	10	270	0.35
Mst-Sst-1	19	700	25	500	750	0.3
Lignite	14	656	45	500	400	0.3
Mst-Sst-2	19	1000	45	700	1073	0.3
Chert	19	3000	45	2000	3647	0.3

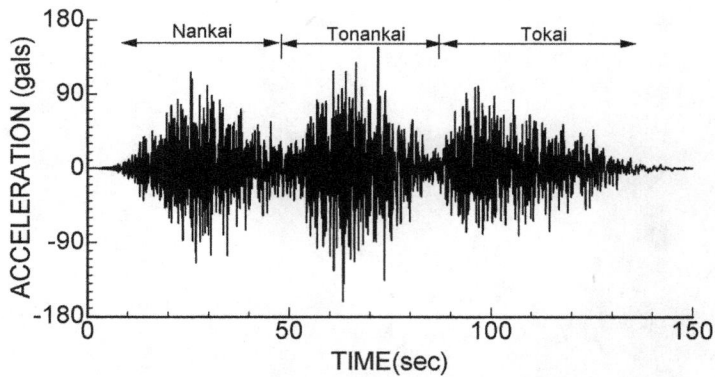

Figure 8.68 Estimated acceleration record due to anticipated Nankai-Tonankai-Tokai earthquake at the site.

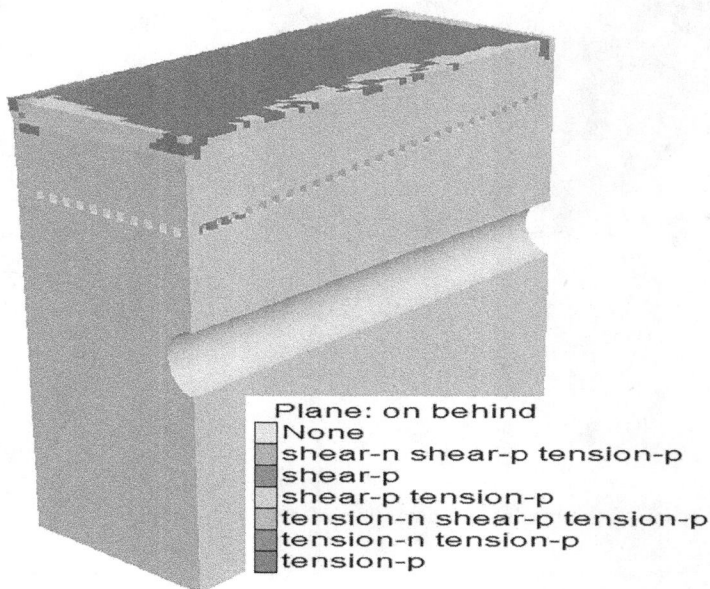

Figure 8.69 The yield zone formation in the computational model (Aydan and Genis 2014).

8.4.5　Seismic design of shafts in rock mass

This section shows the results of the analyses carried out on the ground amplification and frequency characteristics of a deep shaft in rock mass. A series of 3D finite element modal analyses were carried out for four conditions: no shafts, single shaft, double shafts and triple shafts. The software used was 3D MIDAS-FEA. Table 8.16 gives the

Table 8.16 Material properties

Material	UW (kN/m³)	E (GPa)	Poisson's ratio
Rock mass	26.5	0.600	0.37
Concrete	23.5	11.042	0.20

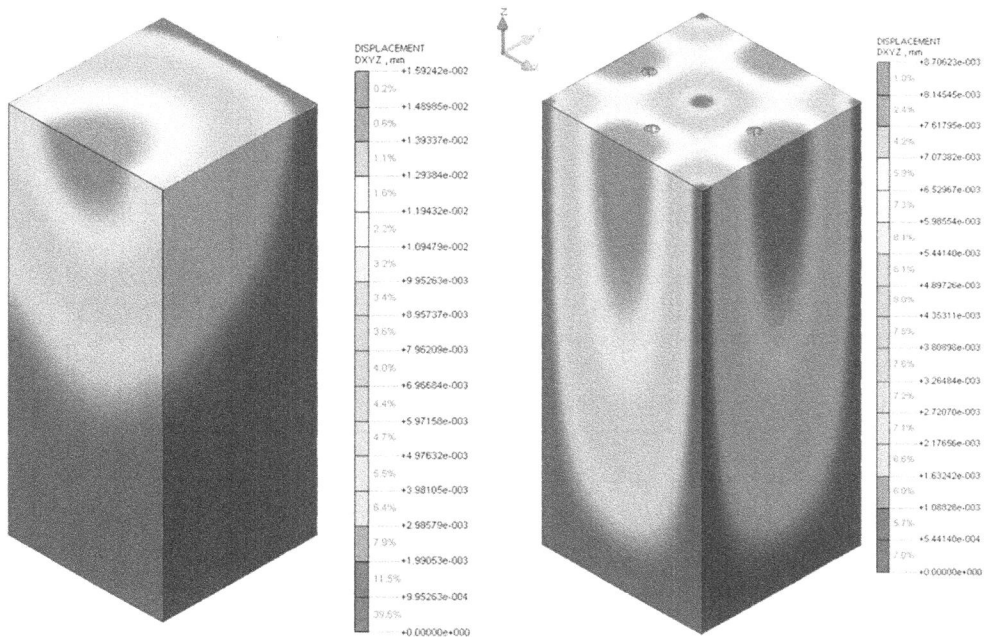

Figure 8.70 Displacement response for Mode 1. (a) No shaft and (b) triple shafts.

Table 8.17 Eigen values for Mode 1

	No shaft	Single	Double	Triple
Mode 1(s)	1.763	1.752	1.203	1.199
Mode 2(s)	1.645	1.635	1.889	1.172
Mode 3(s)	1.564	1.554	1.117	1.111

material properties used in numerical analyses while Table 8.5 compares the Eigen values for four different conditions, and Figure 8.70 shows the displacement response for Mode 1. These results clearly show that there is ground amplification as the depth becomes shallower, which may have some important implications on the dynamic safety of nuclear waste disposal sites at such depths (Table 8.17).

8.4.6 Empirical approaches

Permanent ground deformations may result from relative movement along earthquake faults, earthquake-induced slope failures involving tunnels and plastic deformation of ground due to high ground accelerations. Aydan et al. (2010a,b) compiled case histories and developed databases for three different categories of damages: faulting induced (18 cases histories), shaking induced (98 case history) and slope failure induced (47 cases histories). The parameters of these databases are the name of the tunnel, earthquake parameters (magnitude, hypocenter depth, relative slip), distances from the epicentre and the earthquake fault surface trace (extrapolated surface trace when earthquake fault does not appear on the ground surface), geometry of tunnel, overburden, lining thickness, rock bolt density, rock unit and damage level index (DLI) defined in Table 8.18. Case histories are plotted in Figure 8.71 with different symbols depending upon the damage mode, the moment magnitude (Mw) of the earthquake and the hypocentral distance (R) of the underground structure. As expected, the hypocentral distance of the damaged underground structures increases as the magnitude of the earthquake becomes larger. Furthermore, the limiting relations for fault-induced and ground-shaking-induced damage on tunnels would be different. The available case history data also imply that there is no damage to underground openings by earthquakes when the magnitude is less than six.

Table 8.18 Parameters in Eq. (8.86)

Condition	A	B
Disrupted ground	0.10	0.9
Coherent ground	0.08	0.9

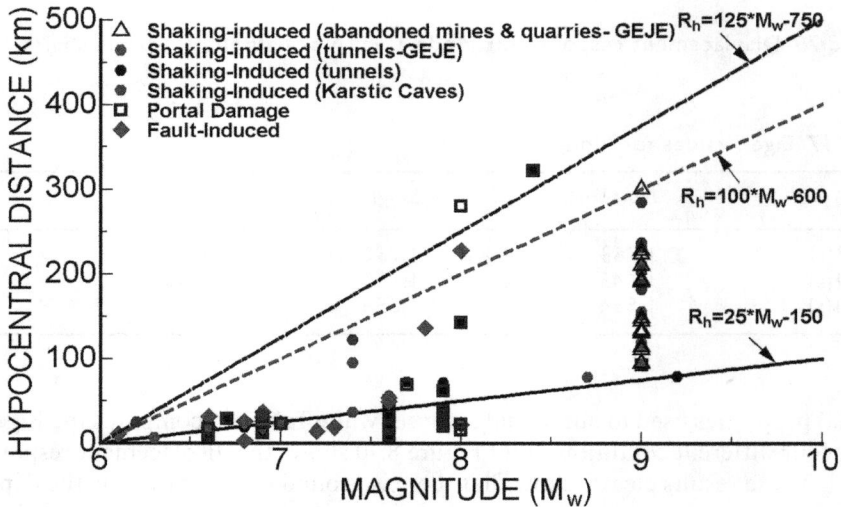

Figure 8.71 Empirical relations between magnitude and limiting damage distance.

The limit of damage for portals of underground openings would be more far-distant and the relation proposed by Aydan (Aydan 2007a; Aydan et al. 2010a,b) for slope failure given below may be used for this purpose:

$$R = A * \left(3 + 0.5\sin\theta - 1.5\sin^2\theta\right) * e^{B \cdot M_w} \tag{8.86}$$

where θ is the angle of the location from the strike of the fault. Constants A and B of Eq. (8.86) are given in Table 8.18 according to ground conditions. Since ground accelerations differ according to the location with respect to fault geometry, the empirical bounds proposed herein can provide some basis for the scattering range of observations.

The definitions of damage to underground structures are generally too broad and a more refined classification of damage is necessary. Aydan et al. (2010a,b) proposed a classification for this purpose, as given in Table 8.19.

Figure 8.72 shows the replotted data shown in Figure 8.71 as a function of distance (R_f) from the surface trace or extrapolated surface trace of the earthquake fault. The vertical axis is the DLI, whose minimum and maximum values are 1 and 7. When the tunnel response is purely elastic, the damage level is assigned as 1.

The functional form for the DLI of underground openings subjected to earthquakes may be given as follows:

$$\text{DLI} = Q(V_s, R_f, \theta, M, \delta_{\max}) \tag{8.87}$$

Table 8.19 Earthquake-induced damage level index (DLI) for underground structures with the consideration of support members

DLI	Remarks
1	No cracking of concrete lining and shotcrete, no plastic deformation of rock bolts or steel ribs, no invert heaving
2	Hair cracking of concrete lining and shotcrete, nonnoticeable deformation of rock bolt platens and steel ribs, no invert heaving
3	Visible cracking of concrete lining, shotcrete, noticeable plastic deformation of rock bolt platens and steel ribs, slight invert heaving
4	Exfoliation of concrete lining and shotcrete, noticeable bending deformation of rock bolt platens and steel ribs, invert heaving; however, is structurally stable
5	Spalling of concrete lining and shotcrete, and considerable plastic deformation of rock bolt platens and bending of steel ribs, invert heaving; is structurally problematic and requires repairs and reinforcement
6	Collapse of concrete lining, shotcrete, and extreme deformation of rock bolt platens and rupturing rock bolts and buckling of steel ribs, buckling and rupturing of invert; collapse of blocks of ground from roof and shoulders; it is structurally unstable and requires immediate repairs and reinforcement
7	Complete closure of the section by failed surrounding ground. Crushing of concrete lining and shotcrete, rupturing of rock bolts and twisted steel ribs and extreme heaving of invert; underground openings are either to be abandoned or re-excavated with extreme precautions

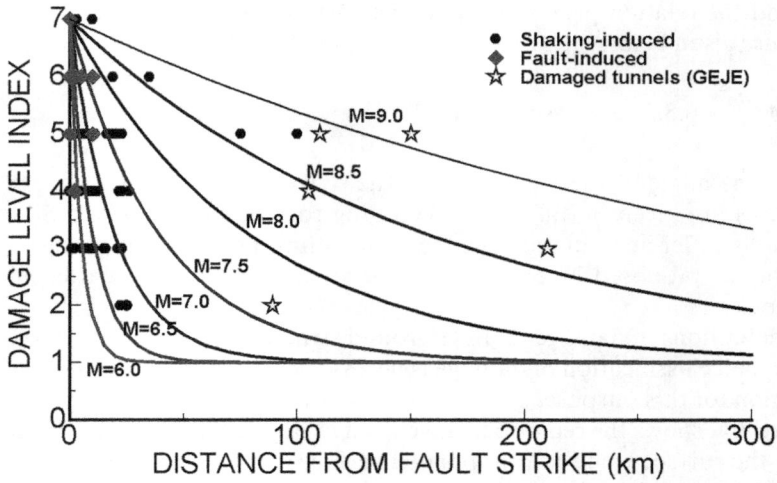

Figure 8.72 Relation between distance (Rf) from surface trace of the fault and DLI.

where V_s, R_f, θ, M and δ_{max} are the shear velocity of ground and the distance from the actual or extrapolated surface fault surface and the angle of the location from the strike of the fault (measured anti-clockwise with the consideration of the mobile side of the fault), earthquake magnitude and maximum relative slip of the earthquake fault. It is an extremely difficult task to select the specific functional form of Eq. (8.87). Aydan (2007a) (see also Aydan et al. 2010a,b) proposed several empirical relations between the various characteristics and moment magnitude of earthquakes. For example, the attenuation relation for maximum ground motion parameters (maximum ground acceleration, or velocity) are given in the following functional form (see Chapter 5):

$$A_{max} \text{ or } V_{max} = F(V_s) * G(R,\theta) * H(M) \tag{8.88a}$$

or more specifically

$$A_{max} \text{ or } V_{max} = A e^{-V_s/B} e^{-R(1-D\sin\theta+E\sin^2\theta)/C}(e^{M_w/F}-1) \tag{8.88b}$$

where parameters A and C depend upon the nature of earthquake (e.g. interplate or intraplate). Taking into account the empirical relation (Eq. 8.88) by Aydan (2007a) with a slight change and the maximum and minimum values of DLI, we propose the following functional form for the DLI and plot it for different magnitudes in Figure 8.72.

$$\text{DLI} = A e^{-V_s/B} e^{-R_f(1-D\sin\theta+E\sin^2\theta)/C^*}+1 \tag{8.89}$$

where C^* is assumed to be a function of moment magnitude as given below:

$$C^* = 10 \cdot 2^{2(M_w-6)} \tag{8.90}$$

It should be noted that the value of $Ae^{-V_s/B}$ must not be greater than six in view of the maximum value of the DLI. Aydan et al. (2011) plotted Eq. (8.89) in Figure 8.72 for different magnitudes by assuming that $\theta = 90°$ and altering the value of $Ae^{-V_s/B}$ to 6 with $D = 0.5$ and $E = 2.5$. As noted from Figure 8.72, the chosen function can closely estimate the observed DLI of underground openings subjected to earthquakes. Nevertheless, there may be a necessity to include a function related to the relative slip of the fault in Eq. (8.89).

8.5 SEISMIC DESIGN OF CONCRETE DAMS

There are different dam types and the seismic considerations of earthfill and rock-fill dams are fundamentally similar to the design of embankments. If the foundation of dams is well-treated against ground liquefaction and/or ground failure, the main considerations would be the seismic stability of earthfill and rockfill dams, and the stability can be undertaken according to the principles of embankments except the scale of such dams could be quite large (e.g. Atatürk Dam in Turkey is 170 m high). Furthermore, earthquake-induced sloshing and dynamic pressure in water reservoirs could be another concern for the seismic design. Particularly, the overtopping of reservoir water during earthquake may cause erosion and it may lead the failure of earthfill and rockfill dams. However, this issue would not be considered herein.

This section presents for the fundamental principles of seismic design of concrete dams and the reader is referred to a textbook by Chopra (2020) for much more detailed presentations and discussions on the seismic design of concrete dams. The force system acting on a dam consists of gravity, seepage and seismic forces. The seismic forces may act on the dam in addition to its deadweight and water level fluctuations and static and dynamic water pressures. The possible failure modes of a dam subjected to various forces can be base shearing (sliding), planar sliding along major discontinuities, flexural/columnar failure or buckling of rock masses, as illustrated in Figure 8.73. Once the acting force system, including the seismic-induced forces, is determined, the SFs of the dam against base shearing and overturning about point O would be easily obtained. For example, the SF against base shearing takes the following form if we utilize the seismic coefficient method (Figure 8.74):

$$\text{SF}_s = \frac{cL_b + N'\tan\phi}{U_s + U_d + E_H}; \quad N' = N - U_b - E_V; \quad N = W \tag{8.91}$$

where U_s, U_d, U_b^{s+d}, W, E_H, E_V are the hydrostatic, hydrodynamic, static and dynamic base uplift, weight, horizontal and vertical seismic forces, respectively. N', T are the effective normal and shear reaction forces.

The major issue in this type of analysis is how to assign the force component given in Eq. (8.91). The vertical seismic coefficient is generally taken as 0.5–0.67 times the lateral seismic coefficient. The horizontal seismic coefficient value used in many countries is generally <0.2, and 0.15 is the most commonly adopted value. However, the recent strong motion monitoring of dams showed that the maximum base acceleration could be very high despite having been designed with a seismic coefficient value of 0.2. For example, Omachi et al. (2003) reported that the maximum acceleration at the top

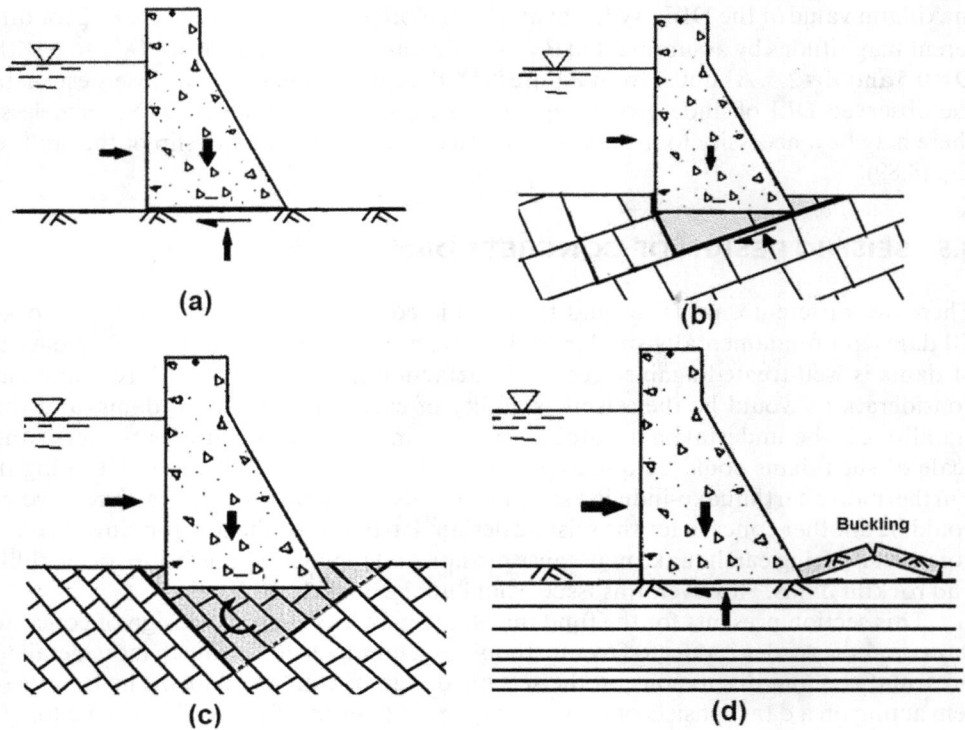

Figure 8.73 Possible failure modes of gravity dams: (a) Base shearing, (b) planar sliding along a thoroughgoing discontinuity plane, (c) flexural/or block toppling failure and (d) buckling failure. (Modified and redrawn from US Army Corps of Engineers (1994).)

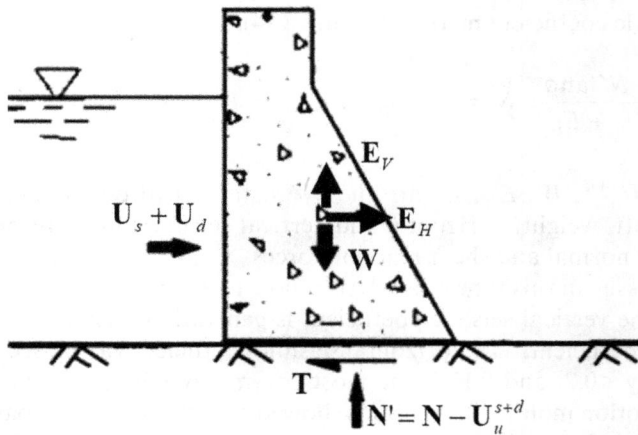

Figure 8.74 Illustration of forces acting on a concrete dam.

of the Kasho gravity dam (46.4 m high) was 2051 gals while the base acceleration was 525 gals. As it is still argued worldwide, Towhata (2008) devoted a chapter to the seismic coefficient and its relation to the observed maximum acceleration. The discussion is beyond this section. Figure 8.92 illustrates a design spectra concept for dams.

On the basis of theoretical studies on the hydrodynamics of dams by Westergaard (1933), hydrostatic and hydrodynamic pressure acting on the upstream side of the dam are given in the following forms:

$$p_s = \rho_w g y, \quad p_s = \rho_w a \sqrt{Hy} \tag{8.92}$$

where ρ_w, g, a, y, H are the density of water, gravitational acceleration, base acceleration, depth from the surface of water reservoir and water depth, respectively. The integrated forces and action locations are

$$U_s = \frac{\rho_w g}{2} H^2; \quad h_s = \frac{1}{3}H; \quad U_d = \frac{7\rho_w a}{12} H^2; \quad h_d = \frac{3}{5}H \tag{8.93a}$$

$$U_u^{s+d} = \left(g + \frac{7a}{6}\right)\frac{\rho_w}{2} HL; \quad l_{s+d} = \frac{2}{3}L; \quad E_H = k_h W; \quad E_V = k_V W \tag{8.93b}$$

Numerical studies would be generally required to evaluate the responses of dams during earthquake motions (Figure 8.75). The simplest approach would be an MDOF-type analysis. A more appropriate approach would be the utilization of numerical techniques such as FEM. The simplest FEM modelling would involve the dam itself and applications of hydrostatic and hydrodynamic pressures with/without mass addition in order to count the water in the reservoir. Figure 8.76 shows the modal analysis results.

A more appropriate finite element modelling would be a representation of the dam's body and foundation with solid elements while representing the water reservoir with fluid elements. In other words it would be a coupling of two domains along the interface through normal traction boundaries. Furthermore, a poro-elastic-type representation of the dam as well as the foundation itself may be more appropriate. Regarding the dam itself, some nonlinear analyses are carried out to check the possibility of cracking, or back-analysing the cracking problems. In addition, some analyses may be concerned with the possibility of stability issues of the foundations of dams. In such cases, the techniques developed in the field of rock mechanics and rock engineering (RMRE) may be necessary, which would be capable of modelling separation and/or, sliding along discontinuities existing in foundation rock masses (Figure 8.77).

8.6   NUCLEAR POWER PLANTS

Nuclear plants are one of the most critical structures and their seismic design is implemented with utmost care; their seismic design loads probably are chosen to be very large compared to other structures. A seismic design is very complex and it covers a wide range of structures such as reactor and turbine buildings, cooling system consisting of large pipes as well as water intake shafts, tunnels and pools and related pumping facilities. Furthermore, the stability of embankments and rock slopes adjacent

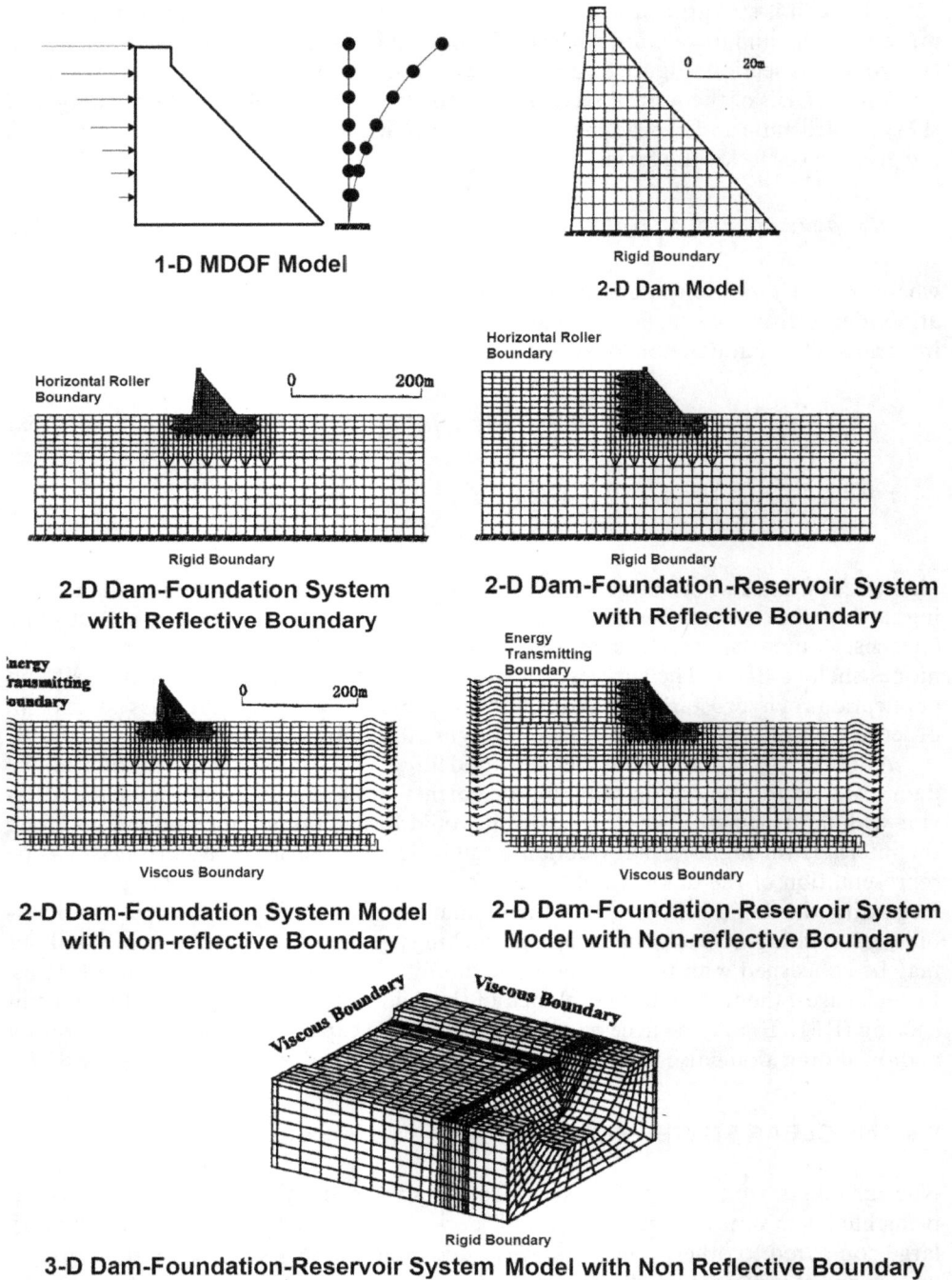

1-D MDOF Model

2-D Dam Model

Rigid Boundary

Horizontal Roller Boundary

Rigid Boundary

2-D Dam-Foundation System with Reflective Boundary

Horizontal Roller Boundary

Rigid Boundary

2-D Dam-Foundation-Reservoir System with Reflective Boundary

Energy Transmitting Boundary

Viscous Boundary

2-D Dam-Foundation System Model with Non-reflective Boundary

Energy Transmitting Boundary

Viscous Boundary

2-D Dam-Foundation-Reservoir System Model with Non-reflective Boundary

Viscous Boundary   Viscous Boundary

Rigid Boundary

3-D Dam-Foundation-Reservoir System Model with Non Reflective Boundary

Figure 8.75 Models utilized for the seismic design of concrete dams. (Modified from Ariga et al. (2000).)

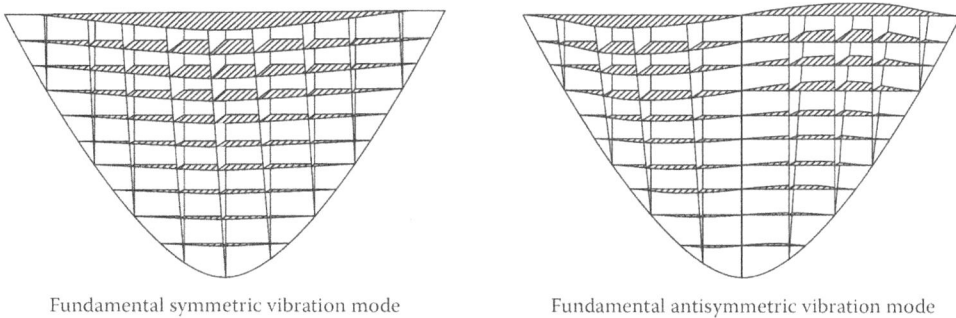

Fundamental symmetric vibration mode   Fundamental antisymmetric vibration mode

Figure 8.76 Results of modal analyses of an arch dam (USBR, 1977).

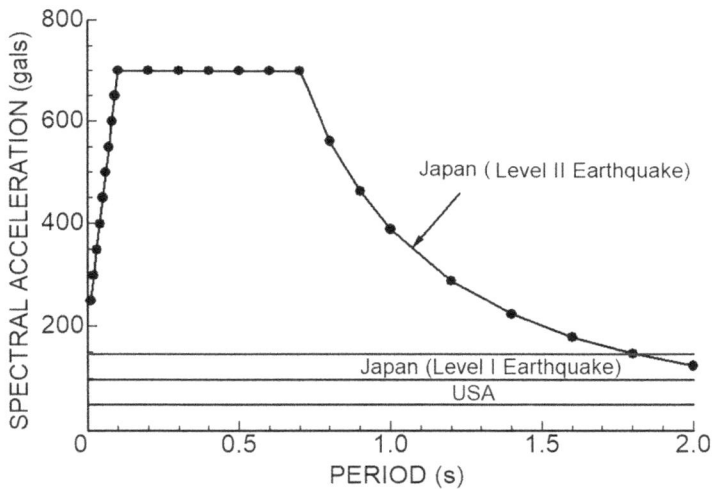

Figure 8.77 Seismic design spectra and seismic design coefficient for dams (Matsumoto et al. 2011).

to the facility are considered. Particularly, the seismic response and resistance of the rock foundation is one of the most critical aspects, and many explorations and investigations are carried out utilizing the principles of RMRE. As the geo-crust always ruptured and there are many faults and fracture zones, the selection of appropriate location is the most critical item in the overall design scheme. Figure 8.78 shows an example of the numerical modelling of the foundation of the Hamaoka Nuclear Station with the considerations of fracture zones at the site. Traces of paleo-events as well as historical and nondocumented earthquakes in the vicinity of the nuclear plants are also other important items of investigations and exploration. These investigations may also on a regional, country or global scale particularly for large geophysical events such as mega-earthquakes, mega-tsunamis and meteorite impacts.

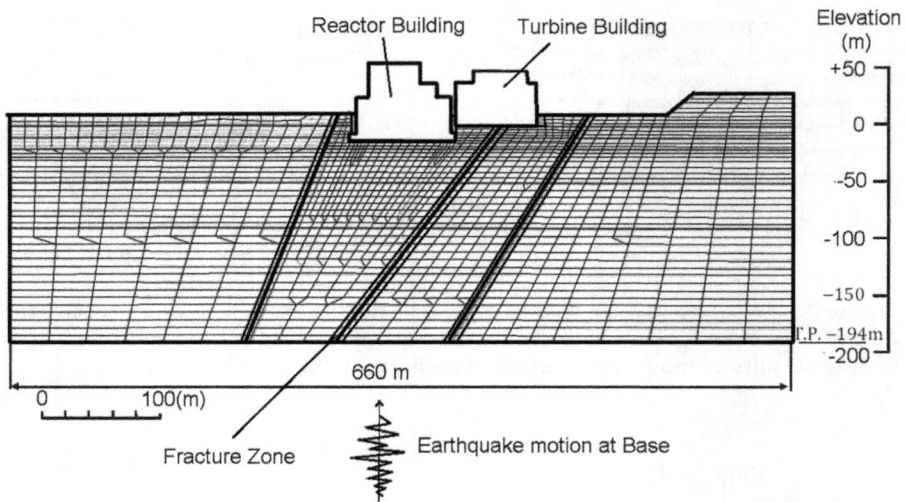

Figure 8.78 An example of numerical modelling of the foundation of the Hamaoka Nuclear Station. (From Hamada and Kuno (2014).)

There are different techniques utilized for different parts of nuclear power plants. The reactor, turbine and associated buildings of nuclear power plants can be modelled using various models ranging from a simple SDOF model to a complex 3D numerical model utilizing FEM, FDM or BEM as illustrated in Figure 8.79. For example, Figure 8.80 shows an example of a simple modelling of a reactor, a reactor building, a turbine and a turbine building through MDOF models with the consideration of a foundation ground support. However, adjacent slopes, tunnels, water intake facilities, retaining walls, shafts and pools would be modelled through the models appropriate for the given structure, most of which have been already described in the previous sections. Particularly, the modelling of adjacent slopes, foundation and tunnels may require proper models for numerical representation of rock mass constituting the foundation, slopes and surrounding tunnels. As rock mass geologically contain many discontinuities in the form of joints, faults, fracture zone, bedding planes and layering, such features must be modelled in numerical analyses. For this purpose, techniques such as finite element method with joint or interface elements (FEM-J), DFEM, discrete element method (DEM) and displacement discontinuity analysis (DDA) method should be used.

8.7 ASSESSMENT OF GROUND LIQUEFACTION AND COUNTERMEASURES

8.7.1 Definition of ground liquefaction

Ground liquefaction was initially known as quick-sand phenomenon, and it has been recognized as "ground liquefaction" since the 1964 Niigata earthquake. Various typical damages due to ground liquefaction (simply liquefaction) was observed

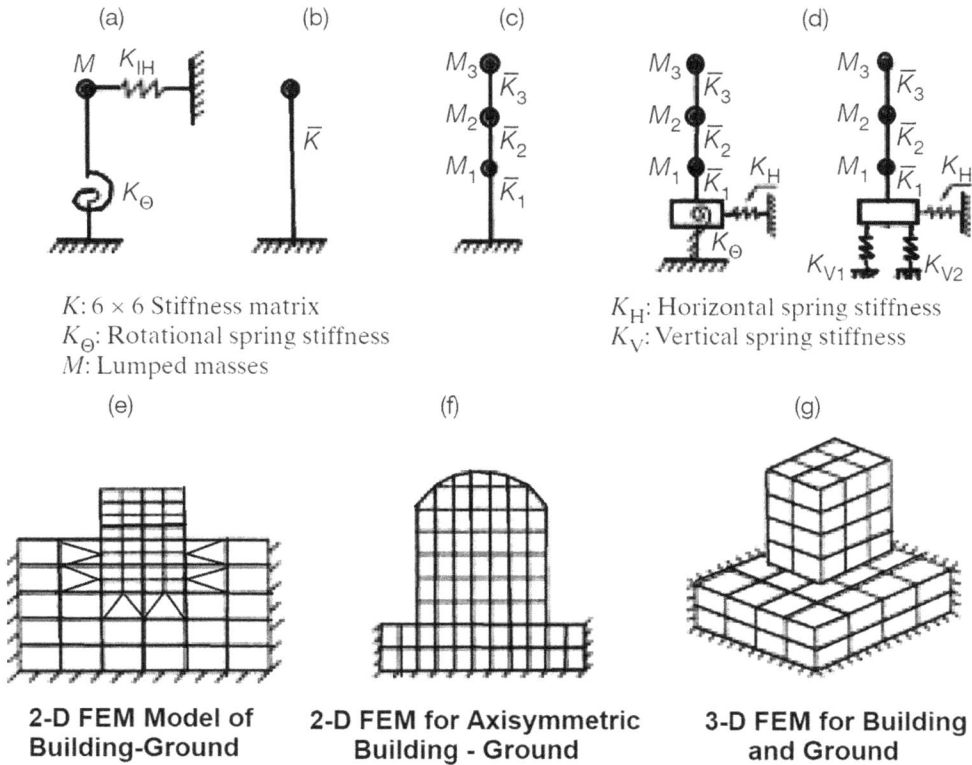

K: 6 × 6 Stiffness matrix
K_Θ: Rotational spring stiffness
M: Lumped masses

K_H: Horizontal spring stiffness
K_V: Vertical spring stiffness

**2-D FEM Model of
Building-Ground**

**2-D FEM for Axisymmetric
Building - Ground**

**3-D FEM for Building
and Ground**

Figure 8.79 Some models used in the seismic design of nuclear power plants. (Modified from International Atomic Energy Agency (2003).)

to be widespread and it has been receiving great attention from researchers as well as engineers and many studies have been carried out since 1964. Ground liquefaction was understood as the behaviour of granular material of ground becoming liquid-like when it is subjected to shaking, as illustrated in Figure 8.81. The main mechanism is understood as the increase in pore pressure resulting in the reduction of effective stress and eventually leading to the liquid-like phenomenon of the ground.

8.7.2   Governing equations of ground liquefaction

Ground prone to liquefaction is considered to be a mixture of solid skeleton and fluid filling the pores. The theoretical bases of ground liquefaction were established by Biot (1956). Based on mixture theory, the governing equations may be shown to be (e.g. Aydan 2001b,c, 2016b, 2021)

Solid phase $\nabla \cdot (1-n)\boldsymbol{\sigma}_s + (1-n)\rho_s g = (1-n)\rho_s \ddot{\mathbf{u}}_s - \boldsymbol{\xi}_{sf}$ (8.94a)

Fluid phase $\nabla \cdot n\boldsymbol{\sigma}_f + n\rho_f g = n\rho_f \ddot{\mathbf{u}}_f + \boldsymbol{\xi}_{sf}$ (8.94b)

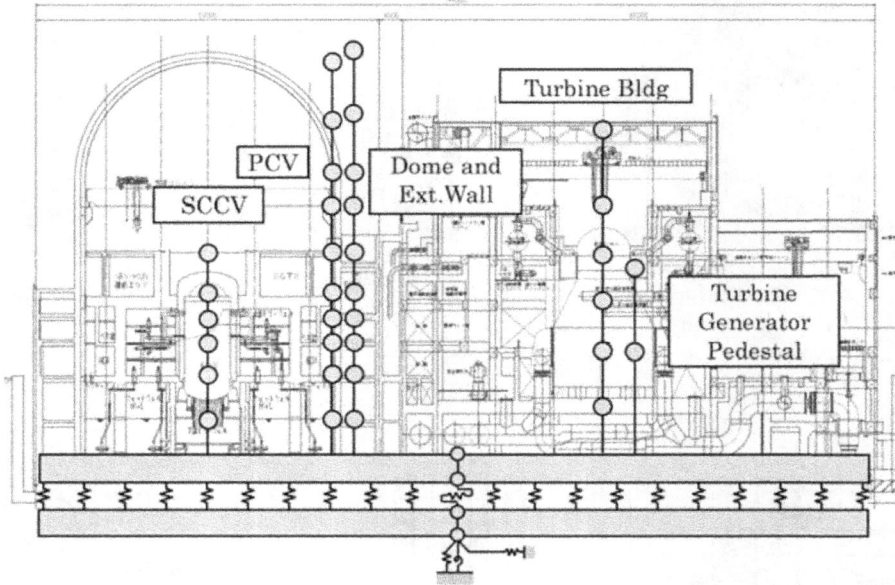

Figure 8.80 Modelling reactor, reactor building, turbine and turbine building using MDOF models together with ground support.

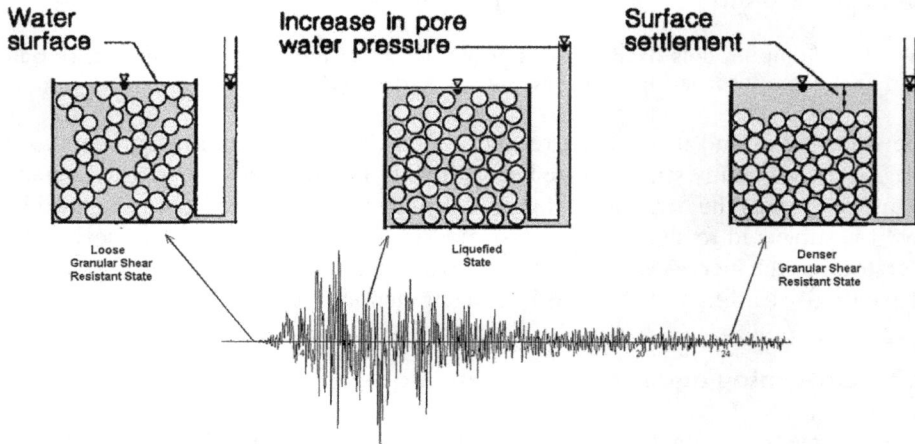

Figure 8.81 Illustration of the mechanism of ground liquefaction.

where g is the gravity; n is the porosity; and $\rho_s, \sigma_s, \ddot{u}_s$ are the density, stress tensor and displacement vector of the solid phase, respectively. $\rho_f, \sigma_f, \ddot{u}_f$ are density, stress tensor and displacement vector of the fluid phase, respectively. These equations are coupled through Darcy's law and effective stress law as given below:

Darcy's law $\xi_{sf} = -n\dfrac{\eta}{k}\mathbf{v}_r$ (8.95)

Effective Stress law $\sigma = \sigma' - \alpha\mathbf{p}$ (8.96)

where α is called the Biot coefficient. When its value is 1, it corresponds to Terzaghi's effective stress law. As noted from the above equations, permeability of the mixture is one of the most fundamental parameters in the overall process of ground liquefaction.

The ground liquefaction state is mathematically defined when the effective stress tensor becomes a nil tensor as given below:

$$\sigma' = 0 \tag{8.97}$$

With this assumption, the ground behaves like liquid with an average density of the mixture. This state is sustained until the pressure starts to decrease. When the excess stress is totally dissipated, the soil would have the properties of a post-liquefaction state.

8.7.3   Solution of governing equations

The coupled equations are solved through some techniques developed for the purpose. The FEM is one of the commonly used numerical solution techniques. For example, if the FEM is chosen, the discretized form of the governing equations take the following form (see Zienkiewicz and Shiomi 1984; Aydan 2021 for details):

$$\begin{bmatrix} \mathbf{M}_1^s & \mathbf{M}_1^f \\ \mathbf{M}_2^s & \mathbf{M}_2^f \end{bmatrix} \begin{Bmatrix} \ddot{\mathbf{U}} \\ \ddot{\mathbf{W}} \end{Bmatrix} + \begin{bmatrix} \mathbf{0} & \mathbf{C}_1^f \\ \mathbf{0} & \mathbf{0} \end{bmatrix} \begin{Bmatrix} \dot{\mathbf{U}} \\ \dot{\mathbf{W}} \end{Bmatrix}$$
$$+ \begin{bmatrix} \mathbf{K}_1^s & \mathbf{K}_1^f \\ \mathbf{K}_2^s & \mathbf{K}_2^f \end{bmatrix} \begin{Bmatrix} \mathbf{U} \\ \mathbf{W} \end{Bmatrix} = \begin{Bmatrix} \dot{\mathbf{F}}_1 \\ \mathbf{F}_2 \end{Bmatrix} \tag{8.98a}$$

or in a compact form

$$[\mathbf{M}^*]\{\ddot{\mathbf{U}}^*\} + [\mathbf{C}^*]\{\dot{\mathbf{U}}^*\} + [\mathbf{K}^*]\{\mathbf{U}\} = \{\dot{\mathbf{F}}^*\} \tag{8.98b}$$

The above equation is discretized in time domain and solved.

In the analyses, the total stress analysis is commonly used as it results in the solution of a single-phase-like formulation. Although it is possible to evaluate the state of liquefaction through the solution equations above, the most difficult aspect is how to evaluate ground behaviour following the liquefaction state. In other words, the post-liquefaction state is quite difficult to handle.

8.7.4　Empirical liquefaction susceptibility methods

Despite many theoretical and numerical studies being available, it is still common to use some empirical liquefaction susceptibility analyses for assessing the ground lique-faction hazard evaluations. The available methods are briefly explained in the follow-ing sections.

8.7.4.1　Geologic criterion

The experiences and observations in past earthquakes showed that saturated quater-nary deposits are vulnerable to ground liquefaction. As an example, Figure 8.82 shows the liquefaction locations observed in Turkey together with quaternary deposit dis-tributions and major active faults. It is also of great interest that the locations almost coincide with quaternary deposits, called Ovas in Turkish, along the active faults of Turkey. Although some liquefaction locations were observed along the shores of Kara Deniz (Black Sea), they were away from the North Anatolian Fault and caused by the Erzincan earthquake of 1939 with a magnitude of 8.0. Similar observations were noted in many earthquakes such as in Nias Island during the 2005 off-Sumatra earthquake and the 2009 Suruga Bay earthquake in Shizuoka prefecture (Aydan et al. 2005b; Ohta and Aydan 2011). It was also interesting to note that many locations of ground lique-faction caused by the 1854 Ansei, 1891 Nobi-Beya, 1944 Mikawa and 2009 Suruga Bay earthquakes in Shizuoka prefecture, Japan, coincided with each other. There are some opinions that ground liquefaction would not occur if the location experienced ground liquefaction previously, but the observations in Turkey and Japan disprove that this

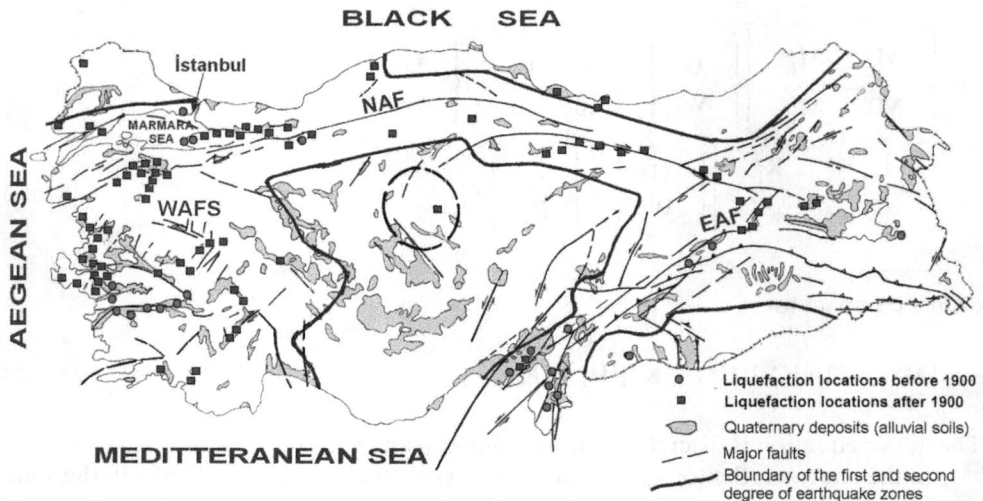

Figure 8.82 Distribution of the Quaternary alluvial deposits and liquefaction sites, and the main structural features of Turkey with first and second degree seismic risk zones.

is not true. For example, the ground that liquefied at Sapanca in the 1967 earthquake liquefied again in the 1999 Kocaeli earthquake.

8.7.4.2 Empirical liquefaction distance-magnitude method

Some empirical relations for assessing ground liquefaction as a function of distance and magnitude of earthquake given in Table 8.20 have been proposed (e.g. Kuribayashi and Tatsuoka 1975; Ambraseys 1988; Wakamatsu 1991, 1993; Aydan et al. 1998; Aydan 2007a, 2012a). When these equations are used, they are wrongly used without differentiating parameters such as magnitude and distance. Consequently, such comparisons become meaningless. If they are compared, additional relations to convert magnitudes and epicentral and hypocentral distances are necessary. The relations between the magnitude of the earthquake and the hypocentral distance to the site of liquefaction are shown in Figure 8.83 together with observational data. Aydan et al. (1998, 2000a) interpreted their upper, mean and lower limits as the bounds of very severe, severe, fair and no liquefaction. Furthermore, observational data implies that the lower limit of moment magnitude is 5.

8.7.4.3 Grain size-based method

Port and Harbour Research Institute of Japan (PARI) (1997) published some bounds of ground liquefaction in *Handbook on Liquefaction Remediation of Reclaimed Land* based on mean grain size of soil. There are basically two bounding lines in Figure 8.84: most liquefiable and potentially liquefiable. Sandy soil is categorized as most liquefiable ground material and silty sand or sandy gravel is categorized potentially liquefiable soils. Figure 8.84a and b shows the grain size distributions of samples collected at various sites affected by recent earthquakes in Turkey and in the vicinity of Sumatra Island

Table 8.20 Estimation of empirical liquefaction distances

Proposed by	Feature	Equation
Kuribayashi-Tatsuoka	Mean	$\log_{10} R_e = 0.87 M_j + 4.5$
	Lower	$\log_{10} R_e = 0.87 M_j + 4.5$
Ambraseys	Epicentral	$M_w = 0.31 + 2.65 \times 10^{-8} R_e + 0.99 \log_{10} R_e$
	Fault	$M_w = 0.18 + 9.2 \times 10^{-8} R_f + 0.9 \log_{10} R_f$
Wakamatsu	1991	$\log_{10} R_e = 2.22 \log_{10}(4.22 M_j - 19)$
	1993	$\log_{10} R_e = 3.5 \log_{10}(1.4 M_j - 6)$
Aydan et al. (1998)	Upper	$R_o = 36 M_s - 160$
	Mean	$R_o = 36 M_s - 200$
	Lower	$R_o = 36 M_s - 240$
Aydan (2007a, 2015d) and Aydan et al. (2012a)		$R_o = 0.08 * (3 + 0.5 \sin\theta - 1.5 \sin^2\theta) * e^{0.9 M_w}$

R_e: epicentral distance; R_o; hypocentral distance; R_f: distance from fault trace; M_j: JMA magnitude; M_s: surface magnitude; M_w: moment magnitude; θ: is the angle between the strike of fault and the site of liquefaction.

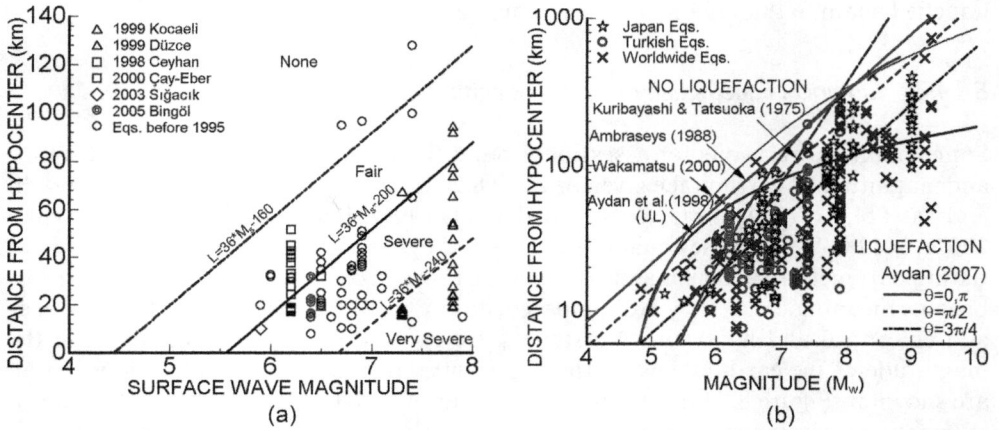

Figure 8.83 The relation between the magnitude of the earthquake and the hypocentral distance to the site of liquefaction. (a) Aydan et al. (2000a) and (b) Aydan (2015d). (From Aydan et al. (2000a) and Aydan (2015d).)

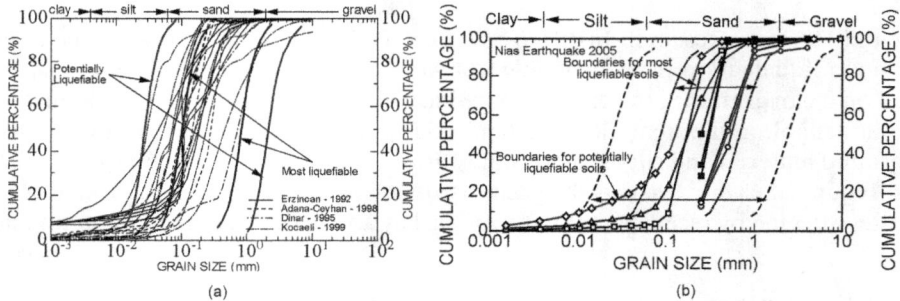

Figure 8.84 Grain size distributions of liquefied soils in Turkey and Sumatra. (a) Mean grain size distribution of liquefied soil in recent earthquakes in Turkey and (b) grain size distribution curves for soils in Nias Island.

of Indonesia together with bounds of liquefaction. Although most of the grain size distribution curves of samples from liquefied sites fall within the empirical bounds, some may also be beyond the bounding curves. Therefore, the bounds should only be regarded as the most likely range of liquefaction for earthquakes with a moment magnitude ≤ 7 and greater than ≥ 5. Clayey or gravelly soils may also liquefy as long as the circumstances are suitable for generating pore pressures to cause liquefaction as observed in the Hyogo-ken Nanbu earthquake of January 15, 1995.

8.7.4.4   Standard penetration test value-based method: the Seed method

Following the 1964 Niigata earthquake, the importance of ground liquefaction was recognized and Seed and his co-workers (Seed 1979; Seed and Idriss 1971, 1979; Seed and DeAlba 1986, Seed et al. 1985) proposed a method for assessing ground lique-faction. The method uses the value of standard penetration, known as N-Value or SPT-Value, maximum ground acceleration and empirical charts for liquefaction as-sessment. Initially, the magnitude of earthquakes was not considered. However, it was understood that the maximum ground acceleration should be related to the magnitude of earthquakes and the fine content should be considered for liquefaction suscepti-bility. The standard penetration value is affected by the depth and also equipment used. Therefore some procedures are proposed to evaluate liquefaction susceptibility of ground and the standard penetration value has to be adjusted (Liao and Whitman 1986). The following formula is used for depth and energy correction of SPT-Values

$$N_{60} = N \cdot C_N \frac{E_m}{0.60 E_{ff}} \tag{8.99}$$

where C_N is the depth correction factor, E_m is the measured energy, and E_{ff} is the free fall energy. Seed and Idriss (1981), Tokimatsu and Yoshimi (1983) and Aydan et al. (1997a, 2000b) suggested the following functions for depth correction:

$$C_N = 0.77 \log\left(\frac{20}{\sigma_v'}\right); \quad C_N = \frac{1.7}{0.7 + \sigma_v'}; \quad C_N = \frac{3}{2 + \sigma_v'} \tag{8.100}$$

In the formula of Seed and Idriss, the effective stress is given in terms of tonf/ft^2 while it is given in terms of kgf/cm^2 in the formula of Tokimatsu and Yoshimi (1983). To con-vert effective stress σ_v' from kPa to the unit of tonf/ft^2, it is multiplied by a coefficient of 0.01044. Figure 8.85 compares the computed results as a function of depth. The value of E_m in terms of E_{ff} will depend upon the type of device used in SPT tests. The value of E_m is said to have a range between 0.6 and 0.72 times E_{ff}. In computations, if the actual value is not known, the values provided by Seed et al. (1985) may be used. Aydan et al. (1997a, 2000b) fitted the following function for the charts of cyclic stress ratio against liquefaction (CSRL) proposed by Seed and Idriss (1981).

$$\frac{\tau_{av}}{\sigma_v'} = 0.0096 * N_{60}(1 + 0.0011 * e^{0.06 N_{60}}) \frac{2}{1 + \left(\frac{M}{7.5}\right)^3} \tag{8.101}$$

The fitted functions with original data points obtained from digitized curves are com-pared in Figure 8.86 as it becomes easier for digital assessments.

The CSRL is interpreted as the resistance of ground against liquefaction. Bartlett and Youd (1992) suggested the following function for CSRL:

$$\frac{\tau_{av}}{\sigma_v'} = 0.0013 * M_s * N_{60}^{0.5} \tag{8.102}$$

It should be noted that this function could not fit the curves shown in Figure 8.86.

The simple procedure to take into account the fine content (FC) is to add a certain constant value to the function given above. Therefore, the following function can be utilized by taking into account the charts given by Seed et al. (1985).

SPT DEPTH CORRECTION FACTOR C_N

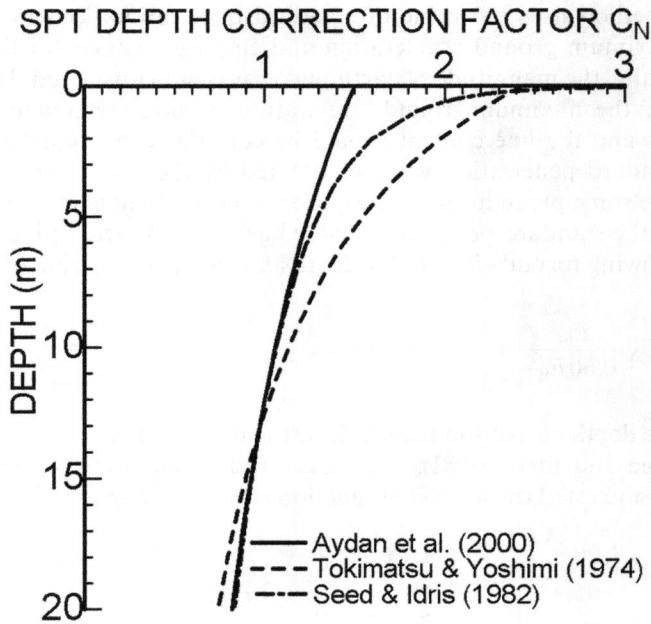

Figure 8.85 Comparison of depth correction factor C_N.

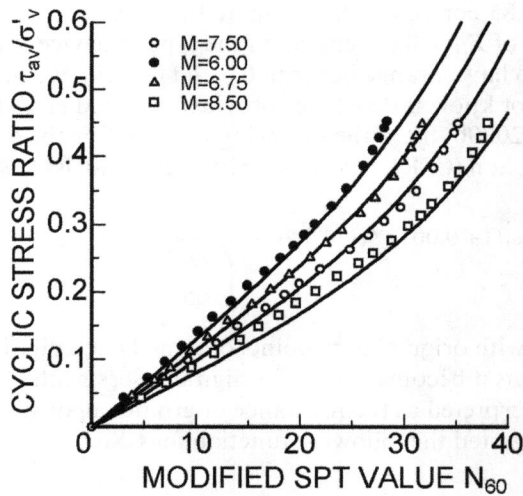

Figure 8.86 Comparison of fitted functions with digitized data.

$$\frac{\tau_{av}}{\sigma'_v} = 0.0096 * N_{60}(1 + 0.0011 * e^{0.06 N_{60}}) \frac{2}{1 + \left(\dfrac{M}{7.5}\right)^3} + 0.02 * \text{FC}^{0.3} \qquad (8.103)$$

Figure 8.87 shows plots of the function (8.103) for different values of FC. The cyclic stress ratio due to earthquake shaking is defined as follows:

$$\text{CSRE} = \frac{\tau_d}{\sigma'_v} = 0.65\frac{a_{\max}}{g}\frac{\sigma_v}{\sigma'_v}r_d \tag{8.104}$$

Reduction factor r_d is related to the depth and it is given in the following form:
   Seed and Idriss (1971)

$$r_d = 1 - \frac{z}{90} \tag{8.105}$$

Tokimatsu and Yoshimi (1983)

$$r_d = 1 - 0.015z \tag{8.106}$$

One of the most important parameters in Eq. (8.101) is how to relate the magnitude of earthquakes. While Seed and his co-workers recommend the utilization of their charts, Tokimatsu and Yoshimi (1983) and Japan Roadway and Bridges Society (1996) suggested the following function:

$$\text{CSRE} = \frac{\tau_d}{\sigma'_v} = 0.1(M-1)\frac{a_{\max}}{g}\frac{\sigma_v}{\sigma'_v}r_d \tag{8.107}$$

where M is viewed as the local magnitude. Another major issue with Eqs. (8.103) and (8.107) is how to assign the maximum ground acceleration. It is possible to relate to

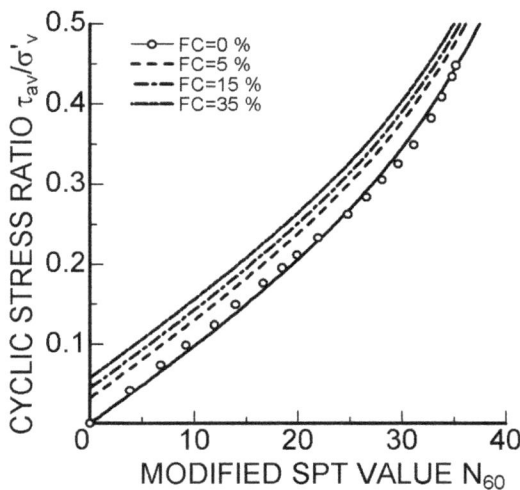

Figure 8.87 Effect of fines content on CSRL.

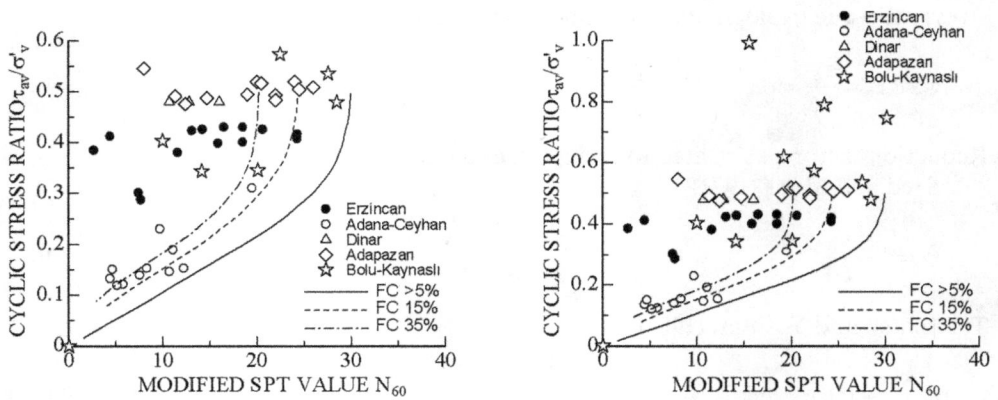

Figure 8.88 The relation between modified SPT-value N_{60} and cyclic stress ratio CSRE together with bounds for liquefaction/nonliquefaction proposed by Seed and DeAlba (1986).

the estimated ground acceleration from some attenuation relations. However, there is a tendency to limit the maximum value of a_{max}/g to 0.3.

Figure 8.88 shows the relation between modified SPT-value N_{60} and cyclic shear stress ratio CSRE together with bounds for liquefaction/nonliquefaction proposed by Seed and DeAlba (1986). The data of liquefied soils fall generally within the range of liquefiable soils. However, the charts of Seed and DeAlba (1986) could not evaluate the liquefaction if the maximum ground acceleration was >0.3 g.

Figure 8.89 illustrates a flow chart for susceptibility assessment of ground liquefaction. This type of assessment can be digitally carried out through some software based on a programming language if the digitized charts for CSRL and CSRE are utilized. Figure 8.90 illustrates some computational results carried out for some sites in Nias Island of Indonesia. As noted from the figure, the susceptibility assessments differ depending upon the method used.

8.7.4.5 Permeability and shear strength based method: method of Aydan-Kumsar

As discussed at the beginning of this section, the liquefaction problem from a mechanical point of view implies that the force equilibrium equations for skeleton and fluid phases are related through Darcy's law, which depends upon a rate-dependent viscous resistance of fluid and the geometrical shape of pores. Therefore, the permeability or hydraulic conductivity of the ground must be one of the most important parameters for liquefaction assessment, although none of the empirical methods have explained it so far. Considering this parameter, Aydan and Kumsar (1997b) proposed a method assuming that *liquefaction inducing stress* results from viscous upward flow of fluid through pores as a result of ground shaking. This method is explained herein.

The force condition acting on a soil grain in saturated ground can be visualized as shown in Figure 8.91. The force equilibrium in vertical direction may be written as

EARTHQUAKE

GROUND

MAXIMUM GROUND ACCELERATION
a_{max}

MAGNITUDE

SPT N VALUE

SPT N VALUE CAN ALSO BE ESTIMATED FROM CPT TESTS AND SWEDISH WEIGHT SOUNDING TEST

$N_1 = C_N N$

EQUIVALENT SHEAR STRESS RATIO CAUSED BY GROUND SHAKING

$$CSRE = \frac{\tau_d}{\sigma_v'} = 0.65 \frac{a_{max}}{g} \frac{\sigma_v}{\sigma_v'} r_d$$

LIQUEFACTION RESISTANCE

$$CSRL = \frac{\tau_l}{\sigma_v'}$$

SAFETY FACTOR AGAINST LIQUEFACTION

$$F_L = \frac{CSRL}{CSRE}$$

YES $F_l > 1$ NO

LIQUEFACTION POTENTIAL SMALL

LIQUEFACTION POTENTIAL HIGH

STOP

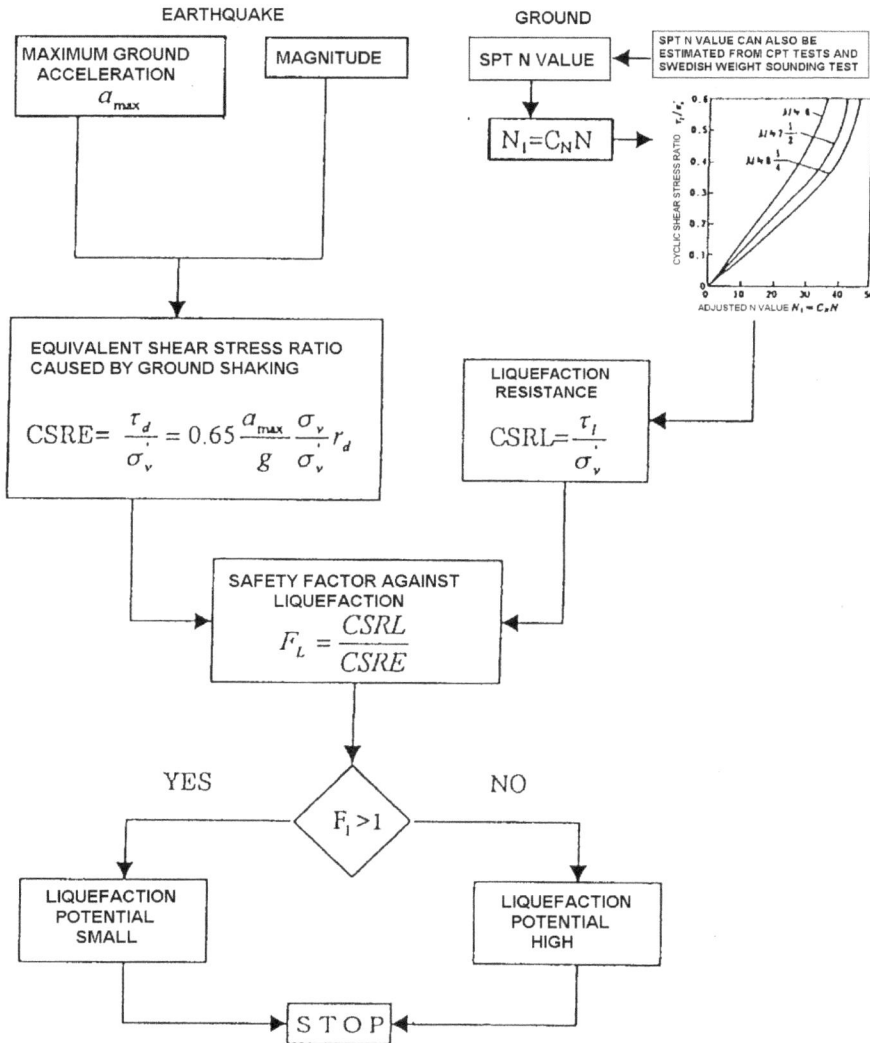

Figure 8.89 Flow chart for ground liquefaction susceptibility analyses.

$$\sum V = dF - dW' - dS = 0 \tag{8.108}$$

where dW' is the effective weight of soil grain, dS is the shear resistance between the grain and surrounding grains and dF is the viscous force associated with ground shaking. The viscous force may be related to grain size (D), velocity (v_r) and viscosity (η) of the fluid as given below (Rouse 1978):

$$dF = 3\pi\eta D v_r \tag{8.109}$$

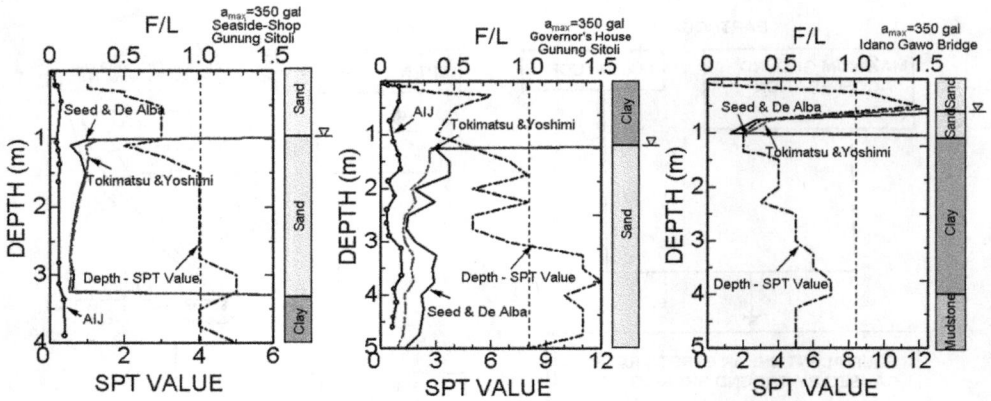

Figure 8.90 Liquefaction assessment for three selected sites. (a) Seaside shop area, (b) Governor's house and (c) Idano-Gawo bridge pier.

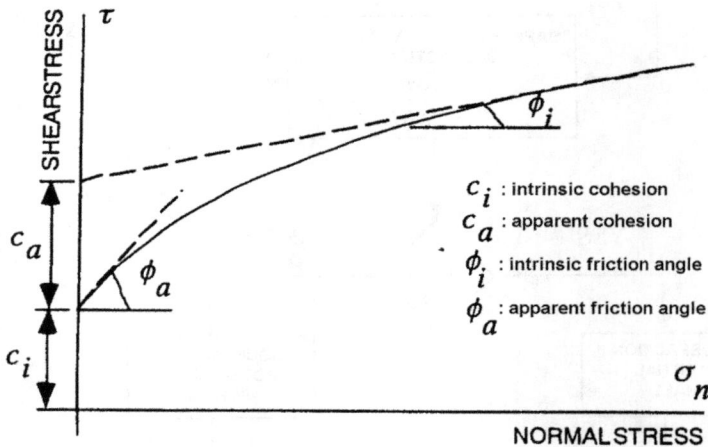

Figure 8.91 Nonlinear yield criterion for soils.

The average velocity ($\bar{v}_r$) of the seeping fluid may be related through Darcy's law as given below:

$$\bar{v}_r = -\frac{\eta}{k}\frac{dp_d}{dz} \tag{8.110}$$

where k, p_d, z are the permeability, pressure induced by ground shaking and vertical axis measured from the ground surface. The average velocity can be related to the true velocity of fluid through the utilization of porosity (n) of the ground in the following form:

$$\bar{v}_r = -nv_r \tag{8.111}$$

Pressure (p_d) may be related to ground acceleration (a_{max}) and fluid density (ρ_f) in the following form

$$p_d = -\alpha \rho_f a_{max} z \tag{8.112}$$

where α is a coefficient and it may be related to the coefficient of Seed and Idriss (1971). The pressure gradient can be obtained by derivation of Eq. (8.112) with respect to z yields

$$\frac{dp_d}{dz} = -\alpha \rho_f a_{max} \tag{8.113}$$

Thus viscous force (dF) takes the following form:

$$dF = 3\pi D\alpha \rho_f a_{max} \tag{8.114}$$

If the grain is assumed to be spherical, the effective weight of soil grain (dW') takes the following form:

$$dW' = \left[1 - \frac{\rho_f}{\rho_s}\left(1 + \frac{a_{max}}{g}\right)\right]dW \quad \text{or} \quad dW' = \left[1 - \frac{\rho_f}{\rho_s}\left(1 + \frac{a_{max}}{g}\right)\right]\rho_s g \frac{\pi D^3}{6} \tag{8.115}$$

The shear resistance of the soil may be given in the following form (Aydan et al. 1996f) (Figure 8.91):

$$\tau = c_i + \sigma'_n \mu_i + c_a\left(1 - e^{-b\sigma'_n}\right) \tag{8.116}$$

where c_i, c_a, μ_i and μ_a are intrinsic cohesion, apparent cohesion, intrinsic friction coefficient and apparent friction coefficient, respectively.

$$b = \frac{\mu_a - \mu_i}{c_a}$$

Effective normal stress may be given as follows:

$$\sigma'_n = \lambda\left(1 - \frac{\rho_f h_w}{\rho_s h}\left(1 + \frac{a_{max}}{g}\right)\right)\sigma_v \quad \text{with} \quad \sigma_v = \rho_s gz \tag{8.117}$$

where λ is the lateral stress coefficient. Thus, the shear resistance may be written by using Eqs. (8.116) and (8.117) as

$$dS = \left[c_i + \sigma'_n \mu_i + c_a\left(1 - e^{-b\sigma'_n}\right)\right]dA \tag{8.118}$$

Area (dA) may be given as

$$dA = \pi D^2 \qquad (8.119)$$

The safety against liquefaction may be defined as

$$SF = \frac{dW' + dS}{dF} \qquad (8.120)$$

If the resistance is assumed to be only frictional, shear resistance (dS) is

$$dS = \mu \lambda dW' \qquad (8.121)$$

The safety factor takes the following form:

$$SF = \frac{(1 + \mu_i \lambda)\left[\dfrac{\rho_s}{\rho_f} - \left(1 + \dfrac{a_{\max}}{g}\right)\right]D^2 n}{18k\dfrac{\alpha a_{\max}}{g}} \qquad (8.122)$$

The properties of granular soils depend upon the porosity and geometry of grains and their arrangement. There are different empirical relations with the mean grain size of soils. It is possible to relate the permeability characteristics with the mean grain size (D_{50}) (e.g. Aydan et al. 1997c). Based on the database on soil properties, Aydan et al. (1997c) suggested the following empirical relation (Figure 8.92):

$$k = 0.135 \cdot D_{50}^{2.145} \qquad (8.123)$$

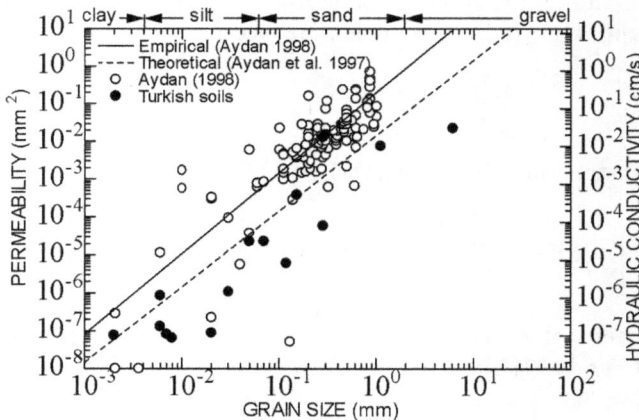

Figure 8.92 The relation between mean grain size and permeability for Turkish soils and comparisons with soils of other countries.

The unit of permeability is mm² and mean grain size is given in mm.

The specific forms of strength parameters defined in Eq. (8.118) are specifically given in terms of the mean grain size as follows, and comparisons with actual data are shown in Figure 8.93:

Intrinsic cohesion (unit is kPa)

$$c_i = \frac{150}{1 + 150 \cdot D_{50}} \tag{8.124}$$

Apparent cohesion (unit is kPa)

$$c_a = 15 \cdot D_{50}^{0.6} \tag{8.125}$$

Intrinsic friction coefficient

$$\mu_i = 0.8 + 0.10 \log D_{50} \tag{8.126}$$

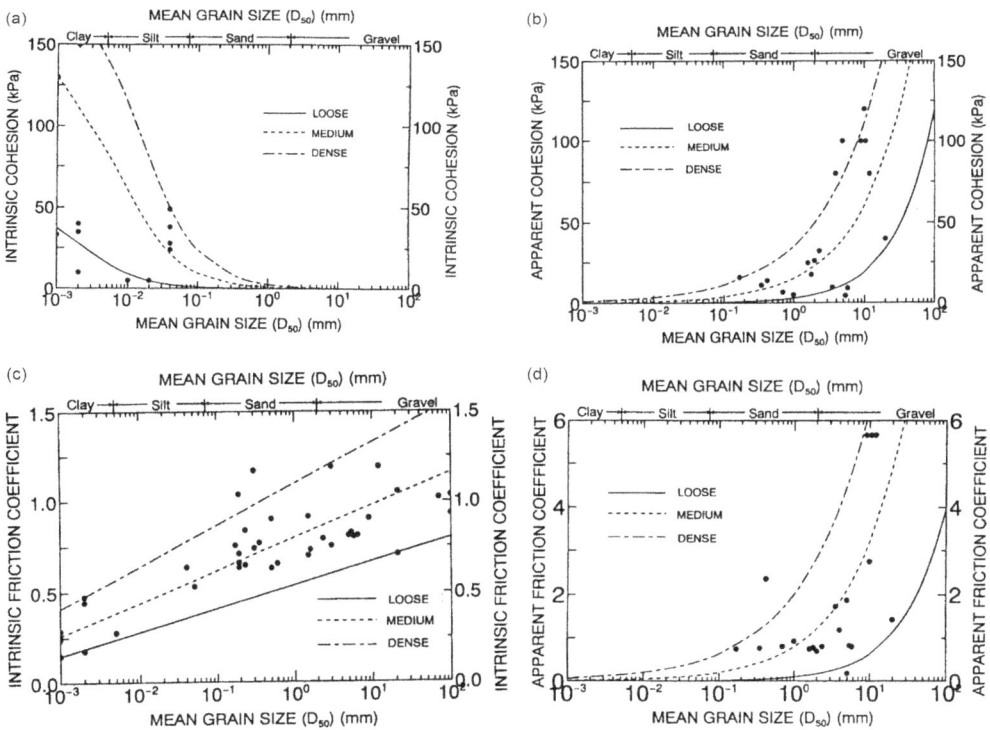

Figure 8.93 Variation in different parameters of the nonlinear yield criterion with mean grain size (D_{50}). (a) Intrinsic cohesion, (b) apparent cohesion, (c) intrinsic friction coefficient and (d) apparent friction coefficient.

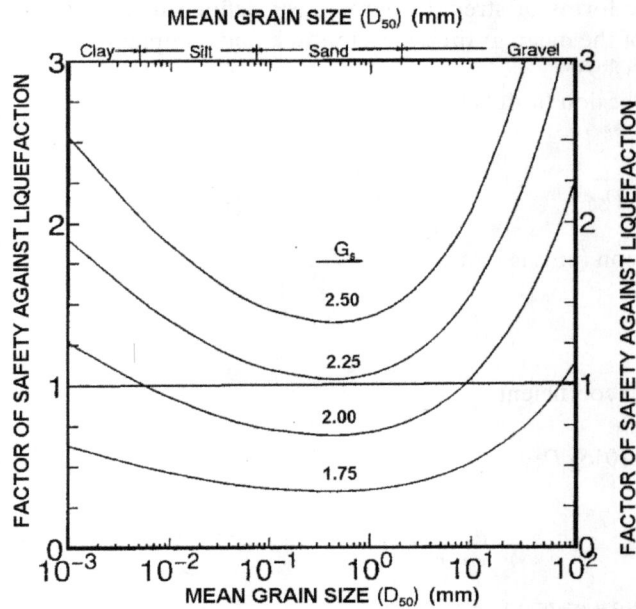

Figure 8.94 Variation in liquefaction potential with mean grain size for different density ratios.

Apparent friction coefficient

$$\mu_a = 0.8 \cdot D_{50}^{0.6} \tag{8.127}$$

Figure 8.94 shows an example of computations for an earthquake with a maximum acceleration of 0.5 g for a given site. G_s in Figure 8.94 stands for the ratio of solid particle density to fluid density. As seen from this example, the proposed method is capable of explaining why clay, silt or gravel is less susceptible to liquefaction compared to sand. Furthermore, as the permeability of the ground reflects the effect of grading of soil and as it is an integrated parameter of grain size distribution, this method, which is based on the permeability of ground as well as its shear resistance, is much more superior to other empirical methods. Figure 8.95 shows a comparison of magnitude versus liquefaction limit distance for the earthquake data shown in Figure 8.83a for some density ratios. The results are quite close to those obtained from empirical relations.

8.7.5   Lateral spreading: deformation estimation

There are basically four different techniques to estimate ground deformations induced by ground liquefaction: (1) empirical methods, (2) sliding body analysis, (3) analytical method and (4) finite element method. These methods are explained herein.

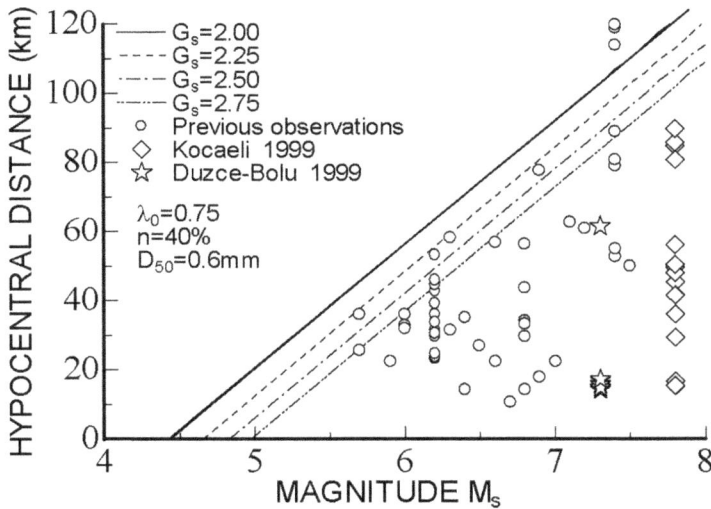

Figure 8.95 Comparison of computed magnitude versus hypocentral distance relations with observations.

8.7.5.1 Empirical methods

Hamada et al. (1986), Bardet et al. (1999), Youd et al. (2002) and Aydan et al. (2005c) have proposed some empirical methods, and methods of Bardet et al. (1999) and Youd et al. (2002) utilize fundamentally experimental and observational results reported by Hamada (1999). The method of Hamada et al. (1986), which is an empirical method based on regression analyses of previous lateral spreading case histories, predicts the amplitude of horizontal ground deformation only in terms of the slope and thickness of the liquefied layer:

$$D = 0.75H^{0.5}\theta^{0.33} \tag{8.128}$$

where D is the horizontal displacement (m), θ is the slope (%) of the ground surface or base of liquefied layer and H is the thickness (m) of the liquefied layer.

Based on the review and comparison of the existing empirical models and data-bases, Bardet et al. (1999) developed a model liquefaction-induced ground displacement using an MLR (multi-linear regression) approach similar to that by Bartlett and Youd (1992). The amplitude D_H of ground deformation (m) is estimated by using the following relation:

$$\begin{aligned} \log(D_H + 0.01) = {} & b_0 + b_{\text{off}} + b_1 M + b_2 \log(R) + b_3 R + b_4 \log(W) + b_5 \log(S) \\ & + b_6 \log(T_{15}) + b_7 \log(100 - F_{15}) + b_8 D50_{15} \end{aligned} \tag{8.129}$$

where M is the moment magnitude, R the nearest horizontal distance (km) to seismic energy source or fault rupture, S the slope (%) of the ground surface, W the free face ratio (%), T_{15} the thickness (m) of saturated cohesionless soil with $(N_1)_{60} < 15$, F_{15} the average fines content (% finer than 75 μm) and $D50_{15}$ the average D_{50} grain size (mm) in T_{15}. The model suggested by Youd et al. (2002) is a modification of the correlation by Bartlett and Youd (1995), which is what is the MLR method. Equation (8.130a) gives the MLR model for free face conditions.

$$\log D_H = -16.713 + 1.532 M_w - 1.406 \log R * -0.012 R + 0.592 \log W$$
$$+ 0.540 \ \log T_{15} + 3.413 \log(100 - F_{15}) - 0.795 \log(D50_{15} + 0.1) \quad \text{(8.130a)}$$

and Eq. (8.130b) is applied to gently sloping ground conditions:

$$\log D_H = -16.213 + 1.532 M_w - 1.406 \log R * -0.012 R + 0.388 \log S$$
$$+ 0.540 \log T_{15} + 3.413 \log(100 - F_{15}) - 0.795 \log(D50_{15} + 0.1) \quad \text{(8.130b)}$$

where D_H is the horizontal ground displacement (m); Mw the moment magnitude; R the horizontal distance to the nearest seismic source or to the nearest fault rupture (km), $R* = R + R_0$, and $R_0 = 10^{(0.89M-5.64)}$; S the gradient of surface topography or ground slope (%); T_{15} the thickness of saturated layers with $(N_1)_{60} < 15$; F_{15} the average fines content (particles <0.075 mm) in T_{15}; $D50_{15}$ the average D_{50} (mm) in T_{15}; and W the free face ratio.

Hamada (1999) has shown that the liquefiable soil behaves as a quasi-plastic fluid and that the Reynold's similitude law holds between the flow of model grounds and that of actual grounds. Based on his studies, horizontal displacements of actual ground in the case studies were predicted from the experimental result by applying the similitude law, which showed a good correlation with the displacements observed during past earthquakes. Then Hamada (1999) suggested the following relation for prediction of displacements of actual grounds from flow tests of the model ground:

$$D_H = \frac{0.0125 \ (H)^{0.5} \ \theta}{\overline{N}^{0.88}} \sum a_i^{0.48} \ T_i \quad \text{(8.131)}$$

where D_H is the total value of ground surface displacement, H the thickness of liquefied soil, θ the surface gradient, N the modified SPT-N values and T_i and a_i are the duration of the time in the ith segment of the time histories of accelerations recorded during the earthquake and the mean magnitude of the acceleration, respectively.

The following empirical equation was proposed by Aydan et al. (2005c) for estimating the displacement of the liquefied ground:

$$\delta = A \frac{\gamma_s H_l^2}{G} \sin \theta \, v_{max} \quad \text{(8.132)}$$

where γ_s is the unit weight of liquefied ground; H_1 is the liquefied layer thickness; G, θ, A and v_{max} are the residual shear modulus of liquefied ground, ground inclination, empirical constant and maximum ground velocity, respectively.

8.7.5.2 Sliding body analysis

The methods based on the sliding body analysis was originally proposed by Newmark (1965) and they are used to estimate the rigid body motion of liquefied layer through the consideration of input-waves and shear strength mobilized along the sliding plane (Dobry and Baziar 1992) (Figure 8.96). The most difficult aspects in this method are how to select the residual shear strength properties and pore pressure variation during shaking and motion of the ground. Despite numerous laboratory tests on the liquefied soils, knowledge of their properties is scarce. The shear strength of liquefied ground under dynamic shaking, which is relevant to actual conditions in situ, is those for undrained state. Some of these data were recently compiled by Ishihara (1993). The normalized shear strength, which may be viewed as the residual friction angle coefficient, may range between 0.1 (5.7°) and 0.2 (11.3°) for pure sand (Figure 8.97). This value decreases as the plasticity index increases (silty or clayey-sandy soil).

Aydan and Ulusay (2002) originally developed a rigid body formulation for assessing landslide problems including post-failure motions together with a Bingham-type yield criterion. This method was extended to the assessment of ground deformation induced by ground liquefaction (Aydan et al. 2005c, 2008b). The formulation is basically the same, and details can be found in Aydan and Ulusay (2002). The static pore pressure coefficient for a gently inclined soil layer is set as follows:

$$\beta_s = \frac{\gamma_w}{\gamma_s} \frac{h_w}{H} \tag{8.133}$$

where γ_w and γ_s are unit weight of water and liquefied ground; h_w and H are water table height above the sliding surface and layer thickness, respectively. However, the dynamic component of pore pressure coefficient is different from that of Aydan and Ulusay (2002), which is given in the following form:

$$\beta_d = \beta_s \left(\frac{a_H}{g} \sin\alpha + \frac{a_V}{g} \cos\alpha \right) \tag{8.134}$$

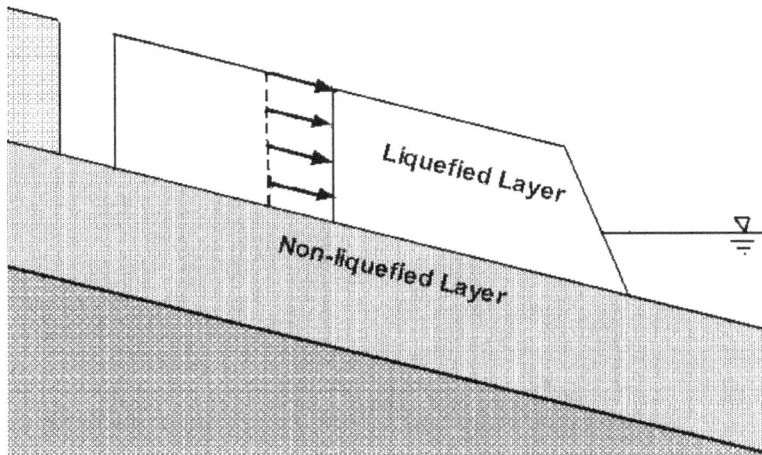

Figure 8.96 Model for sliding body analysis.

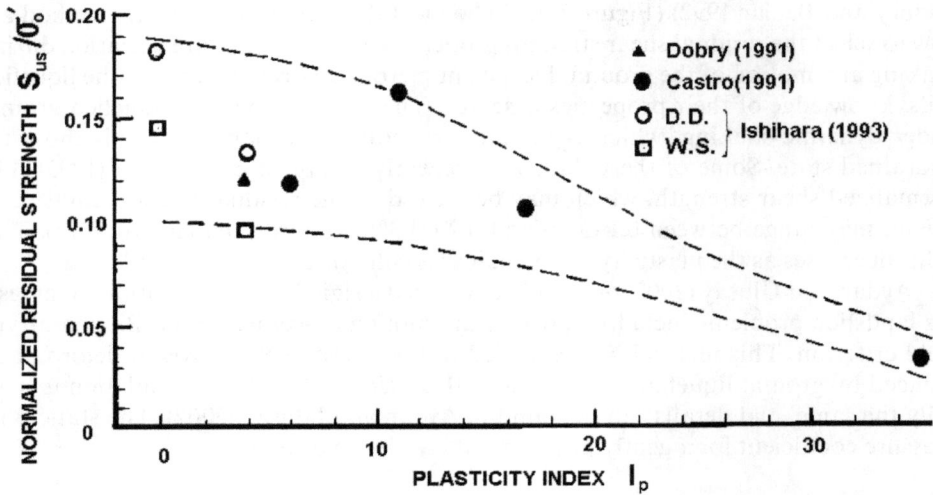

Figure 8.97 Residual characteristics of liquefied soil. (Arranged from Ishihara (1993).)

where α , a_H , a_V and g are the inclination of the sliding plane, horizontal and vertical ground acceleration in the direction of motion and gravitational acceleration. If the above equation is used, the pore pressure may become negative depending upon the amplitude of accelerations. Many experiments on liquefaction showed that the pore pressure is always positive (compression is assumed to be positive). As the effective stress becomes nil at the liquefaction state, the excess pore pressure at the liquefaction state cannot be greater than the effective stress at the initiation of shaking. Thus the following relation may be written:

$$p_d = \sigma - p_s \quad or \quad p_d = \sigma - p_s \quad or \quad p_d = p_s\left(\frac{\sigma}{p_s} - 1\right) \qquad (8.135)$$

If the following identities hold

$$p_s = \gamma_w h_w, \quad \sigma = \gamma_s H, \quad \beta_s = \frac{p_s}{\sigma}, \quad \beta_d = \frac{p_d}{\sigma} \qquad (8.136)$$

One can write the following:

$$\beta_d = \beta_s\left(\frac{\gamma_s H}{\gamma_w h_w} - 1\right) \qquad (8.137)$$

Dynamic variation of pore pressure should be proportional to ground shaking before the ultimate state of liquefaction is achieved. Furthermore, the excess pore pressure will be dissipated in relation to permeability characteristics after ground shaking terminates. The dissipation is very quick for sandy soil without a low-permeability top soil layer. Together with experimental observations on pore pressure variation in experiments, the dynamic pore pressure coefficient is written in the following form:

$$\beta_d = \beta_s \left(\frac{\gamma_s H}{\gamma_w H_w} - 1 \right) \frac{|a|}{g} \tag{8.138}$$

where

$$|a| = |a_H \sin\alpha + a_V \cos\alpha| \tag{8.139}$$

When the liquefaction state is achieved, the sum of static and dynamic pore pressure coefficients cannot be more than 1. This condition is implemented in computations. The introduction of the dynamic component of the pore pressure during ground shaking as well as visco-plastic yield criterion is a further improvement to rigid body sliding approaches.

8.7.5.3   Analytical model for an infinitely long visco-elastic layer

In this section, closed form solutions for the dynamic shear response of an infinitely long visco-elastic layer under gravitational loading are described (Aydan 1994, 1995b). The selection of this problem is particularly important in practice as the motion of liquefied soil is physically similar to the motion of an infinitely long layer. Furthermore, it should be noted that large deformations of liquefied grounds take place long after the main shock wave has passed over.

8.7.5.3.1   CONSTITUTIVE MODEL

The mechanical behaviour of a linear visco-elastic material is associated with the strain ε for a solid-like behaviour and strain rate $\dot{\varepsilon}$ for a fluid-like behaviour of the material. For simple shear behaviour, the constitutive law is given by (Figure 8.98)

$$\tau = G\gamma + \eta\dot{\gamma} \tag{8.140}$$

where G is the elastic shear modulus and η is the viscos shear modulus. This model is known as the Voigt–Kelvin model. When $G = 0$, then it simply corresponds to a Newtonian fluid. On the other hand, when $\eta = 0$, it corresponds to a Hookean solid.

8.7.5.3.2   GOVERNING EQUATION

The governing equation takes the following form by considering the equilibrium of the element by applying Newton's second law (Figure 8.99):

$$\frac{\partial \tau}{\partial y} - \frac{\partial p}{\partial x} + \rho g \sin\alpha = \rho \ddot{u} \tag{8.141}$$

If the thickness of the liquefied layer does not vary with x and the medium consists of the same material, then $\partial p / \partial x = 0$ and the above equation become

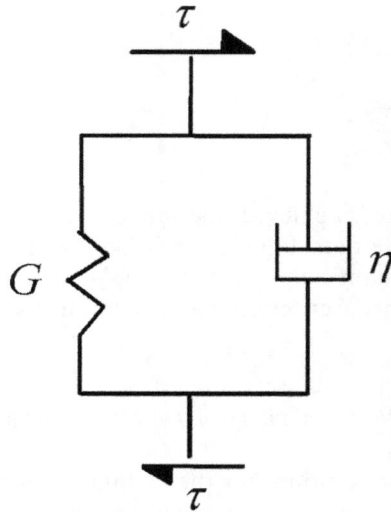

Figure 8.98 Constitutive model.

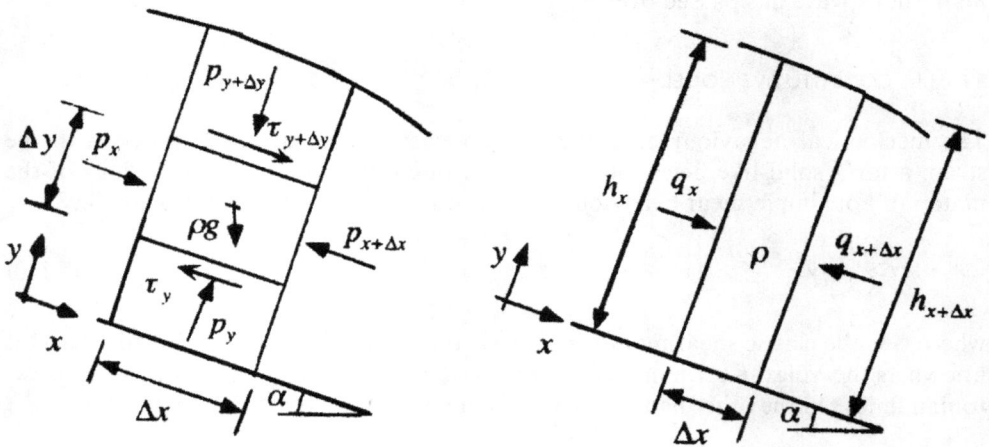

Figure 8.99 Mechanical model.

$$\frac{\partial \tau}{\partial y} + \rho g \sin \alpha = \rho \ddot{u} \tag{8.142}$$

Assuming that the shear stress and shear strain rate can be defined as

$$\gamma = \frac{\partial u}{\partial y}, \quad \dot{\gamma} = \frac{\partial \dot{u}}{\partial y} \tag{8.143}$$

and introducing the constitutive law given by Eq. (8.140) into Eq. (8.142) yields the following partial differential equation:

$$\rho \frac{\partial^2 u}{\partial t^2} + \eta \frac{\partial^2}{\partial y^2}\left(\frac{\partial u}{\partial t}\right) + G\frac{\partial^2 u}{\partial y^2} = \rho g \sin\alpha \tag{8.144}$$

The solution of this partial differential equation may be given as

$$u(y,t) = Y(y) \cdot T(t) \tag{8.145}$$

As a particular case, $Y(y)$ is assumed to be of the following form by considering an earlier solution of the equilibrium equation without inertial term for open-channel boundary conditions:

$$Y(y) = y\left(H - \frac{y}{2}\right) \tag{8.146}$$

Inserting this relation into Eq. (8.138), we have

$$\rho \frac{\partial^2 T}{\partial t^2} y\left(H - \frac{y}{2}\right) + \eta \frac{\partial T}{\partial t} + GT = \rho g \sin\alpha \tag{8.147}$$

Integrating the above equation with respect to y for bounds $y = 0$ and $y = H$ results in the following second-order nonhomogeneous ordinary differential equation:

$$\frac{\partial^2 T}{\partial t^2} - \frac{3\eta}{\rho H^2}\frac{\partial T}{\partial t} + \frac{3G}{\rho H^2}T = \frac{3g\sin\alpha}{H^2} \tag{8.148}$$

The solutions of this differential equation are as follows:

Case 1: Roots are real

$$T = C_1 e^{\lambda_1 t} + C_2 e^{\lambda_2 t} + \frac{1}{\lambda_1 \lambda_2}\frac{3g\sin\alpha}{H^2} \tag{8.149}$$

where

$$\lambda_1 = \frac{1}{2\rho H^2}\left(-3\eta + \sqrt{9\eta^2 - 12G\rho H^2}\right), \quad \lambda_2 = \frac{1}{2\rho H^2}\left(-3\eta - \sqrt{9\eta^2 - 12G\rho H^2}\right)$$

Case 2: Roots are the same

$$T = [C_1 + C_2 t]e^{\lambda t} + \frac{1}{\lambda^2}\frac{3g\sin\alpha}{H^2} \tag{8.150}$$

where

$$\lambda = -\frac{3\eta}{2\rho H^2}$$

Case 3: Roots are complex

$$T = e^{pt}[A\cos qt + B\sin qt] + \frac{1}{p^2 + q^2}\frac{3g\sin\alpha}{H^2} \qquad (8.151)$$

where

$$p = -\frac{3\eta}{2\rho H^2}, \quad q = \sqrt{12G\rho H^2 - 9\eta^2}$$

Integration constants C_1 and C_2 can be determined from the following initial conditions:

$$\begin{aligned} u(y,t) &= 0 \quad \text{at} \quad t = 0 \\ \dot{u}(y,t) &= 0 \quad \text{at} \quad t = 0 \end{aligned} \qquad (8.152)$$

For the above initial conditions, the integration constants for each case are

Case 1: Roots are real

$$C_1 = -\frac{1}{\lambda_1(\lambda_2 - \lambda_1)}\frac{3g\sin\alpha}{H^2}, \quad C_2 = \frac{1}{\lambda_2(\lambda_2 - \lambda_1)}\frac{3g\sin\alpha}{H^2} \qquad (8.153)$$

Case 2: Roots are the same

$$C_1 = -\frac{1}{\lambda^2}\frac{3g\sin\alpha}{H^2}, \quad C_2 = \frac{1}{\lambda^2}\frac{3g\sin\alpha}{H^2} \qquad (8.154)$$

Case 3: Roots are complex

$$C_1 = -\frac{1}{p^2 + q^2}\frac{3g\sin\alpha}{H^2}, \quad C_2 = \frac{p}{q}\cdot\frac{1}{p^2 + q^2}\frac{3g\sin\alpha}{H^2} \qquad (8.155)$$

Integration constants for closed-channel boundary conditions can be obtained in a similar manner.

8.7.5.4   Numerical methods and simplified methods

8.7.5.4.1   REDUCED MATERIAL PROPERTY APPROACH

One of the procedures is to treat the liquefied ground as an elastic medium with residual elastic constants or viscous fluid (Yasuda et al. 1990, 2001, Towhata et al. 1992).

Aydan (1994, 1995b) combined the concepts of residual elasticity and viscosity and proposed the following finite element formulation:

$$\mathbf{M}\,\ddot{\mathbf{U}} + \mathbf{C}\,\dot{\mathbf{U}} + \mathbf{KU} = \mathbf{F} \tag{8.156}$$

where

$$\mathbf{M} = \int_{\Omega_e} \rho \mathbf{N}^{\mathrm{T}} \mathbf{N} d\Omega; \mathbf{C} = \int_{\Omega_e} \mathbf{B}^{\mathrm{T}} \mathbf{D}_v \mathbf{B} d\Omega; \quad \mathbf{K}$$
$$= \int_{\Omega_e} \mathbf{B}^{\mathrm{T}} \mathbf{D}_e \mathbf{B} d\Omega; \mathbf{F} = \int_{\Omega_e} \mathbf{N}^{\mathrm{T}} \mathbf{b} d\Omega + \int_{\Gamma_{te}} \mathbf{N}^{\mathrm{T}} \mathbf{t} d\Gamma$$

The constitutive relations are

$$\sigma = \mathbf{D}^e \varepsilon + \mathbf{D}^v \dot{\varepsilon} \tag{8.157}$$

where $\mathbf{D}^e$ and $\mathbf{D}^v$ are elasticity and viscosity tensors. This type of constitutive law enables one to model media varying from fluid to solid. Furthermore, such a constitutive law will also allow us to treat the liquefied and nonliquefied media simultaneously within the same finite element solution scheme.

8.7.5.4.2 SIMPLIFIED HYBRID ANALYTICAL AND FEM FORMULATION

Momentum conservation law for an infinitely small element of a liquefied ground on a plane with an inclination of α for each respective direction can be written in the following form (Figure 8.99):

x-direction and y-direction

$$\frac{\partial \tau}{\partial y} = \frac{\partial p}{\partial x} - \rho g \sin \alpha; \; \frac{\partial p}{\partial y} = \rho g \sin \alpha \tag{8.158}$$

where τ, p, ρ and g are the shear stress, pressure, density and gravitational acceleration, respectively. Mass conservation law for an infinitely small element of a liquefied ground on a plane with an inclination of α for each respective direction can be similarly written in the following form (Figure 8.99):

$$\frac{\partial(\rho h)}{\partial t} = -\frac{\partial q}{\partial x} \tag{8.159}$$

If ρ is constant and $q = \rho \bar{v} h$, then the equation above becomes

$$\frac{\partial h}{\partial t} = -\frac{\partial(\bar{v} h)}{\partial x} \tag{8.160}$$

Assuming that the liquefied ground is incompressible for volumetric deformation, the shear response of the liquefied ground is given by Eq. (8.140). Integrating Eq. (8.158) with respect to y together with boundary condition $p = 0$ at $y = h$ yields

$$p = \rho g \cos\alpha (h(x) - y) \tag{8.161}$$

Then, its derivative with respect to x may be written as follows:

$$\frac{\partial p}{\partial x} = \rho g \cos\alpha \frac{\partial h}{\partial x} \tag{8.162}$$

Inserting (8.155) into (8.154) yields

$$\frac{\partial \tau}{\partial y} = \rho g (\cos\alpha \frac{\partial h}{\partial x} - \sin\alpha) \tag{8.163}$$

Integrating (9) with respect to y, together with $\tau = 0$ at $y = h$ gives

$$\tau = \rho g \cos\alpha \left(\tan\alpha - \frac{\partial h}{\partial x} \right)(h - y) \tag{8.164}$$

Using Eq. (8.164) in Eq. (8.160) and solving resulting differential equation yields average velocity of laterally spreading soil at a given cross section at x as

$$\bar{v} = \frac{e^{-\frac{G}{\eta}t}}{\eta} \left[\rho g \cos\alpha \left(\tan\alpha - \frac{\partial h}{\partial x} \right) \frac{h^2}{3} \right] \tag{8.165}$$

Inserting Eq. (8.162) into Eq. (8.158) gives

$$e^{\frac{G}{\eta}t} \frac{\partial h}{\partial t} = -\frac{\partial}{\partial x} \left(\frac{\rho g \sin\alpha h^3}{3\eta} \right) + \frac{\partial}{\partial x} \left(\frac{\rho g \cos\alpha h^3}{3\eta} \frac{\partial h}{\partial x} \right) \tag{8.166}$$

This equation is a nonlinear partial differential equation. If the inclination α is very small, the effect of the first term on the RHS is negligible. Thus, we have

$$e^{\frac{G}{\eta}t} \frac{\partial h}{\partial t} = \frac{\partial}{\partial x} \left(\frac{\rho g \cos\alpha h^3}{3\eta} \frac{\partial h}{\partial x} \right) \tag{8.167}$$

Let us denote the following:

$$A = \frac{\rho g \cos\alpha h^3}{3\eta} \tag{8.168}$$

Thus the above equation may be rewritten as

$$e^{\frac{G}{\eta}t} \frac{\partial h}{\partial t} = \frac{\partial}{\partial x} \left(A \frac{\partial h}{\partial x} \right) \tag{8.169}$$

It should be noted that this partial differential is nonlinear. In this study, A will be treated as a constant in at a small time increment in order to linearize the partial differential equation. Taking a variation on δh, the integral form of Eq. (8.169) and integrating the resulting equation by parts gives

$$\int \delta h e^{\frac{G}{\eta}t} \frac{\partial h}{\partial t} dx + \int \frac{\partial \delta h}{\partial x} A \frac{\partial h}{\partial x} dx = \delta h \left(A \frac{\partial h}{\partial x} \right) \Bigg|_{x=a}^{x=b} \tag{8.170}$$

Equation (8.172), which holds for whole domain, must also hold for a typical element if the domain is discretized into n elements, and $h(x)$ can be interpolated element-wise by the following function:

$$h = [N]\{H\}; \quad \frac{\partial h}{\partial x} = \frac{\partial}{\partial x}[N]\{H\} = [B]\{H\} \tag{8.171}$$

Using the above equation in Eq. (8.170) yields

$$[M]_e \{\dot{H}\}_e + [K]_e \{H\}_e = \{F\}_e \tag{8.172}$$

where

$$[M]_e = e^{\frac{G}{\eta}t} \int_{x_{it}}^{xj} [N][N] dx; \quad [K]_e = A \int_{x_{it}}^{xj} [B][B] dx; \quad \{F\}_e = \left[\bar{N} \right]^T \left(A \frac{\partial h}{\partial x} \right) \Bigg|_{x_i}^{xj}$$

Using the Taylor expansion of   with respect to time step t_n, Eq. (8.172) results in the following simultaneous equation system:

$$\left[M^* \right]\{H\}_{n+1} = \{F\}_{n+1} \tag{8.173}$$

where

$$\left[M^* \right] = \frac{1}{\Delta t}[M]; \quad \{F\}_{n+1} = \{F\}_n + \left(\frac{1}{\Delta t}[M] - [K] \right)\{H\}_n$$

The boundary conditions are implemented in the following form:

$$\text{at } x = a, \quad v_a = \frac{\partial h}{\partial t} n_a = -\frac{\partial h}{\partial t} \approx -\frac{h_{n+1}^{x=a} - h_n^{x=a}}{\Delta t} \Rightarrow h_{n+1}^{x=a} = h_n^{x=a} - v_a \Delta t \tag{8.174a}$$

$$\text{at } x = b, \quad v_b = \frac{\partial h}{\partial t} n_b = \frac{\partial h}{\partial t} \approx \frac{h_{n+1}^{x=b} - h_n^{x=b}}{\Delta t} \Rightarrow h_{n+1}^{x=b} = h_n^{x=b} + v_b \Delta t \tag{8.174b}$$

8.7.5.4.3 NUMERICAL EXAMPLES

8.7.5.4.3.1 One-dimensional examples

A series of one-dimensional finite element analyses of an infinitely long layer on an inclined base was first carried out to check the effect of the consideration of the inertia term on displacement, velocity and acceleration responses. The base inclination was 2% and the layer was 4 m thick. Two parametric studies were performed and the parameters used in the computations are given in Table 8.21; more details can be found in a previous publication by Aydan (1994). Figure 8.100 shows the computed acceleration, velocity and displacement responses at the top (4 m) and the middle (2 m) of the layer for two cases. Depending upon the visco-elastic constants of the medium and the geometry of the domain, the initial parts of the response curves are greatly influenced. Nevertheless, as time goes by, the responses computed considering and not considering the inertia term converge toward each other. Considering the visco-elastic constants reported in the literature (Hamada et al. 1993, Towhata and Matsumoto 1992, Yasuda et al. 1990), neglecting the inertia term would not be causing a serious error. Furthermore, it results in stable solutions in multi-dimensional finite element analyses.

8.7.5.4.3.2 Two-dimensional examples

Next, a series of simulations of model tests of liquefied soil shaken in a large box by the proposed method was carried. The box was 5 m long, 1.2 m wide and 1.3 m deep. The soil layer having a surface gradient of 2% was poured into the box, which was 30 cm deep in the middle. The box was shaken in a direction perpendicular to its longitudinal axis using sinusoidal acceleration waves with a frequency of 4 Hz. Once the liquefaction state was attained, the shaking of the box was terminated.

The finite element mesh used in computations is shown in Figure 8.101. In computations, two cases, in which Lame's constant μ was varied, were considered. Mechanical properties of liquefied soil are given in Table 8.22. Constants μ and μ^* are those reported by Hamada et al. (1993), which were obtained from the analysis of displacement and velocity responses measured in model tests using an infinitely long elastic or viscos layer on an inclined plane. Constant λ was chosen such that volumetric strain of liquefied soil is negligible. Computed results are shown in Figure. 8.102. Figure 8.103 shows computed response curves of points at various depths at the mid cross section of the analysed domain. As expected, the variation in Lame's constant μ results in the reduction of magnitude of displacements of the liquefied ground. It is also interesting to note that the ground subsides in the upper part and heaves in the lower part. This is in accordance with observations on model tests and actual liquefied sites in Erzincan, visited by the author (Aydan and Hamada 1992). Furthermore, the displacement distributions are parabolic. Besides the results reported herein, a series of analyses were carried out by varying the magnitude of viscous shear modulus μ^*. Its variation

Table 8.21 Material properties used in analyses

Case no	$\gamma\ (kN/m^3)$	$\mu\ (kPa)$	$\gamma^*\ (kPa\ s)$
1	18	5	20
2	18	5	5

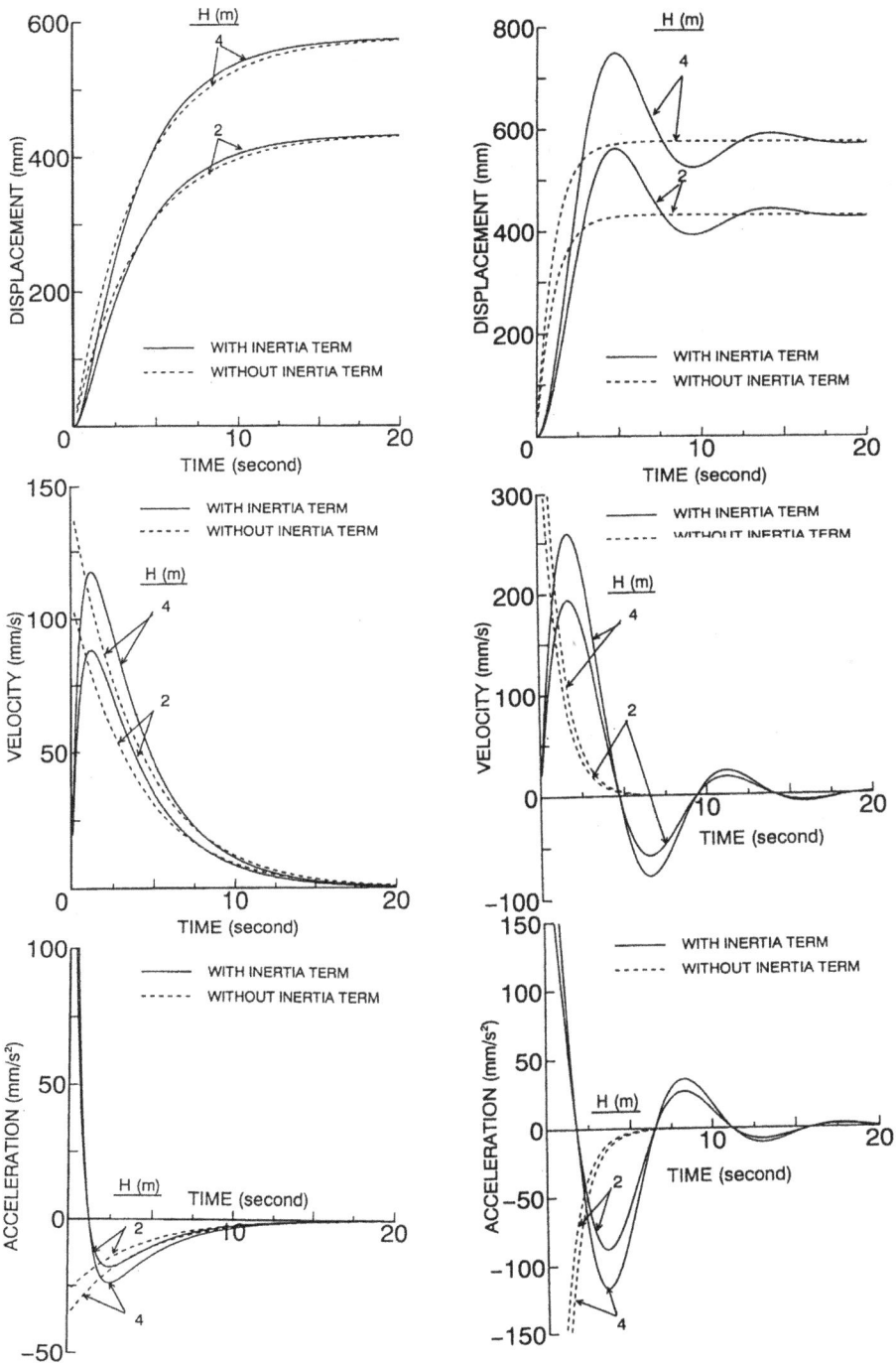

Figure 8.100 Comparison of computed responses. (a) Case I and (b) Case 2.

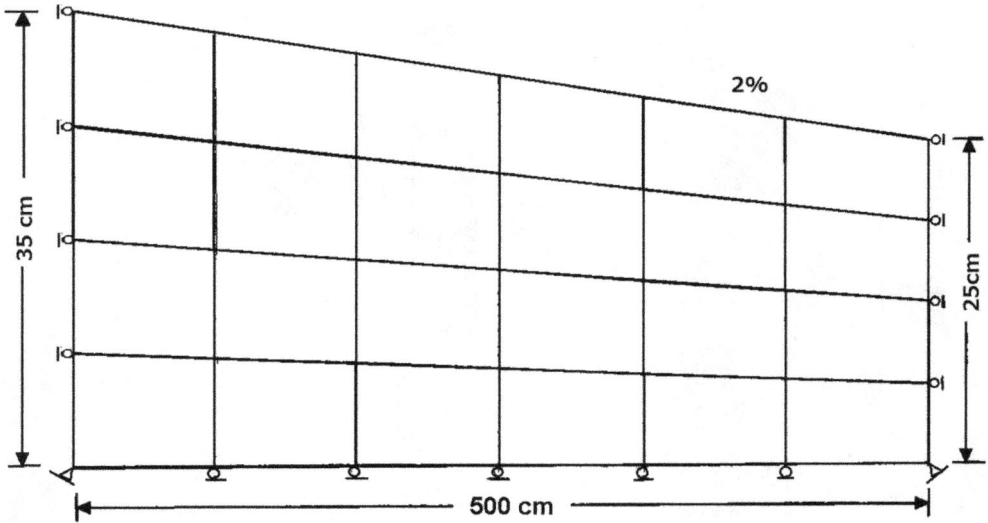

Figure 8.101 Finite element mesh.

Table 8.22 Mechanical properties.

Case no	$\gamma\,(kN/m^3)$	$\lambda\,(Pa)$	$\mu\,(Pa)$	$\lambda^*\,(Pa\;s)$	$\mu^*\,(Pa\;s)$
1	18	5000	50	0	500
2	18	5000	500	0	500

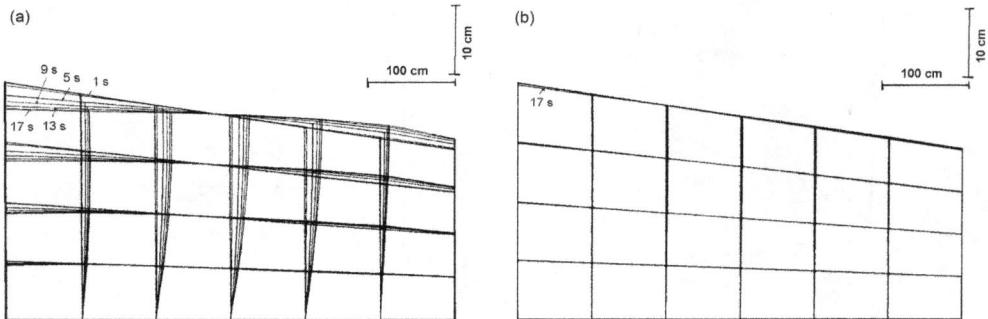

Figure 8.102 Deformed configuration of ground at some selected times. (a) Case 1 and (b) Case 2.

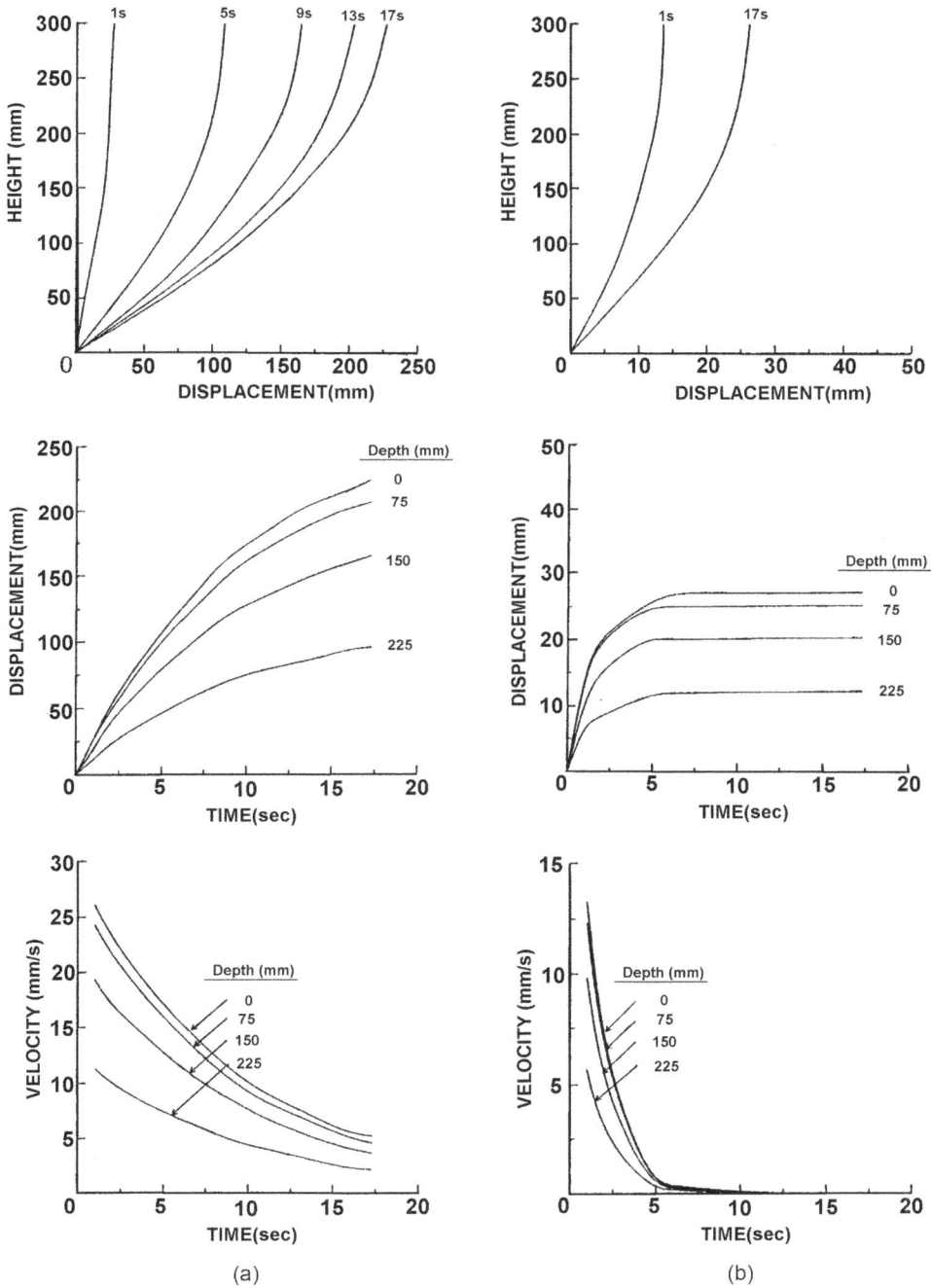

Figure 8.103 Computed responses at selected locations. (a) Case 1 and (b) Case 2.

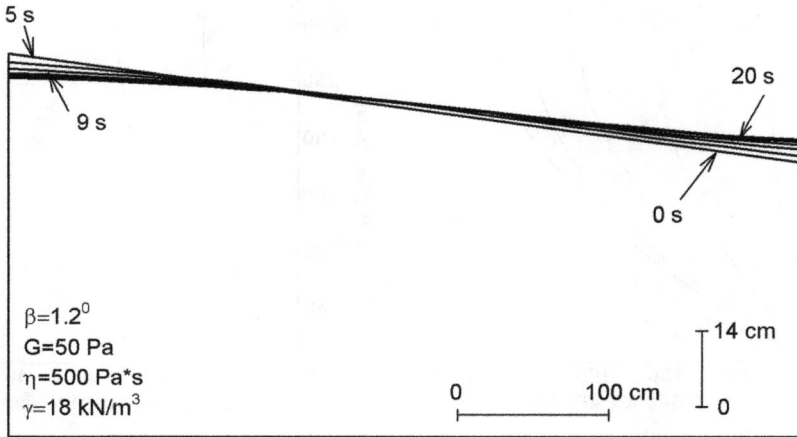

Figure 8.104 Computed surface configurations of the model ground.

influences only the time history of response curves, and the final values depend purely upon the shear modulus μ, which may be viewed as the residual shear modulus of the liquefied ground. This conclusion is similar to that drawn in an earlier publication (Aydan 1994).

The hybrid analytical method and FEM were used for simulating a ground having only visco-elastic resistance (Figure 8.104). The initial ground inclination and geometry were the same as shown in Figure 8.101. Since the ground has only visco-elastic resistance, the ground surface must become asymptotic to a configuration similar to the first mode of sloshing phenomenon.

The final example is concerned with the simulation of model tests at the laboratory using the hybrid method. The inclination of model ground was set as 2.1%, and the physical and mechanical properties used are shown in Figure 8.105. The computed

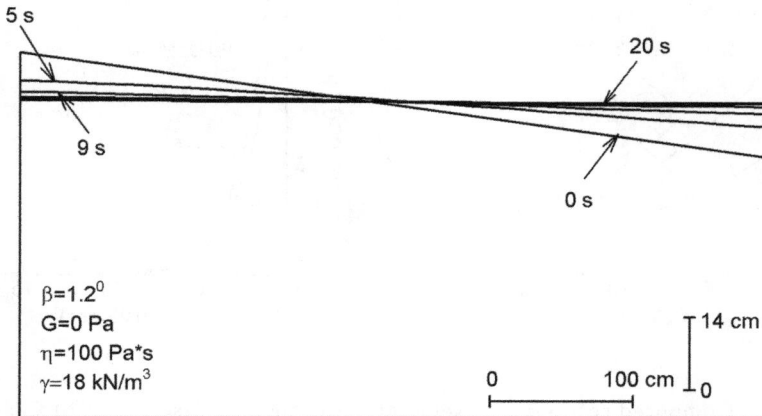

Figure 8.105   *Computed surface configurations of the model ground.*

surface configuration of the ground at 20 seconds is compared with that obtained from an experiment since the variation of computed results after 20 seconds becomes very negligible. As the material is viscous, the ground surface is almost flat at the time step of 20 seconds. The computed surface configuration obtained from the presented approach is almost the same as that observed in experiments. From this comparison, it is once again concluded that the approach is capable of simulating the lateral spreading of liquefied ground in model tests.

8.7.5.5   Experiments on lateral spreading of dry ground

Lateral spreading of ground was generally assumed to be due to ground liquefaction. However, some lateral spreading of dry ground was observed and some shaking table experiments were carried out to check this issue. The sand was poured into the container under gravity and deposited. The angle of the slope of the soil deposits formed through natural deposition. Then the base of table was subjected to shaking. A total of three experiments were carried out as explained in the next section.

Figure 8.106 shows some views of experiment 1 before and after shaking. It is of great interest that the ground surface becomes almost flat after shaking. Figure 8.107 shows the base acceleration and the settlement of the top of the model. As noted from the figure, settlement of the ground increases as the base acceleration becomes larger. As soon as the shaking disappears the settlement terminates. This experiment clearly indicated that ground shaking did cause lateral spreading of dry granular deposits. One of the Mars rovers of NASA induced lateral spreading during the drilling operation and it was mistakenly interpreted as soil liquefaction.

8.7.6   Settlement of structures in liquefiable ground

Settlements of structures, which are heavy compared to the density of surrounding ground, are often observed during earthquakes if the ground liquefies. This problem is of great concern in many engineering projects such as high-rise buildings, piers of bridges, elevated highways and railways and storage tanks. There are some model tests

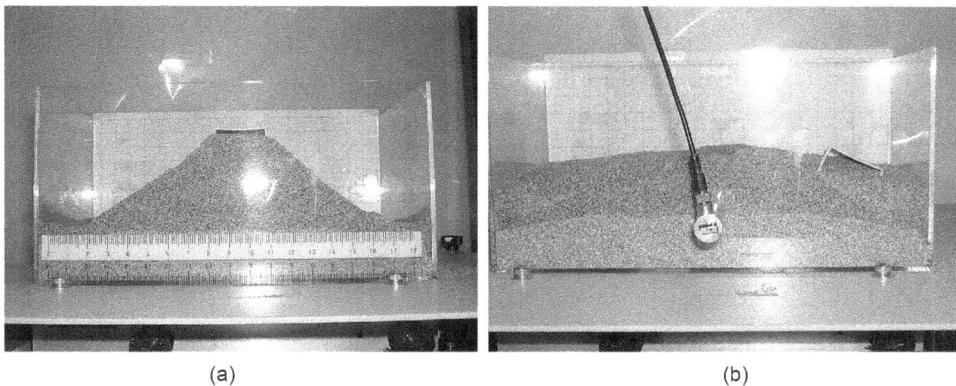

(a)   (b)

Figure 8.106   Views of model experiments (a) before and (b) after shaking.

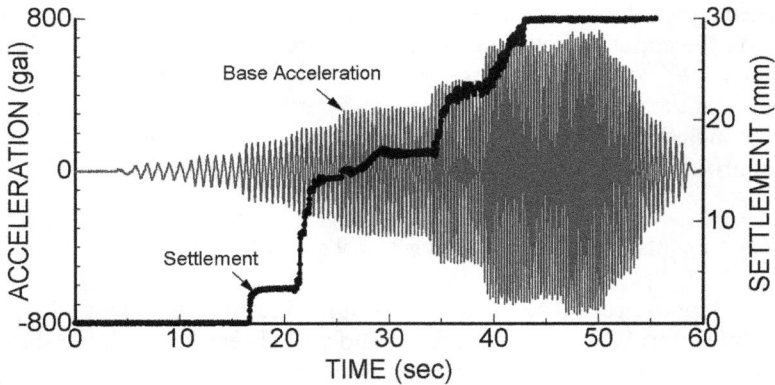

Figure 8.107   Base acceleration and settlement of the ground.

to estimate the behaviour of these structures embedded in liquefaction-prone ground. However, there is no method for estimating the settlement of structures. In this section, a method based on the concept of dynamic limiting equilibrium is described. The basic concept of the method to be described herein is quite similar to that presented by Aydan et al. (1996a, 1999b) and Aydan and Ulusay (2002). A simple structure is assumed to be embedded in a liquefiable ground with an embedment depth of h, as shown in Figure 8.108a.

If the settlement motion of the structure is assumed to be downward, the dynamic force equilibrium condition may be written in the following form for vertical directions (Figure 8.108b):

$$\sum F_V = P - W_b + T + E_V = \frac{W_b}{g}\frac{d^2u}{dt^2} \tag{8.175}$$

where P, W_b, T, E_V and $\dfrac{d^2u}{dt^2}$ are the uplift force of soil, weight of block, side resistance applied by soil, vertical force induced by the earthquake and vertical acceleration of the structure, respectively. The uplift force of soil and ground water and the side resistance may be considered to be consisting of static and dynamic components as given below:

$$P = P_s + P_d \tag{8.176a}$$

$$T = T_d + T_s \tag{8.176b}$$

Under static circumstances, the following relations must hold:

$$P_d = 0 \tag{8.177}$$

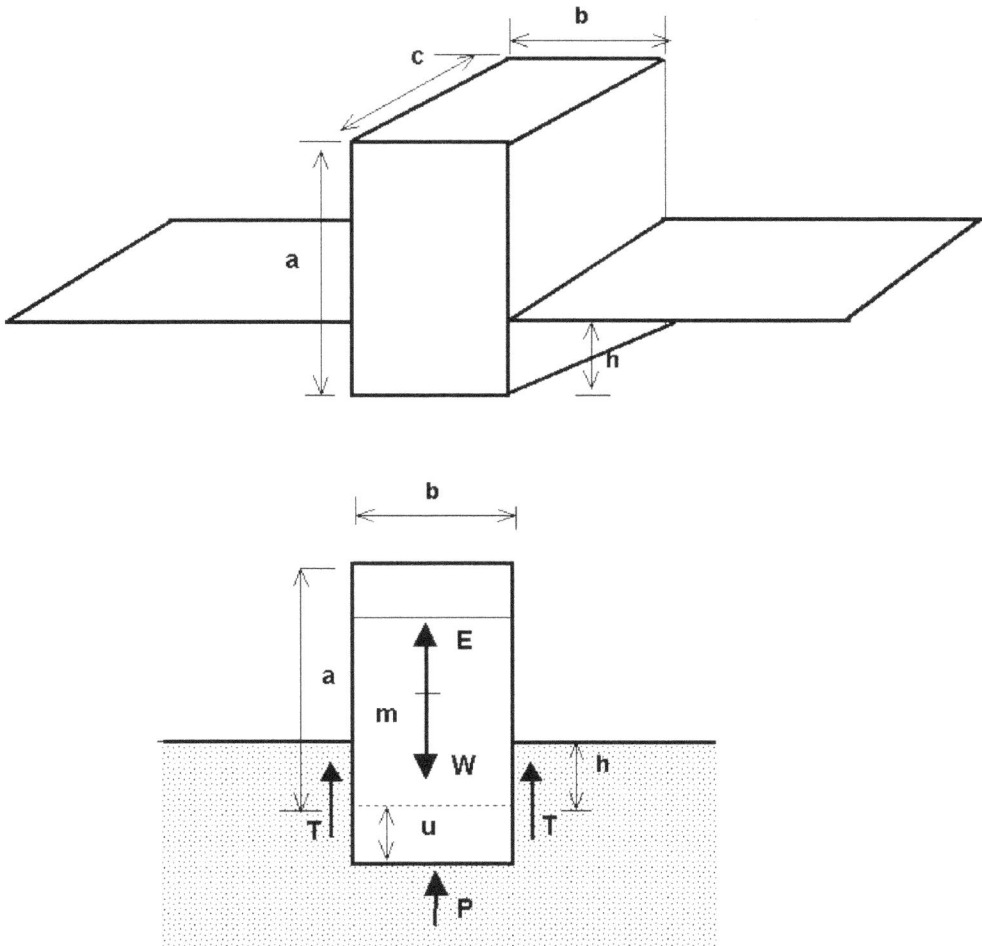

Figure 8.108 Mechanical model for settlement of structures.

As a result, we may assume that the settlement motion of the structure should be due to the remaining components of the force system induced by ground shaking. Thus Eq. (8.175) is rewritten in the following form:

$$\sum F_V = P_d + T_d + E_V = \frac{W_b}{g}\frac{d^2u}{dt^2} \tag{8.178}$$

Let us assume that the liquefaction of ground is due to a horizontal acceleration a_h and let us associate P_d, T_d with the horizontal acceleration acting on the liquefied ground as*

$$P_d = \rho_w a_h (h+u)bc \tag{8.179}$$

$$T_d = 2\rho_w a_h (b+c)\mu K_o (\rho_s - \rho_w) \frac{(h+u)^2}{2} \tag{8.180}$$

where $\rho_s, \rho_w, \rho_b, \mu$ and K_o are the density of ground, water and structure; friction coefficient; and lateral force coefficient of liquefied ground, respectively.

*The dynamic pore pressure may also be evaluated using the procedure explained in Section 8.7.5.2.

The vertical earthquake loading on the structure and its weight may be given as

$$E_V = \frac{a_v(t)}{g} W_b \tag{8.181}$$

$$W_b = \rho_b a\,b\,c \tag{8.182}$$

Using Eqs. (8.179)–(8.182), one may easily obtain the following relation, provided that the dynamic equilibrium condition is satisfied at any time during downward motion.

$$\frac{d^2 u}{dt^2} = \left[\frac{\rho_w}{\rho_b} \frac{(h+u(t))}{a} + \mu K_o \frac{(b+c)(h+u(t))^2}{abc} \frac{(\rho_s - \rho_w)}{\rho_b} \right] a_h(t) + a_V(t) \tag{8.183a}$$

or

$$\ddot{u} - B(u) = 0 \tag{8.183b}$$

where

$$B(u) = \left[\frac{\rho_w}{\rho_b} \frac{(h+u(t))}{a} + \mu K_o \frac{(b+c)(h+u(t))^2}{abc} \frac{(\rho_s - \rho_w)}{\rho_b} \right] a_h(t) + a_V(t)$$

The solution of Eq. (8.183) yields the acceleration, velocity and rigid body displacement of the structure during the settlement motion. During downward motion, the velocity of the structure should remain positive (Newmark 1965, Aydan and Ulusay 2002). Otherwise, the body should not be moving relative to the surrounding soil. It should also be noted that the above equation does not involve the elastic response of the moving body and it is a nonlinear differential equation. Although it may be easy to solve the linearized form of the above equation for simple waveforms, it would be rather difficult to do so when actual earthquake waves are considered. Therefore, some finite difference techniques for the numerical solution of the above equation may be employed. A linear acceleration technique is chosen for solving Eq. (8.183). As Eq. (8.183) holds for any time, it may be written for time step $n+1$ as

$$\ddot{u}_{n+1} - B_{n+1} = 0 \tag{8.184}$$

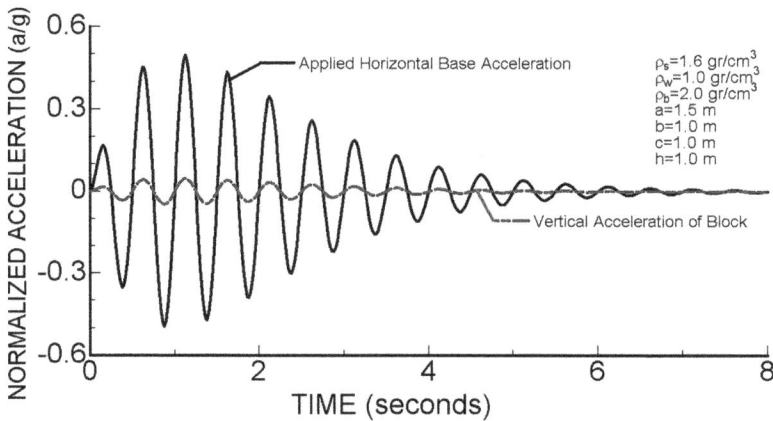

Figure 8.109 Computed acceleration response of the structure in relation to the base acceleration.

This equation is solved using the procedure similar to that in Section 8.3.1.2, specifically Eqs. (8.27) to (8.32). As Eq. (8.184) involves $(h + u_{n+1})^2$, it is a nonlinear differential equation. However, if the time step is kept very small it may be acceptable to assume the following identity holds at time step $n + 1$:

$$(h + u_{n+1})^2 \approx \left(h + \left(u_n + \frac{u_n}{1} \Delta t + \frac{\ddot{u}_n}{3} \Delta t^2 \right) \right)^2 \tag{8.185}$$

An application of the method to the settlement of a rectangular structure is given as an example herein. The structure was 1.5 m high and had a side length of 1 m on either side. The embedment depth was set to 1 m and the density of the structure was 0.6 gr/cm³. The structure was assumed to be subjected a base acceleration with a maximum amplitude of 0.5G as shown in Figure 8.109.

Figures 8.109 and 8.110 show the computed acceleration and settlement of the structure during the motion. The acceleration of the block in the vertical direction is ~0.25 g. Its value will depend upon the material and geometrical properties of structure and ground, and also the base acceleration characteristics. The settlement of the structure tends to become asymptotic as time goes by.

8.7.7 Uplift of structures in liquefiable ground

8.7.7.1 *Dynamic limiting equilibrium method for uplift of structures*

Uplift of structures, which are light compared to the density of the surrounding ground, is often observed during earthquakes if the ground liquefies. This problem is of great concern in many engineering projects such as pipelines, submerged tunnels,

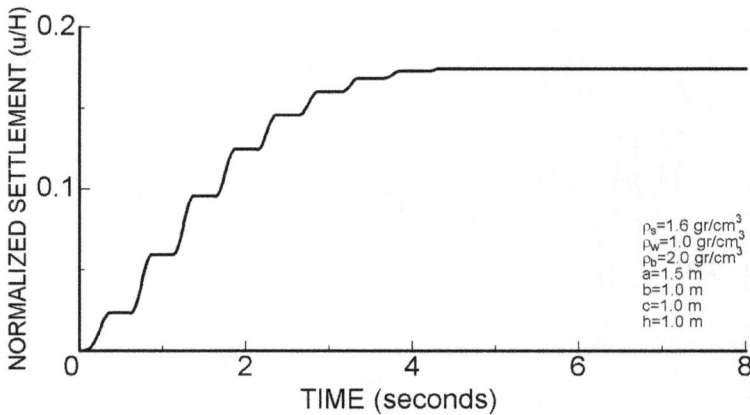

Figure 8.110 Computed settlement displacement response of the structure.

embedded tanks, lifelines such as water mains, sewage pipes and power and telephone cable conduits. There are some model tests to estimate the behaviour of these structures embedded in liquefaction-prone ground. However, there is no method to estimate the uplift displacement of structures. Let us assume a simple structure is situated in a liquefied ground with an embedment depth of h as shown in Figure 8.111a. If the uplift motion of the structure is assumed to be upward compared to the settlement problem, the dynamic force equilibrium condition may be written in the following form for vertical directions (Figure 8.111b):

$$\sum F_v = P - W_b - T + E_V = \frac{W_b}{g}\frac{d^2u}{dt^2} \tag{8.186}$$

where P, W_b, T, E_V and $\dfrac{d^2u}{dt^2}$ are the uplift force of soil, weight of block, side resistance applied by soil, vertical force induced by the earthquake and vertical acceleration of the structure, respectively. Uplift force of soil and ground water and side resistance may be considered to be consisting of static and dynamic components as given below:

$$P = P_s + P_d \tag{8.187}$$

$$T = T_d - T_s \tag{8.188}$$

Under static circumstances, the following relations must hold:

$$P_d = 0 \tag{8.189}$$

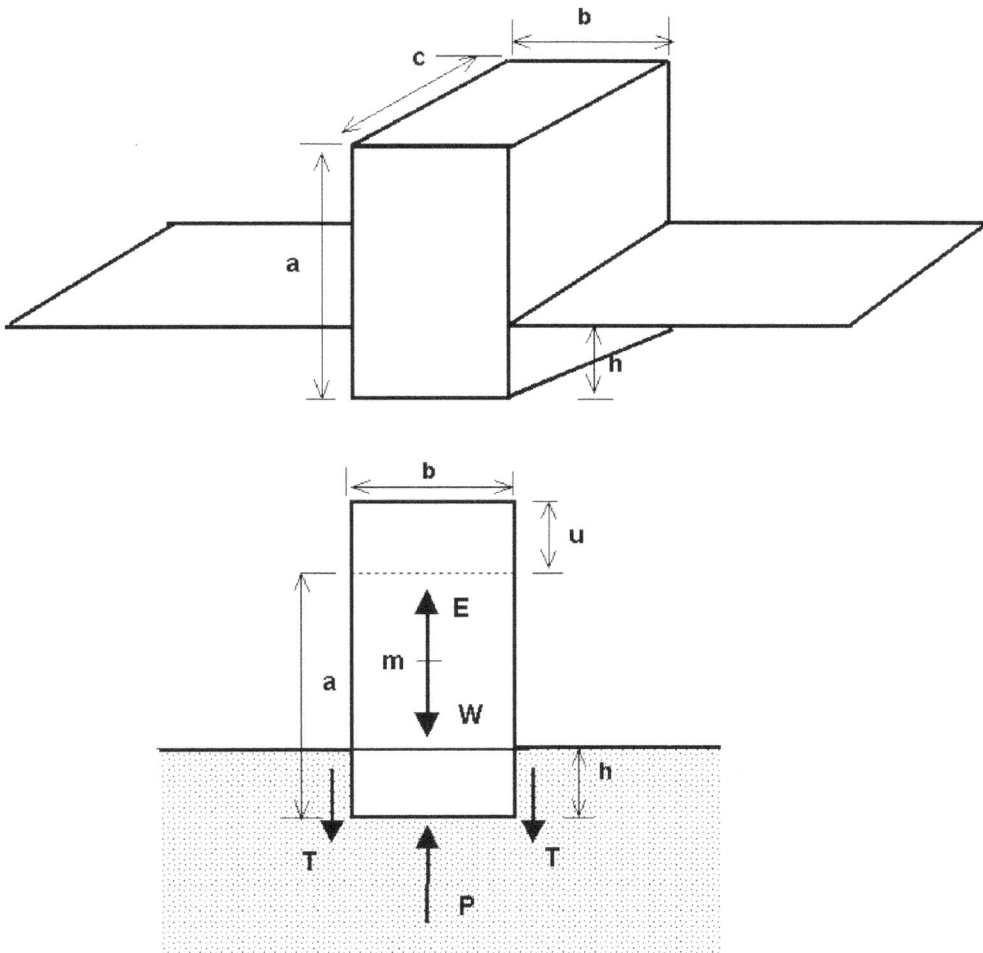

Figure 8.111 Mechanical model for uplift of structures.

As a result, we may assume that the uplift motion of the structure should be due to the remaining components of the force system induced by ground shaking. Thus Eq. (8.186) is rewritten in the following form:

$$\sum F_v = P_d - T_d + E_V = \frac{W_b}{g}\frac{d^2u}{dt^2} \tag{8.190}$$

Let us assume that the liquefaction of ground is due to a horizontal acceleration a_h and let us associate P_d, T_d with the horizontal acceleration acting on the liquefied ground as*

$$P_d = \rho_w a_h (h - u) b \, c \tag{8.191}$$

$$T_d = 2\rho_w a_h (b + c) \mu K_o (\rho_s - \rho_w) \frac{(h - u)^2}{2} \tag{8.192}$$

where $\rho_s, \rho_w, \rho_b, \mu$ and K_o are the density of ground, water and structure, friction angle and lateral force coefficient of liquefied ground, respectively.

*The dynamic pore pressure may also be evaluated using the procedure explained in Section 8.7.5.2.

The vertical earthquake loading on the structure and its weight may be given as

$$E_V = \frac{a_v(t)}{g} W_b \tag{8.193}$$

$$W_b = \rho_b a \, b \, c \tag{8.194}$$

Using Eqs. (8.191)–(8.194), one may easily obtain the following relation, provided that the dynamic equilibrium condition is satisfied at any time during upward motion:

$$\frac{d^2 u}{dt^2} = \left[\frac{\rho_w}{\rho_b} \frac{(h - u(t))}{a} - \mu K_o \frac{(b + c)(h - u(t))^2}{a \, b \, c} \frac{(\rho_s - \rho_w)}{\rho_b} \right] a_h(t) + a_V(t) \tag{8.195a}$$

or

$$\ddot{u} - B(u) = 0 \tag{8.195b}$$

where

$$B(u) = \left[\frac{\rho_w}{\rho_b} \frac{(h - u(t))}{a} - \mu K_o \frac{(b + c)(h - u(t))^2}{a \, b \, c} \frac{(\rho_s - \rho_w)}{\rho_b} \right] a_h(t) + a_V(t)$$

The solution of Eq. (8.195) yields the acceleration, velocity and rigid body displacement of the structure during the uplift motion. During upward motion, the velocity of the structure should remain positive (Newmark 1965, Aydan and Ulusay 2002). Otherwise, the body should not be moving relative to the surrounding soil. It should also be noted that the above equation does not involve the elastic response of the moving body and it is a nonlinear differential equation. Although it may be easy to solve the linearized form of the above equation for simple waveforms, it would be rather difficult to do so when actual earthquake waves are considered. Therefore, instead of this, some finite difference techniques for the numerical solution of the above equation may be employed. In this chapter, a linear acceleration technique is chosen. As Eq. (8.195) holds for any time, it may be written for time step $n + 1$ as

$$\ddot{u}_{n+1} - B_{n+1} = 0 \tag{8.196}$$

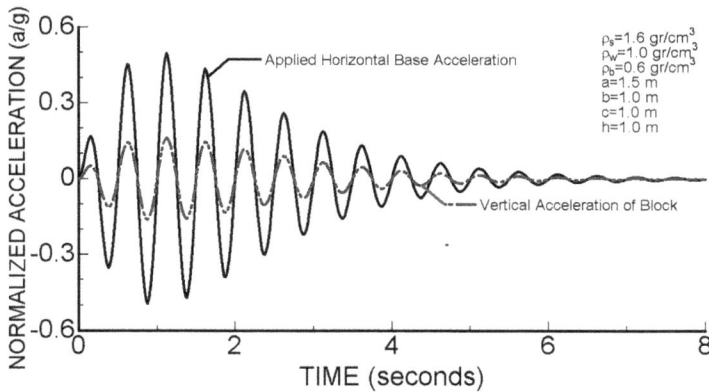

Figure 8.112 Computed acceleration response of the structure in relation to the base acceleration.

This equation is solved using the procedure similar to that in the Section 8.3.1.2, specifically Eqs. (8.27) to (8.32). As Eq. (8.196) involves $(h - u_{n+1})^2$, it is a nonlinear differential equation. However, if the time step is kept very small it may be acceptable to assume the following identity holds at time step $n + 1$:

$$(h - u_{n+1})^2 \approx \left(h - \left(u_n + \frac{\dot{u}_n}{1} \Delta t + \frac{\ddot{u}_n}{3} \Delta t^2 \right) \right)^2 \tag{8.197}$$

In this section, the results of the application of the method to the uplift of a rectangular structure is presented and discussed. The structure was 1.5 high and had a side length of 1 m on either side. The embedment depth was set to 1 m and the density of the structure was 0.6 gr/cm³. The structure was assumed to be subjected a base acceleration with a maximum amplitude of 0.5G as shown in Figure 8.112. Figures 8.112 and 8.113 show the computed acceleration, velocity and uplift displacement of the structure during the uplift motion. The acceleration of the block in the vertical direction is ~0.15G. Its value will depend upon the material and geometrical properties of structure and ground, and also the base acceleration characteristics. As the input base acceleration decrease with time, the uplift displacement of ground tends to become asymptotic as time goes by (Figure 8.113).

8.7.7.2 Pseudo-dynamic design of tunnels, conduits and culverts against uplift

The overall unit weight of immersed tunnels is generally <10 kN/m³. When ground is fully saturated, it will impose uplift pressure on tunnels even under static conditions. This uplift (buoyancy) potential (Φ) under fully submerged conditions may be expressed in the following form according to Archimedes' principle (Aydan 2007a).

$$\Phi = \gamma_f - \gamma_t \tag{8.198}$$

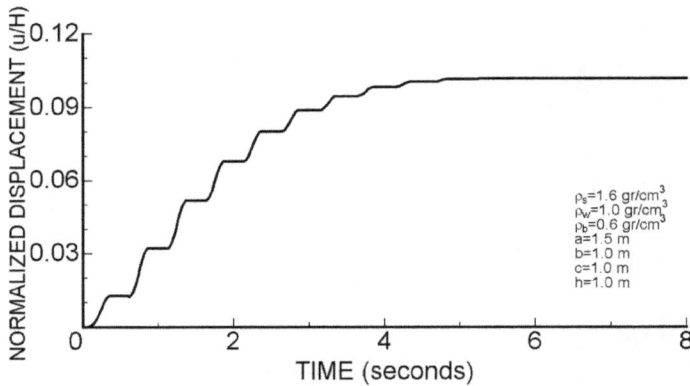

Figure 8.113 Computed uplift displacement response of the structure.

where γ_f and γ_t are the unit weight of fluid and tunnel. If the saturated ground is a particulate media, the dynamic shaking may result in the state of liquefaction of the particulate media. As a result, the unit weight γ_f of fluid will be replaced in Eq. (8.198) by the equivalent unit weight of the liquefied particulate medium, and the uplift potential will increase in amplitude. Figure 8.114 shows a simple design model for estimating the uplift resistance of an immersed tunnel. The SF is defined as

$$\text{SF} = \frac{W_s + W_l + T}{U} \tag{8.199}$$

Figure 8.115 shows an example computation of the SF for no side resistance ($T = 0.0$) as a function of the unit weight of the backfill soil. It should be noted that the dynamic effects and the downward flow of the soil are not taken into account in these types of computations.

8.7.7.3   Numerical analysis of a tunnel in liquefiable ground

A lined tunnel with a diameter of 12.8 m with an overburden of 20 m located in a liquefiable ground was analysed using a coupled finite element method for a given acceleration waveform shown in Figure 8.116. The ground shaking became negligible after 10 seconds. The finite element mesh used in the numerical analysis together with viscous boundaries is shown in Figure 8.117.

First, a gravitational analysis was carried out in order to determine the initial stress state and was then subjected to an input ground motion shown in Figure 8.116. Development of excess pore pressure and deformation occurred within 7 seconds and deformation ceased thereafter. Figure 8.118 shows the deformed configuration and excess pore pressure distribution at 10 seconds. The upward movement was 145 mm at ground surface and 208 mm at the tunnel crown.

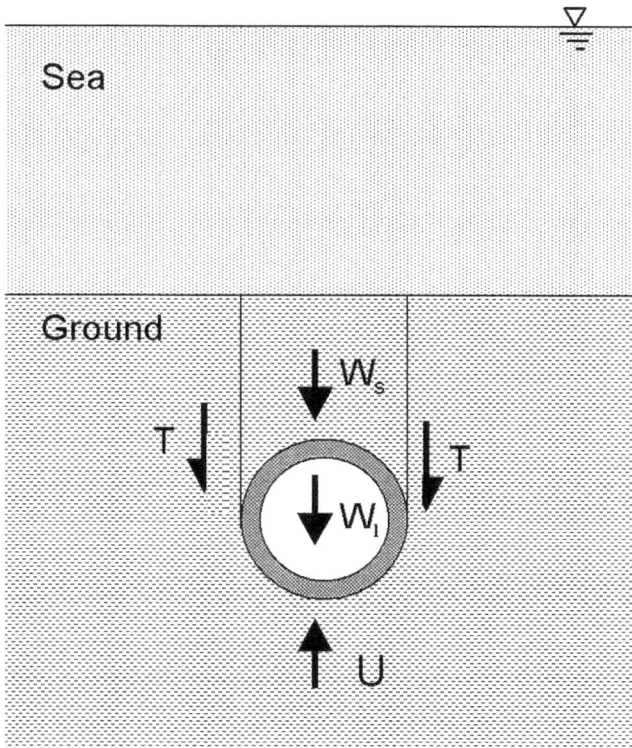

Figure 8.114 Illustration of a simple design concept against uplift (buoyancy) forces.

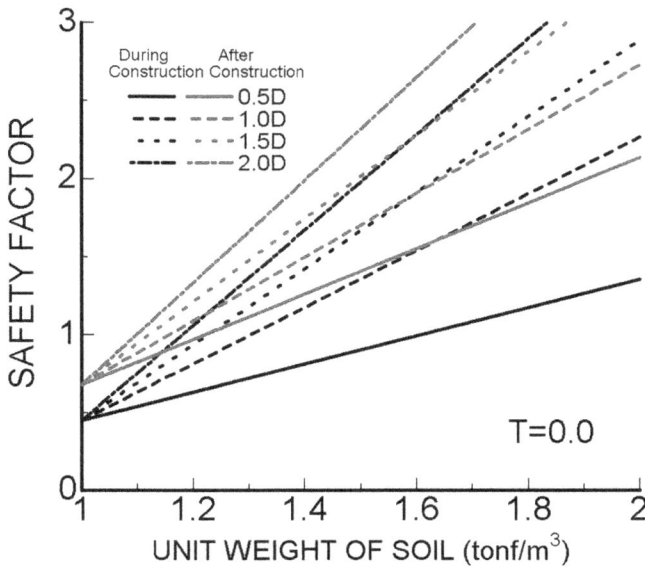

Figure 8.115 Example of computation for design of immersed tunnels against uplift.

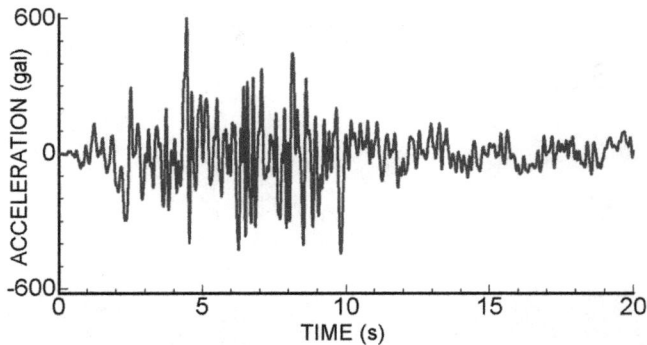

Figure 8.116 Input ground motion used in coupled FEM analyses.

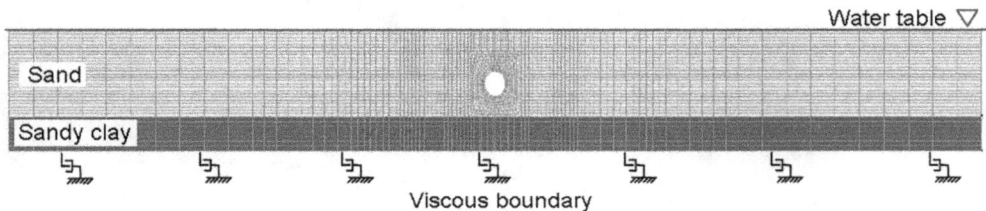

Figure 8.117 Finite element mesh together with viscous boundary conditions.

8.7.8   Shaking table tests on the settlement of wave breaks

Many wave breaks made of natural stone blocks were built along the shores of many countries. Aydan (2009) reported that the off-Sumatra earthquake in the vicinity of West Sumatra Province caused the settlement of water breaks constructed against tsunami attacks (Figure 8.119). In this section, a model study is undertaken to investigate the causes of the settlement of the water breaks. The model setup shown in Figure 8.120 is subjected to vibration using the shaking table. Two accelerometers, two laser displacement transducers and two water pressure sensors were set up as shown in Figure 8.120. Sand No.6 was used as liquefiable ground together with large gravels of sandstone and andesite. Figure 8.121 shows the side and top views of the model water breaks before and after shaking. As noted from pictures, the water breaks sank into the liquefied ground. Figure 8.122 shows the applied acceleration to the shaking table and settlement of the water break. The settlement of the water break ceased soon after shaking was terminated.

8.7.9   Important observations and countermeasures against ground liquefaction

Lateral spreading is not limited to liquefiable ground and it may occur when ground is made of dry granular soil. Measurements for the hilly ground deformations around the

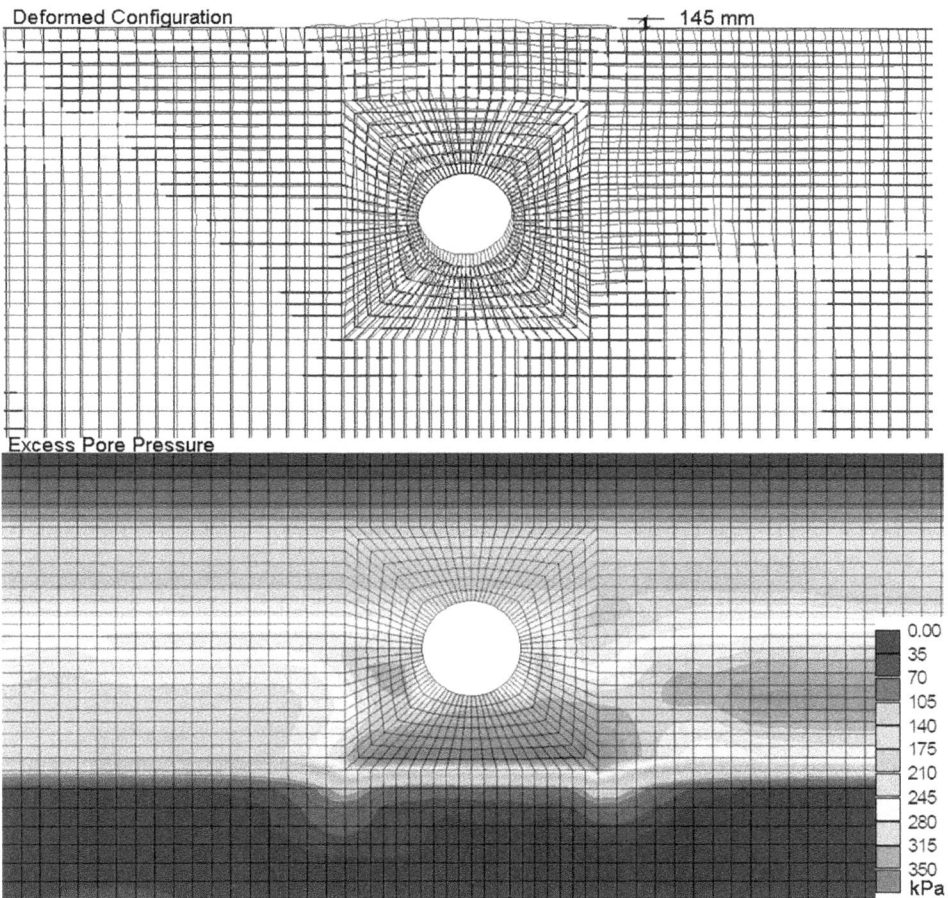

Figure 8.118 Deformed configuration and distribution of excess pore pressure (10 seconds).

(a)   (b)

Figure 8.119 Settlement of wave breaks due to the 2009 Padang-Pariaman earthquakes. (a) and (b) settled wave breaks in Padang shore.

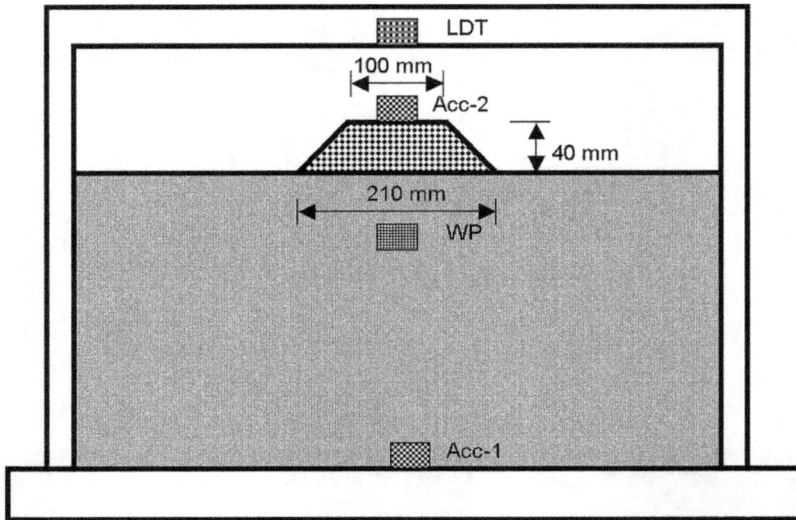

Figure 8.120 Model setup.

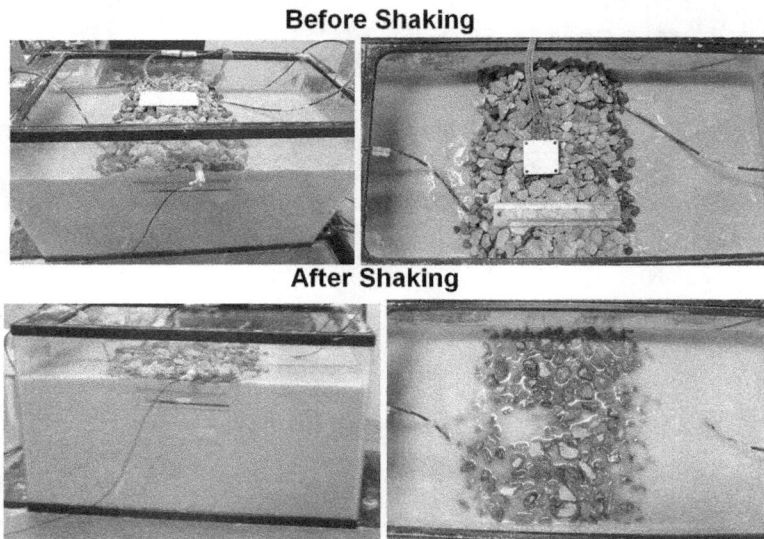

Figure 8.121 Views of model before and after shaking.

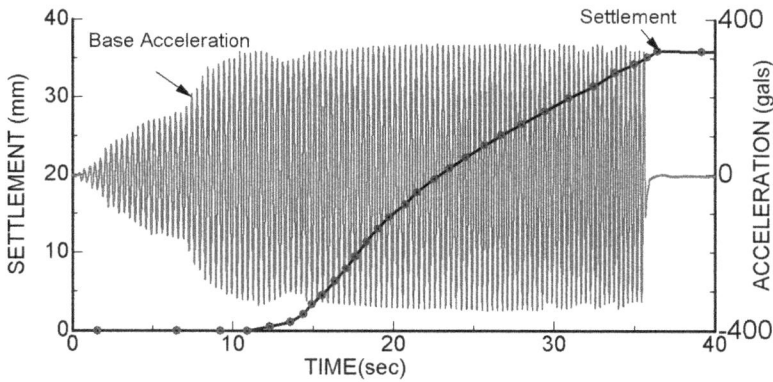

Figure 8.122 Applied base acceleration and settlement of the water break.

Sapanca lake area induced by the 1999 Kocaeli earthquake by the General Command of Mapping of Turkey are in accordance with the observations from the experiments.

Liquefied ground can re-liquefy as observed in earthquakes in Turkey and Japan. In other words, it is incorrect to assume that liquefied ground does not re-liquefy.

To prevent ground liquefaction, the most important countermeasure would be to reduce permeability and increase shear resistance. The specific countermeasures against ground liquefaction are

1. Ground improvement through densification by using vibrations techniques or grouting. This, in turn, results in an increase in shear resistance and decrease in permeability.
2. Structural improvement through piling, anchoring, etc.

Chapter 9

Tsunami

Its effects on structures, and the fundamentals of tsunami-proof design

Tsunami is a Japanese word simply meaning the wave at a harbour or "harbour wave" and it has now become a well-accepted terminology worldwide. The loss of more than 200,000 people along the shores of the Indian Ocean caused by the 2004 Aceh (off-Sumatra) earthquake is one of the well-remembered tsunami events in this century together with the tsunami induced by the 1964 Chile earthquake. In recent years, another very disastrous tsunami was caused by the 2011 Great East Japan earthquake (GEJE) resulting in the loss of about 20,000 people. In this chapter, the mechanism, fundamental equations of tsunami wave propagation, its effect on built and natural environment on the bases of some observations in the 2004 Aceh (off-Sumatra) earthquake and the GEJE, model experiments on tsunami and tsunami boulder occurrence, some empirical estimations of tsunami height as a function of earthquake magnitude and estimations of past mega-tsunami events are presented and discussed. Furthermore, some possible countermeasures for structures against tsunami waves are presented.

9.1 MECHANISM OF TSUNAMIS

The well-accepted causes of tsunami are earthquake faulting, land/submarine slope failure, submarine volcanic eruption/collapse and meteorite impact, as illustrated in Figures 9.1 and 9.2 (e.g. Abe 1979; Aida 1969; Iida 1963; Imamura et al. 2001; Aydan et al. 2005a,b, 2015d; Nakamura 2006; Nistor et al. 2005; Tinti et al. 2006; Ward and Asphaug 2000; Mader and Gittings 2002; Huber and Hager 1997; Semenza and Ghirotti 2000; Slingerland and Voight 1979; Ward 2001; Zeyn et al. 2006; Fritz et al. 2001). In this section, the fundamental mechanism for each cause is explained and some actual examples are described.

9.1.1 Earthquake faulting

Tsunamis occur when an earthquake faulting occurs beneath the sea. The fault should have a vertical component, which may be a normal or thrust-type movement. Pure strike-slip faulting does not cause any tsunami unless the faulting occurs beneath closed water bodies, such as lakes or in-lets. Such small-scale tsunamis occurred

DOI: 10.1201/9781003164371-9

EARTHQUAKE MECHANISM ALONG SUBDUCTION ZONE

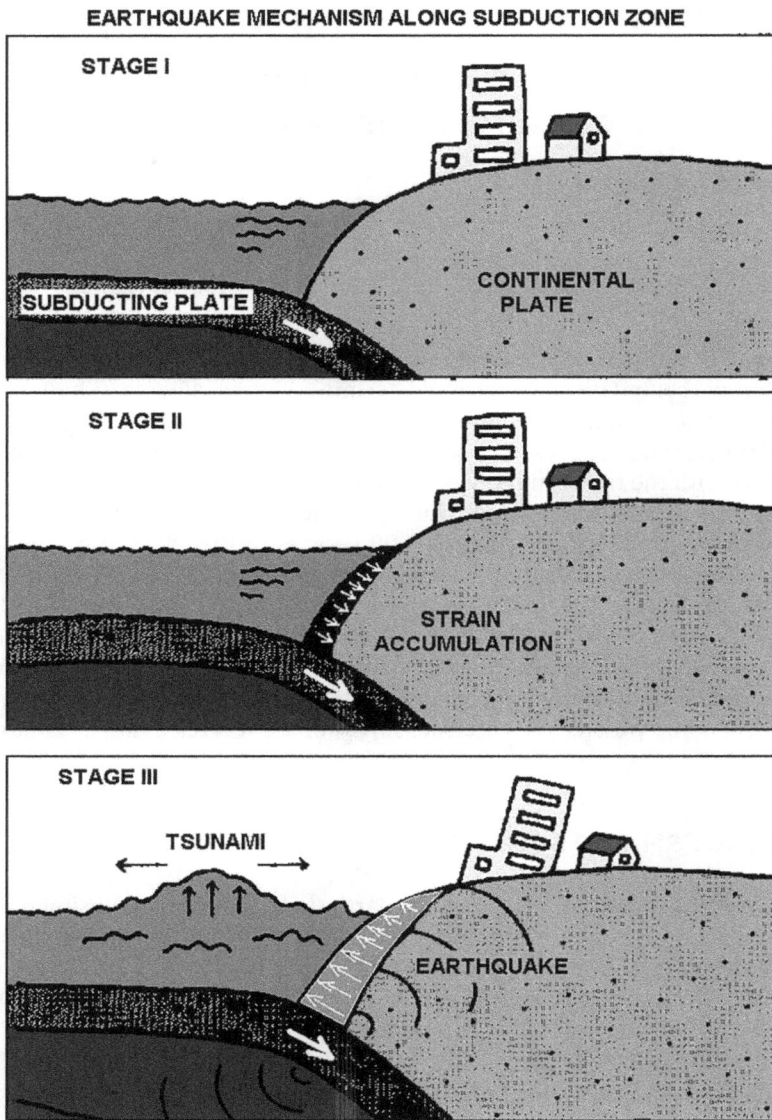

Figure 9.1 (a) Tsunami mechanism in a subduction zone.

during the 1999 Kocaeli earthquake at the eastern end of the Izmit Bay (Tinti et al. 2006; Aydan et al. 1999a). When a tsunami is caused by normal faulting, the sea generally recedes first and then surge waves would appear after a certain period of time depending upon seabed topography and the amplitude of vertical movement, as it has been recently observed in the 2020 Kuşadası Bay earthquake.

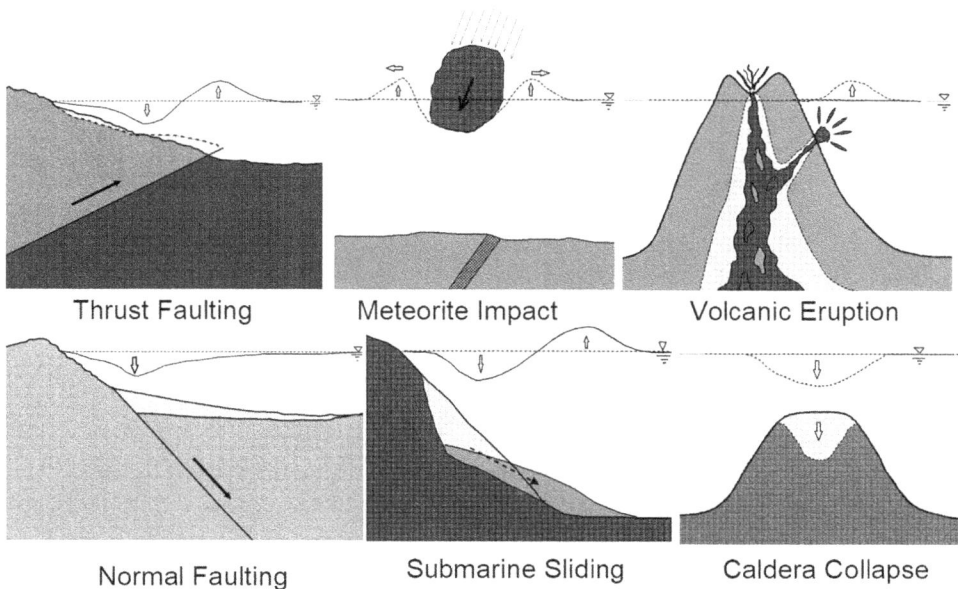

Figure 9.2 Causes of tsunami occurrence and possible initial sea surface profiles.

As for thrust-type faulting, the surge waves would generally appear first. Large tsunamis are generally caused in the close vicinity of subduction zones (Figure 9.1). Figure 9.2 illustrates the mechanism of earthquakes and tsunami along the Sumatra subduction zone. The Aceh earthquake caused a rupture zone for about 1,200 km and killed more than 200 thousand people on the shore of the Indian Ocean, with the largest number of casualties in the Aceh Province of Indonesia.

9.1.2 Land or submarine slides

Rockfalls and mass movements into lakes, fjords and seas are also considered to be causes of tsunami. The failure of Mt. Mayuyama in Kyushu Island in 1792 caused a mega-tsunami, killing more than 14,300 people in Kumamoto prefecture on the opposite side of the bay. The tsunami generated by a submarine landslide during the 1998 Papua New Guinea earthquake killed 2,143 people. The Lituya Bay tsunami was caused by a huge rockfall in 1958 and caused the highest tsunami run-up. Submarine landslides induced by earthquakes have also caused some tsunamis in Marmara Sea despite many past earthquakes in Marmara Sea having a strike-slip focal mechanism. For example, the landslide at Değirmendere resulted in a small tsunami in the 1999 earthquake. Tsunamis induced in the fjords of Norway, Canada and Chile are also well known and they are well documented (i.e. Zeyn et al. 2006). Recently, the rockfall-generated tsunami in Aisen (Aysen) fjord in Chile in the 2007 earthquake (Mw 6.2) was a well-observed event of this type (Pavez et al. 2007).

The 2020 Palu Bay earthquake, which was caused by strike-slip faulting, induced a tsunami with a height of 4m. The main cause of the tsunami was submarine slope failures on the eastern side of the bay.

9.1.3   Submarine volcanic eruption

The volcanic eruptions or the collapse of the volcanoes cause mega-tsunamis. For example, the eruptions and subsequent collapse of the Krakatau volcano in Indonesia in 1883 and the Santorini volcano in Aegean Sea in 1600 BCE resulted in mega-tsunamis. The Krakatau volcano recently became active and caused a small-scale tsunami. Nevertheless, the recent activities in the Krakatau volcano and associated tsunamis clearly illustrated what happened in 1883.

9.1.4   Meteorite falls

The impact of asteroids in oceans and seas is also considered another source of tsunami. The impact of the largest asteroid fall in Yucatan in Central America is thought to be the cause of the extinction of dinosaurs on the earth. Recent studies on the characterization of the size and distribution of the crater through geophysical explorations have been undertaken.

9.2   GOVERNING EQUATIONS OF TSUNAMIS

9.2.1   Fundamental equations in fluid mechanics

The fundamental equations of tsunamis are based on mass conservation law and the equation of motion. It is very common that the fluid obeys the Navier–Stokes type constitutive law (e.g. Eringen 1980; Mase 1970; Aydan 2021; Kowalik 2001). The mass conservation law without any source term is given as

$$\frac{\partial \rho}{\partial t} + \frac{\partial(\rho v_k)}{\partial x_k} = \frac{\partial \rho}{\partial t} + (\rho v_k)_k = 0 \tag{9.1}$$

where ρ, v_k, x_k and t are the density, velocity, physical space and time, respectively. If the fluid is incompressible, that is density is constant, Eq. (9.1) can be re-written as

$$\frac{\partial \rho}{\partial t} + v_k \cdot \frac{\partial \rho}{\partial x_k} = 0 \tag{9.2}$$

Thus, Eq. (9.1) together with Eq. (9.2) is re-written as

$$\frac{\partial v_k}{\partial x_k} = 0 \tag{9.3}$$

In fluid mechanics, the stress tensor is decomposed into two parts as given below:

$$\sigma_{ij} = -p\delta_{ij} + \tau_{ij} \tag{9.4}$$

where σ_{ij} and p, δ_{ij} are the stress tensor, pressure and Kronecker delta tensor, respectively. $\dot{\varepsilon}_{ij}$ is the strain rate tensor. The strain rate tensor and engineering strain rate tensors can be related to velocity vector in the following form:

$$\dot{\varepsilon}_{ij} = \frac{1}{2}\left(\frac{\partial v_i}{\partial x_j} + \frac{\partial v_j}{\partial x_i} \right) \ \ or \ \ \dot{\gamma}_{ij} = \left(\frac{\partial v_i}{\partial x_j} + \frac{\partial v_j}{\partial x_i} \right) \tag{9.5}$$

Equation (3.8) or (3.10) together with Eq. (9.4) can be re-written in the following form:

$$\rho \frac{dv_i}{dt} = -\frac{\partial p}{\partial x_j}\delta_{ij} + \frac{\partial \tau_{ij}}{\partial x_j} + b_i \ \ or \ \ \rho\left(\frac{\partial v_i}{\partial t} + v_j\frac{\partial v_i}{\partial x_j} \right) = -\frac{\partial p}{\partial x_j}\delta_{ij} + \frac{\partial \tau_{ij}}{\partial x_j} + b_i \tag{9.6}$$

where b_i is the body force vector.

9.2.2   Fundamental equations for tsunamis

The starting equation for deriving the fundamental equations of tsunamis is based on Eq. (9.6), while some assumptions are introduced to simplify the physical conditions associated with tsunamis. The waves are assumed to be obeying long wave or shallow water theory. In other words, the three-dimensional problem is simplified to two-dimensional depth-averaged equations to simulate tsunami wave propagation and inundation (Figure 9.3).

Equations (9.3) and (9.6) are re-written using the coordinate system $(Oxyz)$ and velocity vector (u, v, w) with the replacement of the coordinate system $(Ox_1x_2x_3)$ and

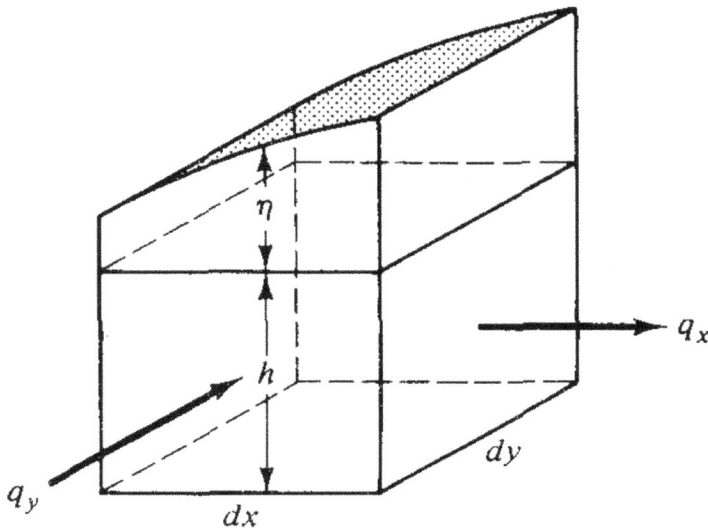

Figure 9.3   Control volume for conservation of mass. The q_x, q_y are replaced by uD and vD, respectively. (From Dean and Dalrymple (1998).)

velocity vector (v_1, v_2, v_3) by assuming that stress tensor is symmetric (e.g. Dean and Dalrymple 1998):

Continuity equation

$$\frac{\partial u}{\partial x} + \frac{\partial v}{\partial y} + \frac{\partial w}{\partial z} = \text{\i}$$

(9.7)

Equations of motion

$$\frac{\partial u}{\partial t} + u\frac{\partial u}{\partial \partial} + v\frac{\partial u}{\partial y} + w\frac{\partial u}{\partial z} = -\frac{1}{\rho}\frac{\partial p}{\partial x} + \frac{1}{\rho}\left(\frac{\partial \tau_{xx}}{\partial x} + \frac{\partial \tau_{xy}}{\partial y} + \frac{\partial \tau_{xz}}{\partial z}\right) + g_x$$

(9.8a)

$$\frac{\partial v}{\partial t} + u\frac{\partial v}{\partial x} + v\frac{\partial v}{\partial y} + w\frac{\partial v}{\partial z} = -\frac{1}{\rho}\frac{\partial p}{\partial y} + \frac{1}{\rho}\left(\frac{\partial \tau_{xy}}{\partial x} + \frac{\partial \tau_{yy}}{\partial y} + \frac{\partial \tau_{yz}}{\partial z}\right) + g_y$$

(9.8b)

$$\frac{\partial w}{\partial t} + u\frac{\partial w}{\partial x} + v\frac{\partial w}{\partial y} + w\frac{\partial w}{\partial z} = -\frac{1}{\rho}\frac{\partial p}{\partial y} + \frac{1}{\rho}\left(\frac{\partial \tau_{xz}}{\partial x} + \frac{\partial \tau_{yz}}{\partial y} + \frac{\partial \tau_{zz}}{\partial z}\right) + g_z$$

(9.8c)

The components of the body force vector is specifically written as

$$\{g_x, g_y, g_z\} = \{0, 0, -g\}$$

(9.9)

The major assumption is that the vertical motion is negligible so that the pressure distribution in the vertical direction is given by

$$p = \rho g(\eta - z)$$

(9.10)

where g, η and z are the gravitational acceleration, vertical displacement of water surface and vertical axis, respectively. Equations (9.7) and (9.8) are integrated over depth $(-h, \eta)$ to obtain the long wave (shallow water) equations using the Leibniz rule of integration. Through linearization of higher-order terms and neglecting Eddy viscosity effects, the final equations are as follows (Imamura et al. 2001):

Continuity equation

$$\frac{\partial M}{\partial x} + \frac{\partial N}{\partial y} + \frac{\partial \eta}{\partial t} = 0$$

(9.11)

Equations of motion

$$\frac{\partial M}{\partial t} + \frac{\partial}{\partial x}\left(\frac{M^2}{D}\right) + \frac{\partial}{\partial y}\left(\frac{MN}{D}\right) + gD\frac{\partial \eta}{\partial x} + \frac{\tau_x}{\rho} = 0$$

(9.12a)

$$\frac{\partial N}{\partial t} + \frac{\partial}{\partial x}\left(\frac{MN}{D}\right) + \frac{\partial}{\partial y}\left(\frac{N^2}{D}\right) + gD\frac{\partial \eta}{\partial y} + \frac{\tau_y}{\rho} = 0$$

(9.12b)

where

$$M = \int_{-h}^{\eta} u\, dz = u(h+\eta) = uD; \quad N = \int_{-h}^{\eta} v\, dz = v(h+\eta) = vD; \quad D = h+\eta \qquad (9.13)$$

$\overline{\tau}_x$ and $\overline{\tau}_y$ in Eq. (9.12) are called the bottom frictions in x and y directions, and they are given in the following forms:

$$\frac{\overline{\tau}_x}{\rho} = \frac{fn^2}{D^{7/3}} M\sqrt{M^2 + N^2} \qquad (9.14a)$$

$$\frac{\overline{\tau}_y}{\rho} = \frac{fn^2}{D^{7/3}} N\sqrt{M^2 + N^2} \qquad (9.14b)$$

where f is the friction coefficient and n is the Manning roughness.

The equations above are subjected to the following boundary conditions:

$$p = 0 \quad \text{at} \quad z = \eta \qquad (9.15a)$$

$$w = \frac{\partial \eta}{\partial t} + u\frac{\partial \eta}{\partial x} + v\frac{\partial \eta}{\partial y} \quad \text{at} \quad z = \eta \qquad (9.15b)$$

$$w = -u\frac{\partial h}{\partial x} - v\frac{\partial h}{\partial y} \quad \text{at} \quad z = -h \qquad (9.15c)$$

However, it should be noted that the behaviour of a tsunami in the close vicinity of shores and during inundation becomes quite nonlinear and complicated.

9.2.3 Applications of fundamental equations for tsunamis

The applications of fundamental equations are given here by considering major tsunamis worldwide, specifically, the 1960 Chile, 2004 Aceh (off Sumatra), 2011 GEJEs for faulting-associated tsunamis, 1871 Mayuyama and 1999 Kocaeli for slope-failure-associated tsunamis, and 1888 and 2020 Krakatau volcano-associated tsunamis.

9.2.3.1 Faulting-induced tsunamis

Large tsunami events are generally associated with mega earthquakes occurring in subduction zones (Figure 9.1). The relative net slip may reach up to 50 metres as reported for the GEJE. Nevertheless, the dip of the subduction zone is generally <20°, which implies that the vertical uplift of the seabed would be generally <18 m. The uplift value of the seabed would be the maximum initial value of the tsunami height. In this section, the characteristics of the tsunamis associated with the 1960 Chile, 2004 Aceh and 2011 GEJEs are briefly described.

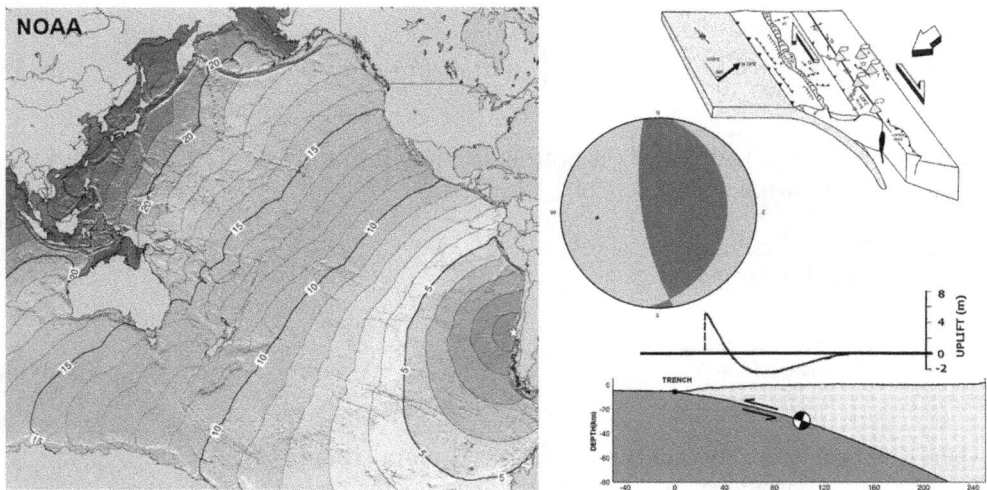

Figure 9.4 The mechanism and tsunami propagation of the 1960 Valdivia earthquake. (Tsunami propagation by NOAA, Mechanism by Barrientos and Ward, (1990).)

9.2.3.1.1 THE 1960 VALDIVIA CHILE EARTHQUAKE

The largest instrumentally recorded earthquake occurred on May 22, 1960, with a moment magnitude Mw of 9.4–9.6 in the southern Chile, named Valdivia earthquake (Figure 9.4). The earthquake occurred in the subduction zone of the Nazca Plate subducting beneath the South American Plate. The earthquake caused heavy damage along nearly 1,000 km off the Chilean coast. The main shock generated a destructive tsunami that hit the nearest coast with 8–10 m waves. The maximum waves with a height of 15 m were observed along the 350 km section of the coast between Corral in the south and Conception in the north. In 15 hours, 8–10 m high waves reached Hawaii and caused 61 casualties in Hilo. In 22 hours, the waves reached the east coast of Japan with a height of 5–6 m. More than 10,000 houses were destroyed and 122 people died.

9.2.3.1.2 DECEMBER 26, 2004, SUMATRA, INDONESIA

Off the coast of Sumatra, a devastating mega-thrust earthquake occurred on December 26, 2004, at 250 km south-west of Banda Aceh. The estimated moment magnitude varies between 9.1 and 9.3. This earthquake is the fourth-largest earthquake instrumentally recorded and it caused a devastating tsunami in the Indian Ocean region (Figure 9.5). The estimated rupture fault length is about 1,300–1,500 km with an estimated slip of 20–25 m. The earthquake generated a destructive tsunami that caused severe damage in Aceh Province of Indonesia and along the coasts of 14 countries and caused almost 280,000 casualties (Aceh Province alone is 204,000). The worst hit country in terms of fatalities was Indonesia. The remaining fatalities occurred in Sri Lanka (36,081), India (16,423), and Thailand (8,567). In addition to these four most affected countries, there were 300 reported fatalities in Somalia, 82 in the Maldives,

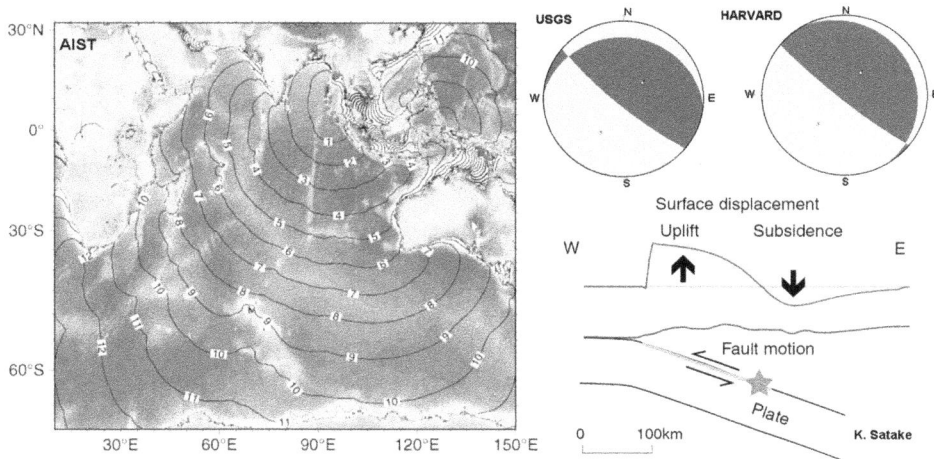

Figure 9.5 (right lower figure) Seabed deformation model, (left figure) computed tsunami arrival time in Indian Ocean by Satake (2005) and (top right figure) the focal mechanism solutions by USGS and HARVARD.

Figure 9.6 Some views in Lhonga area. (a) A view from the highest run-up site and (b) a view of the cement factory.

68 in Malaysia, 61 in Myanmar, 11 in Tanzania, 2 in Seychelles, 2 in Bangladesh and 1 in Kenya. The Pacific Tsunami Warning Center (PTWC) in Beach, Hawaii, had timely determined the earthquake with an operational magnitude as high as 8.0 (soon upgraded to 8.5), but since its source was located well outside of the PTWC area of responsibility, the warning was not issued to the affected areas. The tsunami wave heights varied from 15 to 20 m along the coast of Aceh Province in north-western Sumatra. The maximum measured run-up was 49 m near Lhonga village, some 15 km east-west of Banda Aceh (Figure 9.6). In the far-field, the largest run-up (9.3 m) was measured along the coast of Somalia.

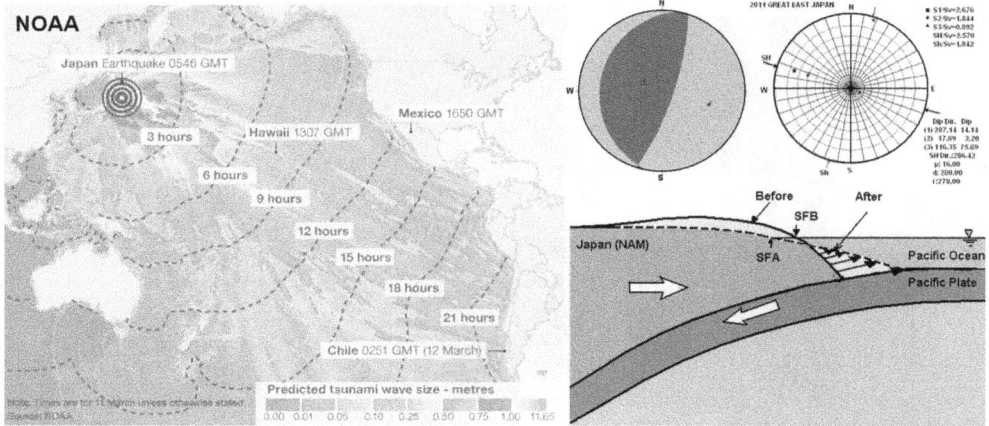

Figure 9.7 (a) Tsunami arrival time estimated by NOAA and (b) crustal deformation model and focal mechanism of the GEJE. (From Aydan (2015d).)

9.2.3.1.3 THE 2011 GREAT EAST JAPAN EARTHQUAKE

The GEJE with a moment magnitude 9.0 occurred at 14:46 (JST) on March 11, 2011 (JMA 2011). The earthquake was a subduction plate–boundary earthquake and the rupture area was ~450 km long and 200 km wide (i.e. Aydan 2015c; Aydan et al. 2011; PARI 2011; Hamada 2011). However, the estimated values of the parameters of the earthquake vary depending upon institutes and researchers. The earthquake occurred along the subduction zone between the North American Plate (NAM) and the Pacific Plate (PAC) (Figure 9.7) and the inclination of the rupture plane was estimated to be 14°–16° (Koketsu 2012). The faulting was due to thrust faulting with a maximum offset estimated to be ranging between 25 and 56 m depending upon institutes and/or researchers. The seabed was uplifted by 5 m while the land sank down by about 0.8 m with a horizontal movement of 4.6 m measured by the GEONET GPS monitoring system of Japan (GSI 2011). This earthquake caused gigantic tsunami waves, which destroyed many cities and towns along the shores of the Tohoku and Kanto regions of Japan. The casualties caused by this tsunami exceed 20,000 people. The tsunami destroyed and heavily damaged buildings of various types, transportation facilities and infrastructure.

The crustal deformation caused the intrusion of the sea front into the land as illustrated in Figure 9.7. If the strained overriding plate slips along the plate boundary, it rebounds to its unstrained state. As a result, the upper surface (before) configuration changed to an "after" configuration. This configuration change caused ground uplift near the plate front and subsidence of ground away from the plate front. As a result, the sea front moved towards the land, causing inundated areas. This caused inundation problems particularly in the settlements during high tides and erosion along the seashore.

There are two different definitions for characterizing tsunami height: height at the shoreline and inundation height, or run-up height. As it was previously difficult to measure tsunami height at shoreline, the inundation tsunami height obtained from

(a)

(b)   (c)

Figure 9.8 Some views from the tsunami-affected areas. (a) Views of structures in Ri-kuzen-Takada (note the only remaining pine tree), (b) toppled slate block in Onagawa and (c) small temple protected by trees in Natori.

traces of tsunami front on the land was commonly quoted as the tsunami height in the past. The publicized tsunami heights may have tremendous values depending upon how it is measured. For example, the newspapers reported a tsunami height of 38.9 m at Aneyoshi (Miyako) measured by Tokyo University of Marine Science and Technology, which probably includes the splash zone. The maximum tsunami height at shoreline in this earthquake was measured to be 15.3 m at Onagawa town by the author (Figure 9.8).

9.2.3.2   Volcanism-induced tsunami

Tsunamis may be caused by the eruption of submarine volcanoes, pyroclastic flows into the sea or collapse of volcanoes. The collapse of the Krakatau volcano during its volcanic activities in 1883 is a most well-known and disastrous event (Figure 9.9). It happened in the morning of August 27, 1883, generating a 25–30 m tsunami in the Sunda Strait, with a maximum reported run-up of 41 m. The waves swept away many villages along both coasts of the strait and killed more than 36,000 people. It also resulted in an atmospheric pressure wave that was globally recorded by barographs even at very remote locations such as France, England, Alaska and Hawaii. Among 38 available sea level records, the largest wave (1.5 m) was recorded by the tide-gauge in Galle (Sri Lanka, formerly known as Ceylon). It should be noted that the heavy damage and casualties were confined to the close neighbourhood of the source area and

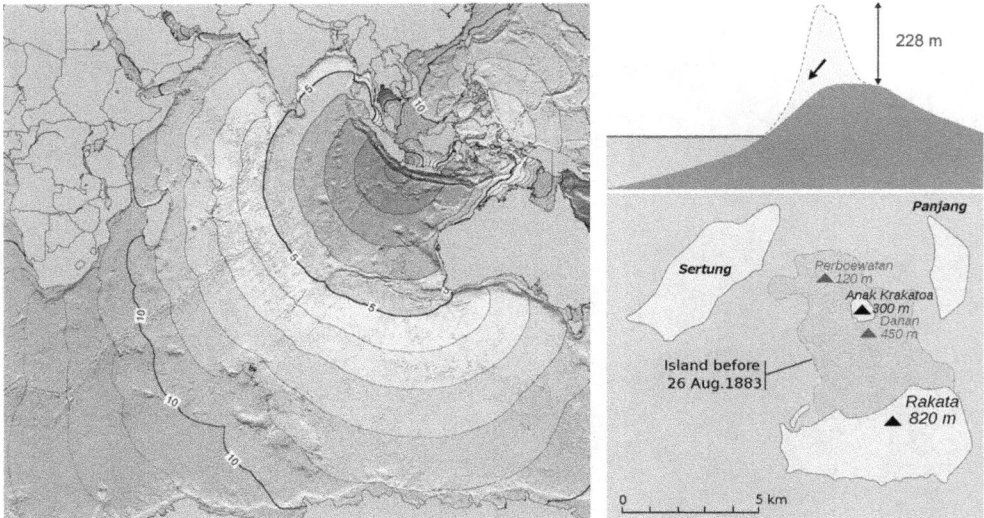

Figure 9.9 Tsunami arrival time estimated by NOAA and configuration of Anak Kratau before and after collapse and the collapse area and islands.

occurred within one hour propagation time. Anak Krakatau (also spelled as Krakatoa) recently became active again and caused eruptions and pyroclastic flows into the sea since June 19, 2018, from time to time. On December 22, 2018, it caused some submarine eruption and induced a tsunami with a height of up to 5 m, with 437 casualties.

9.2.3.3 Slope-failure-induced tsunami

As mentioned previously, tsunamis may be caused by slope failures as has happened in many parts of the world. Particularly, tsunamis induced by submarine slope failures during earthquakes may not be differentiated. For example, the 1771 Meiwa tsunami that occurred in the close vicinity of Ishigaki Island could have been induced by large-scale submarine slope failure (Imamura et al. 2001); there was another claim that it was induced by a normal faulting between Ishigaki and Miyako Islands (Nakamura 2006). In this section, these two case histories are discussed.

9.2.3.3.1 THE 1871 MAYUYAMA SLOPE FAILURE AND ASSOCIATED TSUNAMI

Mt. Mayuyama (formerly called Mt. Maeyama) failed following an earthquake at about 8 p.m. during the volcanic activity of Unzen volcanoes in 1792 as seen in Figure 9.10, which shows the present situation of Mayuyama (Mt. Mayu). On May 21 at 8 P.M., the mountain failed following an earthquake (Misawa et al. 1993). The volume of the failed region is said to be ~340,000,000 m^3 and the coast line progressed towards the sea by 800 m at maximum. As a result of the failure of the mountain, a big tsunami with a

Figure 9.10 A recent view of Mt. Mayuyama.

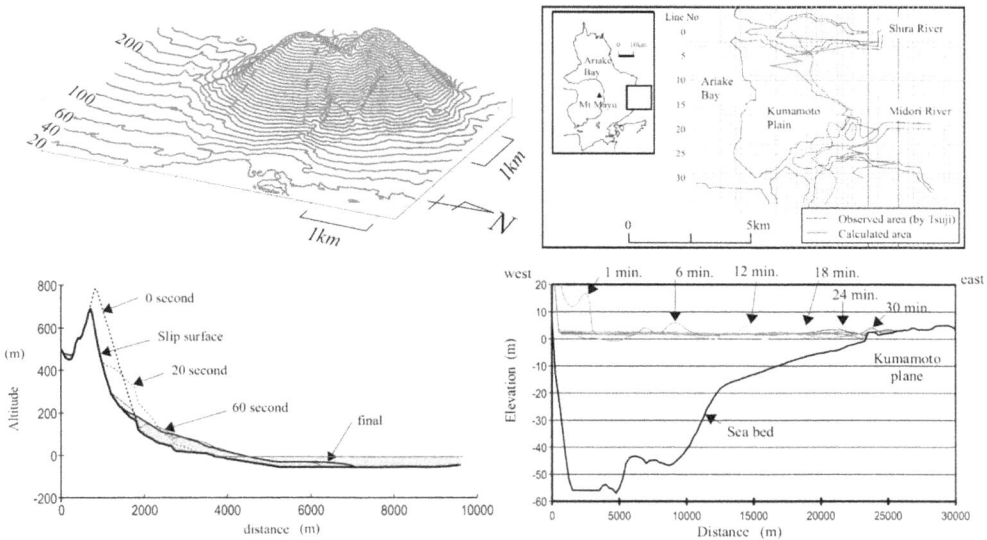

Figure 9.11 Computational results of the tsunami. (Arranged from Miyako (2010).)

wave height of 10 m occurred and 15,000 people lost their lives both in Shimabara and Kumamoto facing the Shimabara peninsula. Recently, this incident has been investigated using present-day computational techniques. Figure 9.11 shows the evaluations reported by Miyako (2010). In this particular case, the collapsed part of Mayuyama moved into the Ariake Sea and induced a huge tsunami, causing a high number of casualties in Shimabara and Kumamoto.

9.2.3.3.2   DEĞIRMENDERE SUBMARINE SLOPE INDUCED TSUNAMI DURING THE 1999 KOCAELI EARTHQUAKE

Although the 1999 Kocaeli earthquake was caused by a strike-slip faulting, a tsunami occurred in the Izmit Bay. The main causes of the tsunami are the slope failures that occurred mostly on the mobile side (the southern side of the North Anadolu fault) and the large-scale ground subsidence due partly to liquefaction and partly to a secondary normal fault at Kavaklı and Seymen, as seen in Figure 9.12. The ground subsidence was up to 2 m. The bathymetric surveying following the Bay showed that the submarine slope failure was of a large scale. The results of tsunami heights obtained from the field survey reported by Aydan et al. (2008b) are shown in Figure 9.13 together

(a)   (b)

Figure 9.12 Aerial views of Değirmendere (a) and Kavaklı (b).

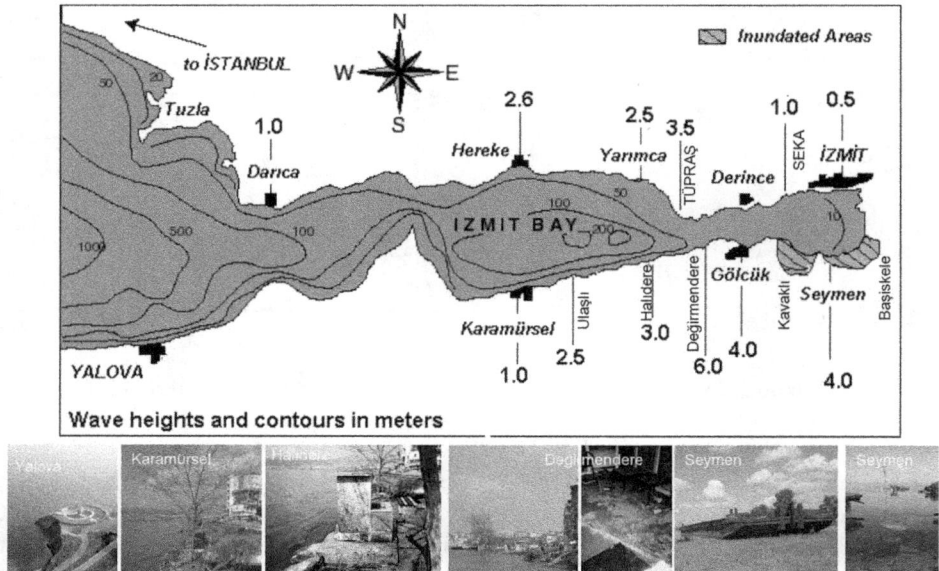

Figure 9.13 Observed run-up heights along the shore of Izmit Bay and some views. (Modified from Aydan et al. (2008b).)

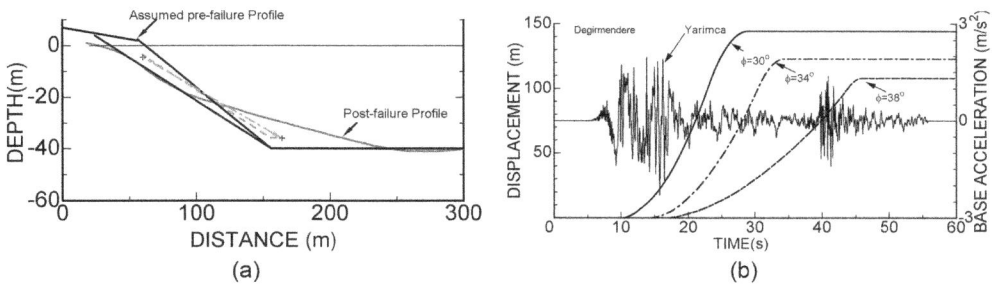

Figure 9.14 Mechanical model and computed displacement responses of the failed slope at Değirmendere. (a) Mechanical model and (b) computed displacement of mass centre. (Modified from Aydan et al. (2008b).)

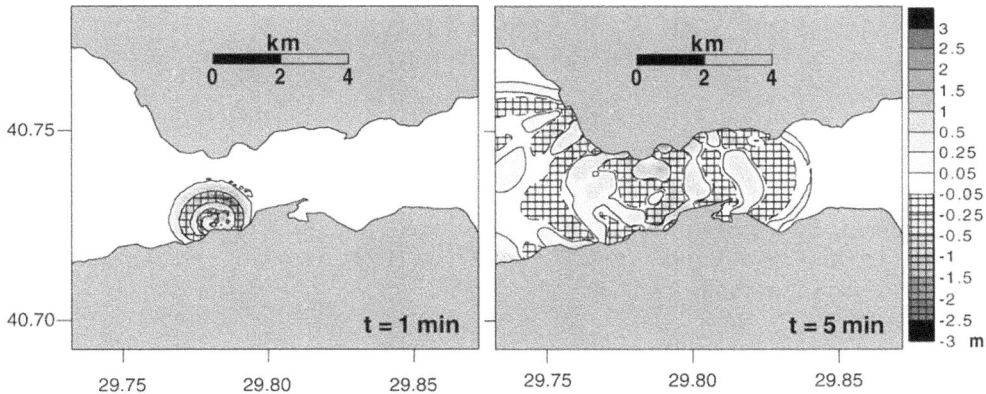

Figure 9.15 Computed tsunami propagation. (Modified from Tinti et al. (2006).)

with some views at different locations. As seen in Figure 9.13, the inside of a bank at Değirmendere had been engulfed by the waves as mud and traces of seawater were still seen 2 weeks after the earthquake. The windows of a coffee shop next to this bank were broken and similar traces were also seen here. The distance from the shore at this particular location is about 12 m. In the same figure, traces of crude oil slicks can be seen on the fountain along the shore of Halıdere.

Aydan et al. (2008b) analysed the submarine slope failure at Değirmendere, and found that the maximum velocity of the mass centre was about 93 cm/s (Figure 9.14). Tinti et al. (2006) analysed the tsunami propagation caused by the slope failure at Değirmendere. The results of tsunami propagation are shown in Figure 9.15. The computational results are generally in agreement with the observations. The difference might be due to nonconsideration of other large ground failures and settlements observed in the southern shore of the Izmit Bay.

9.2.3.4 Tsunami occurrence by meteorite impacts

Following the tsunami by the 2004 off-Sumatra earthquake and the recent Chely-abinsk meteorite impact in Russia, there is a growing interest on the possibility of tsunami occurrence induced by meteorite impacts on oceans (i.e. Weiss et al. 2006; Wünnemann et al. 2010). While there is no historic examples of meteorite impacts to have produced a tsunami, the impact of a meteorite at the end of the Cretaceous Period, about 65 million years ago near the tip of the Yucatan Peninsula of Mexico, produced tsunami that left deposits all along the Gulf coast of Mexico and the United States (e.g. Ward and Asphaug 2000). Ward and Asphaug (2000) have simulated the tsunami induced by the impact of a meteorite with a diameter of 200 m hitting the sea surface at a velocity of 20 km/s. There is recently a growing interest to determine the size of the impact and its effects on the environment. Figure 9.16 shows their computational results. The major issue with tsunami occurrence associated with meteorite impacts is their size and location.

9.2.4 Estimation of tsunami arrival time

Tsunami wave velocity in shallow waters is given in the following form (e.g. Dean and Dalrymple 1998):

$$v = \sqrt{gh} \tag{9.16}$$

where g and h are gravitational acceleration and sea depth, respectively. Tsunami arrival time may be obtained from the integration of the following relation as a function of seabed topography as

$$T = \int_0^\ell \frac{1}{v(x)} dx \tag{9.17}$$

For integration, seabed topography data would be necessary and the integration can be carried out using numerical integration techniques. Figure 9.17 shows an example of computation for an offshore earthquake along the west coast of Sumatra Island near Padang in Indonesia. The deformation front of the fault is assumed to be about 260 km away from the coast and the seabed profile along a line perpendicular to the shoreline passing through the channel between Siberut and Sipura Islands is used for computations. The tsunami arrival time is about 58 minutes for Padang and about 14 minutes for Siberut and Sipura Islands. The existence of a sea-mound between Siberut and Sipura Islands and a steep slope drastically decrease the propagation of the tsunami wave. If the offshore earthquake causes sufficient shaking to alert people, the arrival time for Padang is sufficiently long for people to move to a higher ground. However, it should be noted that if the epicentre of the earthquake happens to be in the Mentawai fault zone between Sumatra Island and islets, the duration would be reduced to 20–30 minutes.

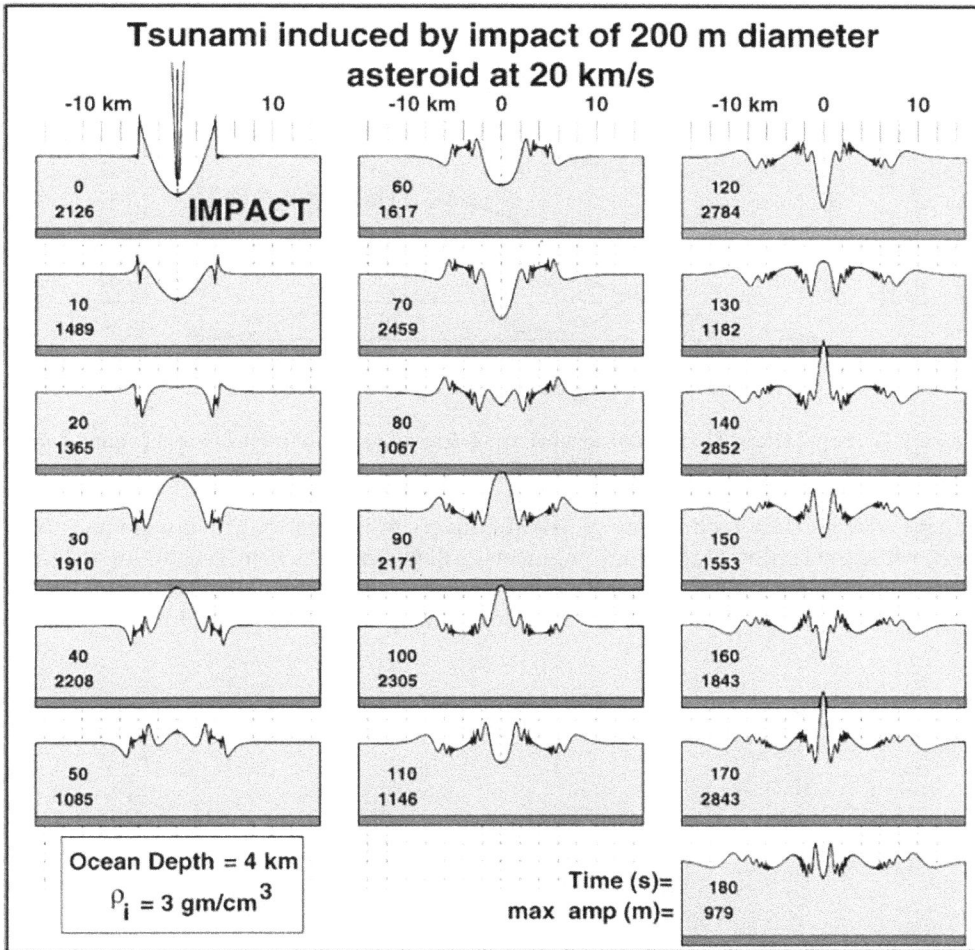

Tsunami induced by impact of 200 m diameter asteroid at 20 km/s

Figure 9.16 Tsunami induced by the impact of a 200 m diameter asteroid at 20 km/s. The waveforms (shown at 10 s intervals) trace the surface of the ocean over a 30 km cross section that cuts rings of tsunami waves expanding from the impact site at $x = 0$. Maximum amplitude in metres is listed on the left. (From Ward and Asphaug (2000.)

9.3 MODEL TSUNAMI TESTS

There are different experimental procedures to simulate tsunamis in the laboratory. The simplest tsunami generation device consists of a water tank and a channel with the other end open to prevent reflection. Such devices are found in many laboratories worldwide as well as in the hydraulic laboratories of Tokai University and the University of the Ryukyus (e.g. Aydan 2018c; Nakaza et al. 2015; Rahman et al. 2016; Shimohira et al. 2019), where the author was affiliated before his retirement. The facility at the

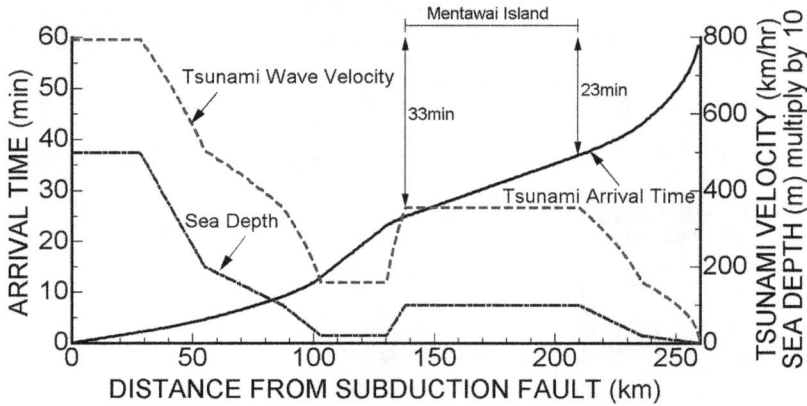

Figure 9.17 Estimation of tsunami arrival time for a scenario offshore earthquake near Padang. (From Aydan (2008b).)

Marine Science and Technology of Tokai University is 30 m long, 1 m wide and 0.6 m high, while the facility at the Civil Engineering department is 10 m long, 0.6 m wide and 0.4 m high (Figure 9.18). This type of device creates water flow depending upon the slit height of the sluice gate at the time of water release from the tank and it can simulate

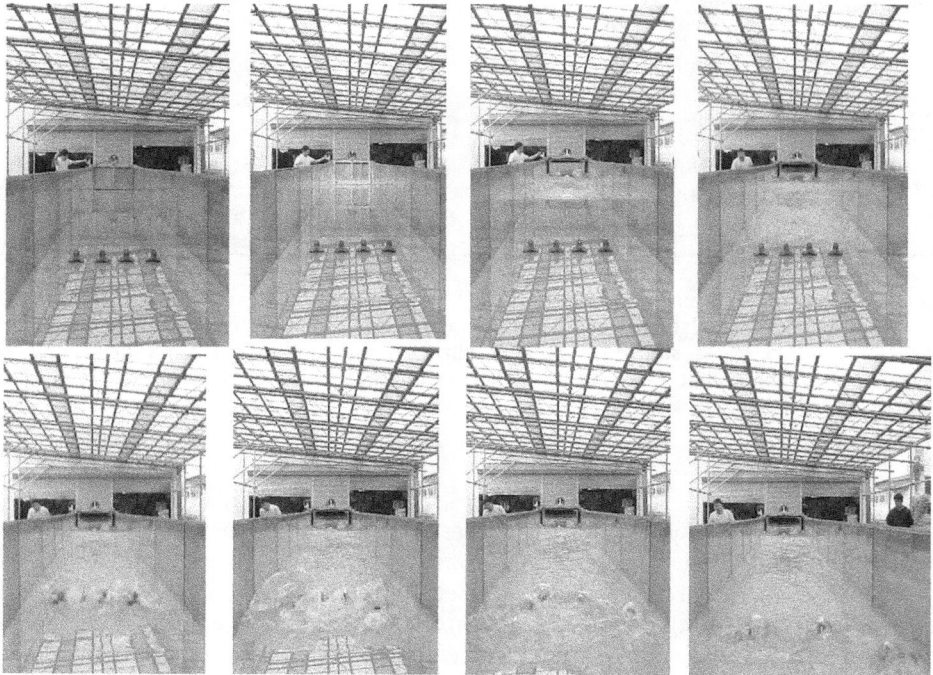

Figure 9.18 A tsunami experiment at the Marine Science and Technology of Tokai University.

Figure 9.19 Tsunami demonstration facility of the Marine Science Museum of Tokai University.

various types of effects on structures. However, the major issue with this device is that the main cause of water flow is not similar to the actual mechanism of tsunamis.

The Marine Science Museum of Tokai University has a unique demonstration facility to illustrate the mechanism and effects of tsunamis on the built-in and natural environment (Figure 9.19). This demonstration facility was established in 1970 and it can simulate thrust-type and normal-type seabed movements, which makes the facility unique in illustrating the mechanism of tsunamis associated with faulting movement of the seabed. The facility is 10 m long, 9 m wide and 1.2 m high. A 7 m long, 1.8 m wide and 50 cm high box at one end of the facility is uplifted or sank for 20 cm to cause thrust-type and normal faulting movement of the seabed. The maximum water depth is 60 cm. This facility is also capable of simulating the effect of wave-breaks, rivers and topography, physically.

9.3.1 Faulting-induced tsunami (normal and thrust)

The author performed two series of experiments using model tsunami generation devices. The first series of experiments were done at Tokai University and the second at the University of the Ryukyus. The experimental data were presented by Shimohira et al. (2019), and brief summaries of these experiments are outlined in this section.

9.3.1.1 Experimental facility and experiments at Tokai University

9.3.1.1.1 EXPERIMENTAL DEVICE

The tsunami generation model was made of 2,000 mm long, 300 mm wide and 400 mm high acrylic box. Figure 9.20 shows the concept of rising sea wave and receding wave

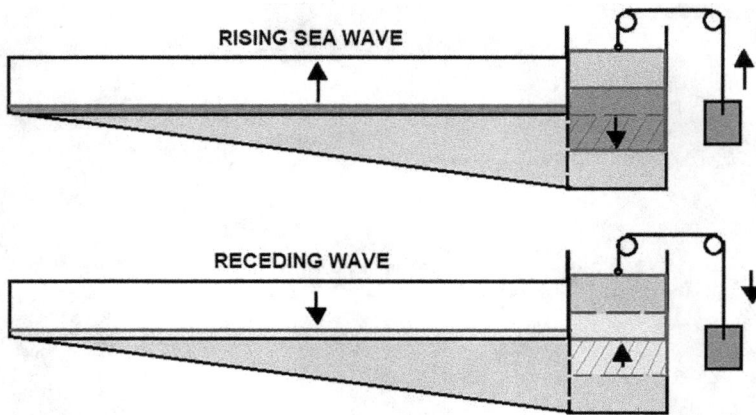

Figure 9.20 Illustration of the basic concept of tsunami generation device at Tokai University.

Figure 9.21 Views of Ryukyu limestone blocks before and after the test. (a) Water level is 40 mm and (b) water level is 70 mm.

tsunami modelling. The inclination of the sea bottom was 1/10. The water level change at stationary state was about 100 mm. Single, rectangular, prism-type blocks made of Ryukyu limestone or porous concrete were tested. Several overhanging configurations were also tested. Furthermore, the performances of boulder-type wave-breaks were also tested.

9.3.1.1.2 EXPERIMENTS

9.3.1.1.2.1 Ryukyu limestone blocks

First, Ryukyu limestone blocks with a height of 100 mm and width of 40 mm immersed to a depth of 30 mm initially were subjected to tsunami waves with different heights. When the final water level variation was <40 mm, the blocks slid upward as seen in Figure 9.21a.

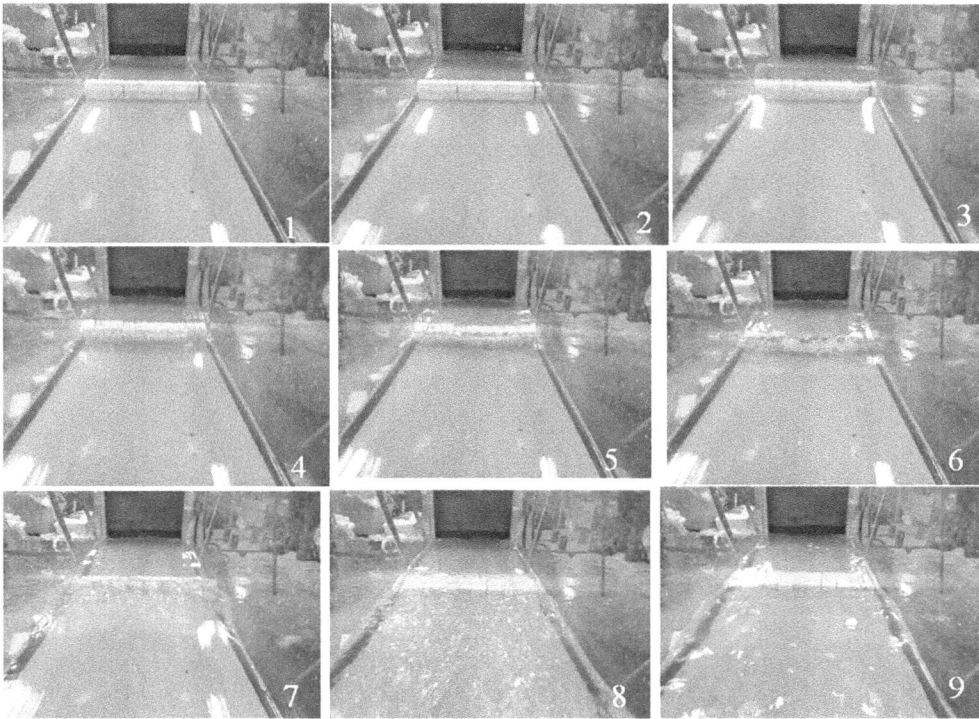

Figure 9.22 Views of the test shown in Figure 9.21b at different time steps.

However, if the final water level variation was more than 60 mm, the blocks were toppled landward in the direction of tsunami wave propagation as seen in Figure 9.21b. Figure 9.22 shows the experiments at several time intervals. The movements of the blocks occurred mainly during the water level rise stage and there was almost no movement during the receding wave stage.

9.3.1.1.2.2 Porous concrete blocks

Next, porous concrete blocks having dimensions of 150 mm height and 50 mm width were tested as shown in Figures 9.23 and 9.24. As the density of the blocks is lower than that of Ryukyu limestone and the height over width ratio is higher, they easily toppled, as seen in the Figures 9.23 and 9.24.

9.3.1.1.2.3 Experiments on boulder-type wave-breaks

The final experiment was carried out on the performance of boulder-type wave-breaks. The stone was siliceous sandstone from Abe River in Shizuoka Prefecture, Japan. The height of the wave-break was 50 mm and it was immersed in water up to 25 mm. The final water level rise was about 25 mm after the test. Figure 9.25a shows side and top views of the wave-break before and after the experiment. As seen from the figure, the

Figure 9.23 View of porous concrete blocks before and after the test.

Figure 9.24 Views of the test shown in Figure 9.23 at different time steps.

Figure 9.25 (a) Side and top views of wave-breaks before and after testing and (b) during testing.

wave-break subsides and spreads due to the impact of tsunami waves. In other words, the boulders are displaced by the generated tsunami waves. Figure 9.25b shows the experiment at time intervals. As noted from the figure the boulders are displaced during the overflow process of the tsunami waves. The receding tsunami waves did not cause major movements during the experiment.

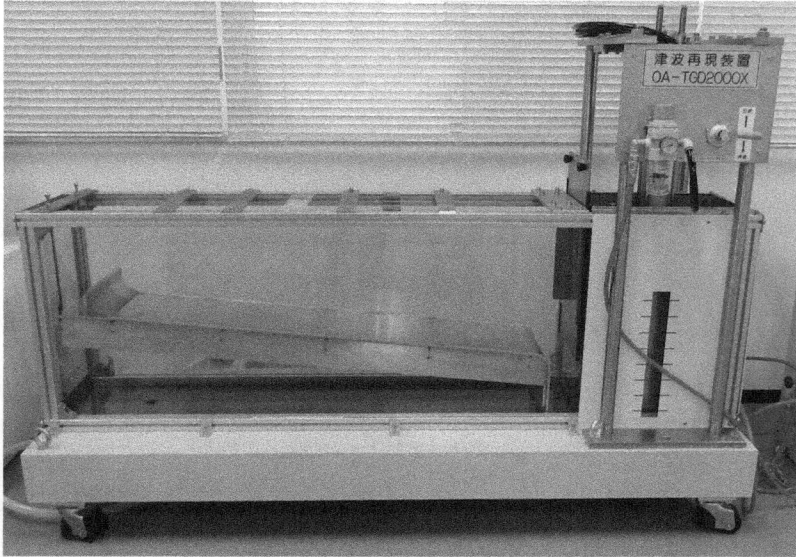

Figure 9.26 A view of the tsunami generation device OA-TGD2000X.

9.3.1.2 *Experimental facility and experiments at the University of Ryukyus*

9.3.1.2.1 EXPERIMENTAL DEVICE

Following the tests at Tokai University, the author designed a unique tsunami generation device named OA-TGD2000X to study the tsunami waves due to thrust and
normal faulting events shown in Figure 9.26. The dimensions and characteristics of the
device were quite similar to those used in Tokai University except the wave-induction
system. A tank was lowered or raised through pistons with a given velocity to generate
rising or receding tsunami waves. The pressure and wave velocity at specified locations were measured using pressure sensors and the amount of tank movement was
measured using laser transducers. Figure 9.27 shows an example of record during the
movement of tanks inducing rising and receding tsunami waves.

9.3.1.2.2 EXPERIMENTS

9.3.1.2.2.1 Triangular Ryukyu limestone blocks

Rectangular prism blocks were tested and the results were quite similar to those
tested in Tokai University. The triangular prismatic block shown in Figure 9.28 was
tested under the same conditions. The longest side of the triangular prismatic block
shown in Figure 9.28a was downward while that shown in Figure 9.28b was upward.
The downward block was almost nondisplaced, while the upward one was considerably displaced. One of the main reasons is that the tsunami force acts on the block

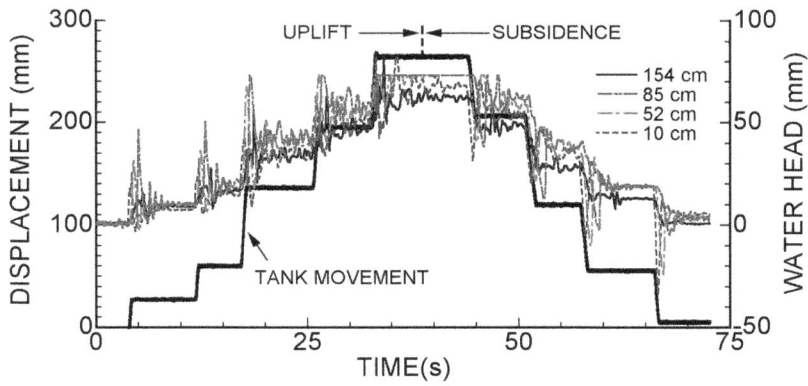

Figure 9.27 Water head response at specified locations in relation to the tank movement.

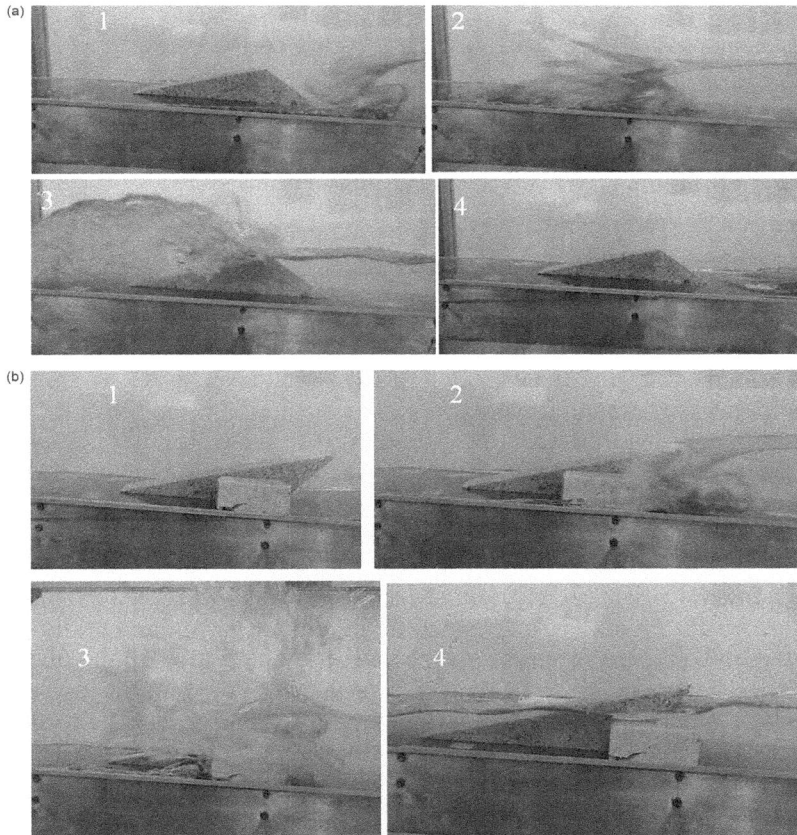

Figure 9.28 Views of downward (a) and upward (b) triangular prism block at different time steps.

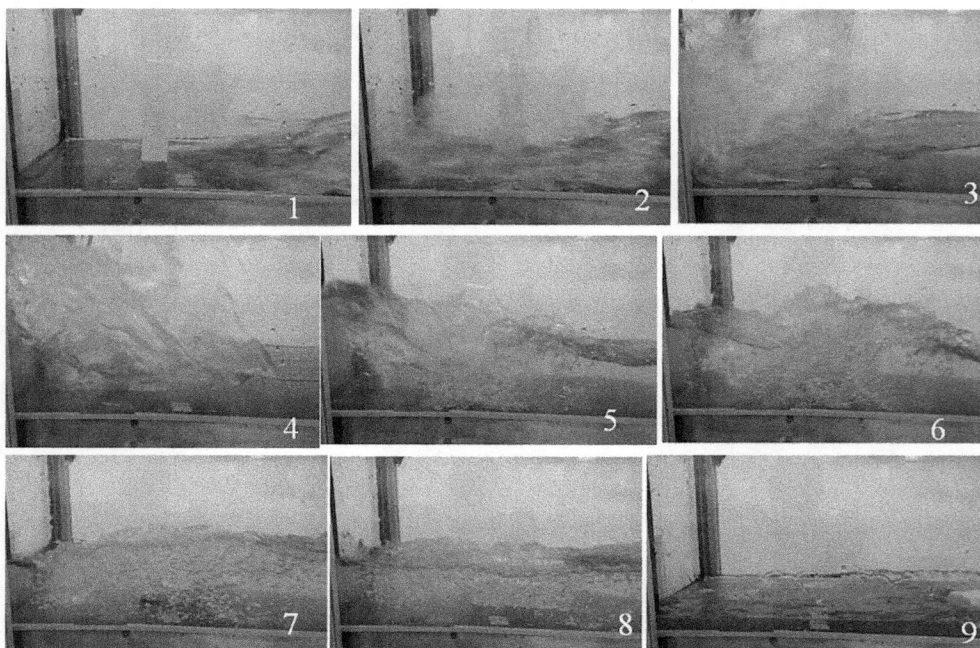

Figure 9.29 Views of the plaster block test at different time steps.

in a different way. For the upward triangular block, there is upward force acting on the side of the block facing the direction of the tsunami wave. As for the downward triangular prism the tsunami force increases the normal force on the block. A rectangular prism of Ryukyu limestone was placed next to the triangular prismatic block. The displacement of the rectangular block was quite small compared to previous experiments.

9.3.1.2.2.2   Plaster blocks

First a rectangular prismatic block made of plaster was subjected to rising tsunami waves as shown in Figure 9.29. The overall behaviour is fundamentally similar to those tested in Tokai University. Nevertheless, the block toppled downstream and was displaced horizontally in the direction of the receding tsunami waves, as seen in Figure 9.29. Next, two plaster prismatic blocks were laid over the Ryukyu limestone blocks as shown in Figure 9.30. The density of the plaster blocks is almost half of that of the Ryukyu limestone blocks. As seen from images 2 and 3 in Figure 9.30, the plaster blocks were thrown upward and displaced in the direction of the tsunami waves. This experiment clearly demonstrates the importance of the density and overhanging degree of blocks when they are subjected to tsunami waves in nature.

Figure 9.30 Top views of the plaster block hanging over the base of the Ryukyu limestone blocks at different time steps.

9.3.1.2.2.3   Breakable overhanging cliffs

The next series of experiments involve breakable blocks in relation to tsunami boulders seen in nature. Finding appropriate material for breakable blocks under the tsunami forces induced by the experimental device was quite cumbersome. Although the materials had a very small density compared those in nature, they provided an insight into the mechanism of the formation of tsunami boulders. Figure 9.31 show images of the models at different time steps. The surging tsunami wave enters beneath the overhanging blocks and applies upward forces. As a result, the overhanging block starts to bend upward and they are broken after a certain amount of displacement. In other words, the failure of the overhanging blocks is quite close to cantilever beams. However, the failure of the overhanging blocks is against the gravitational force. Once the block is broken, it is dragged by the overflowing tsunami waves. This observation is in accordance with the mechanism proposed by Aydan and Tokashiki (2019) for the formation of tsunami boulders. The experiments clearly indicated that if the inclination of the lower side of the overhanging block ranges between 10° and 20°, they are quite vulnerable to failure.

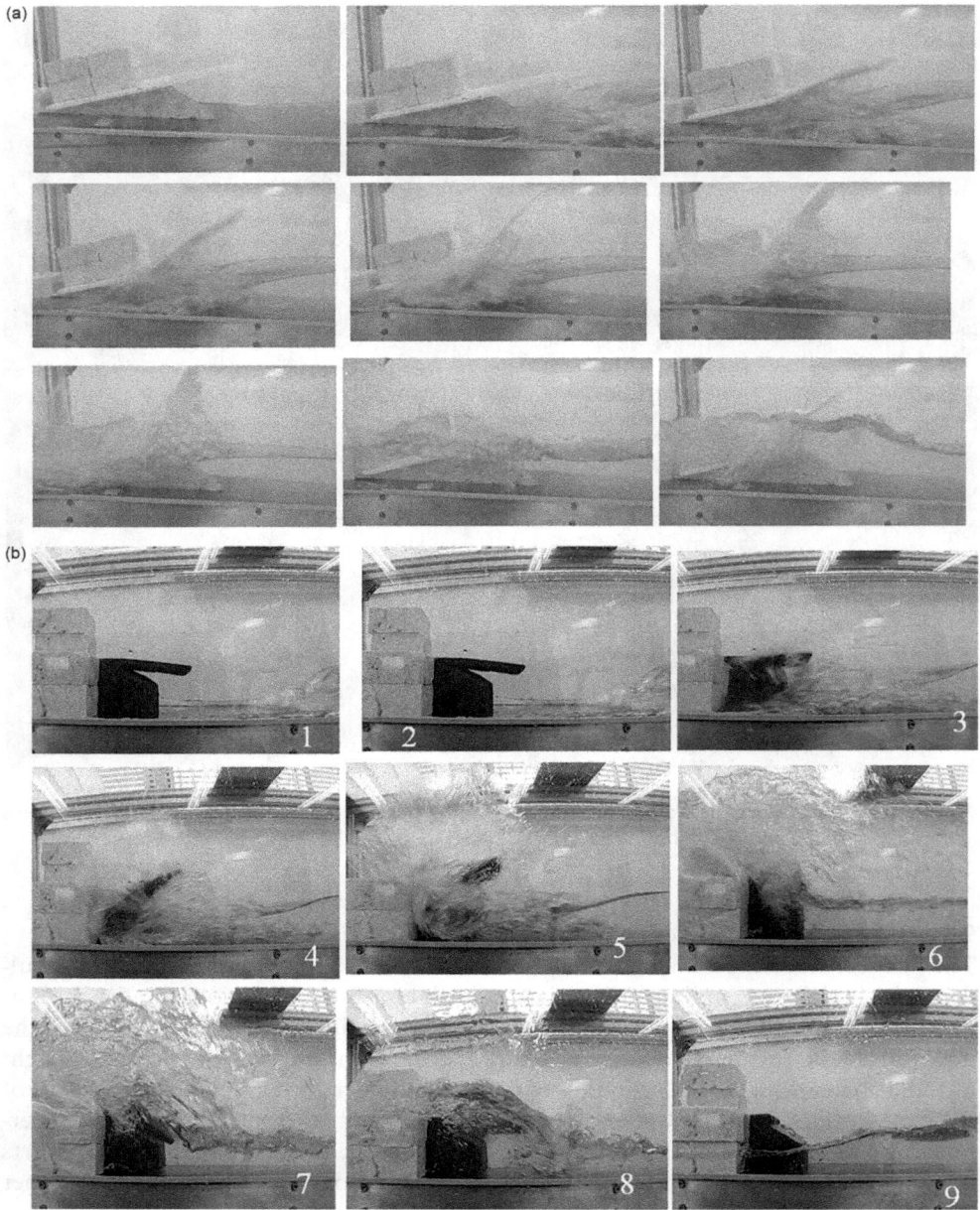

Figure 9.31 Views of the experiments using a breakable overhanging block at different time steps. (two models with different configurations)

Figure 9.32 Small-scale tsunami induced by the landslide in Aratozawa Dam induced by the 2008 Iwate-Miyagi intraplate earthquake. (From Aydan (2015a).)

9.3.2   Water surface changes due to impactors

9.3.2.1   Experiments on water level variations due to impactor in closed water bodies

The disastrous effects of tsunami are well known worldwide, and tsunami is generally caused by earthquakes of normal or thrust faulting types. They may also be caused in closed water bodies even by strike-slip faulting. The rockslide into Vaiont dam in 1960 into the reservoir triggered a huge tsunami and caused inundation of a settlement below and killed more than 2,000 people. The 2008 Iwate-Miyagi intraplate earthquake also caused a huge landslide resulting in a small-scale tsunami in the reservoir of Aratozawa Dam as seen in Figure 9.32 (Aydan 2015a). The landslide acted like an impactor on the reservoir.

An experimental study on tsunami generation in reservoirs was performed by Aydan (2017). An object was dropped from different heights and the induced tsunami waves were recorded. For this purpose, a cylindrical acrylic container with an internal diameter of 144 mm and water height of 75 mm was used as a reservoir and a plastic prismatic block ($50 \times 50 \times 40$ mm^3) was the object being dropped. The wave height was measured using a water pressure sensor attached to the side wall of the reservoir, and the acceleration of the base was measured in order to evaluate the time of the impact and its vibration on the surrounding (Figure 9.33). The unit weight of the prismatic block was 10.6 kN/m^3. The drop height of the block was selected as 0, 50, 100 and 150 mm. The variation of drop height (h) changes the initial impact velocity (v_{in}) of the

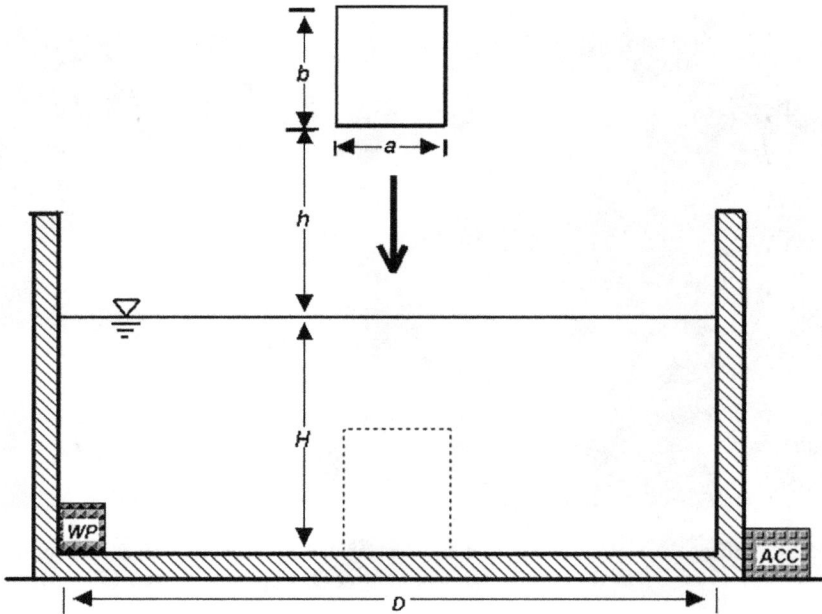

Figure 9.33 Illustration of the experimental setup.

falling body, which may be given in the following form by considering the transfer of potential energy into kinetic energy:

$$v_{in} = \sqrt{2gh} \tag{9.18}$$

where g is gravitational acceleration.

Four experiments were carried out by varying the dropping height of the object. Figure 9.34 shows the time histories of water level variations and acceleration for 10 seconds. The fluctuations in the tsunami waves almost disappeared after 10 seconds in an exponential manner. As expected the maximum wave height increased as the drop height increased. The recorded acceleration at the base of the model occurred just after the peak tsunami wave height and remained almost constant thereafter. The water level of the reservoir increased in proportion to the volume of the falling object. The water level increases for the given geometry of the experimental setup when the fluctuations disappeared can be given by the following expression:

$$\Delta H = \frac{4 \times a \times b \times c}{\pi D^2} \tag{9.19}$$

The experimental results were consistent with the above relation. Figure 9.35 shows the relation between drop height and induced maximum wave height together with experimental results. The experimental results fit the following relation very well:

$$\Delta h = \left(37.5 + 6.9 h^{0.7}\right) \frac{r_1}{r_2} \tag{9.20}$$

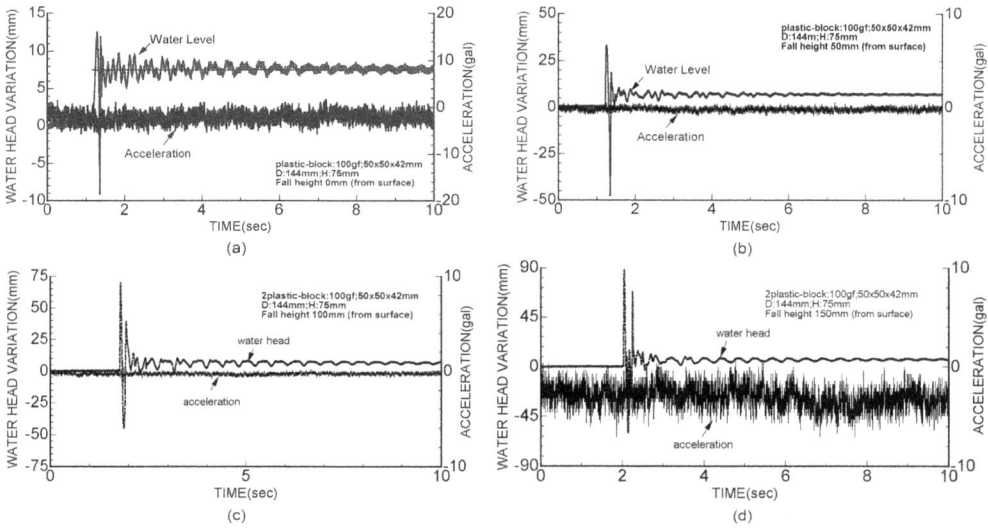

Figure 9.34 Acceleration and water height responses for different drop height from the water surface (a) 0 mm, (b) 50 mm, (c) 100 mm, and (d) 150 mm.

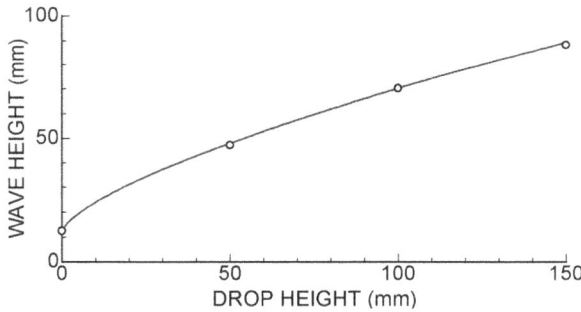

Figure 9.35 Comparison of the empirical relation with experimental results.

where r_1 and r_2 are the equivalent radius of the falling object at the time of impact and the radius of the leading edge of the wave. r_2 / r_1 is about 3 in the experiments.

Figure 9.36 shows the relation between normalized drop height and induced wave height by the height of the falling object together with experimental results. The experimental results fit the following relation very well:

$$\frac{\Delta h}{b} = \left(1.0 + 2.25\left(\frac{h}{b}\right)^{0.7}\right)\frac{r_1}{r_2} \tag{9.21}$$

It is also interesting to note that the power of the function remains the same.

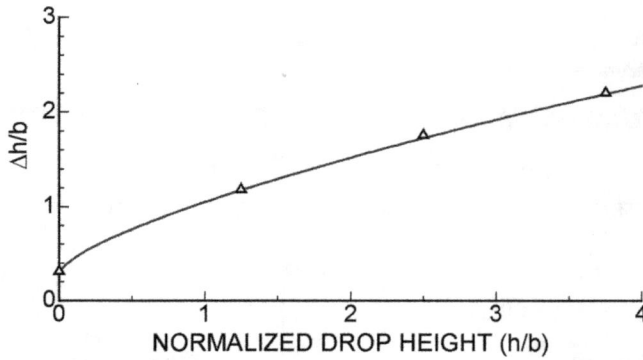

Figure 9.36 Comparison of the empirical relation with experimental results.

9.3.2.2   Theoretical modelling on water level variations due to impactor in closed water bodies and its applications

Water level variations in closed water bodies can be related to volume change due to volume change of the impactor (Aydan 2017). Let us consider a very simple geometry of both closed water body and impactor. Their volumes are given as

$$V_w = H \cdot L \cdot B; \; V_i = h \cdot l \cdot b \tag{9.22}$$

H, h, L, l, B and b are the dimensions, respectively. By considering the mass conservation law, the final water level change (u_f) may be given in the following form:

$$u_f = \frac{l}{L} \cdot \frac{b}{B} h \tag{9.23}$$

The vertical force equilibrium relation may be written as

$$\sum F_v = W - F_v - P - m \frac{d^2 s}{dt^2} = 0 \tag{9.24}$$

where

$$W = \rho_w g \left(H \cdot B \cdot L - u_b \cdot l \cdot b + u \cdot L \cdot B \right); \; P = \rho_w g \cdot H \cdot L \cdot B; \; m = \rho_w H \cdot L \cdot B$$

ρ_w is the density of water; u_b is the displacement of water body by impactor; u is the water body level change; and g is the gravitational acceleration.

The rate-dependent resistance (F_v) of the water body against motion may be assumed to given in the following form:

$$F_V = \rho_w \cdot H \cdot L \cdot B \cdot g \cdot \beta \cdot v^n \tag{9.25}$$

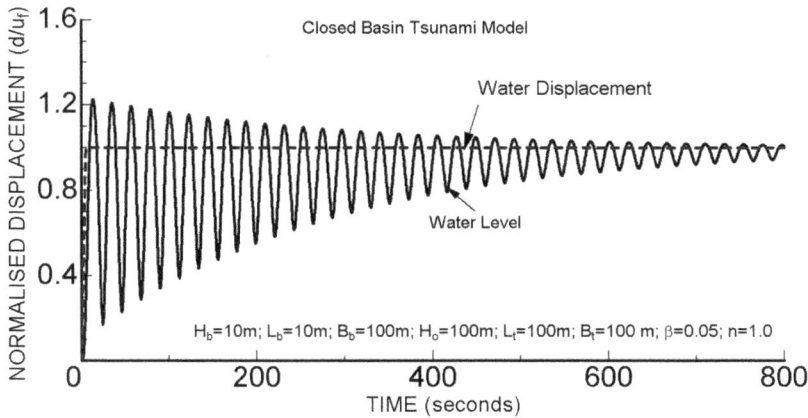

Figure 9.37 Water level fluctuations following impactor entry.

where β is the viscous coefficient, v is the velocity and n is a power coefficient. Thus the following differential equation is obtained for the water level change in the closed water body:

$$\frac{d^2u}{dt^2} + \beta g \left(\frac{du}{dt}\right)^n + g\frac{u}{H} = g\frac{l}{L}\cdot\frac{b}{B}\cdot\frac{u_b}{H} \tag{9.26}$$

The variation in the displaced water body may be assumed to be

$$u_b = \frac{t}{T_r}u_f \tag{9.27}$$

where T_r is called "rise time".

The ordinary differential equation given by Eq. (9.26) is nonhomogenous and nonlinear. Its solution can be obtained only through numerical methods. The procedure described in Chapter 8 can be used. Figure 9.37 shows computational results for a water body displaced by a prismatic impactor. The entry period of the impactor was assumed to be 5 seconds in this particular example.

9.3.2.3   Experiments on water level variations due to sliding or toppling bodies into closed water bodies

In addition to the tests presented in the previous section, some tests were carried out to see the effect of rock movements into reservoirs on water level variation. For this purpose, the configurations shown in Figures 9.38 and 9.39 were used and the inclination of the base plate was 39°. The size of the block was $40 \times 80 \times 100\,mm$, and it resulted in 2.5 mm water level change when it was fully immersed in the reservoir.

Figures 9.40 and 9.41 show the measured acceleration and water level changes for each model test setup. As seen in the figures, the water level changes are quite high when compared with the responses without initial velocity. It should be noted that the

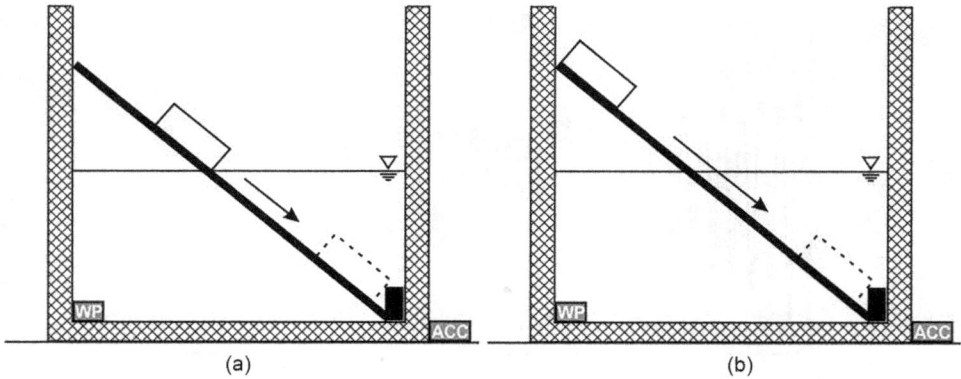

Figure 9.38 Model test setups for sliding mode. (a) Sliding with no initial velocity and (b) sliding with initial velocity.

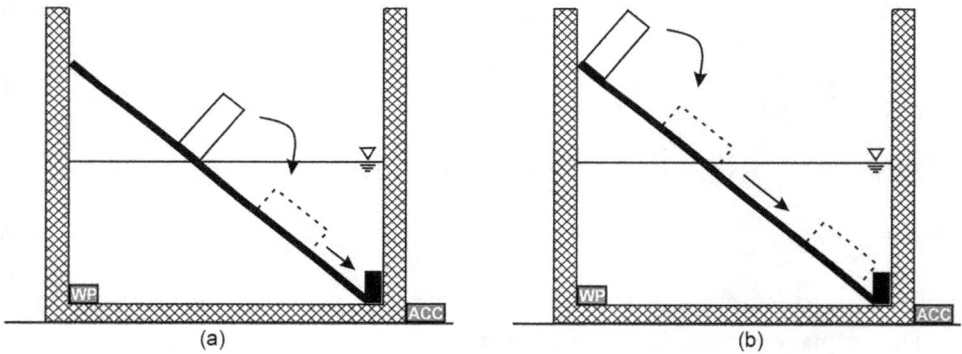

Figure 9.39 Model test setups for toppling mode. (a) Toppling with no initial velocity and (b) toppling with initial velocity.

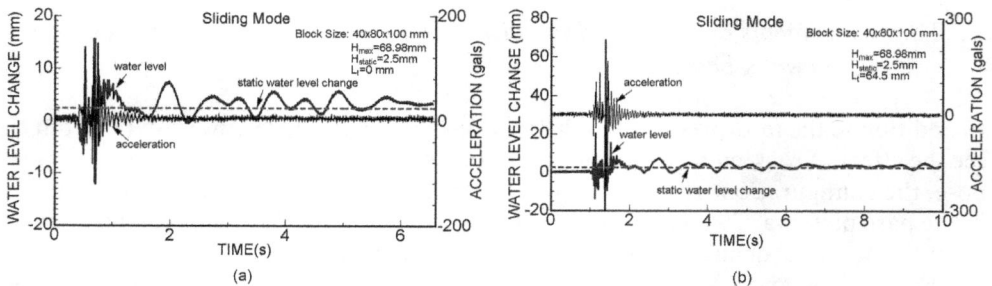

Figure 9.40 Water level and acceleration responses for the setups for sliding mode. (a) Sliding with no initial velocity and (b) sliding with initial velocity.

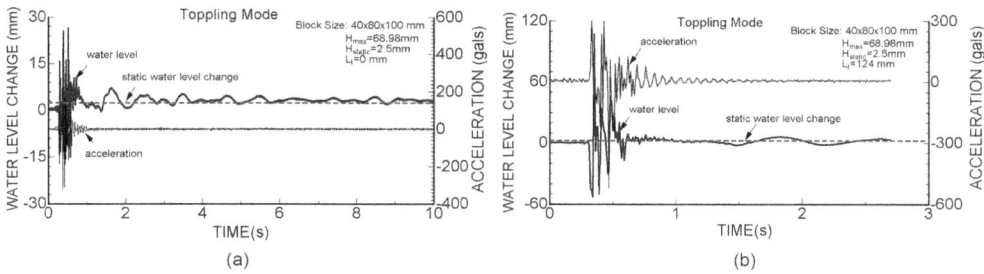

Figure 9.41 Water level and acceleration responses for the setups for toppling mode. (a) Toppling with no initial velocity and (b) toppling with initial velocity.

block has angular velocity and it is converted to translation velocity upon toppling failure. Another important conclusion from these experiments is that the blocks failing by toppling would induce much higher water level fluctuations, which may have important implications for evaluating the risk of block movements into the reservoirs of dams in practice.

9.4 EFFECTS OF TSUNAMIS ON STRUCTURES AND THE ENVIRONMENT

As mentioned in the introduction, the 2004 Aceh earthquake tsunami inflicted 200,000 casualties in countries around the Indian Ocean with highest casualties in Aceh Province of Indonesia. Nevertheless, there were many casualties and damage to built-in and natural environments since earlier times. The most recent disastrous tsunami was caused by the 2011 Great East Japan earthquake despite Japan probably having the most advanced precautions against tsunami disaster, and it surprised the world. This tsunami caused a secondary disaster due to the incident of the Fukushima Nuclear Power Plant. Furthermore, reinforced concrete (RC) buildings were overturned or uplifted and dragged several tens of metres in Onagawa town. The effects on structures and environment observed mainly in the affected areas by the 2004 Aceh earthquake tsunami and the GEJE are described herein together with some additional observations in other tsunami events.

9.4.1 Tsunami damage to industrial facilities

Aceh Province is one of the major oil- and natural-gas-producing regions in Indonesia. These facilities are on the east coast of the province. None of these facilities were damaged either by ground shaking or induced tsunami. However, a cement plant at Lhonga was heavily damaged by tsunami waves. A French company originally built the plant and an Indonesian company now operates it. The steel-framed structures, kiln and fuel storage tanks were heavily damaged by tsunami while RC silos remained intact (Figure 9.42a). It seems that tsunami waves reached a height of 15–20 m in some sections and the steel beams and columns either bent or buckled by the impact of tsunami

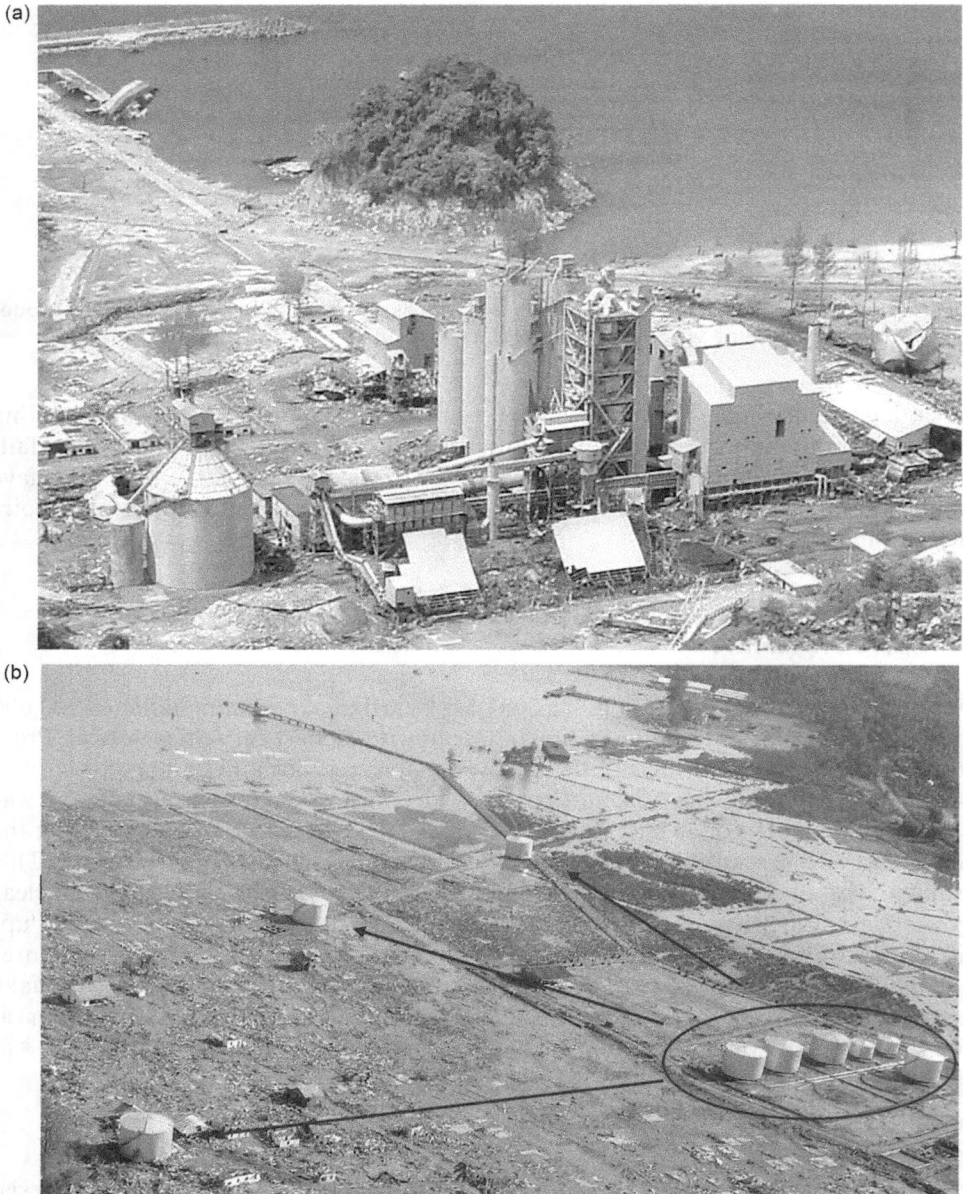

Figure 9.42 Damage by the tsunami on industrial facilities. (a) Damaged cement factory and (b) damaged oil tank farm and displaced tanks.

waves and dragged objects. There were two cylindrical fuel tanks in the plant. One of them was completely destroyed while the other one buckled and offset from its original location. Storage tanks from an oil tank farm in the eastern part of Banda Aceh city were displaced by the tsunami waves for a considerable distance from their original position, as seen in Figure 9.42b. Figure 9.43 shows the damage to a crusher plant and an oil tank in the cement factory. The crusher plant was a steel-framed structure. The panels were smashed and steel beams and columns were either bent or buckled. A palm oil tank along the seashore in Meulaboh was hit by tsunami waves and one side of the tank was buckled as seen in Figure 9.44. Furthermore, the settlement of the base of the tank occurred probably due to the liquefaction of the foundation ground.

(a)   (b)

Figure 9.43 Damage to facilities in Lhonga cement plant. (a) Crusher plant and (b) oil tank.

Figure 9.44 Buckled and settled tank at Meulaboh.

(a)   (b)

Figure 9.45 Satellite views of Banda Aceh port (a) before and (b) after the earthquake.

9.4.2   Tsunami damage to ports and coastal facilities

Tsunami-induced heavy damage to ports and coastal facilities along the west and north coast of Sumatra Island. Figure 9.45 shows satellite views of Banda Aceh harbour before and after the earthquake. As noticed from the two satellite views, a huge area was damaged by the tsunami as a result of settlement and erosion due probably to ground liquefaction induced by ground shaking as well as due to the tsunami. The ground consists of sandy soil in this area. It is also of great interest that some parts of the dykes of the harbour disappeared. Besides the effects of liquefaction, the flow direction of tsunami waves might have some damaging effects on the missing section of the dykes.

The RC building of the port facility collapsed at the ground floor. However, the main cause of collapse was ground shaking rather than the tsunami waves. Large stone blocks were thrown by the tsunami waves over the wharf of the port. Although the wharf of a barge with a power generator was not damaged by the tsunami as seen in Figure 9.46, the barge was displaced from the wharf to a distance of 3 km inland.

The port facility for the Lhonga cement factory was also damaged by the tsunami. The piles of wharf of the port were fractured by the impact forces of the capsized ship moored to the wharf as seen in Figure 9.47. Furthermore, the tetrapods of the wave-break were displaced for a considerable distance due to the whirling of the tsunami waves.

9.4.3   Tsunami damage to transportation facilities

Tsunami damage to transportation facilities in Aceh Province were mainly associated with roadways and bridges along the west coast between Banda Aceh and Meulaboh cities. The road between Banda Aceh and Meulaboh cities for a length of 384 km was disrupted at 123 locations according to United Nation records. The only airport in the earthquake-affected region is the Sultan Iskandar Muda International Airport.

(a)

(b)

Figure 9.46 (a) Displaced barge with a power generator and (b) its wharf.

Bridges are either of truss or RC type. Twenty-one truss-type steel bridges were mainly uprooted from their bearings and 47 concrete bridges collapsed or were heavily damaged. The causes of the damage were the impact, drag and uplift forces of the tsunami as well as from the impact of dragged objects by the tsunami waves. Figure 9.48 shows some examples of bridge failures in Aceh Province. The following statements can be made on the damage of bridges:

- When bridges have no shear keys, the girders are easily displaced due to horizontal forces.
- When the bridges are sufficiently elevated for the unobstructed passage of tsunami and dragged materials, they are not damaged.

Figure 9.47 (a) Damaged wharf and (b) displaced tetrapods of wave-break at Lhonga port.

- When there are some obstructions such as small hills next to the bridges against tsunami waves, there is almost no damage or very limited damage to bridges.

Damage to roadways was mainly due to erosion resulting from tsunami waves, ground liquefaction or embankment failure (Figure 9.49). The damage was quite extensive in lowland areas, nearby rivers and seashores next to steep slopes.

The tsunami of the GEJE caused severe damage to railways and railway bridges along railway lines running close to the seashore. The piers of railway bridges were toppled, rails were dragged and embankments were scoured (Figure 9.50). Furthermore, trains were uplifted and dragged for considerable distances.

Fundamentally, damage to roadways and roadway bridges were similar to that of railways and railway bridges and it occurred in roadways running close to the seashore. The bridge decks were uplifted and dragged and embankments were scoured (Figure 9.51).

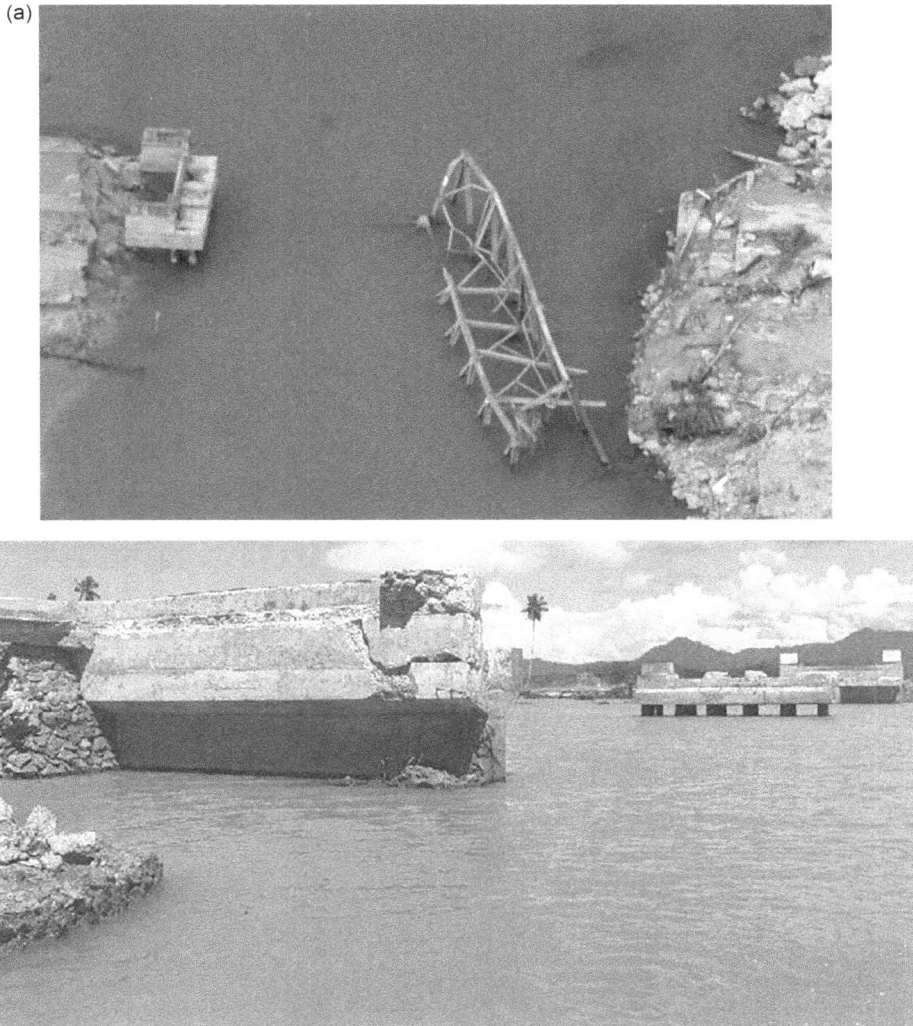

Figure 9.48 Examples of (a) collapsed truss bridge and (b) concrete bridge.

9.4.4 Responses of airports

The only airport in the Aceh Province was the Sultan Iskandar Muda International Airport to which commercial flights are in operation. This airport was not damaged either by ground shaking or by tsunami waves. However, the airstrip nearby Meulaboh was damaged by tsunami waves and by ground shaking.

The Sendai Airport, which is about 1 km away from the seashore and at an elevation of 2 m above sea level, was inundated by the tsunami of the GEJE. Tsunami

(a)

(b)

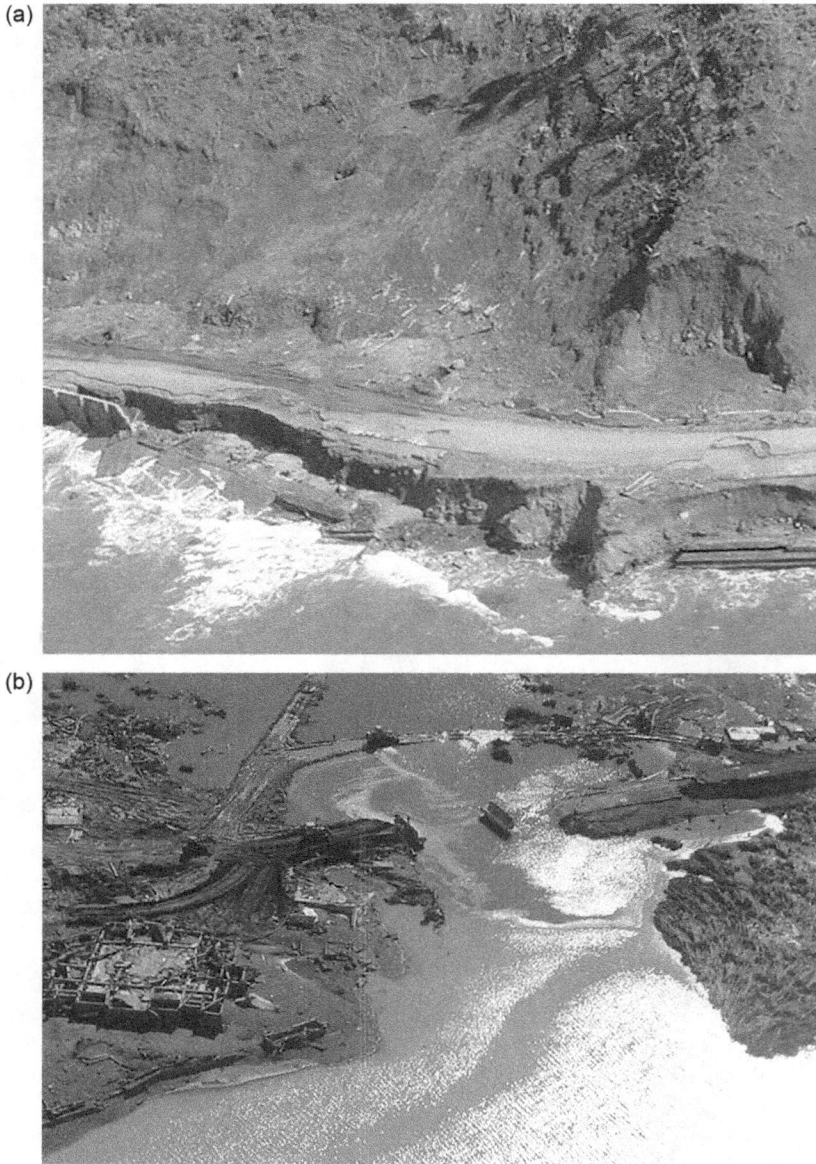

Figure 9.49 (a) Roadway damage due to erosion and (b) failures of embankments.

height at the airport was about 4.9–5.7 m. The runway and ground floor facilities were all damaged by the inundation, mud and debris of the tsunami (Figure 9.52a). The airport was cleaned up and it restarted its function on April 13, 2011, with some domestic flights. The piers and embankments of railway and roadway bridges were also scoured, and at some locations there were deck falls (Figure 9.52b).

Figure 9.50 Damage to railways and railway bridges.

9.4.5 Tsunami damage to buildings

Building stocks in Aceh Province can be broadly classified as follows: (1) timber houses, (2) masonry (brick) houses, (3) RC buildings and (4) mosques and minarets. The causes of damage to buildings were ground shaking, tsunami or both. While the number of stories of buildings in the populated cities such as Banda Aceh and Meulaboh could be greater than six, most of the buildings along the west coast of Sumatra were mainly single- or two-story buildings.

Timber houses are generally single- or two-story buildings. These buildings were almost nondamaged in the regions which were not affected by the tsunami. Therefore the main cause of the damage was the tsunami. The tsunami may impose at least four types of loading: impact force, drag force, hydrostatic water loading and buoyancy (uplift) force on wooden buildings. While impact and drag forces are directly related to the velocity of tsunami, the hydrostatic and buoyancy forces depend upon the tsunami height and relative density differences between the building material and tsunami waves. Figure 9.53 shows some examples of damage to timber houses.

Masonry (brick) houses are generally single-story buildings, with some being two story. Solid red clay bricks or hollow concrete blocks were used for constructing the masonry (brick) houses. Although concrete column and slabs are used during the construction, they are merely for achieving structural integrity. Such buildings were not damaged in the non-tsunami-affected areas. However, they were completely destroyed when the tsunami waves hit these structures. The impact forces induced by the tsunami were quite high as noted from the state of debris and the fallen tree trunks. The

(a)

(b)

Figure 9.51 Views of damage to (a) roadways and (b) roadway bridges.

(a)

(b)

Scouring at Piers and bridges in Sendai Airport

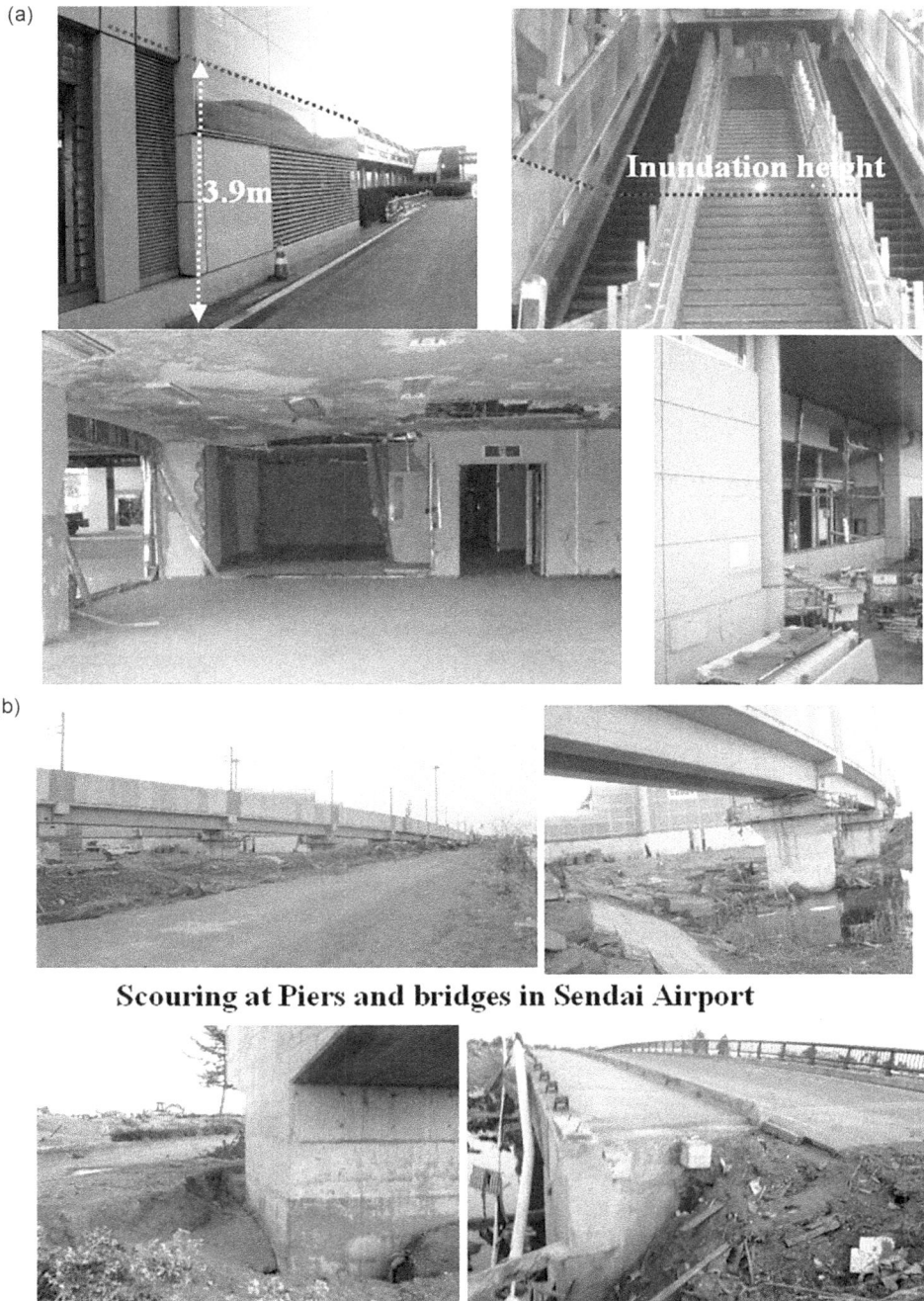

Figure 9.52 Views of damage to Sendai Airport and scouring at piers and embankment damage of roadways and railways leading to Sendai Airport. (a) Sendai Airport and (b) piers and embankments.

(a)

(b)

Figure 9.53 Damage to timber houses (a, b).

other force components of the tsunami waves could be drag forces and hydrostatic water loads. While the masonry (brick) houses were still standing, their walls were punched out. Figure 9.54 shows some examples of damage to masonry houses.

Most of the RC buildings have two to three stories in Aceh Province. Nevertheless, new buildings for governmental offices and shopping malls have more than three stories. Commonly they are five-storied. Almost all new buildings in the non-tsunami-affected areas collapsed or were heavily damaged by ground shaking. In addition, the collapsed or heavily damaged RC buildings were constructed nearby rivers or swampy areas. The RC structures in tsunami-affected areas have two–three-story buildings. Some of their columns were broken and in-fill walls were punched out by the impact

(a)

(b)

Figure 9.54 Damage to masonry (brick) buildings (a, b).

forces of tsunami waves and tsunami-dragged objects (Figure 9.55). Nevertheless, they survived against the ground shaking and also the forces resulting from tsunami waves. The columns were generally ruptured or fractured at their mid-height, which implies that they were subjected to high bending forces. Furthermore, the broken columns and punched-out walls were facing the flow direction of the incoming tsunami waves.

Figure 9.55 Damage to RC buildings in tsunami-affected area (a, b).

Figure 9.56 Damage to mosques (a, b).

Mosques are generally built as single-story RC structures without in-fill walls. Mosques survived against the ground shaking and tsunami waves even in the severely hit areas (Figure 9.56). The possible reason for such a good performance against tsunami may be associated with its columnar structure without in-fill walls. The only mosque damaged in this earthquake is in Ulee Lheue district and the central RC column of the wall facing the sea was fractured by bending at mid-height. Many mosques were built as masjids with no minarets. The grand mosques in Banda Aceh and Meulaboh have only minarets. Their minarets were lightly damaged, as they are slender structures.

The damages caused by the GEJE are briefly discussed herein. Timber buildings are the less-resistant building type among all other types of buildings (Figure 9.57a). As they are very light, there are easily uplifted and carried away and crushed. Therefore, such structures should not be built in tsunami-prone areas. Although stone masonry buildings are rare in tsunami-affected areas, they are much more resistant against tsunami waves (Figure 9.57b). Steel-framed structures also performed poorly (Figure 9.58a). The in-fill panels of the steel-framed structures were punctured by impact forces of tsunami and debris, and steel frames bent or buckled. RC structures or steel-reinforced concrete (SRC) structures performed much better among all building

(a)

(b)

Figure 9.57 Views of damage to (a) timber and (b) masonry buildings.

Figure 9.58 Views of damage to steel-framed buildings (a) and RC buildings (b).

types (Figure 9.58b). Except the damage to some of RC or SRC buildings in Onagawa town, almost all of these types buildings were structurally sound and they were the only remaining structures in all visited towns and cities.

In Onagawa town, even the RC or SRC buildings were damaged and some of them were toppled and dragged for several 10 m (Figure 9.59). It was interesting to note that one building, which had piles with a diameter of 30 cm and length of 4 m, was uplifted from its piled foundations and dragged for about 70 m. The back-analysis of damage to RC and SRC buildings in Onagawa should be the starting point for future tsunami-resistant building designs.

Figure 9.59 Views of damage to RC and SRC buildings.

9.4.6   Effect of tsunami on slopes

The effects of tsunamis on rock slopes along seashores are almost unknown. The tsunami waves apply shock waves on rock slopes and lower the effective stress during pull-back. These are thought to be the major causes of rock slope failures induced by tsunamis. Aydan (2013, 2015b) was first to point out the effects of tsunamis on rock slopes along seashores on the basis of his observations during the reconnaissance on the effects of the tsunamis of the 2004 Aceh (off-Sumatra) earthquake and the GEJE. Although there is no report yet that tsunamis caused any deep-seated slope failures so far, the necessity for further studies on the effect of tsunamis on rock slopes along seashores with the consideration of their structural geological features was pointed out by Aydan (2013).

In Aceh Province, the seashore between Banda Aceh and Lamno is particularly mountainous and slopes are steeply inclined. Furthermore, the rock layers are folded and inclined towards the sea. The failure of rock slopes was mainly due to sliding on bedding planes of hard sedimentary rocks (i.e. limestone, sandstone). Mainly the slopes facing the causative fault plane failed (Figure 9.60). Some toppling failures were also observed near Lhonga.

In the region affected by the tsunami of the GEJE, the seashore between Ishinomaki (Miyagi Prefecture) and Noda (Iwate Prefecture) is particularly mountainous and slopes are steeply inclined. Furthermore, the rock layers are folded and inclined towards the sea. The failure of rock slopes was mainly due to sliding on bedding planes of hard sedimentary or metamorphic rocks and the slopes facing the causative fault

Figure 9.60 Slope failures induced by the 2004 Aceh (off-Sumatra) earthquake.

plane (Figures 9.61–9.63). Tsunami waves first apply shock-type forces with tremendous amplitude on rock slopes along shores and climb up. Then they pull back and apply drag forces on already disturbed rock mass by shaking due to the preceding earthquake and shock waves by the tsunami, which results in the failures of rock slopes in different modes depending upon their structural geologic features. Particularly, the effective stress is drastically reduced due to the rapid draw-down of the seawater and insufficient drainage of saturated rock mass by the seawater. The situation is very similar to the effective stress variations when the water level of reservoirs is rapidly lowered. The author also observed this situation in his experiments on rock blocks subjected to planar failure subjected to rapid rising and subsequent lowering of water in reservoirs.

9.4.7 Damage to embankments

Damage to embankments by tsunami waves is due to tsunami-induced liquefaction and erosion resulting from drag forces. Such damages were observed along river embankments, at railway and roadway embankments (Figure 9.64) along the shores of Tohoku region, Japan. Furthermore, erosion of approach embankments of railway and roadway bridges was also common in the tsunami-affected areas.

Figure 9.61 Effects of the tsunami on rock slopes with horizontal layering.

Figure 9.62 Views of the effects of the tsunami on rock slopes with inclined layering.

Figure 9.63 Views of the effects of the tsunami on jointed rock slope failures.

Figure 9.64 Views of scouring in embankment of roadways and railways in Sendai Airport.

9.4.8 Responses of gigantic breakwaters and causes of their damage

Japan constructed gigantic breakwaters at various locations. Breakwaters at Kamaishi, Ofunato, Rikuzentakata and Taro were all damaged and they could not function as they were supposed to be. The Kamaishi gigantic breakwater was 61 m high from the sea bottom. Figure 9.65 shows its cross section. The caissons of the breakwater were sank down and tilted after the tsunami. The video records taken during the arrival of the tsunami clearly demonstrated that the tsunami surf appeared at locations where it should not be. This simply implied that the damage to caissons of the breakwater already took place before the arrival of the tsunami due to ground shaking induced by the earthquake. It is most likely the rubble mound below the caissons sank down due to ground liquefaction or failure of seabed soil. This is actually a commonly observed phenomenon in past earthquakes as well as in this earthquake (Figure 9.66). Similar damage was also observed in model tests.

There are also many wave-breaks along the shores and river banks in the tsunami-affected region. When tsunami waves are quite high and flow over the wave-breaks, the overflow may induce their toe erosion, which leads further instability and eventual failure. This was a quite common situation in this earthquake. In addition, some erosion took place between superstructures and foundations, as seen in Figure 9.67.

Figure 9.65 Illustration of the cross section of the Kamaishi port gigantic breakwater.

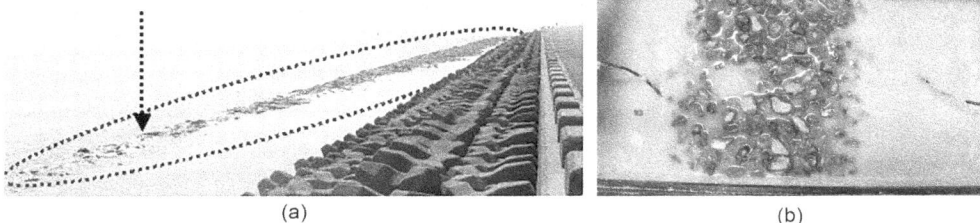

(a) (b)

Figure 9.66 (a) Settlement of tetrapods in Shin-Urayasu caused by the GEJE and (b) a wave-break model test on liquefiable soil.

Figure 9.67 Views of the damaged state of countermeasures after the tsunami.

9.5   INFERENCE OF TSUNAMIS HEIGHTS

Tsunamis may sometimes cause tremendous damage as observed during the 2004 Aceh earthquake and the GEJE and other major tsunami events in the past. Any tsunami warning system would involve the estimation of the hypocenter, magnitude of earthquake, faulting mechanism and seabed deformation. As stated previously, normal or thrust faulting may cause tsunami depending upon the amount of uplift or subsidence and seabed topography. Except closed basins and lakes, strike-slip faults do not generate tsunami unless they result in submarine landslide as often observed in the earthquakes of Marmara Sea in Turkey (Aydan et al. 1999b, 2008b). Before discussion, tsunami height must be defined. Figure 9.66 illustrates some definitions. The parameters shown in Figure 9.68 may be defined as

H_o : Original tsunami height at the source area, and it is directly associated with the vertical deformation of seabed due to faulting. If the sea depth is >4,000 m, this value remains the same. Otherwise it will differ during the propagation.

H_t : Shelf tsunami height. This height would change as tsunami propagates.

H_s : Tsunami height at shoreline. This is the most important parameter for tsunami-proof design.

H_r : Tsunami run-up height. Mass media often quote this value and its value would depend upon the topography of the inundation area.

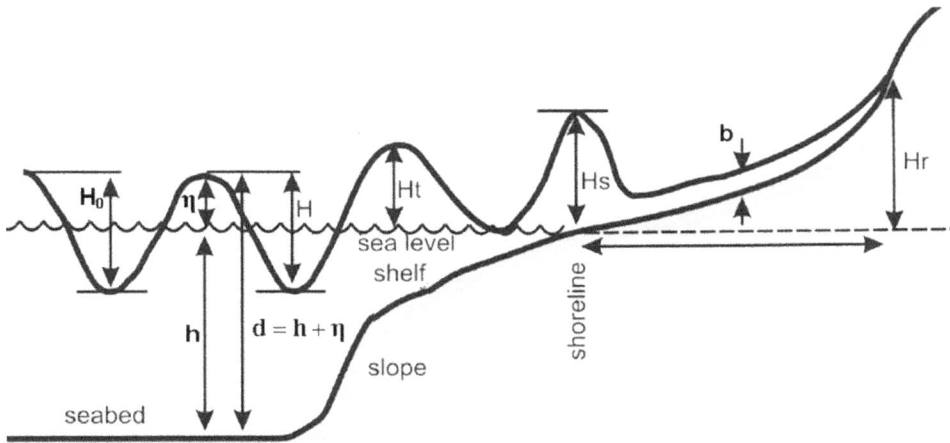

Figure 9.68 Illustrations of tsunami heights. (Modified from Alpar et al. (2003).)

Iida (1963) and Abe (1979) proposed some empirical relations for estimating tsunami wave height at the shoreline and run-up. Iida's formula is given in the following form:

$$M = 0.383 \log_2 H + 7.065 \qquad (9.28)$$

Abe (1979) proposed two relations between tsunami magnitude (M_t) (roughly equivalent to moment magnitude (M_w)) and tsunami wave height (H_m) at the shoreline for near-field earthquakes.

$$M_t \approx M_w = 2 \log_{10} H_m + 6.6 \qquad (9.29)$$

As for far-field earthquakes, H_m is replaced with wave height (H_{tide}) at tidal gauge and the following equation was proposed:

$$M_t \approx M_w = \log_{10} H_{tide} + \log_{10} X + 5.5 \qquad (9.30)$$

Abe (1979) also proposed that maximum run-up height of the tsunami could be taken twice that at the shoreline:

$$H_r = 2H_m \qquad (9.31)$$

Aydan (2008b) recently compiled available data on run-up heights and shoreline height of tsunamis that occurred in Indonesia and neighbouring areas. These data are plotted in Figure 9.69 together with the following empirical relations for shoreline height and run-up heights of tsunamis:

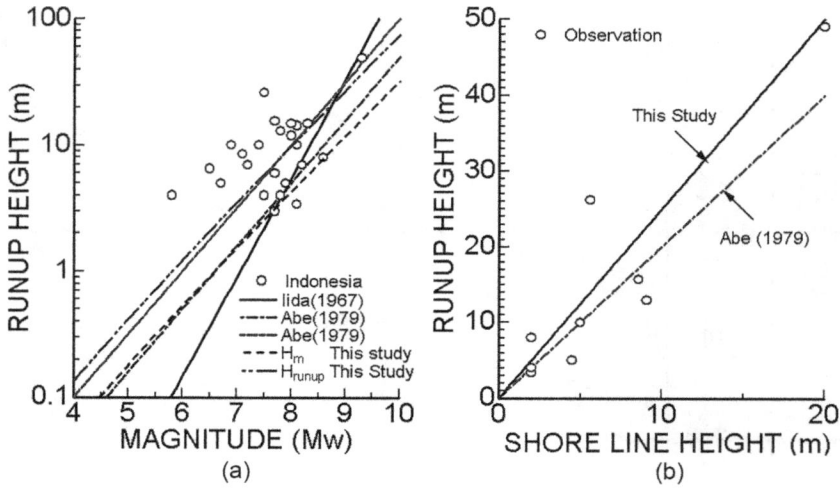

Figure 9.69 Relation between tsunami wave height and earthquake magnitude (a) runup height-magnitude , (b) shoreline height vs runup height.

$$H_s = A M_w \exp(b M_w) \tag{9.32a}$$

$$H_r = B H_s \tag{9.32b}$$

where A, b and B are constants. In this chapter, H_m is replaced by H_s.

For tsunamis in Indonesia and neighbouring areas, the values of constants A, b and B are 0.0004, 0.9 and 2.5, respectively. As noted from the figure, the observation data is highly scattered around empirical relations. In other words, the empirical relations can only serve as guidelines, and this fact must be considered in any assessment of tsunami potential. Furthermore, tsunami run-up height can be taken 2.5 times that at the shoreline for the Indonesian tsunami data set.

Table 9.1 compares tsunami height estimations according to empirical equations proposed by Abe (1979) and Iida (1963) and Eq. (9.32) for the last three earthquakes. As noted from the table, it seems that the tsunami height given by Iida's empirical relation is an average of estimations by Abe (1979) and Eq. (9.32). While estimations from Eq. (9.32) and Abe's empirical relations are quite reasonable for tsunamis of the 2004 Aceh earthquake and the 2006 South Java earthquake, they fail to predict those for the tsunami induced by the 2005 Nias earthquake and the 2007 Bengkulu earthquake, which may be due to some peculiar topography and seabed deformation that occurred in these earthquakes.

Aydan (2007a, 2012a) recently proposed the following relations between fault length and moment magnitude of earthquakes (see also Chapter 5). The empirical relation proposed by Aydan (2007a) for thrust, strike-slip and normal faults is written in the following form and coefficients for each fault type are given in Table 9.2:

$$L = C \cdot M_w e^{M_w/D} \tag{9.33}$$

Table 9.1 Comparison of estimation by empirical equations with observations

Earthquake	Mw	Iida (m)	Abe(1979)		Equation (9.32)		Observation	
			H_m (m)	H_r (m)	H_s (m)	H_r (m)	H_s (m)	H_r (m)
2004 Aceh	9.3	56	22	44	16	40.1	20	49
2005 Nias	8.6	19.5	11	22	7.9	19.8	2.5	8.0
2006 South Java	7.7	3.2	3.3	8.3	3.15	6.3	8.6	15.7
2007 Bengkulu	8.4	11.2	7.94	15.89	6.45	16.13	3.6	5.0

Table 9.2 Values of constants in Eq. (9.33) for each faulting type

Fault type	Rupture length L (km)	
	C	D
Normal faulting	0.0014525	1.21
Strike-slip faulting	0.0014525	1.19
Thrust faulting	0.0014525	1.25

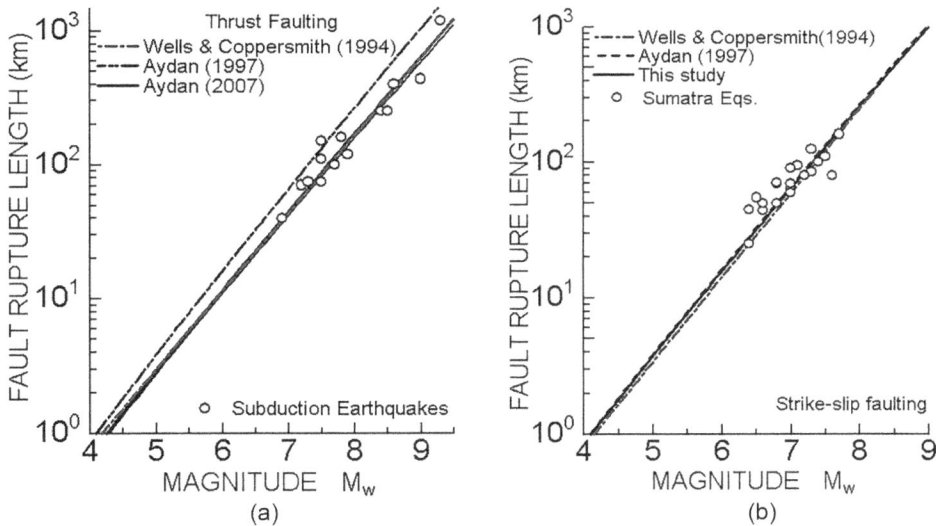

Figure 9.70 Comparison of empirical relations with observations from Indonesia. (a) Thrust faults and (b) strike-slip faults.

Figure 9.70 compares the estimations by Eq. (9.33) with observational data for Sumatra Island and its close vicinity. The empirical relations, originally proposed for the earthquakes of Turkey by Aydan (2008b) seem to hold for Indonesia and its close vicinity. The magnitude of the earthquake for each scenario fault and associated tsunami heights are computed and given in Table 9.3.

Table 9.3 Estimated tsunami heights for scenario earthquakes near Padang

Fault length (km)	Mw	Iida (m)	Abe (1979)		Equation (9.32)	
			H_m (m)	H_r (m)	H_m (m)	H_r (m)
33	6.8	0.62	1.26	2.52	1.24	3.09
140	7.8	3.78	3.98	7.96	3.49	8.73
275	8.4	11.2	7.94	15.89	6.45	16.13
420–450	8.7	19.3	11.22	22.44	8.75	21.88

Figure 9.71 Examples huge tsunami boulders in major islands of the Ryukyu archipelago (base map from 11th Regional Coast Guard Headquarters of Japan).

9.6   TSUNAMI BOULDERS AND THEIR UTILIZATION FOR INFERENCE OF MAGNITUDE OF PALEO MEGA EARTHQUAKES

As explained in the previous sections, tsunami boulders are associated with tsunami height. Aydan and Tokashiki (2019) proposed a procedure to estimate the magnitude of paleo mega earthquakes from tsunami boulders. This procedure is described and utilized with the considerations of tsunami boulders in Ryukyu archipelago. Tsunami boulders on several major islands in the Ryukyu archipelago are shown in Figure 9.71. There is also a great interest if there could be some large events like the 2004 off-Sumatra earthquake and tsunami along the Ryukyu Trench and Okinawa Trough.

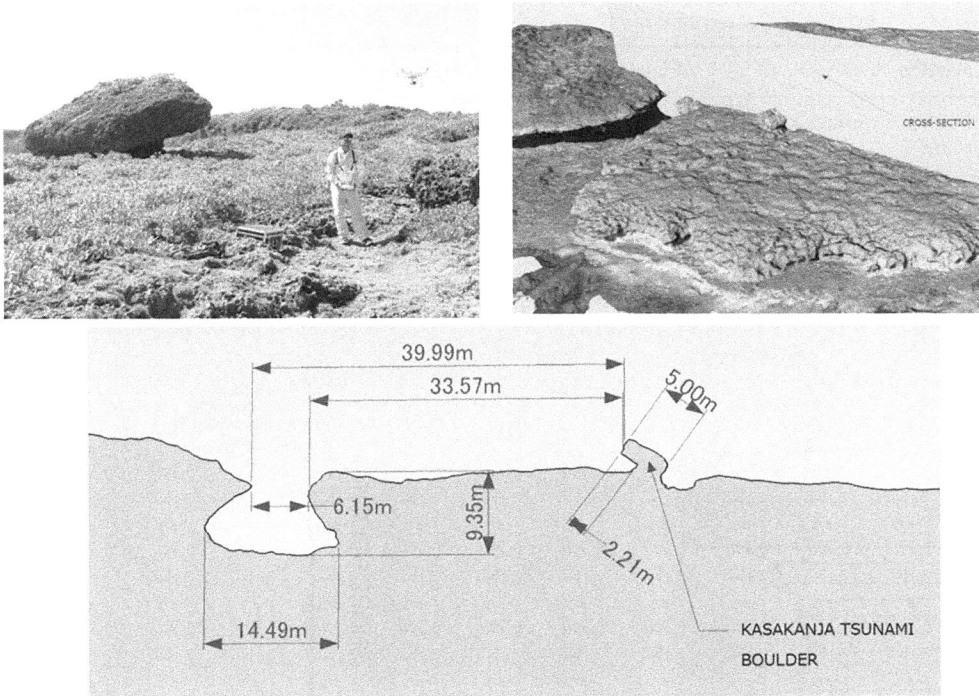

Figure 9.72 Aerial photogrammetry technique utilizing a drone at Kasakanja site.

One of the major issues with tsunami boulders is how to differentiate boulders caused by tsunami and cliff failures due to toe erosion and storm-waves during typhoons. Sometimes, boulders may be overthrown to higher elevations by strong storm-waves. For example, the Kasakanja boulder (see Figures 9.69 and 9.70) along the southern shore of Okinawa Island, Japan, was wrongly interpreted as a typhoon-transported boulder.

In view of the principles of hydro-mechanics, it would be unlikely to induce and transport such a huge boulder to the elevation of 12 m above the mean sea level by however strong storm-waves. Some observations were made on large tsunami boulders in Okinawa (Kasakanja), Miyako (Higashi-Hennasaki), Shimojiri (Obiwa) and Ishigaki (Ohama) Islands (Figure 9.71). For this purpose, the author and his group have been utilizing aerial photogrammetry and laser scanning techniques. Figure 9.72 shows the utilization of a drone at Kasakanja site and the results of aerial photogrammetry. Although the tsunami boulders in Ishigaki, Miyako and Shimoji islands were initially believed to be due to the 1771 Meiwa earthquake with an estimated magnitude of 7.4, recent studies indicated that they were much older (Goto et al. 2010). Particularly, the tsunami boulder in Shimoji Island is probably the largest in the world (Figure 9.71). Table 9.4 gives the height and elevation of the tsunami boulders in selected locations. In addition, some large boulders of metamorphic origin and sandy tsunami deposits were observed within a Ryukyu limestone layer during an excavation of a large

Table 9.4 Estimated magnitude of earthquakes from elevation and height of tsunami
         boulders

Location	Elevation (m)	Height (m)	Mw (LB)	Mw (UB)
Miyako-Hennazaki	20	4	9.5	9.7
Shimoji	12.5	9.0	9.0	9.5
Okinawa-Kasakanca	12	3	9.0	9.2
Ishigaki-Ohama	8.0	5.9	8.6	9.1

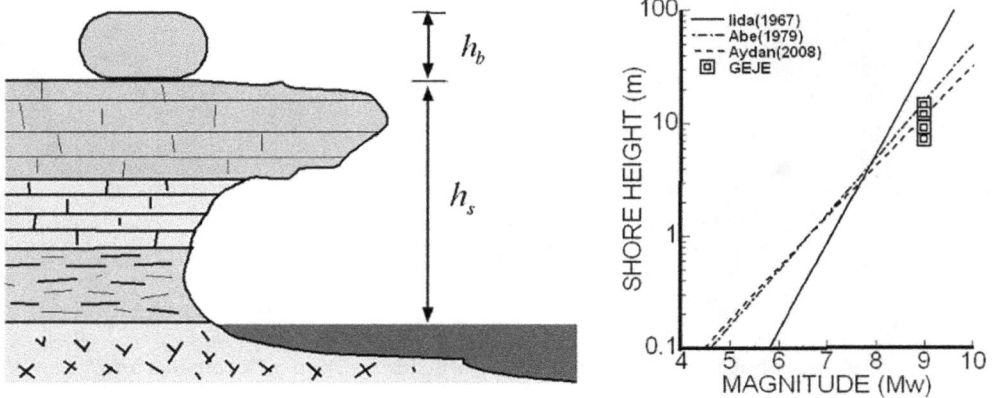

Figure 9.73 (a) Illustration of shore height versus tsunami boulder height and (b) empirical
          relation between magnitude and shoreline height (Aydan and Tokashiki 2019).

engineering structure in Ishigaki Island. These observations also imply that the events
were cyclically occurring in the Ryukyu archipelago.

   Aydan and Tokashiki (2019) proposed a method of inference to estimate the magnitude of the mega-earthquake resulting in great tsunamis utilizing tsunami shoreline
height defined in Figures 9.68 and 9.73a and some empirical relations illustrated in
Figure 9.73b and given by Eqs. (9.32) and (9.33). This method is applied to estimate the
magnitude for present elevation and height of tsunami boulders given in Table 9.4. The
estimated moment magnitudes of earthquakes with the consideration of the position
of tsunami boulders are also given in Table 9.4. Lower-bound (LB) values correspond
to those estimated from present elevation height and upper-bound (UB) values obtained from the present elevation plus block height. The magnitude of the mega earthquakes along the Ryukyu archipelago for the lower bound ranges between 8.6 and 9.5
and for the upper bound between 9.1 and 9.7. The results clearly indicate that mega
earthquakes are also possible along the Ryukyu archipelago and that disaster-preven-
tion measures must take this fact into account.

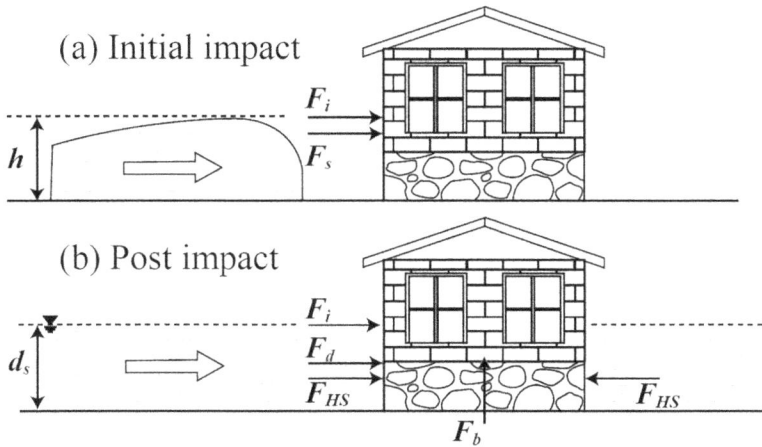

Figure 9.74 Loading conditions. (From Palermo and Nistor (2008).)

9.7 TSUNAMI-PROOF STRUCTURAL DESIGN PRINCIPLES

9.7.1 Tsunami-induced forces on structures

The most important issue for the design of structures against tsunami-induced forces requires accurate information on density, height and velocity of the tsunami acting on structures. These forces are named as (Figure 9.74)

1. Buoyant force
2. Hydrodynamic (drag) force
3. Hydrostatic force
4. Surge force
5. Impact force

It is claimed that these forces do not act on the structure simultaneously. This situation is illustrated in Figure 9.75.

These forces are specifically are given as follows:

Buoyant force is given in the following form based on Archimedes' principle:

$$F_b = \rho_f g V \tag{9.34}$$

Hydrodynamic (drag) force is a function of fluid density (ρ), flow velocity (v) and structure geometry ($A = hB$). The hydrodynamic force is given in the following form:

$$F_d = F_{hd} = \frac{1}{2}\rho_f C_d A v^2 \tag{9.35}$$

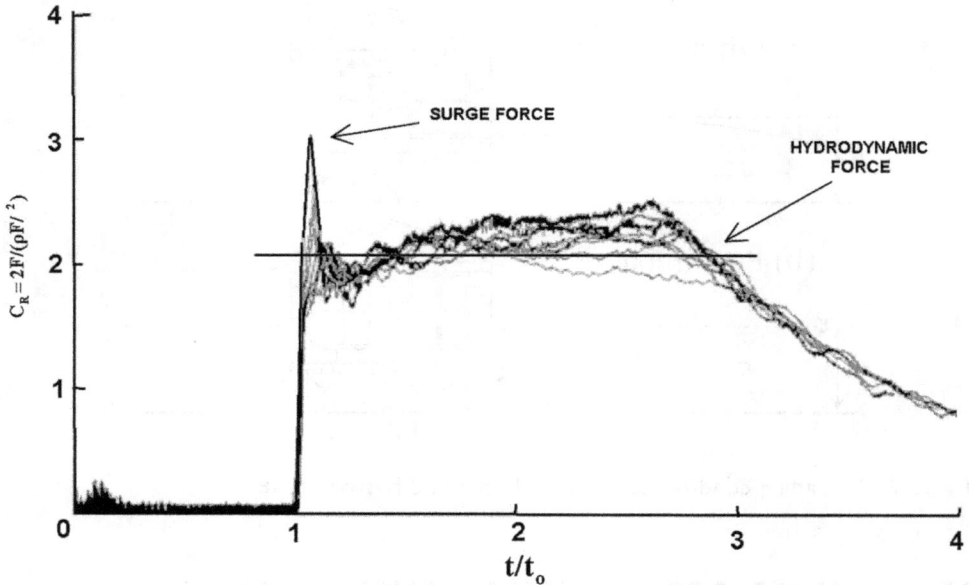

Figure 9.75 Action of tsunami-induced forces on structures with time. (From Yeh (2007).)

where h is the flow depth at the location of interest when there is no flow obstruction; hv^2 represents the momentum flux per unit mass per unit breadth; B is the breadth of the structure in the plane normal to the flow direction. FEMA-CCM (2012) recommends the drag coefficient C_d2 for square or rectangular objects and 1.2 for cylindrical objects. Velocity is generally related to the transient tsunami height through the following equation:

$$v = \sqrt{2gh} \tag{9.36}$$

Hydrostatic force acting laterally on a structure with an overflowing fluid is given as

$$F_{hs} = \frac{1}{2} B\rho_f g \left(h + \frac{v^2}{2g} \right)^2 \tag{9.37}$$

Surge force is caused by the leading edge of running-up water impinging on a structure. The following equation is proposed by Dames and Moore (1980) for surge force F_s per unit width:

$$F_s = F_{imp} = 4.5\rho_f gh^2 B \tag{9.38}$$

FEMA (2012) suggests that it may be taken 1.5 times the force given by Eq. (9.36).

Impact force resulting from objects such as lumbers is the most difficult force component to be evaluated, and it is given by

$$F_i = m\frac{dv}{dt} \approx m\frac{v_I}{\Delta t}$$ (9.39).

where m, v_I and Δt are mass, impact velocity and duration of impact. There are some empirical relations for evaluating the parameters given by Eq. (9.39). See the discussion provided by Yeh (2007).

9.7.2 Recommendations for measures against tsunami

Building dykes, gates and water breaks and planting trees along the coastline can also be implemented as hardware measures against tsunami disasters.

Educating children and people in general is of great importance to create public awareness about earthquake and tsunami disasters. The activities of voluntary groups and other related establishments must be further promoted through educational materials and financial support for their activities.

The following conclusions and lessons may be drawn from observations of earthquake-induced tsunamis:

1. Shore height (inundation height at seashore) is the most important parameter for the design of structures against potential tsunamis. Shore heights induced by numerous tsunamis in Indonesia and elsewhere are in accordance with the estimations from the empirical relations by Abe (1979) and Aydan (2008b).
2. Although Japan has been well prepared for tsunamis and their disastrous effects, the anticipated tsunami height for implementing hardware measures was much less than that induced by the tsunami induced by the GEJE.
3. The observations on the damage induced by the tsunami of the GEJE were basically similar to those caused by the 2004 Aceh (or off Sumatra).
4. Timber buildings cannot stand against high tsunami waves. Steel-framed structures are also weak against tsunami waves due to the fragility of their in-fill panels. RC buildings perform best to resist gigantic tsunami waves provided that they are well built against ground shaking. Therefore, timber buildings must not be allowed in tsunami-prone areas.
5. The GEJE also showed that vertical evacuation in relatively flat areas was important to save lives.
6. The effectiveness of planting trees against tsunami must be re-evaluated.
7. Very strong impact, surge, buoyancy and dragging forces of tsunami waves were the primary factors in damage to bridges despite their being well-designed against ground shaking.
8. The failure of breakwaters to protect settlements may have been caused by failure of seabed ground due to loss of strength resulting from the reduction of effective stress under prolonged tsunami waves and/or ground liquefaction.

Figure 9.76 Conceptual model and actual crustal changes along shores. (a) Graphical illustration, (b) Banda Aceh, (view from ground) (c) Banda Aceh (aerial view) and (d) Nias.

9. The inundation of ground along seashores is due to rebound of strained overriding plates (Figure 9.76). It would be quite difficult to preserve previous shorelines for a considerable period of time. Shorelines may further retreat due to erosion by high waves.

Chapter 10

Earthquake prediction

Earthquake prediction is a hot topic in earthquake science. Despite too many studies, there is no single reliable method to predict earthquakes. It is also very important to warn people before an earthquake strikes. It is well known that various anomalous phenomena occur before, during and after earthquakes. The anomalous phenomena are generally associated with the behaviour of animals, lightning, fireballs and variations in various gas emissions, groundwater level, gravity, geomagnetic field and electric potential before, during and after earthquakes. Some earth-scientists from the former USSR, China and Japan have been the pioneers in using these phenomena as precursors of earthquakes in order to predict them. Earthquake prediction research in Japan, United States, USSR and China gained considerable acceleration in the early 1970s. Particularly, the successful prediction of the Haicheng earthquake in 1975 made many seismologists all over the world optimistic about earthquake prediction. Nevertheless, the failure to predict the 1976 Tangshan Earthquake, which killed more than 250,000 people, in the following year made many geoscientists understand that earthquake prediction was still in its infancy. This resulted in the disappearance of the enthusiasm for earthquake prediction studies and projects seen in 1970s among scientists and politicians. Japan gave up hope on the success of earthquake prediction in 1997 after the Hyogo-ken Nanbu earthquake, which devastated Kobe City.

Although humankind is still premature in predicting earthquakes, it is believed that the accumulation of anomalous behaviours observed in each earthquake should be carefully documented and examined for future generations, who might be successful in doing so. There is no doubt that the correct information on observation should provide some hints for such people to develop the methods for predicting earthquakes in spite of current pessimistic views. The most important factors in earthquake prediction are the prediction of time of occurrence, location and magnitude. Earthquake prediction may involve short-, intermediate- and long-term predictions. Particularly, short-term prediction should be within 3 days when the patience of laypeople is taken into account.

In this chapter, a brief outline is given on possible physical backgrounds for various anomalous phenomena observed before, during and after earthquakes on the basis of findings from various disciplines. Then, presently available earthquake prediction methods are summarized. Finally, some attempts at prediction are described.

DOI: 10.1201/9781003164371-10

10.1 PHYSICAL BACKGROUND ON ANOMALOUS PHENOMENA OBSERVED IN EARTHQUAKES

Earthquakes are produced as a result of rupturing of Earth's crust due to the stress state acting on it. It is generally believed that the so-called tectonic stresses are the principal actors. The questions are why such stresses exist in Earth's crust and how they are generated.

The plate tectonics theory has been presumed to be able to answer the causes of tectonic stresses and to explain why earthquakes occur along some regions. However, this theory is also insufficient to explain intra-plate earthquakes as the theory is based on rigid body kinematics. The driving force for plate tectonics is assumed to be mantle convection, which is thought to be resulting from nonuniform temperature distribution in the upper mantle caused by subducting plates. There is no doubt that such a temperature difference could cause the convection. Then, the questions are why subduction of plates occurs and why Earth's surface is divided into several plates. There are probably no answers to these questions in the field of geophysics, presently. As explained in Chapter 2, Aydan (1995a) analysed the stress state of Earth by modelling it as a spherical object consisting of layers exhibiting thermo-elasto-plastic behaviour under pure gravitational acceleration. He was able to show that the whole of Earth could not be an elastic object at all and it must have already been in a plastic state. This simply implied that the fracturing and plastic yielding of the whole mantle must have taken place in its geological past together with tangential and radial stresses being the compressive maximum and minimum principal stresses, respectively. This finding also indicated that subduction or overriding phenomenon should occur within Earth's crust, and the mantle so that the conditions for mantle convection may be generated. It should not be forgotten that Earth is a part of the solar system. Earth rotates around the Sun with a varying speed between 29.3 and 30.321 km/s, and it wobbles. These facts should certainly cause some special circumstances to disturb its spherical symmetry, resulting in material inhomogeneity. Within this perspective, we have to consider the rupturing of Earth's crust, resulting in earthquakes.

Earthquakes are simply the products of rupture process of rocks composing Earth's crust. The stored mechanical work done of Earth's crust resulting from its deformation is transformed into various forms throughout its rupturing process if the energy conservation law of continuum mechanics holds. The forms of transformation of the mechanical work done can be observed as heat flux, electric current (magnetic current), kinetic energy, etc. Without any doubt, these transformations result in various phenomena, which may be called anomalous phenomena. With the birth and advancement of rock mechanics in the 1960s, some physical backgrounds for various phenomena were established from laboratory tests on rocks, which are directly relevant to the rupturing process of Earth's crust. If an idealized compression test of a rock specimen submerged in a fluid under a triaxial stress state is considered and the results are plotted, diagrams similar to those shown in Figure 10.1 are generally obtained. As seen from the figure, the rock specimen starts to behave in a nonlinear manner after a certain stress threshold. After this threshold, some fracturing starts to take place. Each time new fractures occur, various forms of transformation of the work done would take place and the imposed stress level increase would be stored as mechanical work done in the specimen. These transformations may be seen as (Aydan 2003a)

a. sound waves (acoustic emissions)
b. electric (magnetic) pulses
c. increase in permeability and porosity, implying decrease in pore pressure, and fluid flow
d. temperature increase and heat flux
e. gas emissions
f. degradation of elastic properties, subsequently reduction in P and S wave velocities
g. decrease in electrical resistivity
h. secondary creep

It is experimentally known that the fracturing process becomes unstable after a certain stress level. From that level onward, the so-called secondary creep and subsequently tertiary creep processes take place and result in the failure of the specimen. Although the boundary conditions are different in actual earthquakes, the stages in the rupture process and transformation forms of the mechanical work done should be very similar to those observed on rock specimens tested under laboratory conditions.

10.2   IMPLICATIONS OF RESPONSES OF ROCKS AND DISCONTINUITIES DURING FRACTURING AND SLIPPAGE

As said previously, an earthquake is an instability problem of Earth's crust and it is a subject of geoscience and geoengineering. Earthquake is caused by the varying crustal stresses and it is a product of rock fracturing and/or slippage of major discontinuities such as faults and shear/fracture zones.

When rock starts to fail, the stored mechanical energy in rock tends to transform itself into different forms of energy. Experimental studies have been conducted by Aydan and his group to understand multi-parameter variations including electric potential, electrical resistivity, magnetic field, acoustic emissions during deformation and fracturing process of geomaterials, which ranges from crystals, gouge-like materials to rocks under different loading regimes and environments (Aydan 2004a; Aydan et al. 2001 2005d,e, 2007c, 2010d,e).

Various responses measured during some of these experiments are shown in Figures 10.1–10.3. Detailed discussions can be found in previous articles (Aydan 2003a, 2004a, 2006c; Aydan et al. 2001, 2005d,e, 2007c, 2011, 2015; Ohta et al., 2008; Aydan and Tano, 2003; Aydan and Daido 2002). Nevertheless, one can easily notice the distinct variations in multi-parameters during the deformation and fracturing of rocks. As seen from the experimental results, the deformation and fracturing of rock cause distinct variations in electric potential, electrical resistivity, magnetic field and acoustic emissions in addition to conventional parameters such as displacement (strain) and force (stress), which may be useful in geoengineering and geoscience. The author drew the following conclusions from his previous experimental studies and those shown in Figures 10.2 and 10.3:

*   The experimental results clearly indicate that the deformation, fracturing and sliding processes induce electric potential in geomaterials.
*   The magnitude of induced electric potential depends upon both the piezo-electric characteristics of minerals or grains and the moment caused by the separation of

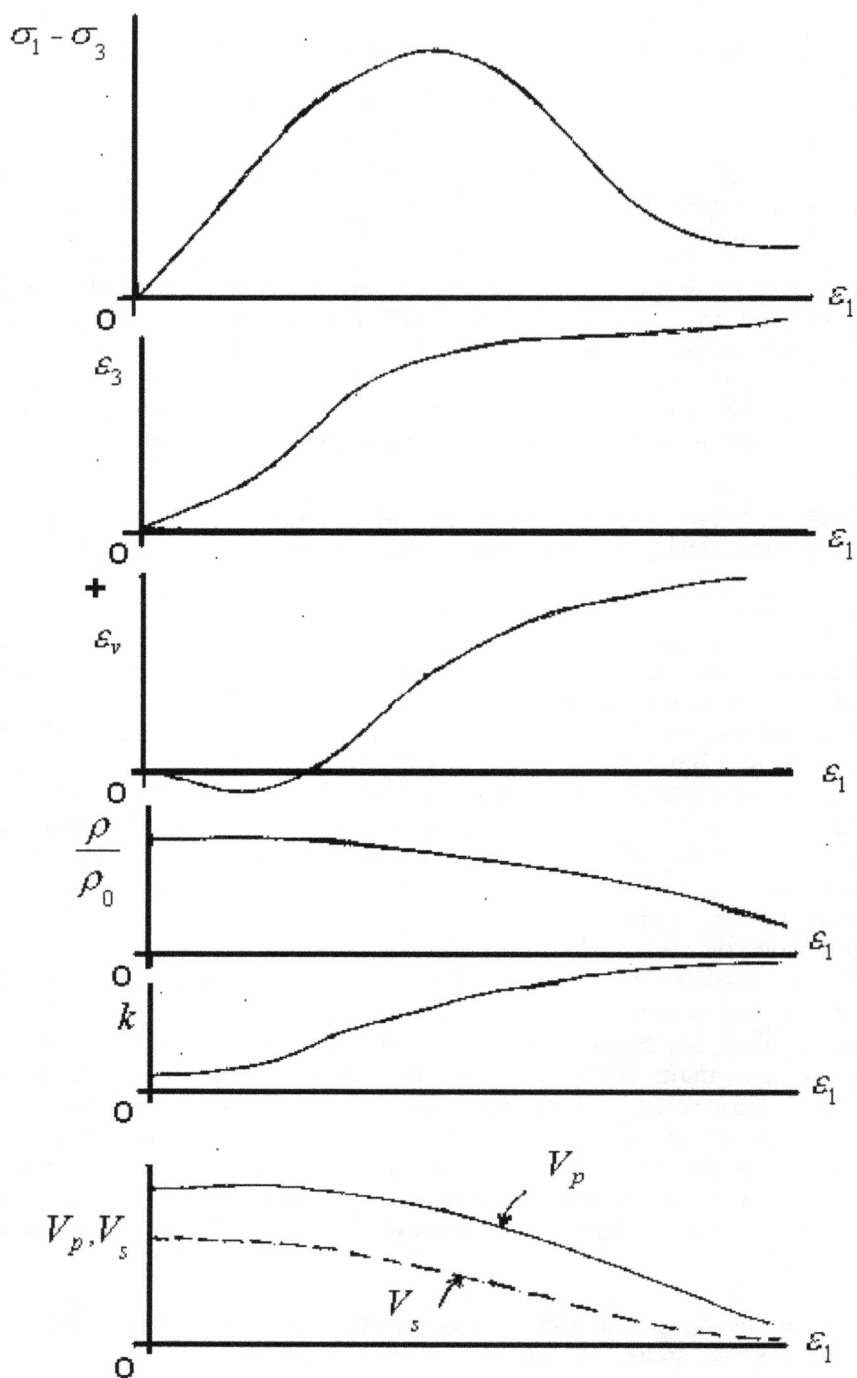

Figure 10.1 Variation in the properties of a rock under triaxial compression test.

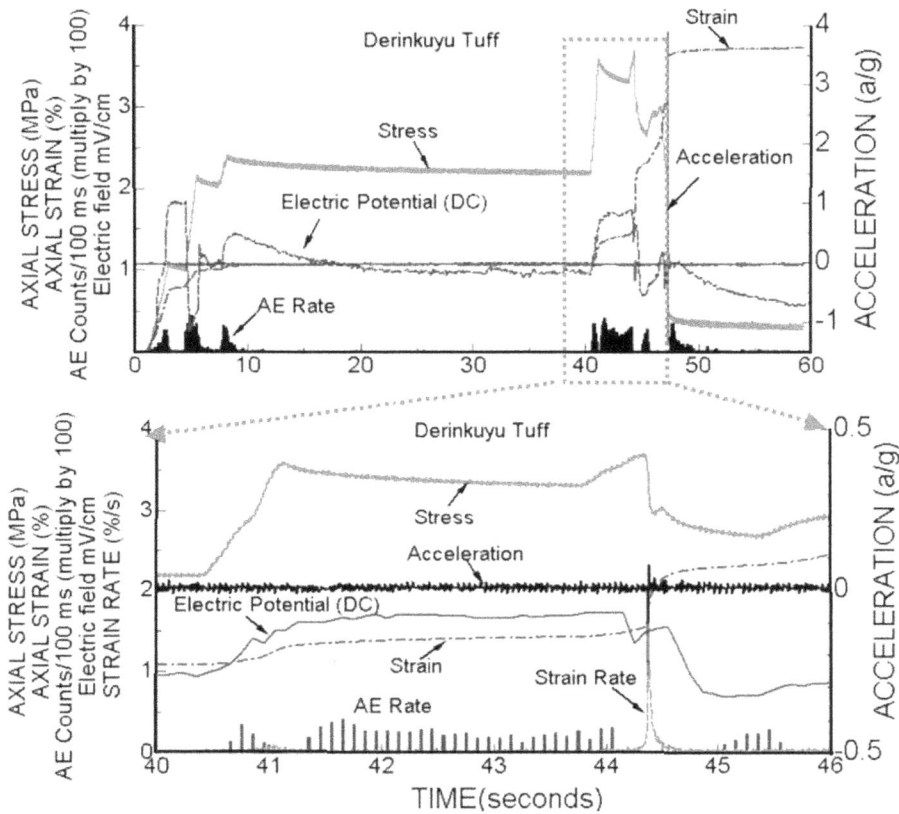

Figure 10.2 Multi-parameter response of Derinkuyu tuff during deformation and rupturing. (From Aydan et al. (2011).)

electrons of minerals and momentum as a result of deformation and inter-crystal
or inter-grain separation and/or sliding during dislocations as a result of fracturing or sliding due to momentum imposed.
• The amplitude of accelerations of the mobile part of the loading system is higher
than that of the stationary part. This feature has striking similarities with the
strong motion records nearby earthquake faults observed in recent high-magni-
tude in-land earthquakes.
• The amplitude of accelerations during the fracturing of rocks is directly proportional to the energy stored in samples before the fracturing.
• The experimental results are considered to be very important to both engineers,
for predicting potential rock bursting in deep high-level waste-disposal projects
and mining, and scientists who are closely associated with earthquake prediction
projects based on multi-parameter monitoring systems.

As described in Chapter 2 (Section 2.4.3.2), a stick-slip experimental device consisting of
an endless conveyor belt and a fixed frame was used to investigate the multi-parameter

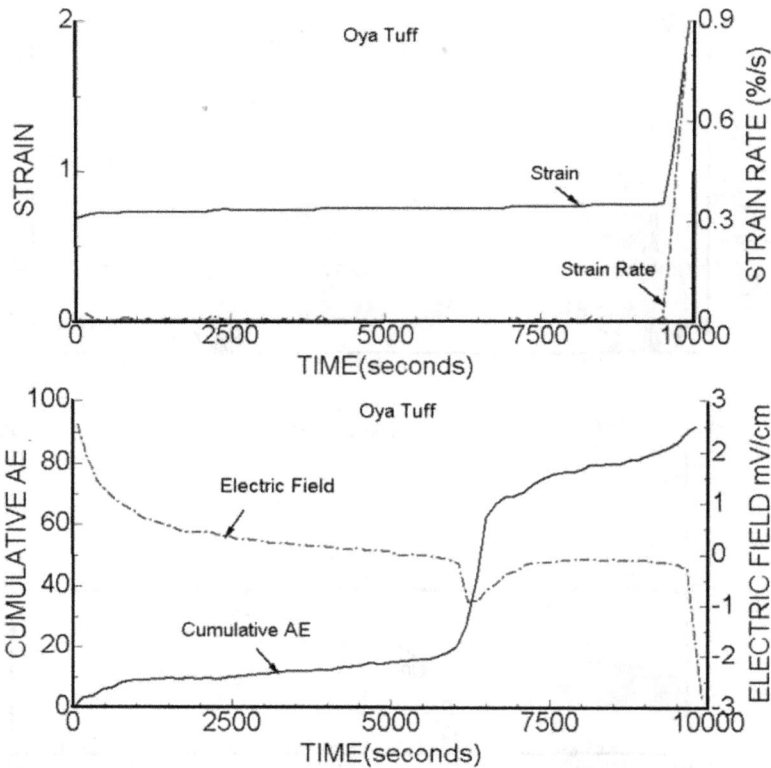

Figure 10.3 Multi-parameter response of Oya tuff during creep deformation and rupturing. (From Aydan et al. (2011).)

response of rock discontinuities (Ohta and Aydan 2010, Aydan 2017). The stick-slip experiment shown in Figures 2.20–2.23 given in Chapter 2 (Section 2.4.3.2) can provide very valuable insight into what happens during stick and slip phases during experiments. The responses shown in Figures 2.20–2.23 may be of great significance during the interpretation of crustal deformations by the GPS method, particularly (Aydan et al. 2000b).

10.3   AVAILABLE METHODS FOR EARTHQUAKE PREDICTION

Currently available earthquake prediction methods are mainly observational and they are basically too empirical (e.g. Aggarwall et al. 1973; Barsukov and Sorokin 1973; Buskirk et al. 1981; Derr 1973; Finkelstein et al. 1973; Geller 1997; Igarashi et al. 1995; Ikeya et al. 1997; Kondo 1968; Mizutani et al. 1976; Pierce 1976; Sato et al. 1995; Sultankhodzhaev 1984; Toksöz et al. 1975; Tsunogai and Wakita 1995; Uyeda et al. 1999; Yasui 1973). Figure 10.4 shows the basic concepts of the models for earthquake prediction adopted in the 1970s.

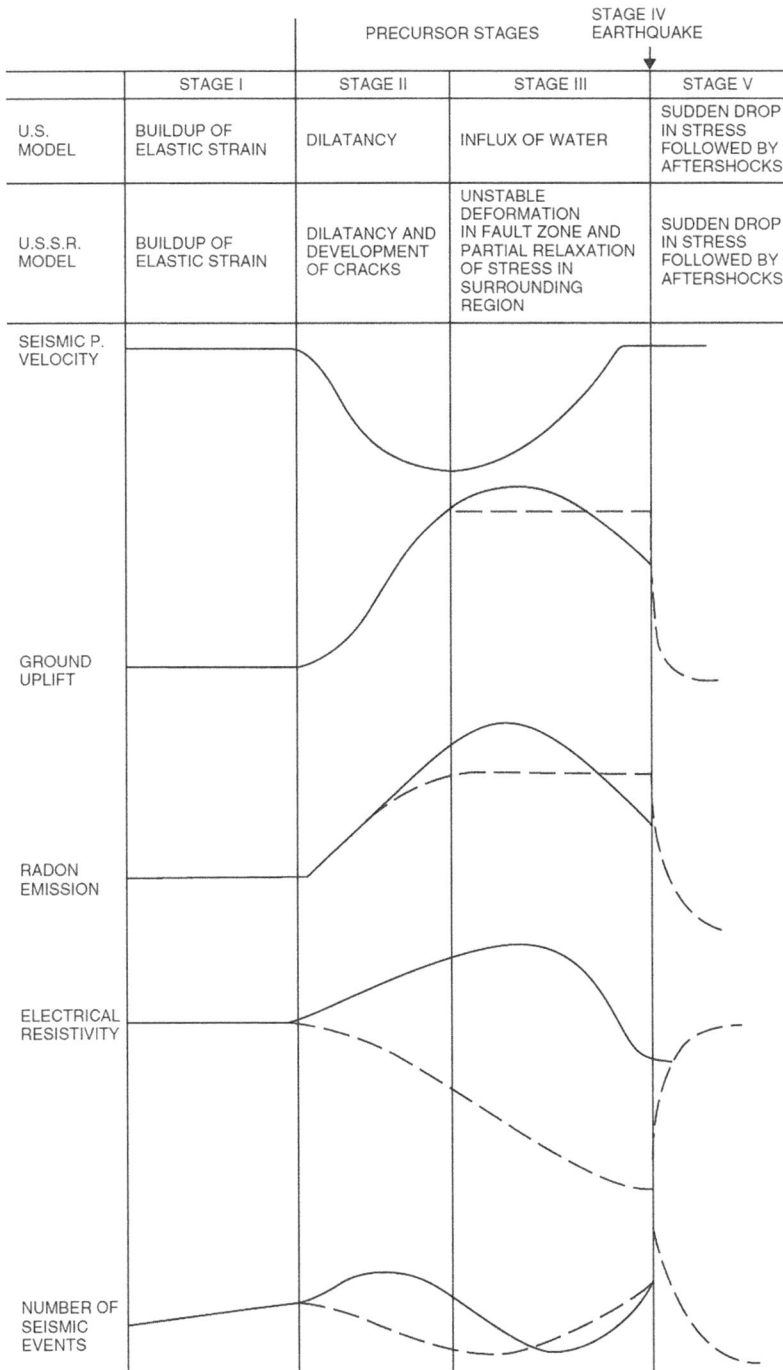

	STAGE I	STAGE II	STAGE III	STAGE V
U.S. MODEL	BUILDUP OF ELASTIC STRAIN	DILATANCY	INFLUX OF WATER	SUDDEN DROP IN STRESS FOLLOWED BY AFTERSHOCKS
U.S.S.R. MODEL	BUILDUP OF ELASTIC STRAIN	DILATANCY AND DEVELOPMENT OF CRACKS	UNSTABLE DEFORMATION IN FAULT ZONE AND PARTIAL RELAXATION OF STRESS IN SURROUNDING REGION	SUDDEN DROP IN STRESS FOLLOWED BY AFTERSHOCKS

PRECURSOR STAGES

STAGE IV
EARTHQUAKE

SEISMIC P. VELOCITY

GROUND UPLIFT

RADON EMISSION

ELECTRICAL RESISTIVITY

NUMBER OF SEISMIC EVENTS

Figure 10.4 The basic concepts of earthquake prediction models of United States and the former USSR in the 1970s. (After Toksöz (1977).)

The methods available may be categorized as follows:

1. Tilting or ground deformation anomaly method
2. Creep method
3. Groundwater level anomaly method
4. Elastic wave velocity anomaly method
5. Electrical resistivity anomaly method
6. Electric field anomaly method
7. Magnetic field anomaly method
8. Seismic gap method
9. Gas emission anomaly method
10. Gravity anomaly method
11. Anomalous animal behaviour method

Although Nur (1972) tried to unify some of these methods into a dilatancy-diffusion method, there is presently no worldwide accepted approach, based on a sound universal theory. In many sites, such as Parkfield in the United States and Tokai region in Japan, some of these methods are simultaneously used. The fundamental concepts of the methods listed above and their applications are briefly presented next.

10.3.1   Tilting or ground deformation anomaly method

This method is fundamentally based on the elastic rebound theory and it assumes that the ground rebounds after a certain amount of tilting or deformation as illustrated in Figure 10.5. The rebound may be uniform or nonuniform cyclic or random. This model is generally adopted for locations along subducting plate boundaries such as Japan, United States and Taiwan.

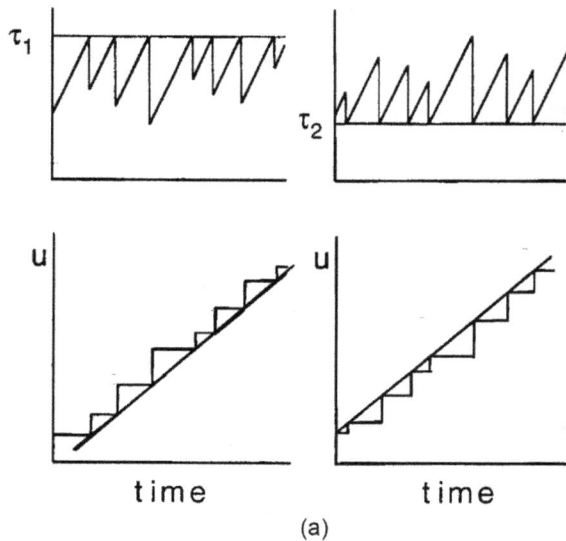

Figure 10.5 (a) Elastic rebound theory model and (b) ground tilting in the 1964 Niigata earthquake. (Arranged after Shimazaki – Nakata (1980) and Mogi (1985).)

(Continued)

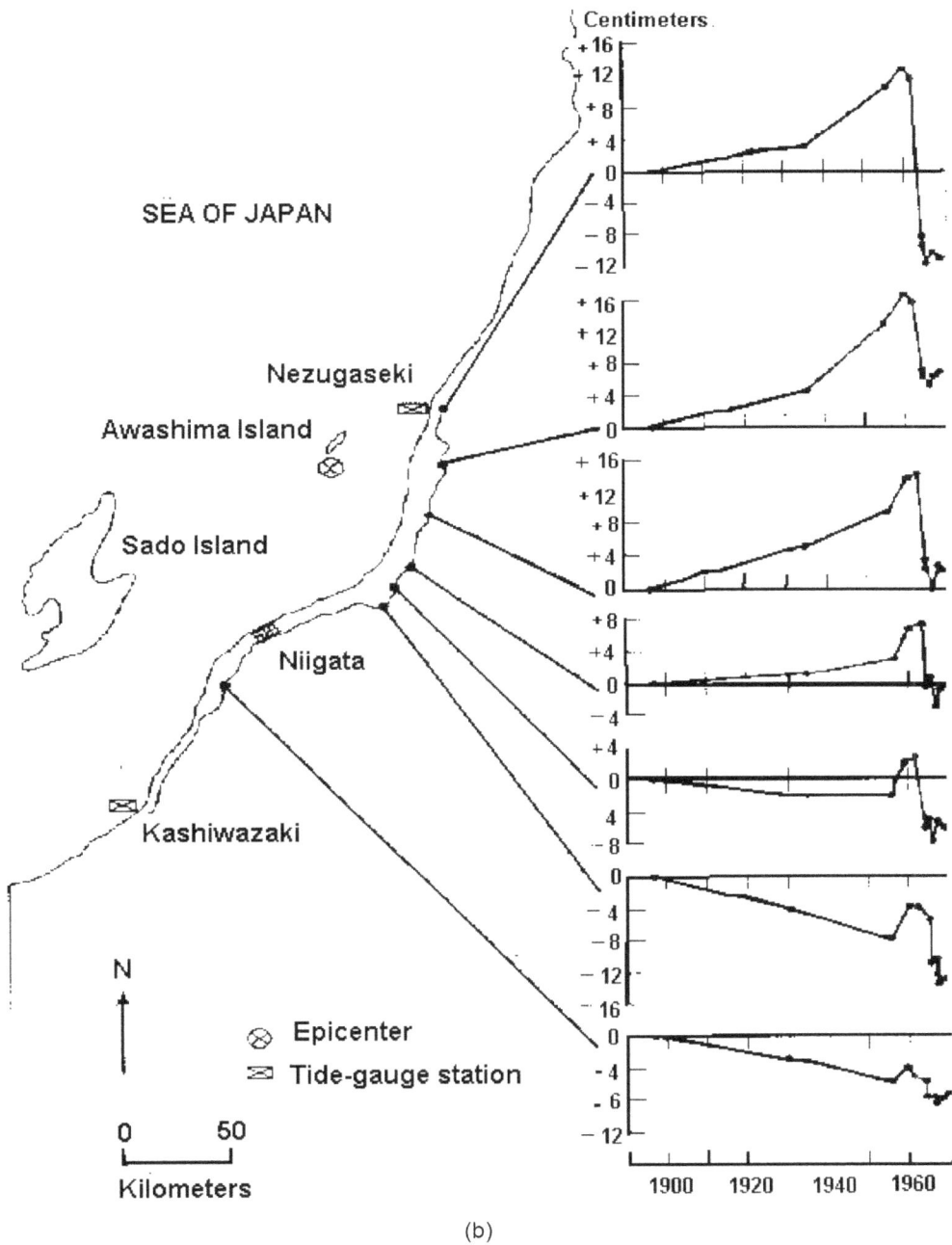

(b)

Figure 10.5 (Continued) (a) Elastic rebound theory model and (b) ground tilting in the 1964 Niigata earthquake. (Arranged after Shimazaki – Nakata (1980) and Mogi (1985).)

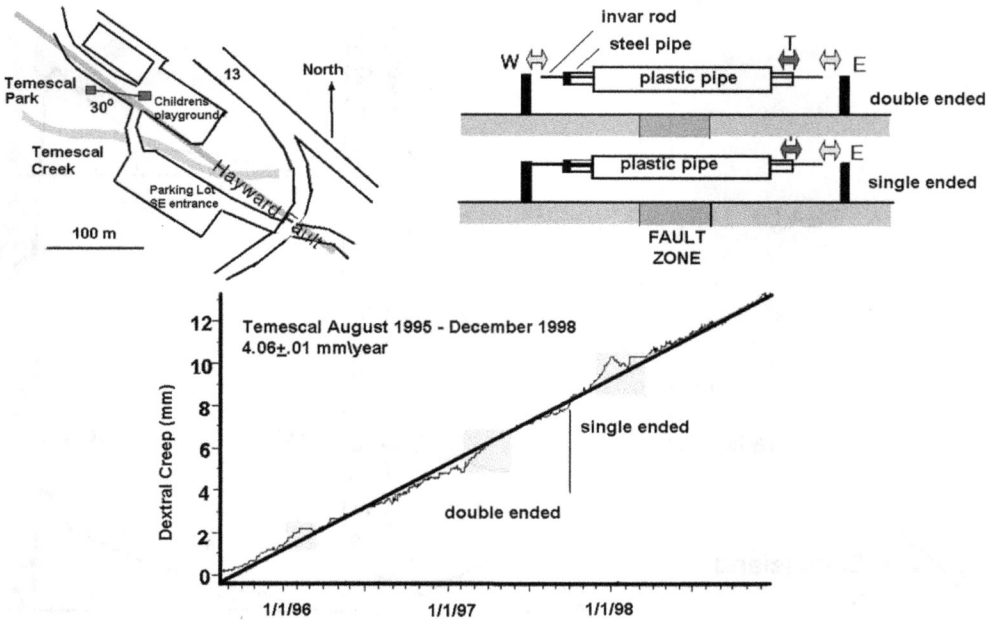

Figure 10.6 Creep response along Hayward Fault at Temescal Park, Oakland. (After USGS (1999a).)

10.3.2   Creep method

This method basically utilizes the creep models developed in laboratory, and it is applied to observe the behaviour of Hayward fault within the San Andreas Fault zone through the San Francisco Bay Area in the United States. Figure 10.6 shows a creep response of the Temescal Park station in Oakland (USA) along the Hayward fault. The creep behaviour of faults is also reported elsewhere (Aytun 1973, 1982; Barka 1996; Cakir et al. 2005). Figure 10.7 shows two examples of creep behaviour of Gediz fault at Sarıgöl and North Anadolu Fault at İsmetpaşa. The creep behaviour was already reported by Aytun (1982) for İsmetpaşa with an average displacement rate range of 0.9–1.1 cm/year at İsmetpaşa for a 44-year period. The displacement rate at Sarıgöl was also estimated to be about 1 cm/year, which is very close to that measured at İsmetpaşa along the North Anadolu Fault.

10.3.3   Groundwater level anomaly method

It is one of the most widely used methods in many earthquake prediction studies. Groundwater level may change due to tilting, straining and permeability change, and a wide area around the potential epicentre may respond to groundwater level variations. Figure 10.8 shows the groundwater level response of Cholame Hills along the San Andreas Fault to the large earthquakes between 1988 and 1995. As seen from the

Figure 10.7 Deformation of walls, fences and railways due to fault creep. (a) Gediz Fault at Sayısal Grafik (2004) and (b) North Anadolu Fault at İsmetpaşa.

figure, there is a good correlation between groundwater level variation and earthquakes. Following an earthquake, it is also known that some wells dry up or some new springs develop as a result of permeability change of the ground and new path developments. Sato et al. (1995) reported a new springs appeared in Awaji island after the 1995 Hyogo-ken Nanbu earthquake. Aydan and Hamada (1992), Aydan and Kumsar (1997a), Aydan et al. (1998) and Ulusay et al. (2003a) made similar observations in the epicentral areas of the 1992 Erzincan, 1995 Dinar and 2002 Çay-Eber earthquakes as seen in Figure 10.9. In addition, petroleum was discharged from ground in a field near Soysallı village of Ceyhan during the 1998 Adana-Ceyhan earthquake as also seen in Figure 10.9.

10.3.4 Elastic wave velocity anomaly method

This method is based on the anomaly of V_p/V_s ratio of the epicentral region. Before the earthquake, the ratio decreases and then it tends to recover its previous value following the earthquake. It was first applied to the earthquakes in the Garm region of Tajikistan (Figure 10.10). The same method was also applied in the prediction of the 1971 Blue Mountain Lake earthquake by Aggarwall et al. (1973). It should be noted that to measure the elastic wave velocities of a region under investigation, some artificial or natural seismic sources are required. In other words, the real-time measurement of the variations in wave velocities of the observation stations would not be easy.

CHOLAME HILLS WATER LEVEL

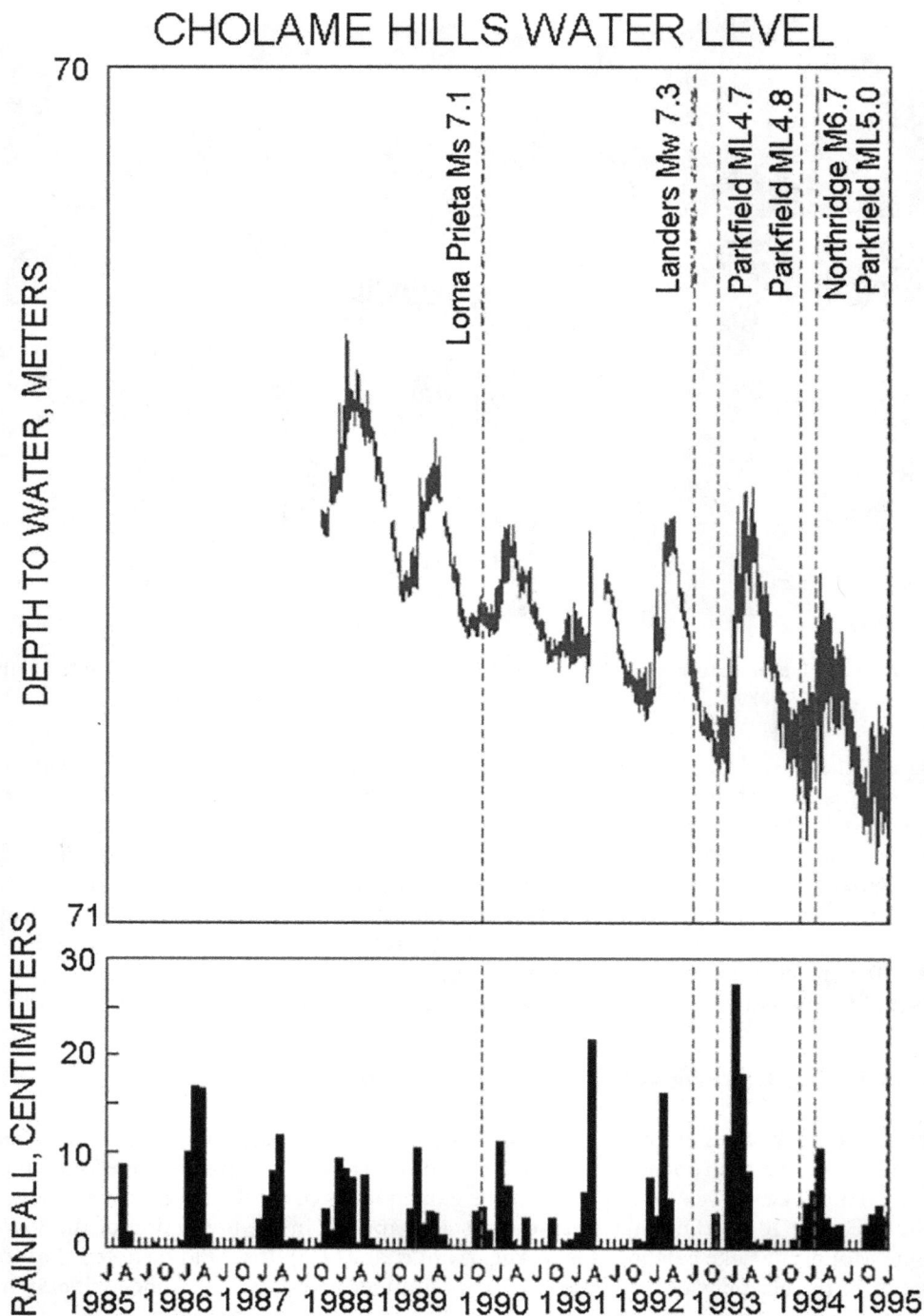

Figure 10.8 Groundwater level response of Cholame Hills along the San Andreas Fault to the large earthquakes between 1988 and 1995 (USGS 1999b).

Figure 10.9 New springs and petroleum discharge due to recent earthquakes in Turkey.

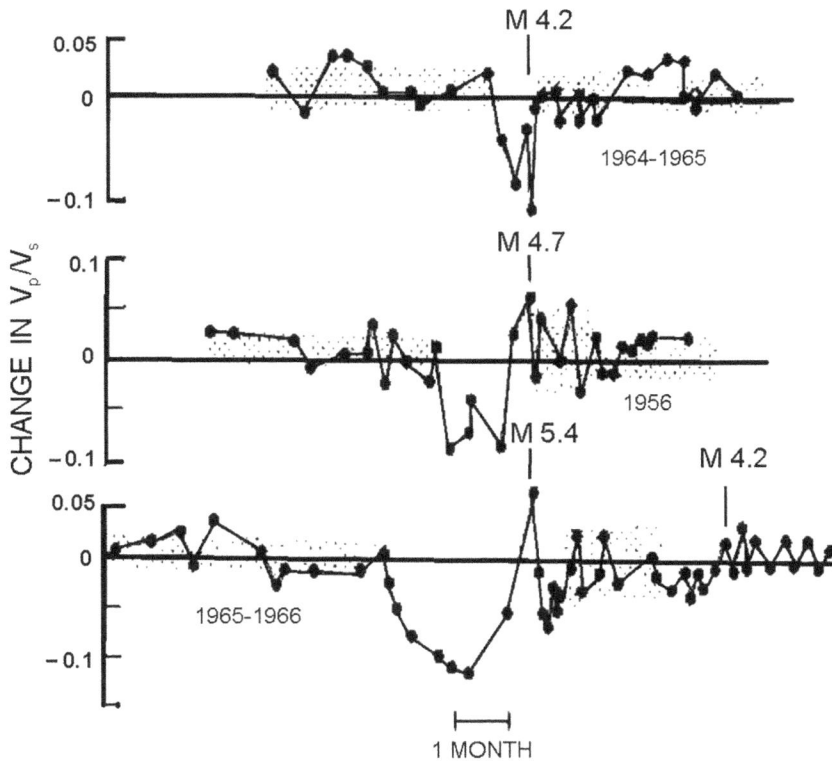

Figure 10.10 Application of elastic wave velocity anomaly method to earthquake prediction in Garm region of Tajikistan. (After Barsukov and Sorokin (1973).)

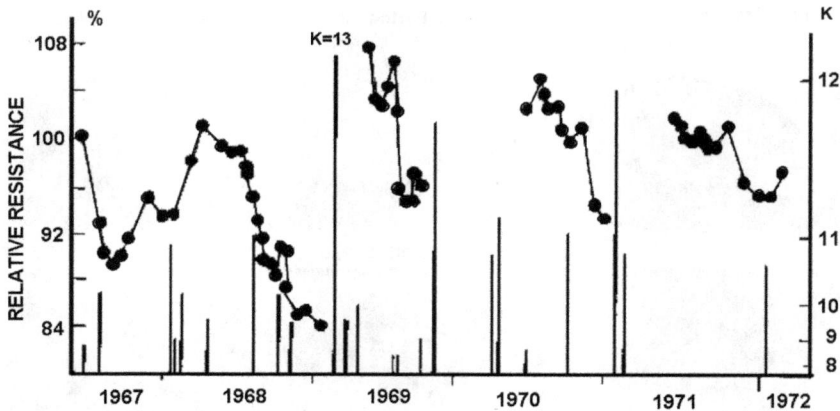

Figure 10.11 Relation between resistivity variation and earthquake occurrence in Garm region in Tajikistan. (After Barsukov and Sorokin (1973).)

10.3.5 Electrical resistivity anomaly method

It is also assumed that the electrical resistivity of Earth's crust decreases and then recovers its previous value in a similar manner to the elastic wave velocity anomaly. This method was first applied in the Pamir region of Tajikistan by Barsukov and Sorokin (1973) (see Figure 10.11). The cause of the reduction in electrical resistivity is assumed to be associated with the fracturing of rock and the filling of the fracture by less-resistant groundwater.

10.3.6 Electric field anomaly method

When rock fractures, some of the work done is transformed to pulsed electric charges due to piezoelectric characteristics of quartz crystals in rock, streaming fluid flow and/or frictional heating (Kondo 1968; Finkelstein et al. 1973; Mizutani et al. 1976; Lockner et al. 1983). This phenomenon is also regarded as the cause of earthquake lightning and fireballs before, during and after earthquakes (Yasui 1973; Derr 1973; Pierce 1976). Figure 10.12 shows examples of earthquake lightning and earthquake clouds observed before the 1965 Matsushiro earthquakes and the 1995 Hyogo-ken Nanbu (Kobe) earthquake.

Such pulsed electric charges are well known and they were recently named seismic electrical signals (SES) in the VAN method developed for earthquake prediction (Varotsos and Alexopolous, 1984). However, the validity of the VAN method is severely criticized by some geoscientists (see for example, Geller 1997) as its signal discrimination from background noise caused by other means is not objective. Nevertheless, this method is now under trial for application to earthquake prediction by Uyeda et al. (1999) in Japan.

(a)

(b)

(c)

(d)

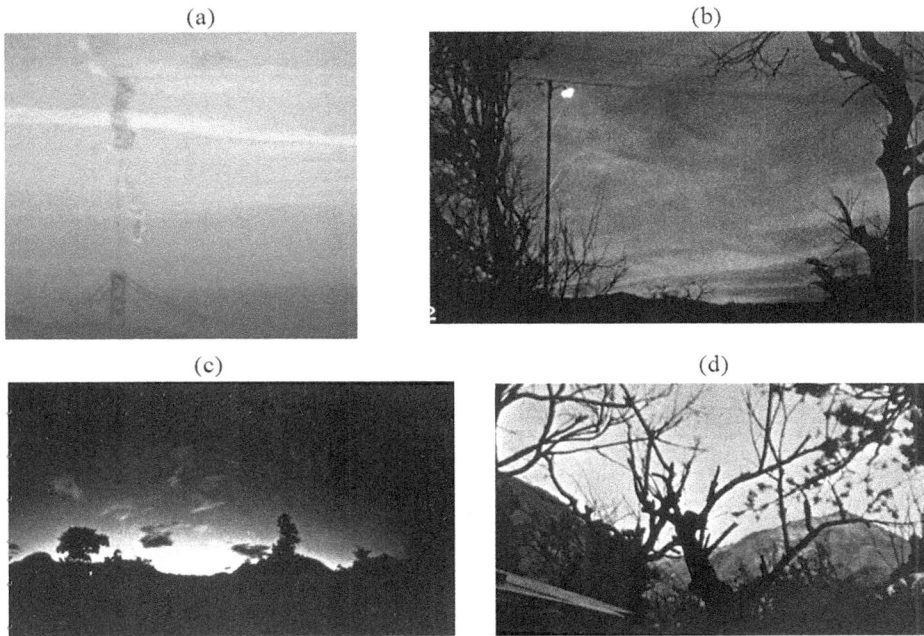

Figure 10.12 Earthquake lights and clouds observed in Japan: (a) 1995 Kobe earthquake; (b–d) 1965 Matsushiro Earthquake. (After Yasui (1973).)

10.3.7   Magnetic field anomaly method

Magnetic and electric field anomalies are closely related to each other, and both fields are interrelated through Maxwell's equations. Therefore, causes of magnetic field anomalies should be the same as those of the electric field. Figure 10.13 shows the variation in a magnetic anomaly observed in the 1989 Loma Prieta earthquake. Such magnetic field anomalies may have numerous effects on modern electronic devices. Ikeya and his group investigated the effects of this phenomenon on various electronic devices from both experimental and theoretical points of view (Ikeya and Matsumoto 1997; Matsumoto et al. 1998).

10.3.8   Seismic gap method

This method is based on the quiescence of the seismic activity of a known fault zone. To apply this method, the seismic activity of the region must be observed for a long period of time. The Parkfield site in the United States and Tokai region in Japan were selected for the seismic watch studies. For example, Toksöz et al. (1975) discussed earthquake migration from Eastern Turkey towards Western Turkey along the North Anadolu Fault. The 1999 Kocaeli and Düzce earthquakes were recognized as evidence of the seismic gap hypothesis (Figure 10.14). However, the duration of earthquake prediction

Earthquake

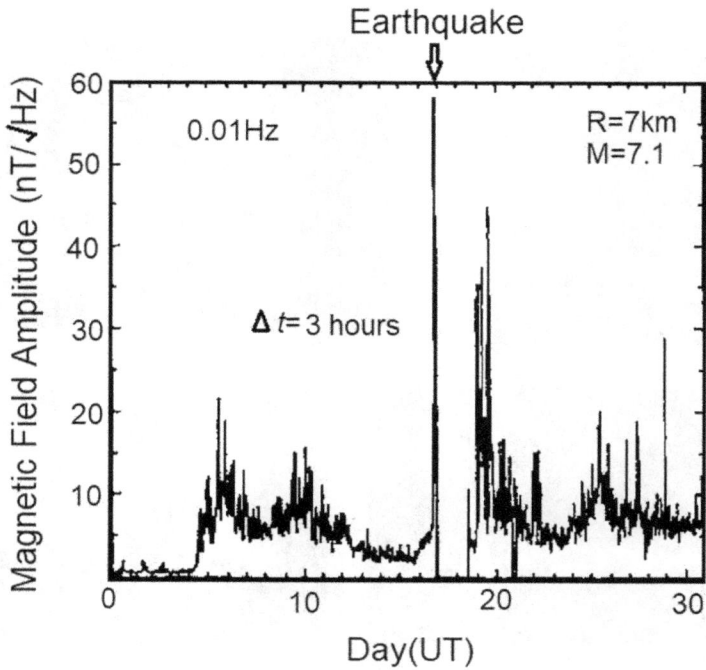

Figure 10.13 Magnetic field variation in the 1989 Loma Prieta earthquake.

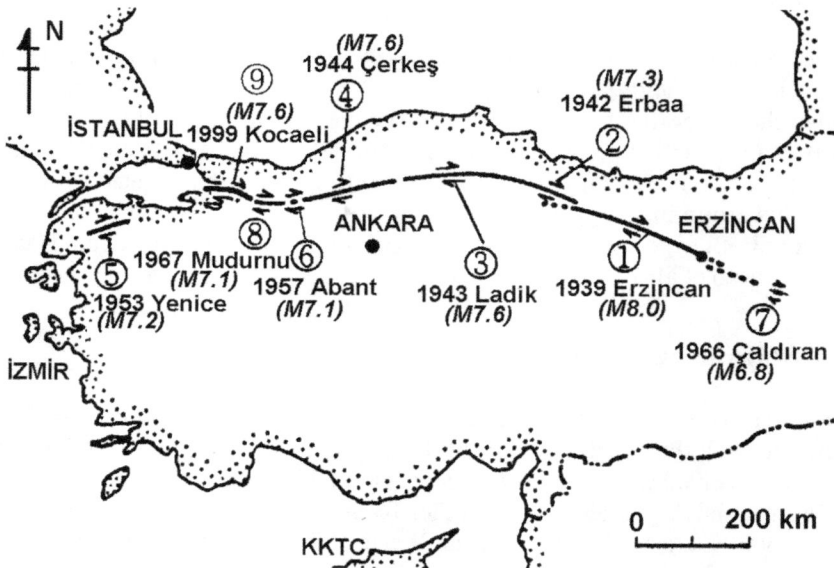

Figure 10.14 Earthquakes with surface ruptures between 1910 and 1999 in Turkey.

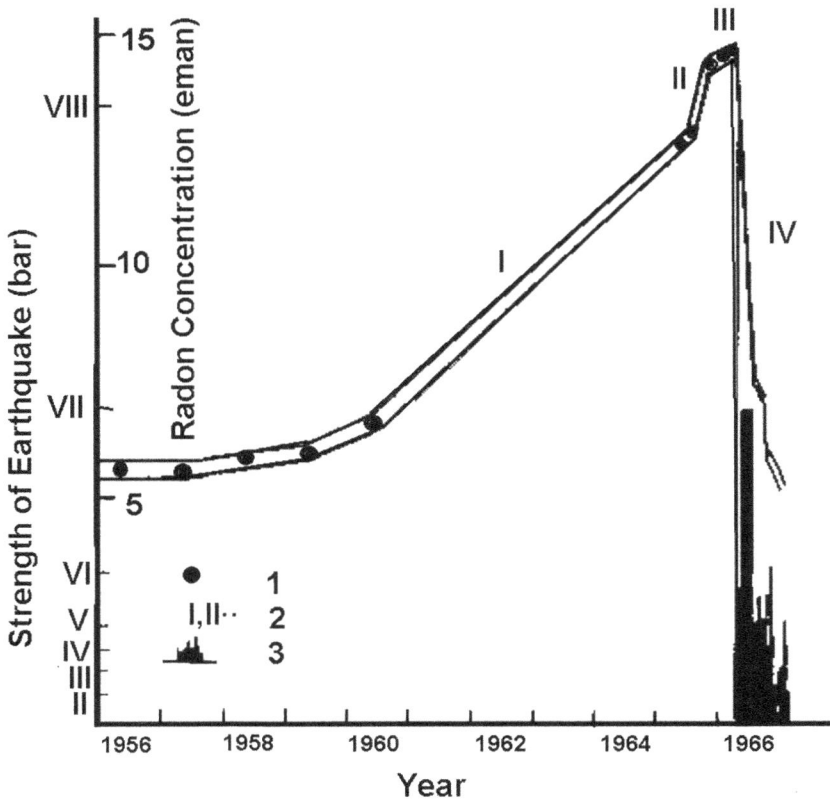

Figure 10.15 Radon concentration in the 1966 Taşkent earthquake.

may be categorized as long-term prediction. The other methods would generally be used after a site is designated as a seismic gap by this method.

10.3.9   Gas emission anomaly method

It is known that various gases are emitted during the rupture process of rocks. The most easily measured gas is radon, which is a radioactive gas with a half-life decay of 3.8 days. Radon emitted during the rupture process of rock is conducted to the free surface mostly through groundwater. The first known application of radon gas measurements for earthquake prediction was for the 1966 Taşkent (Tashkent) earthquake (Sultankhodzhaev 1984) in Uzbekistan (Figure 10.15). It is also known that gases such as He, Ar, NO, CO_2 and ions such as Cl^{-1}, SO_4^{2-} may also be emitted (Sultankhodzhaev 1984; Tsunogai and Wakita 1995).

10.3.10   Gravity anomaly method

As Earth's crust is strained, its density changes. As a result of this phenomenon, Walsh (1975) theoretically showed that gravity anomalies may take place. In addition to these, if there are material variations beneath the potential fault zone, gravity changes would be added to that resulting from ground deformation.

10.3.11   Anomalous animal behaviour method

Many scientists were suspicious of various reports on anomalous animal behaviour observed before earthquakes. However, it was one of the decisive elements in the successful prediction of the timing of the 1975 Haicheng earthquake. Although it has been a well-known phenomenon for centuries, it did not receive enough attention from scientists. The USGS organized a symposium on anomalous animal behaviour in 1976. Since then some additional studies were undertaken in China, Japan and the United States. Recent studies on this theme conducted by Ikeya and his group following the 1995 Kobe earthquake are quite promising (Ikeya et al. 1997, 1998). They were able to explain why animals behave anomalously before earthquakes through the disturbance caused by electromagnetic waves, to which animals are most sensitive. Another good piece of work on this theme was carried out by Buskirk et al. (1981). Figure 10.16

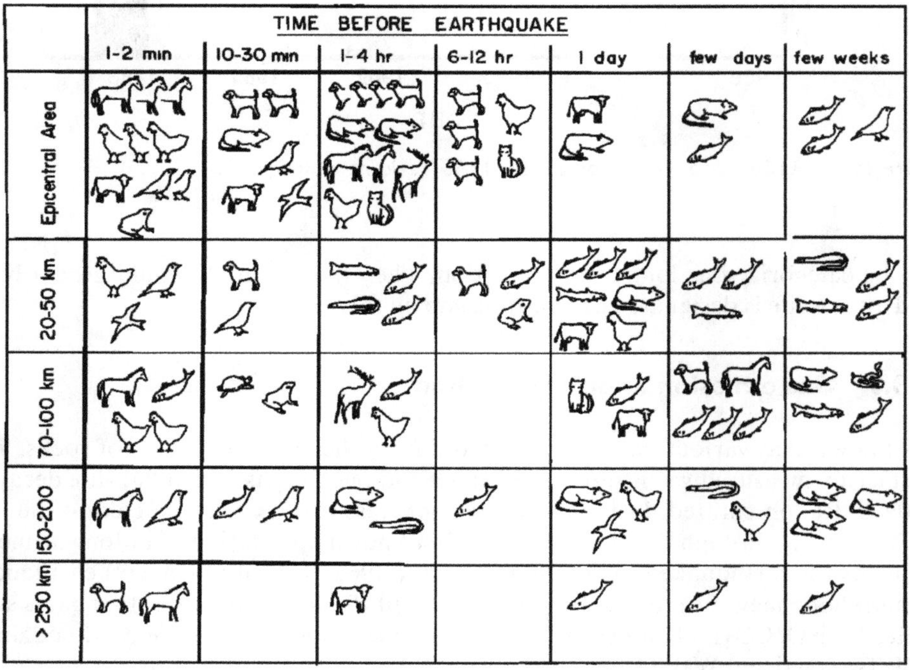

Figure 10.16 Distribution of animal behaviour incidents according to the distance from the epicentre and the time before the main shock. (After Buskirk et al. (1981).)

shows how and which animals display anomalous behaviour before earthquake in a time span.

It should be realized that any earthquake prediction method must be able to predict the location, time and magnitude of the earthquake. If these requirements are not met it cannot be called a proper earthquake prediction method. Therefore, the methods listed above do not still satisfy the requirements. Nevertheless, they may prove useful in developing an integrated earthquake prediction method.

10.4 GLOBAL POSITIONING METHOD FOR EARTHQUAKE PREDICTION

As stated previously, if the stress state and the yielding characteristics of Earth's crust are known at a given time, one may be able to predict earthquakes with the help of some mechanical, numerical and instrumental tools. The GPS method may be used to monitor the deformation of Earth's crust continuously with time. From these measurements, one may compute the strain rates and probably the stress rates. The stress rates derived from the GPS displacement rates can be effectively used to locate areas with high seismic risk as proposed by Aydan et al. (2000e). Thus, daily variations of derived stress-strain rates from dense, continuously operating GPS networks in Japan and the United States may provide high-quality data to understand the behaviour of Earth's crust preceding earthquakes.

10.4.1 Theoretical background

First we describe a brief outline of the GPS method proposed by Aydan (2000b, 2003a, 2004b, 2006c). The crustal strain rate components can be related to the displacement rates at an observation point (x, y, z) through geometrical relations (i.e. Eringen 1980; Aydan 2021) as given below:

$$\dot{\varepsilon}_{xx} = \frac{\partial \dot{u}}{\partial x}; \; \dot{\varepsilon}_{yy} = \frac{\partial \dot{v}}{\partial y}; \; \dot{\varepsilon}_{zz} = \frac{\partial \dot{w}}{\partial y}; \; \dot{\gamma}_{xy} = \frac{\partial \dot{v}}{\partial x} + \frac{\partial \dot{u}}{\partial y}$$
$$\dot{\gamma}_{yz} = \frac{\partial \dot{w}}{\partial y} + \frac{\partial \dot{v}}{\partial z}; \; \dot{\gamma}_{zx} = \frac{\partial \dot{w}}{\partial x} + \frac{\partial \dot{u}}{\partial z}$$

(10.1)

where $\dot{u}$, $\dot{v}$ and $\dot{w}$ are displacement rates in the x, y and z directions, respectively. $\dot{\varepsilon}_{xx}$, $\dot{\varepsilon}_{yy}$ and $\dot{\varepsilon}_{zz}$ are strain rates normal to the x, y and z planes and $\dot{\gamma}_{xy}$, $\dot{\gamma}_{yz}$, $\dot{\gamma}_{zx}$ are engineering shear strain rates. GPS measurements can only provide the displacement rates on Earth's surface (x (EW) and y (NS) directions) and do not give any information on displacement rates in the z direction (radial direction). The strain rate components in the plane tangential to Earth's surface would be $\dot{\varepsilon}_{xx}$, $\dot{\varepsilon}_{yy}$ and $\dot{\gamma}_{xy}$. Additional strain rate components $\dot{\gamma}_{yz}$ and $\dot{\gamma}_{zx}$, which would be interpreted as tilting strain rate in this chapter, are defined by neglecting some components in order to make use of the third component of displacement rates measured by GPS as follows:

$$\dot{\gamma}_{zx} = \frac{\partial \dot{w}}{\partial x}; \; \dot{\gamma}_{zy} = \frac{\partial \dot{w}}{\partial y}$$

(10.2)

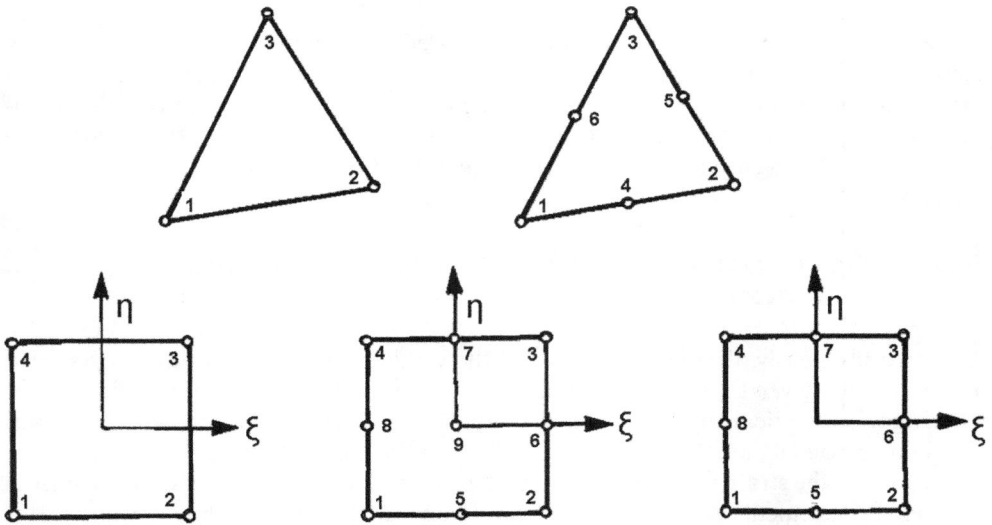

Figure 10.17 Finite elements for GPS method.

Let us assume that the GPS stations are rearranged in such a manner so that a mesh is constituted similar to the ones used in the finite element method. It is possible to use different elements as illustrated in Figure 10.17. Applying the interpolation technique used in the finite element method, the displacement in a typical element may be given in the following form for any chosen order of interpolation function:

$$\{\dot{u}\} = [N]\{\dot{U}\}$$
(10.3)

where $\{\dot{u}\}, [N]$ and $\{\dot{U}\}$ are the displacement rate vector of a given point in the element, shape function and nodal displacement vector, respectively. The order of shape function $[N]$ can be chosen depending upon the density of observation points. The use of linear interpolation functions is presented elsewhere (Aydan 2000b and 2003a).

From Eqs. (10.1), (10.2) and (10.3), one can easily show that the following relation holds among the components of the strain rate tensor of a given element and displacement rates at nodal points:

$$\{\dot{\varepsilon}\} = [B]\{\dot{U}\}$$
(10.4)

Using the strain rate tensor determined from the Eq. (10.4) for a given time interval, the stress rate tensor can be computed with use of a constitutive law such as Hooke's law for elastic materials, Newton's law for viscous materials and Kelvin's law for visco-elastic materials (Aydan and Nawrocki 1998). For simplicity, Hooke's law is chosen and is written in the following form:

$$
\left\{
\begin{array}{c}
\dot{\sigma}_{xx} \\
\dot{\sigma}_{yy} \\
\dot{\sigma}_{xy}
\end{array}
\right\}
=
\left[
\begin{array}{ccc}
\lambda+2\mu & \lambda & 0 \\
\lambda & \lambda+2\mu & 0 \\
0 & 0 & \mu
\end{array}
\right]
\left\{
\begin{array}{c}
\dot{\varepsilon}_{xx} \\
\dot{\varepsilon}_{yy} \\
\dot{\gamma}_{xy}
\end{array}
\right\}
\tag{10.5}
$$

where λ and μ are Lame's constants, which are generally assumed to be $\lambda=\mu=30\,\text{GPa}$ (Fowler 1990). It should be noted that the stress and strain rates in Eq. (10.5) are for the plane tangential to the Earth's surface. From the computed strain and stress rates, principal strain and stress rates and their orientations may be easily computed as an eigen value problem.

To identify the locations of earthquakes, one has to compare the stress state in Earth's crust at a given time with the yield criterion of the crust. The stress state is the sum of the stress at the start of GPS measurement and the increment from GPS-derived stress rate given as follows:

$$
\{\sigma\} = \{\sigma\}_0 + \int_{T_0}^{t} \{\dot{\sigma}\}dt
\tag{10.6}
$$

If the previous stress $\{\sigma\}_0$ is not known, a comparison for the identification of the location of the earthquake cannot be done. The previous stress state of Earth's crust is generally unknown. Therefore, Aydan et al. (2000b) proposed the use of maximum shear stress rate, mean stress rate and disturbing stress for identifying the potential locations of earthquakes. The maximum shear stress rate, mean stress rate and disturbing stress rate are defined below:

$$
\dot{\tau}_{\max} = \frac{\dot{\sigma}_1 - \dot{\sigma}_3}{2}; \quad \dot{\sigma}_m = \frac{\dot{\sigma}_1 + \dot{\sigma}_3}{2}; \quad \dot{\tau}_d = |\dot{\tau}_{\max}| + \beta\dot{\sigma}_m
\tag{10.7}
$$

where β may be regarded as a friction coefficient. It should be noted that one (vertical) of the principal stress rates is neglected in the above equation as it cannot be determined from GPS measurements. The definition of disturbing stress rate is analogous to the well-known Mohr–Coulomb yield criterion in geomechanics and geoengineering. The concentration locations of these quantities may be interpreted as the likely locations of earthquakes as they imply an increase in disturbing stress (Figure 10.18). If the mean stress has a tensile character and its value increases, it simply implies the reduction of resistance of the crust.

Figure 10.19 shows the flowchart for the implementation of the procedure described above. The computation programs are coded in FORTRAN and True BASIC programming languages. The computed results can be visualized through the embedded graphical programs in the codes or other visualization software.

10.4.2 Applications

Earthquake prediction involves three fundamental parameters: location, time and magnitude. If any of these parameters cannot be predicted, the prediction cannot be

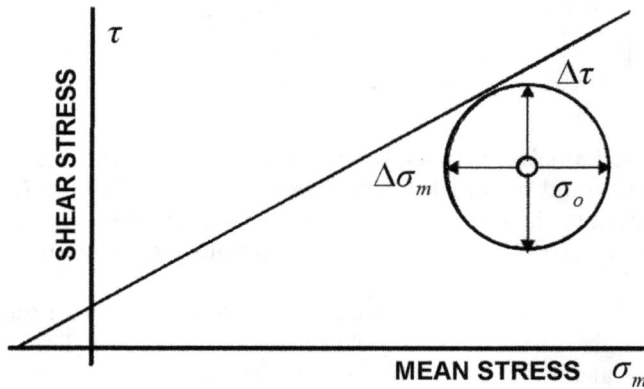

Figure 10.18 Illustration of stress rates in the space of mean and shear stresses.

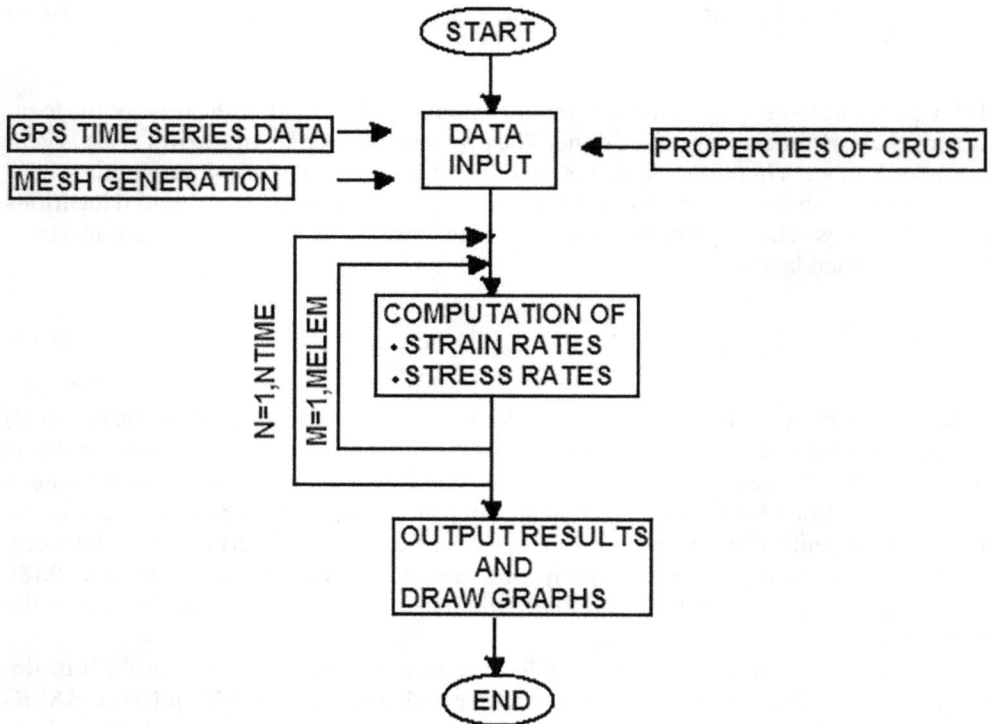

Figure 10.19 Flowchart of computation codes.

claimed to be true. In this section, applications of the GPS technique to some specific countries are described in view of the three fundamental parameters of earthquake prediction. The earthquakes plotted in this section are mainly from the catalogue of the USGS.

10.4.2.1 Prediction of earthquake epicentres

10.4.2.1.1 TURKEY

As Aydan (2000b) has shown previously, recent earthquakes in Turkey fall within the maximum shear stress concentration regions. Similarly, close correlations exist between mean stress rate and disturbing stress rate concentrations and epicentres of the earthquakes. Therefore, the concentrations of maximum shear stress rate and disturbing stress rate may serve as indicators for identifying the location of potential earthquakes. The high mean stress rate of a tensile characteristic may also be used to identify likely earthquakes due to normal faults (Aydan 2000b). Figure 10.20 shows the contours of disturbing stress rate together with the epicentres of earthquakes with a magnitude greater than 4 that occurred during 1995 and 1999 using the GPS data reported by Reilinger et al. (1997). Particularly the epicentres of the 1999 Kocaeli, 1999 Düzce-Bolu, 2000 Orta-Çankırı and 2000 Honaz-Denizli earthquakes coincide with the regions of concentration of these stress rates. Therefore, the GPS method implies that it is possible to locate earthquakes.

10.4.2.1.2 TAIWAN

Yu et al. (1997) reported the annual crustal displacement rate (velocity) as shown in Figure 10.21a. Figure 10.21b shows the disturbing stress rate contours together with earthquake magnitudes greater than 6 until 2000. As noted from the figure, earthquake activity is very high in areas where stress concentrations occur. The M7.4 1999 Chi-Chi earthquake occurred in one of such areas as indicated in Figure 10.21b.

10.4.2.1.3 JAPAN

Japan has the most extensive network of continuous GPS, called GEONET. An evaluation of GPS measurements by this network for 2003 is shown in Figure 10.22. As noted from this figure, high stress concentrations occur along the eastern shore of Japan compared to that along the western shore. The seismic activities along the east coasts of Hokkaido Island in 2003 (M8.3 Tokachi earthquake), Honshu Island in 2004 and 2011 (M7.6 Tokaido-oki earthquake; M9.0 East Japan Mega Earthquake) and Suruga Bay earthquake in 2008 in the area of the anticipated Tokai earthquake coincide with largest concentrations.

10.4.2.1.4 INDONESIA

Indonesia has suffered many large earthquakes along Sumatra Island and Java Island since 2004. Figure 10.23 shows the seismic activity and locations of the major large earthquakes. Figure 10.24 shows the contours of disturbing stress rate obtained from GPS stations in the region bounded by Latitudes 15S–15N and Longitudes 90E–140E (Aydan 2008b). As noted from the figures, stress rate concentrations are clearly observed in the Moluccas area (Banda Sea). Concentrations in the vicinity of Sunda

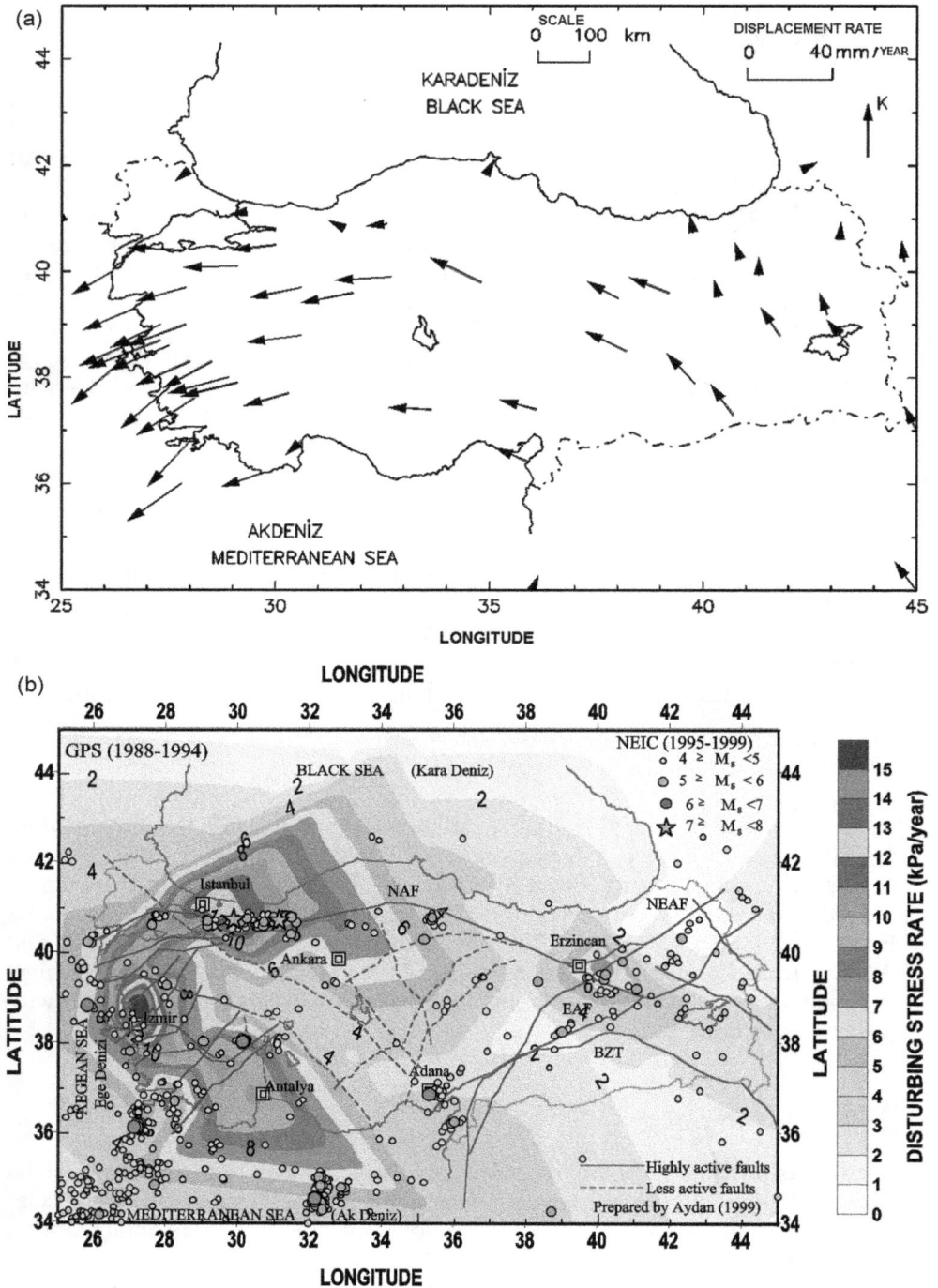

Figure 10.20 Computed disturbing stress rate and earthquakes. (a) Displacement rate and (b) disturbing stress rate.

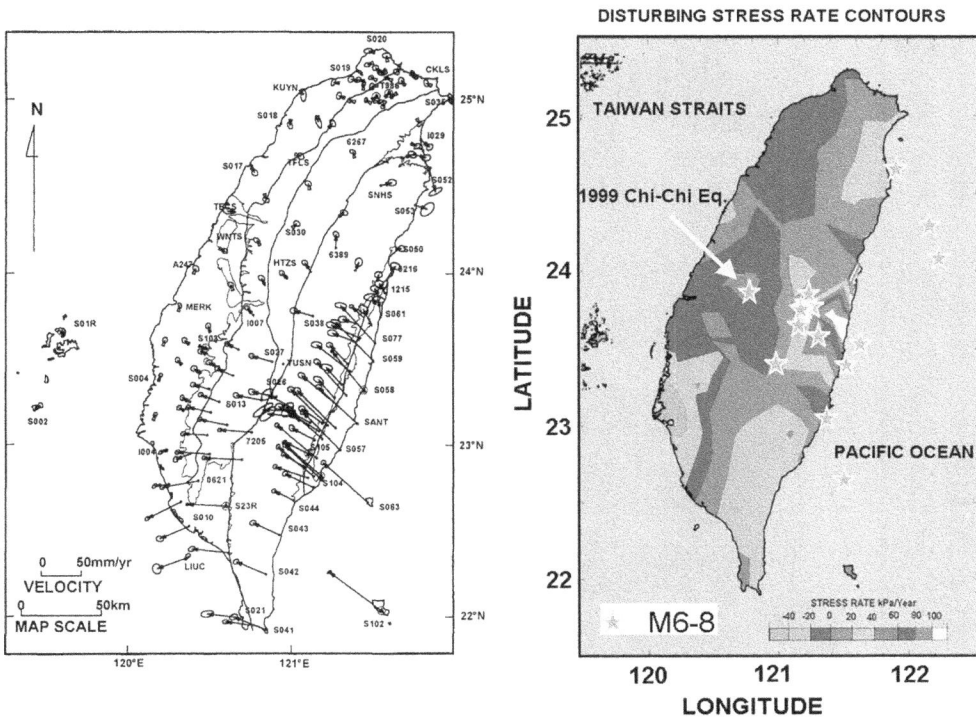

Figure 10.21 Displacement rate vectors and disturbing stress rate contours with earthquakes.

Strait and west of Sumatra Island are worth noticing. However, it should be noted that the GPS stations in the west of Sumatra Island are sparse. Therefore it is expected that the actual concentrations may be larger than those seen in Figure 10.24. Nevertheless, it is of great interest that the stress rate concentrations are closely associated with regional seismicity.

10.4.2.1.5   GPS STATIONS IN AFGHANISTAN AND ITS VICINITY

Compared to other countries, Afghanistan has only single GPS station located in Kabul. However, it is almost impossible to make a detailed analyses of crustal strains in Afghanistan. Nevertheless, the available data of the GPS station in Kabul and other GPS stations in the neighbouring countries would be used in this section to evaluate crustal deformation and straining in Afghanistan and its close vicinity. Furthermore, their relation to seismicity is discussed.

10.4.2.1.5.1   GPS stations and deformation rates
As pointed out, there is only one GPS monitoring station in Afghanistan. Table 10.1 gives the code, coordinates and annual horizontal deformation velocity of GPS stations

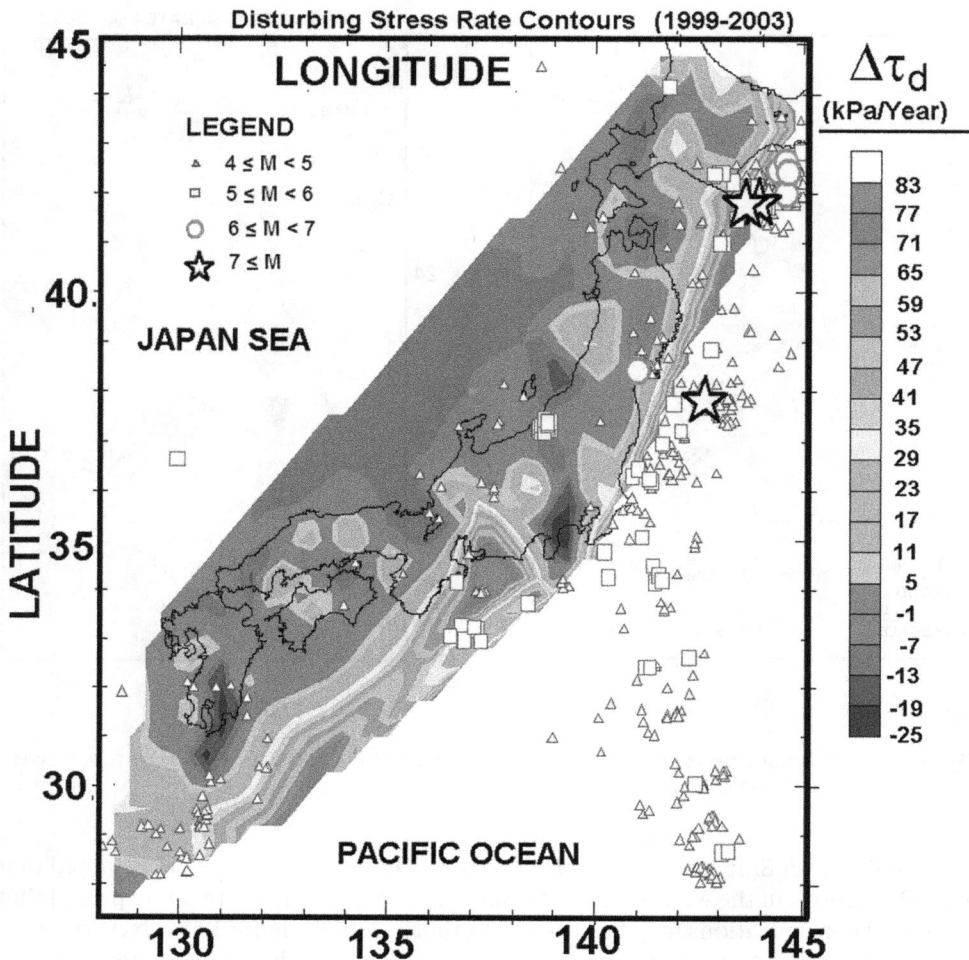

Figure 10.22 Comparison of disturbing stress contours of 2003 with earthquake activity. (a) Displacement rate vectors and (b) disturbing stress rate contours.

together with sources. Figure 10.25 shows the locations of the stations and annual deformation velocities at the respective stations. The velocities given in Table 10.1 are computed with respect to the Euro-Asian plate.

10.4.2.1.5.2   Strain rates
The method presented is utilized to compute the strain rates. Figure 10.26 shows the annual strain rates in Afghanistan and neighbouring regions (blue: compression; red: extension). Crustal straining is high along the major tectonic structures such as the Pamir and Hindu Kush Subduction zones, the Chaman Fault, The Makran subduction zone and Sistan Suture Zone (SSZ), while the annual strain in central Afghanistan (Helmand Block) is quite small.

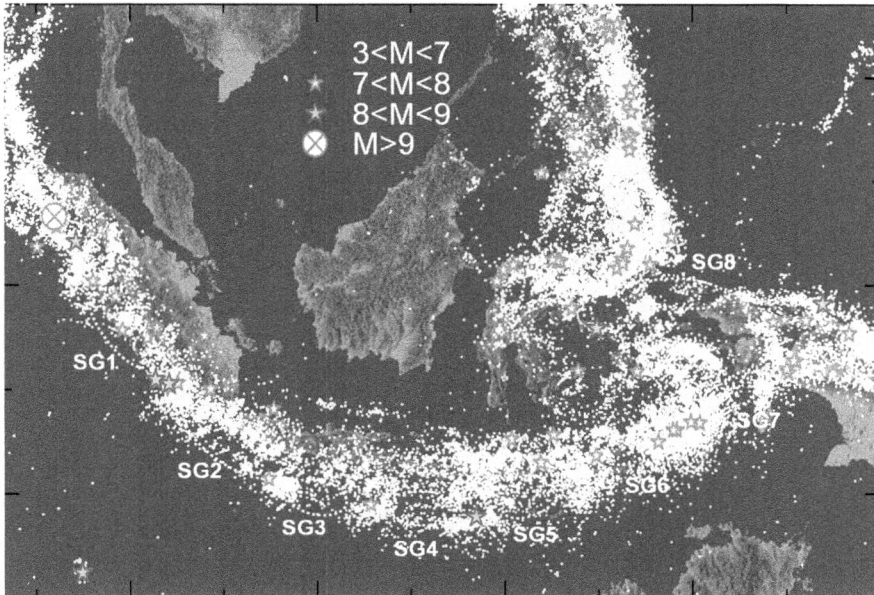

Figure 10.23 Distribution of epicentres of earthquakes and seismic gaps.

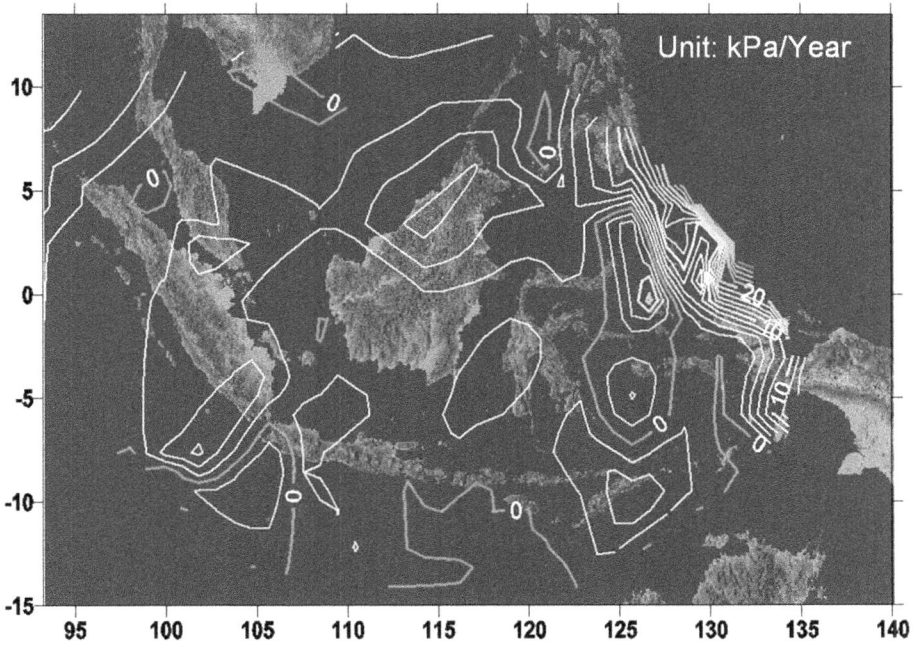

Figure 10.24 Distribution of contours of the disturbing stress rate and seismic activity.

Table 10.1 The code, coordinate and annual horizontal deformation velocity of GPS stations in Afghanistan and its vicinity used in this study

Site	lon	lat	E (mm)	N (mm)	References
YAZT	61.034	36.601	3.14	0.91	Masson et al. (2007)
BAKH	60.36	35.02	0.08	−0.01	Walpersdorf et al. (2014)
DESL	59.297	31.96	1.33	6.24	Walpersdorf et al. (2014)
ZABO	61.57	31.049	1.72	0.91	Walpersdorf et al. (2014)
BAZM	60.18	27.865	5.22	3.77	Walpersdorf et al. (2014)
JASK	57.767	25.636	2.78	14.56	Masson et al. (2007)
CHAB	60.694	25.3	1.14	7.96	Masson et al. (2007)
KCHI	67.113	24.931	5.6	28	Mohadjer et al. (2010)
QTAG	66.991	30.166	1.4	18.8	Mohadjer et al. (2010)
BAHAWALPUR	71.6833	29.4	5.04	28.6	Walpersdorf et al. (2014)
S2004	72.45	32.59	6.82	36.27	Walpersdorf et al. (2014)
RSCL	77.6	34.128	−5.7	20.3	Mohadjer et al. (2010)
NCEG	71.487	34.004	−0.8	29	Mohadjer et al. (2010)
MANM	71.68	37.542	−10.7	15.8	Mohadjer et al. (2010)
GARM	70.317	39.006	−1.2	2.4	Mohadjer et al. (2010)
SHTZ	68.123	37.562	22.7	3.9	Mohadjer et al. (2010)
KIT3	66.885	39.135	0.1	−1.4	Mohadjer et al. (2010)
KBUL	69.13	34.574	−0.1	10	Mohadjer et al. (2010)

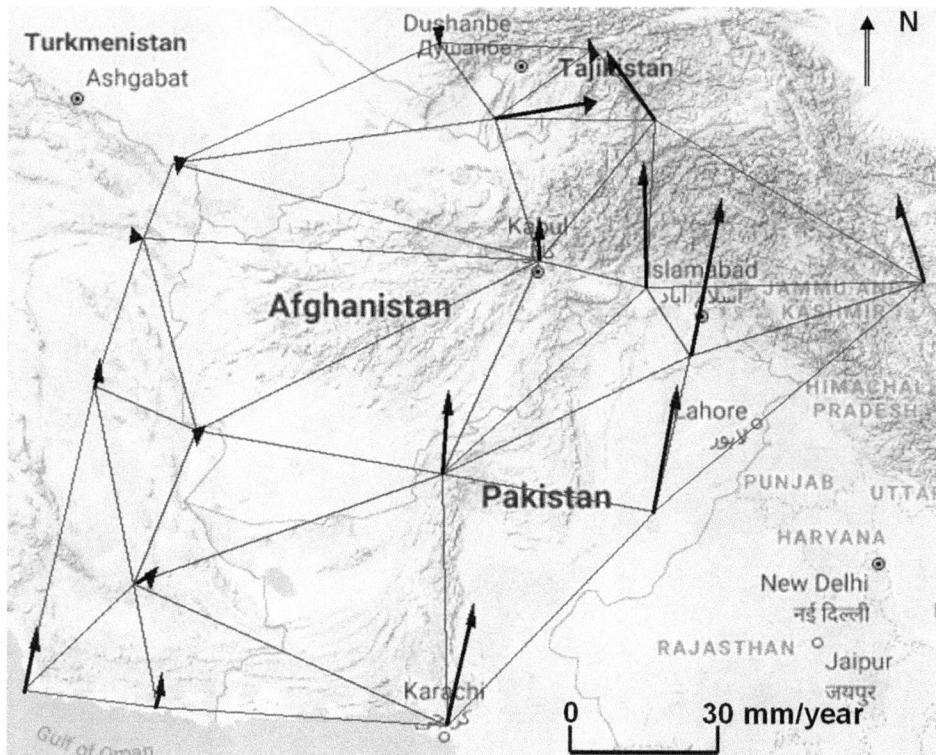

Figure 10.25 Annual deformation rates of GPS stations in Afghanistan and its vicinity.

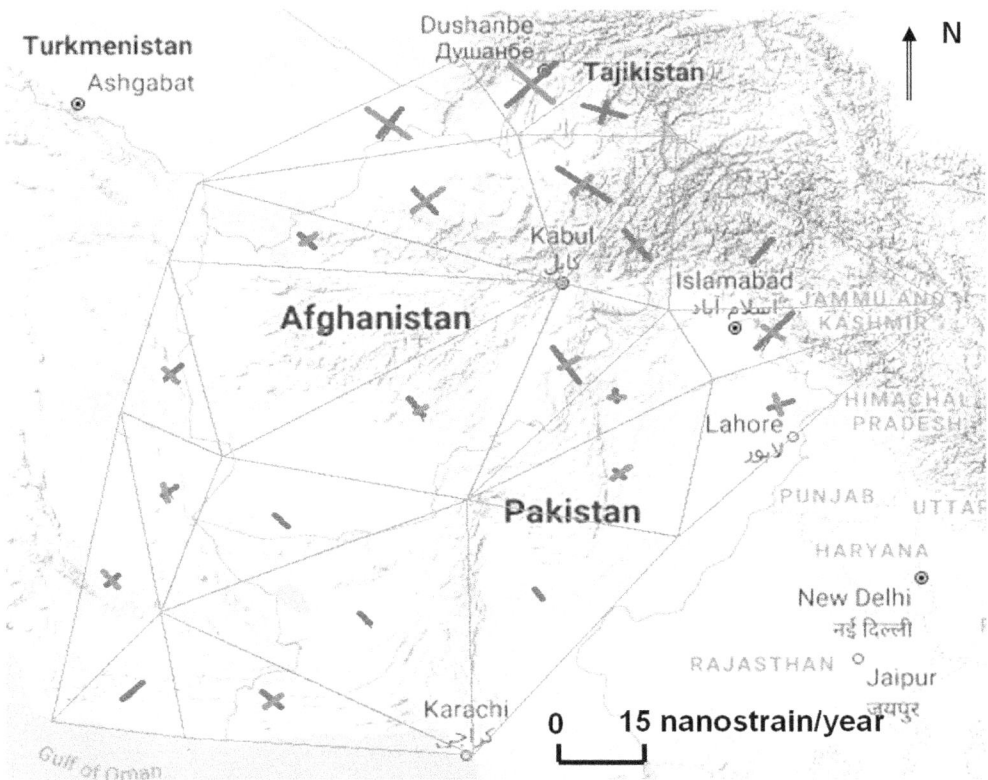

Figure 10.26 Annual strain rates in Afghanistan and its vicinity.

As noted from Figure 10.26, compressive principal strain rates are perpendicular to the plate boundary between the Indian and Euro-Asian plates. It is interesting to note that the principal strain rates in the vicinity of Kabul are high and are close to the strain rate variations in the vicinity of sinistral faults. Furthermore, the strain rate to the north of the Hari-rud (Herat) Fault implies dextral-type straining in accordance with the deformation sense of this fault.

10.4.2.1.5.3   Stress rates
The method presented is used to compute the principal stress, maximum shear stress, mean stress and disturbing stress rates. Figures 10.27–10.29 show the annual maximum shear stress, mean stress and disturbing stress rates in Afghanistan and neighbouring regions. Crustal stresses are high along the major tectonic structures, such as the Pamir and Hindu Kush Subduction zones, the Chaman Fault, the Makran subduction zone and SSZ, while the annual stress rate in central Afghanistan (Helmand Block) is quite small. Furthermore, both the contours of maximum shear stress and disturbing stress rates have a striking similarity to the seismic zoning map of Shareq (1981).

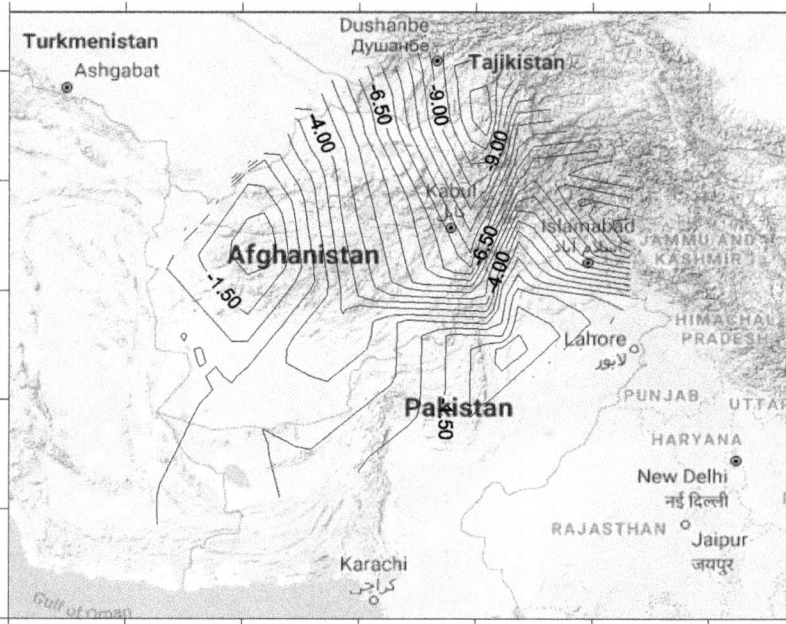

Figure 10.27 Maximum shear stress rate variations in Afghanistan and its neighbouring countries.

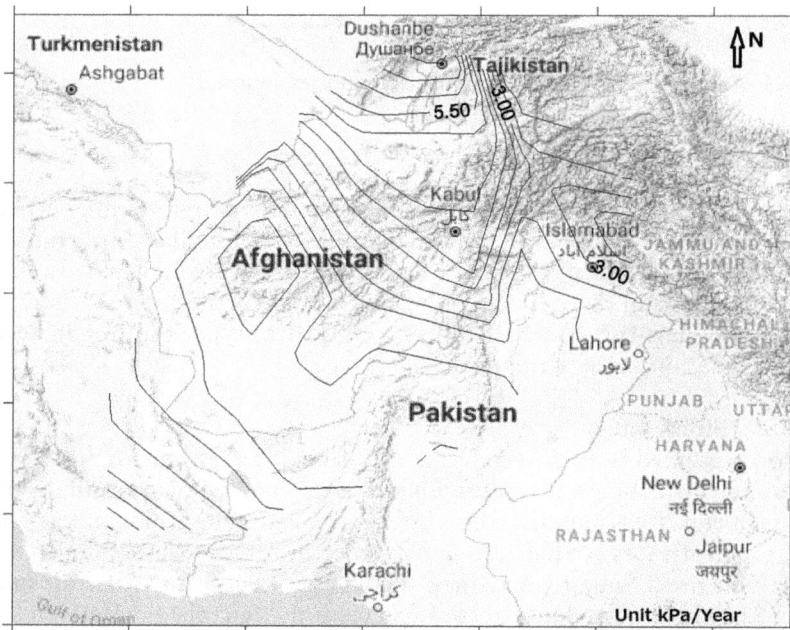

Figure 10.28 Mean stress rate variations in Afghanistan and its neighbouring countries.

Figure 10.29 Disturbing stress rate variations in Afghanistan and its neighbouring countries.

10.4.2.1.5.4 Interrelation between strain and stress rates and seismicity

Aydan et al. (2011) showed that the strain and stress rates are closely related to regional seismicity. Figures 10.30 and 10.31 compare the computed strain and stress rates and seismicity of Afghanistan and its vicinity. Seismicity is quite high in Hindu Kush, Kashmir Quetta and SSZ. It is interesting to note that high seismicity is observed particularly in the regions undergoing high extension-type straining rate. There is no doubt that the GPS instrumentation in relation to the main tectonic features would yield better crustal straining and stress variations in Afghanistan and its relation to regional seismicity. Nevertheless, the strain and stress rates computed in this study are sufficient enough to have a clear image of crustal straining and stress changes in Afghanistan.

10.4.3 Prediction of time of occurrence and recurrence

Aydan (2003a, 2004a) also showed that prediction of the time of occurrence of earthquakes in terms of weeks may be possible using GPS measurements recorded during

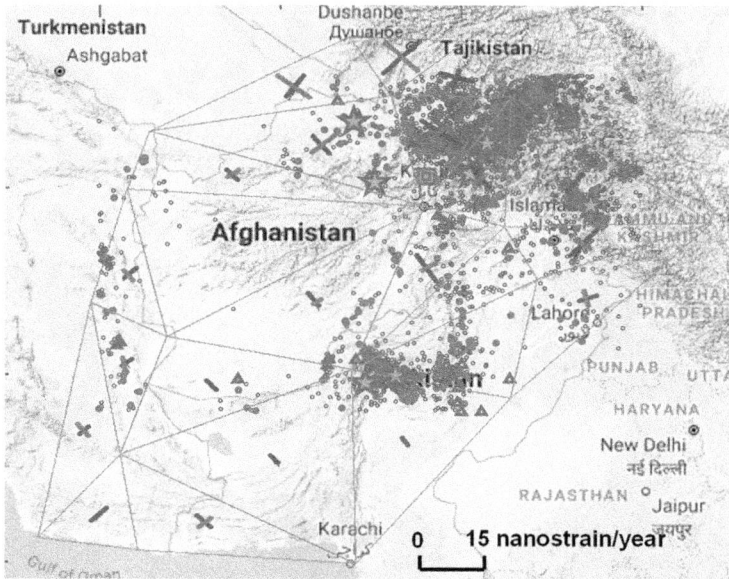

Figure 10.30 The relation of annual strain rates in Afghanistan and its vicinity to its seismicity.

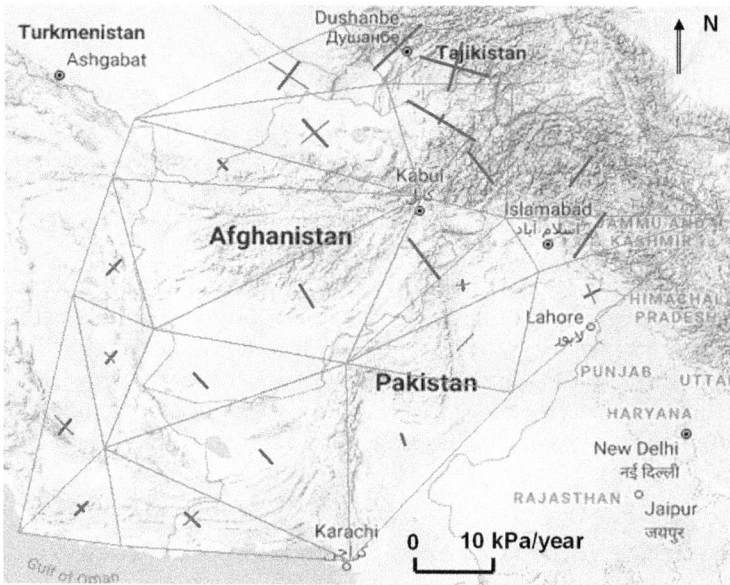

Figure 10.31 The relation of annual principal stress rates in Afghanistan and its vicinity to its seismicity.

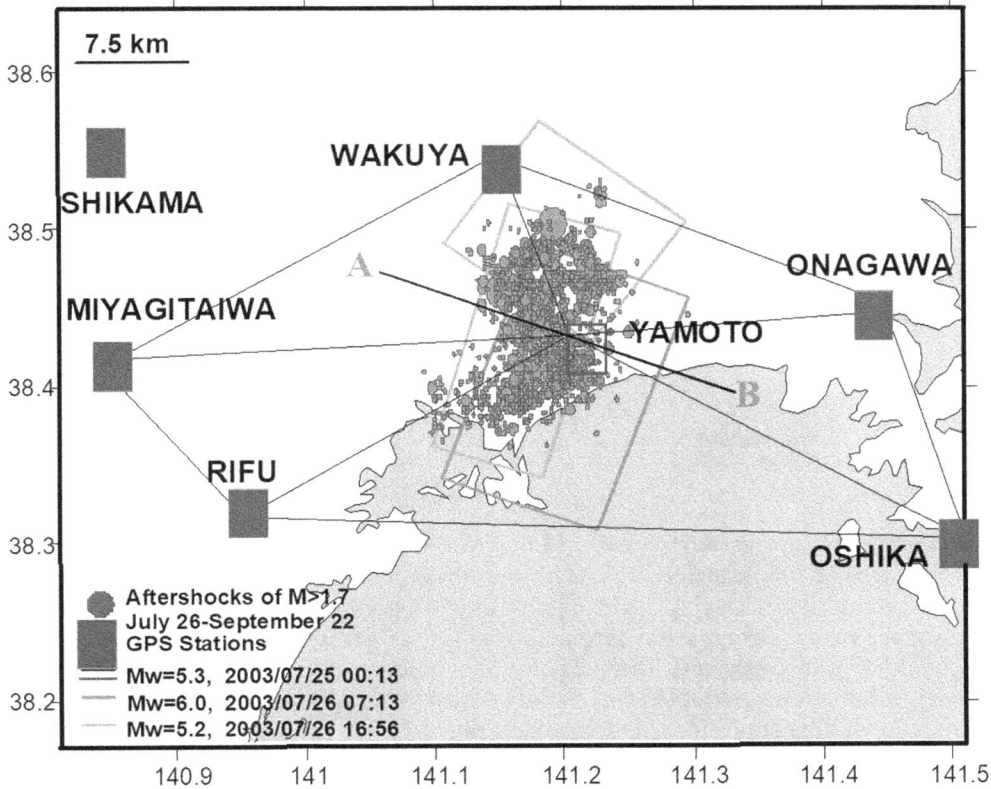

Figure 10.32 GPS stations and configuration of GPS mesh. (From Aydan (2004a).)

the 2003 Miyagi-Hokubu earthquake (Figures 10.32 and 10.33). The parameter MRI shown in Figure 10.33 is defined as

$$MRI = \frac{M}{R} \times 100 \tag{10.8}$$

where M and R are the magnitude and hypocenter distance of an earthquake, respectively. The MRI is a measure of the effect of earthquakes in a given point. As noted from Figure 10.33, the stress rate components of the Yamoto-Rifu-Oshika element indicated that remarkable stress variations started in October 2002. However, the strain rate components of the elements of Yamoto-Oshika-Onagawa, Yamoto-Onagawa-Wakuya and Yamoto-Wakuya-Miyagi-Taiwa started to change remarkably at the beginning of May 2003, about 1 month before the M7.0 Kinkazan earthquake that occurred on May 26, 2003. The high rate of variations continued after the M7.0 earthquake and resulted in the July 26, 2003, Miyagi-Hokubu earthquakes. Variations before the earthquake resembles those observed in creep tests. As the variations of the

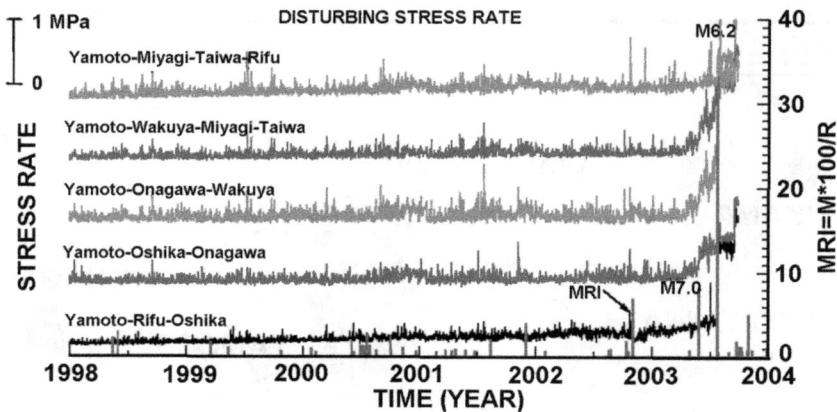

Figure 10.33 Time series of disturbing stress rates of GPS elements. (From Aydan (2004a).)

disturbing stress rates were greater than those of mean and maximum shear stress rates, Aydan (2004a) concluded that the disturbing stress rate may be a good indicator of regional stress variations in and precursors of following earthquakes. Therefore, the time of the earthquake may be obtained from GPS measurements.

As shown in Chapter 2, the stick-slip phenomenon of preexisting discontinuities (faults, subduction zones, etc.) can explain why earthquakes occur periodically under sustained loading conditions. The strain is released when the shear strength of discontinuities is exceeded and the next event of slip would take place roughly at the same time interval unless the properties of discontinuities and adjacent rock masses and the overall imposed loading conditions do not change. Stress-strain changes during a seismic event should correspond to the most likely stress-strain changes. During a seismic event, surface expression of total crustal deformation may occur simultaneously or slightly delayed due to redistribution of stress state and viscous characteristics of Earth's crust. On the basis of stick-slip experiments reported in the literature (i.e. Ohta and Aydan 2010), the stress-strain rate per year may be assumed roughly linear during the pre-slip period. Thus, the recurrence interval (T_r) of an earthquake with a given magnitude may be put forward as follows (Aydan 2015c):

$$T_r = \frac{|\Delta\sigma_t|}{|\Delta\sigma_y|} \tag{10.9}$$

where $|\Delta\sigma_t|$ and $|\Delta\sigma_y|$ are the measures of total stress change and annual stress change. The stress measure may be chosen as principal stresses, maximum shear stress or disturbing stress. This approach was first applied to the 2011 Great East Japan Earthquake (GEJE) for the area shown in Figure 10.32 and an element consisting of Oshika, Wakuya and Rifu GPS stations of the GEONET. Figure 10.34 shows the computed principal stresses, maximum shear stress and disturbing stress and the orientation of

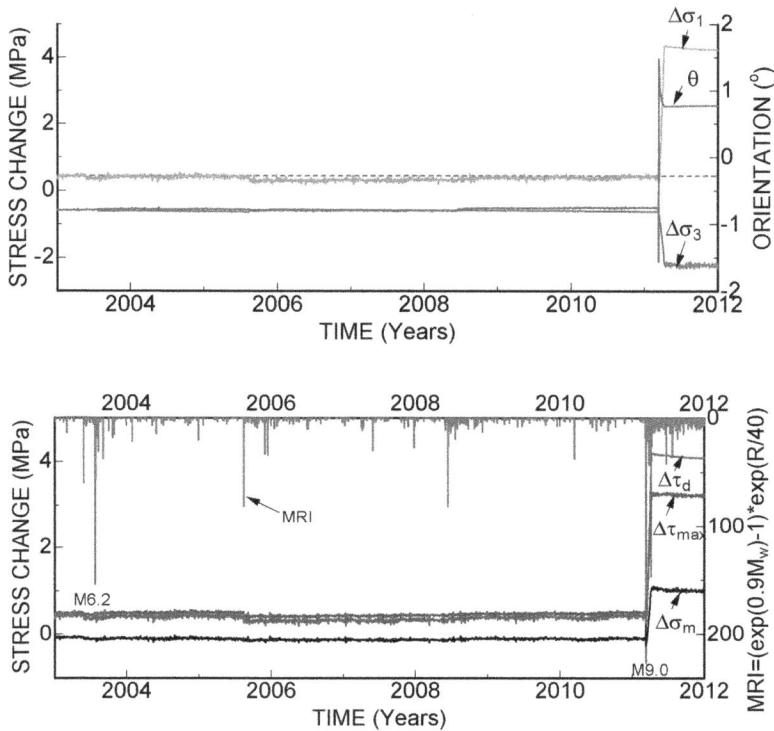

Figure 10.34 Stress changes between the beginning of 2003 and the end of 2011. (From Aydan (2015c).)

the principal stress change from east together with another definition of MRI, which was suggested by Aydan et al. (2010b) as:

$$\text{MRI} = \left(e^{0.9M_w} - 1\right)e^{-R/40} \tag{10.10}$$

where the unit of hypocentral distance R is given in km. This definition of MRI also considers the effect of earthquake magnitude much more clearly compared to the one given by Eq. (10.8).

Table 10.2 gives the computed recurrence interval for each annual stress change averaged for the period between January 1, 2003, and December 31, 2009. It is known that the Jogan earthquake occurred in 869 in this region and its estimated surface wave magnitude was initially presumed to be 8.6 and it was recently discussed if its magnitude was higher than this. The computed recurrence interval changes range from 597 to 1219 years depending upon the stress measures. If the Jogan earthquake were taken as the previous event, the recurrence interval would be 1,142 years. The recurrence interval based on the maximum shear stress measure yields is 1,219 years and it is close to that if the 869 Jogan earthquake is assumed to be the previous event.

Table 10.2 Computed recurrence interval for the 2011 Great East Japan Earthquake

Measured parameter	$\Delta\varepsilon_1$ (μs)	$\Delta\varepsilon_3$ (μs)	$\Delta\sigma_1$ (kPa)	$\Delta\sigma_3$ (kPa)	$\Delta\tau_{max}$kPa)	$\Delta\tau_d$ (kPa)
Annual	0.09	−0.07	6	−3.7	2.7	4.79
Total	63.9	−45.9	4369	−2218	3294	4208
Recurrence (years)	701	656	728	599	1220	879

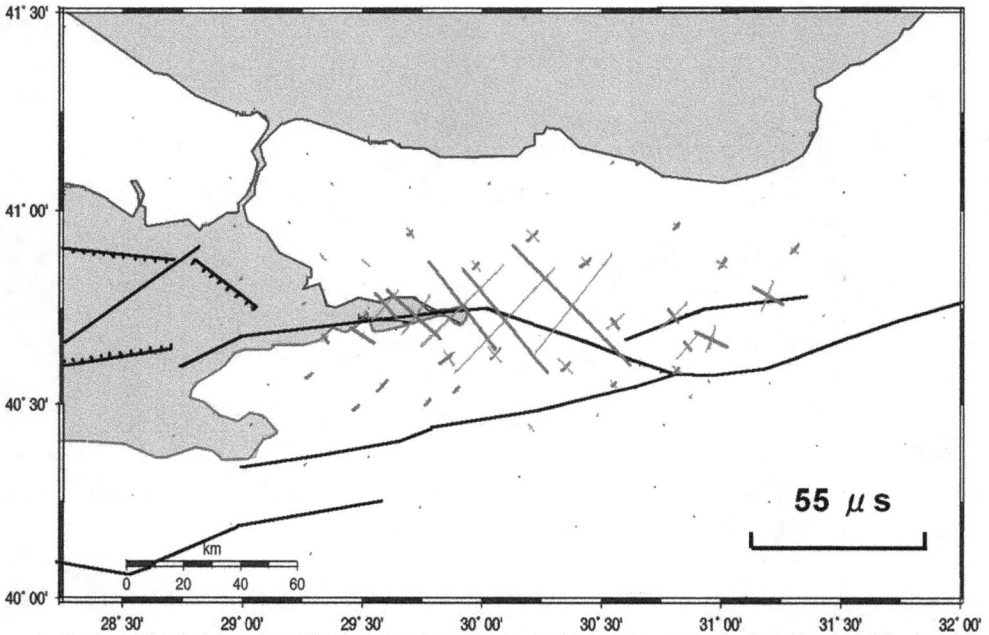

Figure 10.35 Principal strain variations associated with co-seismic deformations of the 1999 Kocaeli earthquake. (From Aydan et al. (2011).)

The same approach was also applied to the co-seismic GPS measurements of the 1999 Kocaeli earthquake by Reilinger et al. (2000). Aydan (2000b) computed the average yearly stress changes for all Turkey earthquakes including the epicentral area of the 1999 Kocaeli earthquake. Aydan et al. (2011) computed strain and stress changes following the 1999 event, and Figure 10.35 shows the principal strain changes associated with the 1999 event.

Table 10.3 gives the recurrence interval for each stress measure using the yearly and total stress changes during the 1999 event. The computed recurrence interval changes between 108 and 163 years. The largest event prior to the 1999 event occurred at 1894. Thus the recurrence interval is 104 years. However, it was reported that there were several large events with an estimated magnitude of 7 or more in 1509, 1719, 1754 and 1894 (Ergin et al. 1967). It seems that the recurrence interval computed from the disturbing stress rate is quite close to the actual recurrence interval. From these two

Table 10.3 Computed recurrence interval for the 1999 Kocaeli earthquake

Measured parameter	$\Delta\varepsilon_1$ (μs)	$\Delta\varepsilon_3$ (μs)	$\Delta\sigma_1$ (kPa)	$\Delta\sigma_3$ (kPa)	$\Delta\tau_{max}$ (kPa)	$\Delta\tau_d$ (kPa)
Annual	0.36	−0.26	24	−12.8	18.4	23.1
Total	41.2	−37.0	2602	−2092	2347	2564
Recurrence (years)	114	142	108	599	128	111

specific examples, it may be concluded that the maximum shear stress or disturbing stress changes may be used to estimate the recurrence interval with some confidence for long-term prediction.

10.4.4   Prediction of magnitude

Prediction of the magnitude of an earthquake is still difficult. Nevertheless, the area of stress rate concentrations with a chosen value may be used to determine the magnitude. As a result, the fundamental parameters of earthquake prediction, i.e. location, time and magnitude, may be determined from the evaluation of GPS measurements. However, there are still some technical problems associated with GPS observations and artificial disturbances as pointed by Aydan (2000b, 2004a).

10.4.5   Effect of the 2011 Great East Japan earthquake on the epicentral area of the anticipated Tokai earthquake

Tokai earthquake was anticipated since 1970s on the basis of some researches by Ishibashi (2004). Recently, a successive rupture of the subduction zones from Nankai, Tonankai to Tokai regions was proposed, and this earthquake has been anticipated to have a similar magnitude to that of the GEJE. As the rupture area of the GEJE was 450 km long and 150 km wide, it is expected to alter the stress state in Earth's crust in the archipelago of Japan. Aydan (2006) computed strain and stress changes for Suruga Bay, which is the epicentral area of the anticipated Tokai earthquake by utilizing the monitoring results at Minami Izu, Fuji and Omaezaki GPS stations of the GEONET. As the Omaezaki GPS station was damaged by the M6.4 Suruga Bay earthquake on August 11, 2009, the author selected the Hamaoka GPS station instead and repeated the same computations for the period between January 1, 2003, and December 31, 2011. Figure 10.36 shows the computed stress changes in the epicentral area of the anticipated Tokai earthquake. As noted from the figure, the GEJE undoubtedly caused stress changes in Earth's crust beneath the Suruga Bay. While the amplitude of the maximum compressive stress increased, there was stress reduction in other components.

Similarly great variations occurred in other stress measures. It is also worth noticing that maximum shear stress response started to change from the middle of 2008 and has been increasing since then. The 2009 Suruga Bay earthquake may be a consequence of the change in the stress state in Earth's crust, and the GEJE further increased such stress changes. This may imply that the region may have entered a critical stage of crustal straining, and continuous hourly evaluations of the GPS monitoring would be needed.

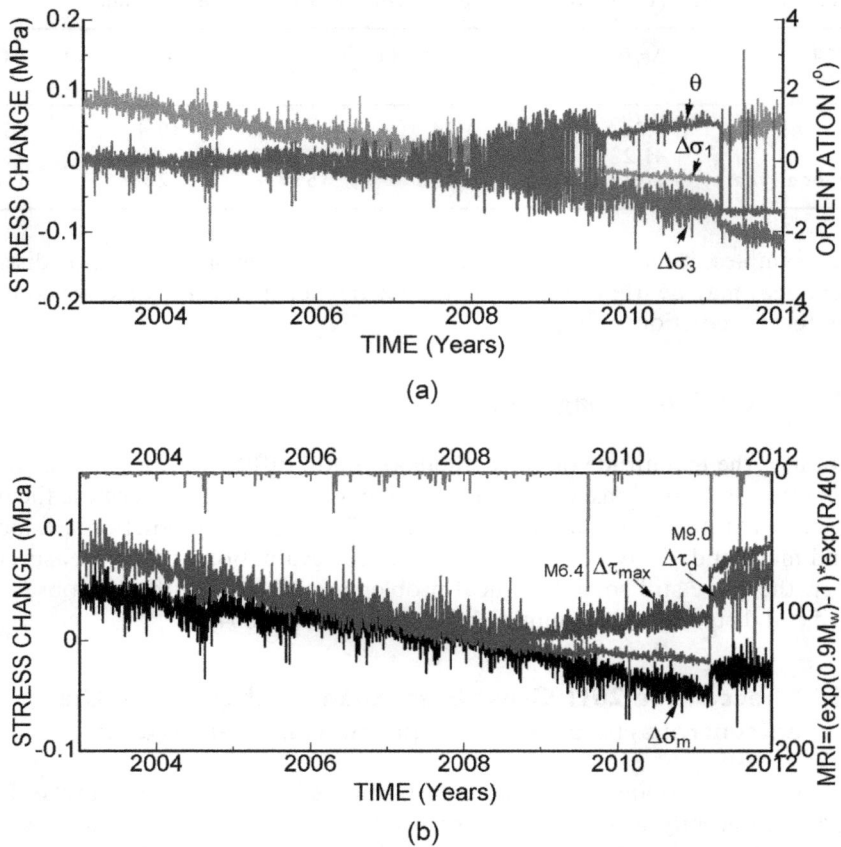

Figure 10.36 Changes in stress measures in Earth's crust beneath Suruga Bay.

10.5 ANOMALOUS PHENOMENA OBSERVED IN THE 1999 DÜZCE EARTHQUAKE AND OTHER EARTHQUAKES IN TURKEY

Several anomalous phenomena were observed before, during and after the Düzce-Bolu earthquake as well as other earthquakes in Turkey. Most of such anomalous phenomena observations are not well documented. Even if they are, those sources are not easily accessible. Besides various anomalous phenomena observations on atmosphere, wells, springs and animals by individuals and institutes unrelated to earthquake prediction, some well-instrumented projects on various parameters mentioned in Section 10.2 for earthquake prediction have been going on in Turkey for some time.

Turkey is a real in situ seismic laboratory and anyone can directly access the faults without any physical hindrance compared to the extreme natural and physical difficulties in observation and measurement of seabeds at great depths for the same purpose. As a result, Turkey attracts the attention of many geoscientists all over the world, and some non-seismic, technologically advanced countries, whose sole purpose is to develop technologies which may be tested against the actual data obtained on the economical, sociological, physical and psychological sufferings of the people of Turkey. Such real-time data pertain to:

(a) Seismic velocity measurements
(b) Ground tilting and borehole strain measurements
(c) Groundwater level and temperature recordings
(d) Gravity field measurements
(e) Monitoring of the total magnetic field and broadband magneto-telluric signals
(f) Recording of meteorological parameters
(g) Radon emissions and hydro-chemical parameters

Most of these measurements are done in the area where two recent largest events, namely, the 1999 Kocaeli earthquake and the Düzce earthquake, occurred.

10.5.1 Gas emissions

Co-seismic spectacular gas emissions were observed near Hamamüstü and Cevizli villages along the southern shore of Efteni Lake where the fault break was observed. Although the bubbling of the lake was observed on November 10, 1999 (Figure 10.37a), the burning of emitted gas appeared over the lake surface immediately after the earthquake (Figure 10.37b). Figure 10.37c shows pictures taken by the MTA and others (Şimşek and Yıldırım 2000a; Emre et al. 1999). The chemical component of the burning gas was methane (CH_4).

The shallow boreholes drilled into the ground nearby the lake are known to be emitting methane gas for a long time, and it could be burned as shown in Figure 10.38a. Similar types of gas emissions were also observed in the 1999 Kocaeli earthquake. The geophysical investigations by the ARAR seismic exploration ship of İstanbul University observed some co-seismic CH_4 gas emissions in the seabed, as shown in Figure 10.38b.

In addition to the concentration of CH_4, the MTA also found out that the concentration of CO_2 was quite high. This finding is consistent with observations by Sultankhodzhaev (1984) in other countries, such as the Ulugbek well during the 1976 Gazlı, 1978 Isfara and 1978 Tavaksai earthquakes in Uzbekistan. He reported that the concentration of CO_2 during earthquakes could go up to 14% and gradually decrease to its normal level after 15 days.

Figure 10.37 Gas emission in Efteni Lake (Şimşek and Yıldırım 2000a; Emre et al. 1999).

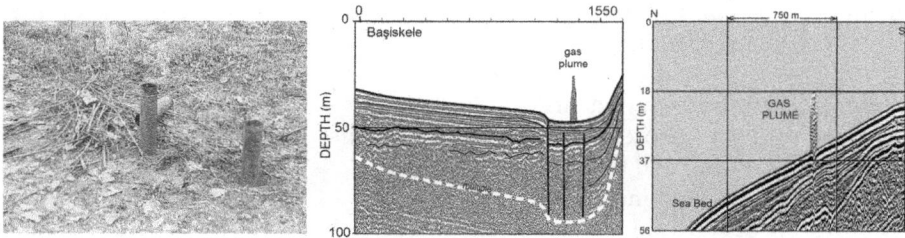

Figure 10.38 (a) Burning of CH$_4$ gas emitted from boreholes drilled into the ground near Cevizli village (picture by MTA; Emre et al., 1999) and (b) CH$_4$ gas plume observed along the southern shore of İzmit Gulf. (Istanbul).

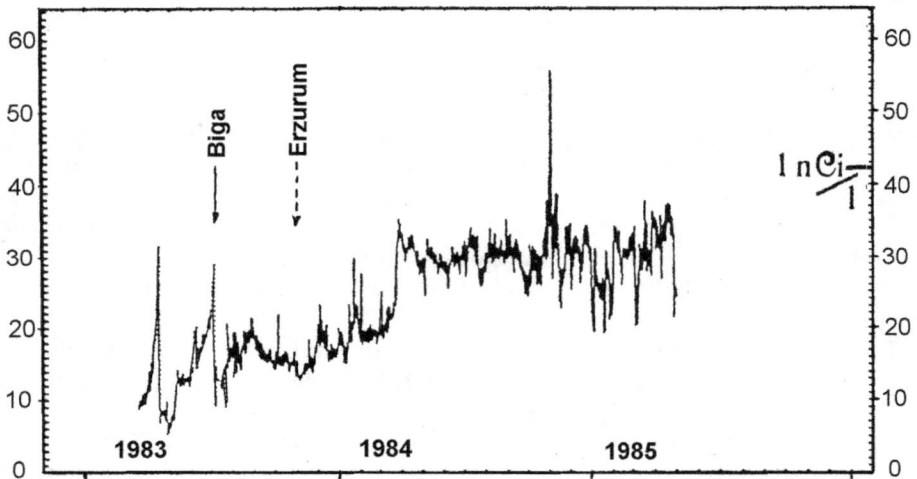

Figure 10.39 Radon concentration recorded in a hot spring nearby Bolu. (After Friedman et al. (1988).)

Radon gas emission was observed during the last two earthquakes according to the information given by the Çekmece Nuclear Research and Education Institute. The MTA (Emre et al. 1999) also measured radon concentration after the earthquake and reported that it was lower than the usual concentration levels observed in nearby faults. Although their interpretation of this result was mistaken, the results are again consistent with the co-seismic variation in the concentration level of radon gas observed in other earthquakes in the world such as the 1976 Gazlı, 1978 Isfara, 1978 Tavaksai (Sultankhodzhaev, 1984) and the 1995 Hyogo-ken Nanbu (Kobe Earthquake) earthquakes (Igarashi et al. 1995). Figure 10.39 shows radon concentration recorded at a hot spring nearby Bolu between 1983 and 1985 (Friedman et al. 1988). During the measurements, the 1984 Biga earthquake ($M = 5.7$) and the 1984 Erzurum earthquake ($M = 6$) occurred. In spite of great distances between the hot spring and the epicentre of these earthquakes, it seems that there is some correlation between earthquakes and radon gas emission responses.

10.5.2   Groundwater level observations

The MTA observed some co-seismic new springs along the fault break (Emre et al. 1999). These new springs were considered to be a result of variation in permeability of the ground and development of new water channels as a result of ground deformation. Similar observations were also made by the authors during the 1992 Erzincan earthquake, the 1995 Dinar earthquake and the 1998 Adana-Ceyhan earthquake (Hamada and Aydan 1992; Aydan and Kumsar 1997a; Aydan, et al. 1998).

It was also reported that the hot spring water become greyish (turbidity) 2 days before the earthquake in Hamamüstü village and the temperature of the hot spring was unusually higher than its usual value. Furthermore it was also reported that new hot springs appeared nearby existing hot springs after this earthquake and the 1999 Kocaeli earthquake (Figure 10.40) (Şimşek and Yıldırım 2000a). A similar phenomenon was also observed in Ekşisu hot spring in the 1992 Erzincan earthquake (Hamada and Aydan 1992).

The MTA also measured the water temperature of the lake where the burning of CH_4 gas was observed. The lake water temperature at the spot was 25.2°C, which was higher than the usual lake water temperature of 9.1°C at the time of measurement. The temperature of the hot spring was about 42.3°C. This result implied that the hot spring water ratio of the water at that location was about 50%.

Tezcan et al. (2000) reported a very interesting example of groundwater variations observed at wells in Sivrihisar and Günelli in Eskişehir Province (190 km away from the epicentre) after the 1999 Kocaeli earthquake (Figure 10.41a,b). The water level of the well in Günelli decreased at an almost constant rate before the earthquake. Since no precipitation data were available for the region, the rate of decrease was quite similar to the response of the Cholame well shown in Figure 10.8. Furthermore the responses of wells are quite remarkable despite great distances between the epicentres of the earthquakes and their locations.

Another remarkable feature is that the variations in the co-seismic groundwater levels at Günelli and Sivrihisar are contrasting. While the groundwater level at Günelli decreased, that at Sivrihisar increased. It is generally said that the groundwater level should increase in compression zones and decrease in dilation zones obtained from focal plane solutions. Considering the focal plane solution for the 1999 Kocaeli earthquake, both wells must be in the dilation zone, implying that the groundwater level should decrease. However, it was also pointed out by Fleeger and Goode (1999) that the variation in permeability of the ground after the earthquake may result in different responses in wells if they are topographically situated at different levels, regardless of whether they are in the zones of dilation or compression obtained from the focal plane solutions of earthquakes.

10.5.3   Earthquake lights

It is reported that the colour of the sky is unusually red just before and after an earthquake occurred. Earthquake lights (EQL) were observed at the time of the main shock (Figure 10.12). Considering that the Sun had set at the time of the earthquake, it is quite possible that EQL should be observed, which are caused during the rupturing

(a)

(b)

Figure 10.40 New springs at Efteni Kaplıca after Düzce-Bolu earthquake (a) and Yalova Termal after Kocaeli earthquake (b). (After Şimşek and Yıldırım (2000a).)

process of the Earth's crust described in the previous section. The same phenomenon was also observed by the author during the 1999 Kocaeli earthquake, who experienced the earthquake in İstanbul, and the luminosity of the sky continued for about 5–7 minutes. A very interesting example of EQL, which was seen even at a distance of 300 km away from the epicentre was during the 1976 Çaldıran earthquake, as reported by Toksöz (1977).

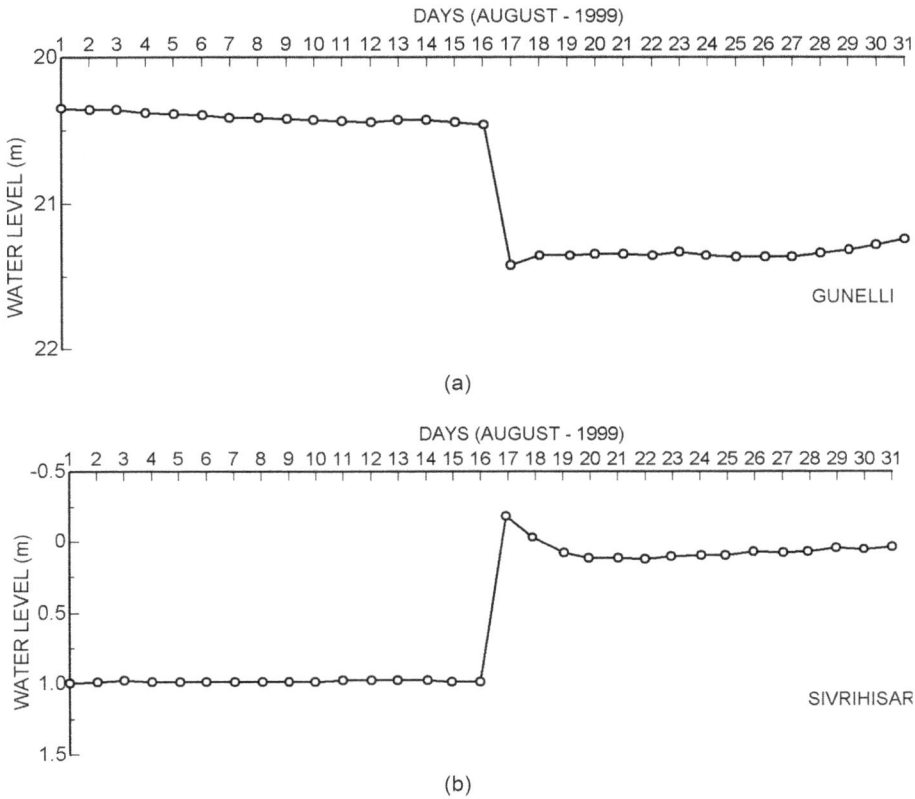

Figure 10.41 Groundwater level variation at wells in Günelli (a) and Sivrihisar (b). (From Tezcan et al. (2000).)

10.5.4 Geomagnetic and gravity anomalies

So far there have been no reports on geomagnetic and gravity anomalies. However, it is known that geomagnetic anomalies occurred in previous earthquakes in the Marmara region as shown in Figure 10.42 (İspir et al. 1976). It was also reported by Büyüksaraç et al. (1998) that high magnetic anomalies exist along the surface fault traces near the site of this earthquake. While magnetic anomalies in the northern part of the main strand of the NAFZ have negative values up to 700 nT, those in the southern part have positive values up to 1,000 nT. The gradient of magnetic anomalies across the NAFZ is very high. These studies further showed that magneto-elastic stress rates have a tensile character in the direction of 63°–66° and compressive character in the direction of 145°–155°.

Gerstenecker et al. (1999) measured gravity changes across the NAFZ between 1988 and 1998. Their results are replotted in Figure 10.43. As seen in the figure, the gravity decreases in the northern part of the NAFZ while it increases in the southern part of the NAFZ. It is quite interesting that these tendencies are quite similar to those of geomagnetic anomalies.

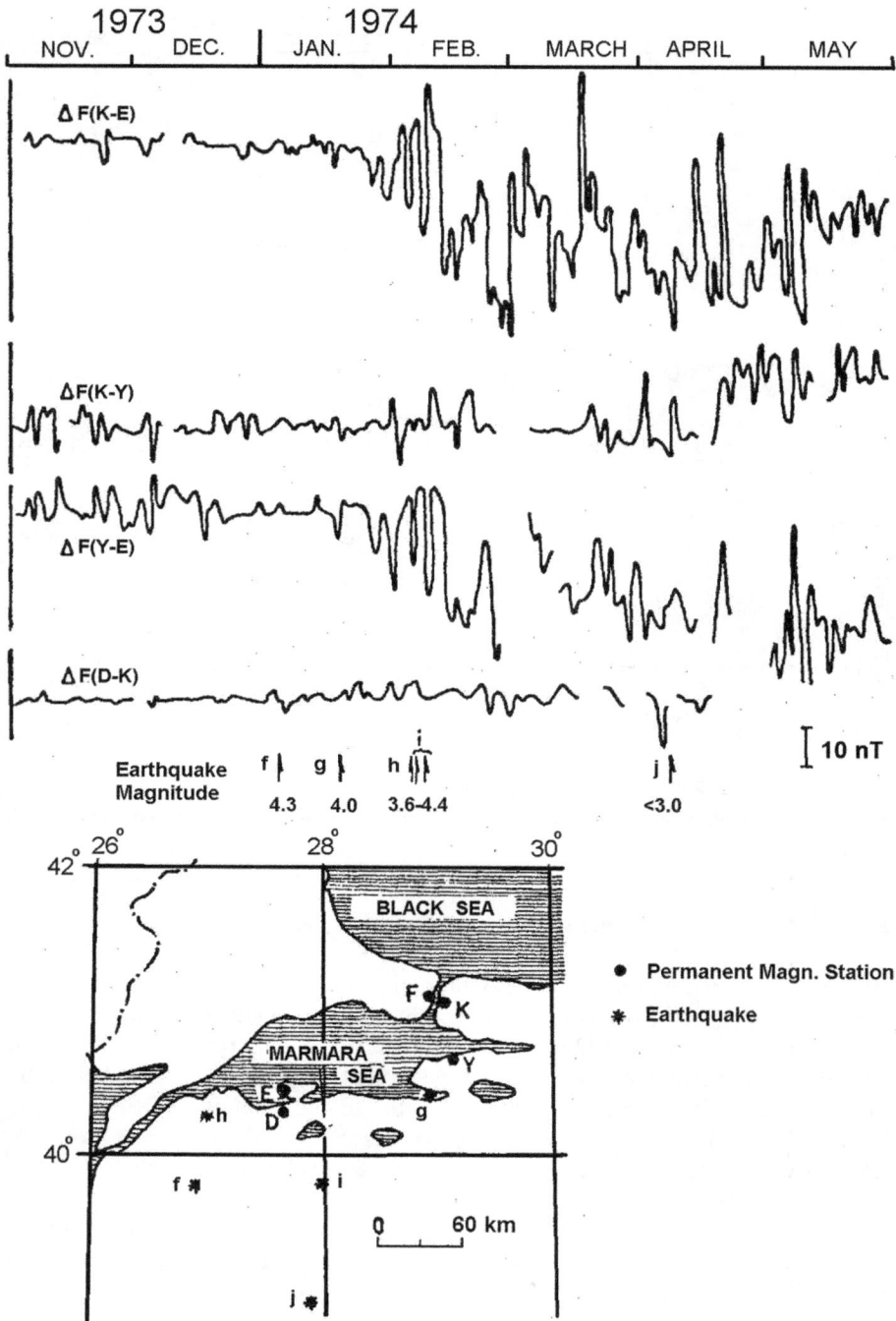

Figure 10.42 Magnetic anomalies measured during some earthquakes in Marmara region by İspir et al. (1976).

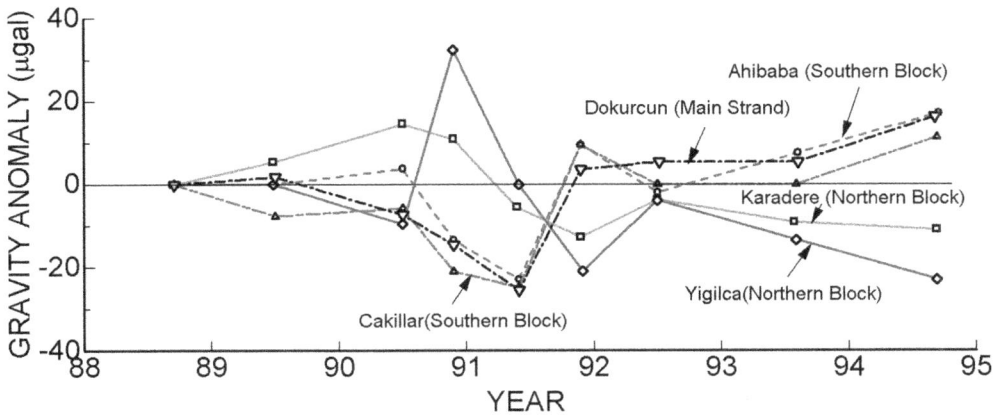

Figure 10.43 Gravity variations across the NAFZ. (Data from Gerstenecker et al. (1999).)

10.5.4.1 Ground tilting and deformation

Although ground deformation and tilting measurements are carried out in the region, these results are presently inaccessible. Instead, the measurements carried out in Ankara and Gebze using GPS would be referred here. Figure 10.44 shows time histories of the components of deformation of both stations (EUREF 2000). The Gebze station of TÜBİTAK, which is regarded to be situated on the Eurasian plate (or Black Sea plate) was included in the European GPS network on August 8, 1999. The Ankara station is on the Anatolian plate. The Ankara station moves NW horizontally and subsides vertically. While the westward motion of Ankara is almost linear, its northward motion starts to be reversed ~3 months before the 1999 Kocaeli Earthquake. After the earthquake, it returns to its original trend. However, this trend of motion again stops and becomes opposite ~6 weeks before the Düzce-Bolu earthquake, after which it again returns to its original trend of motion. As for the vertical component, the movement starts to be upward ~12 weeks before the Kocaeli earthquake and then it becomes downward 6 weeks before the earthquake. The same cycle was repeated before the Düzce-Bolu earthquake. Although the westward motion jumped at the time of both earthquakes, the general trend remained the same. The sudden jumps at the Ankara station at the time of the Kocaeli earthquake, which are also seen in the motion of the Gebze station, are extremely difficult to explain. The jumps may be due to some other causes on the stations. The horizontal movement of the Gebze station is almost stationary except the responses after the Kocaeli earthquake.

The vertical movement of the Gebze station is almost stationary. Nevertheless, an uplift behaviour of this station is noticed. The movement of the Ankara station is of great interest in relation to the deformation and earthquake occurrence in the region. The increased rate of the westward motion of the Ankara station may have further implications on future earthquake occurrences in western Anatolia.

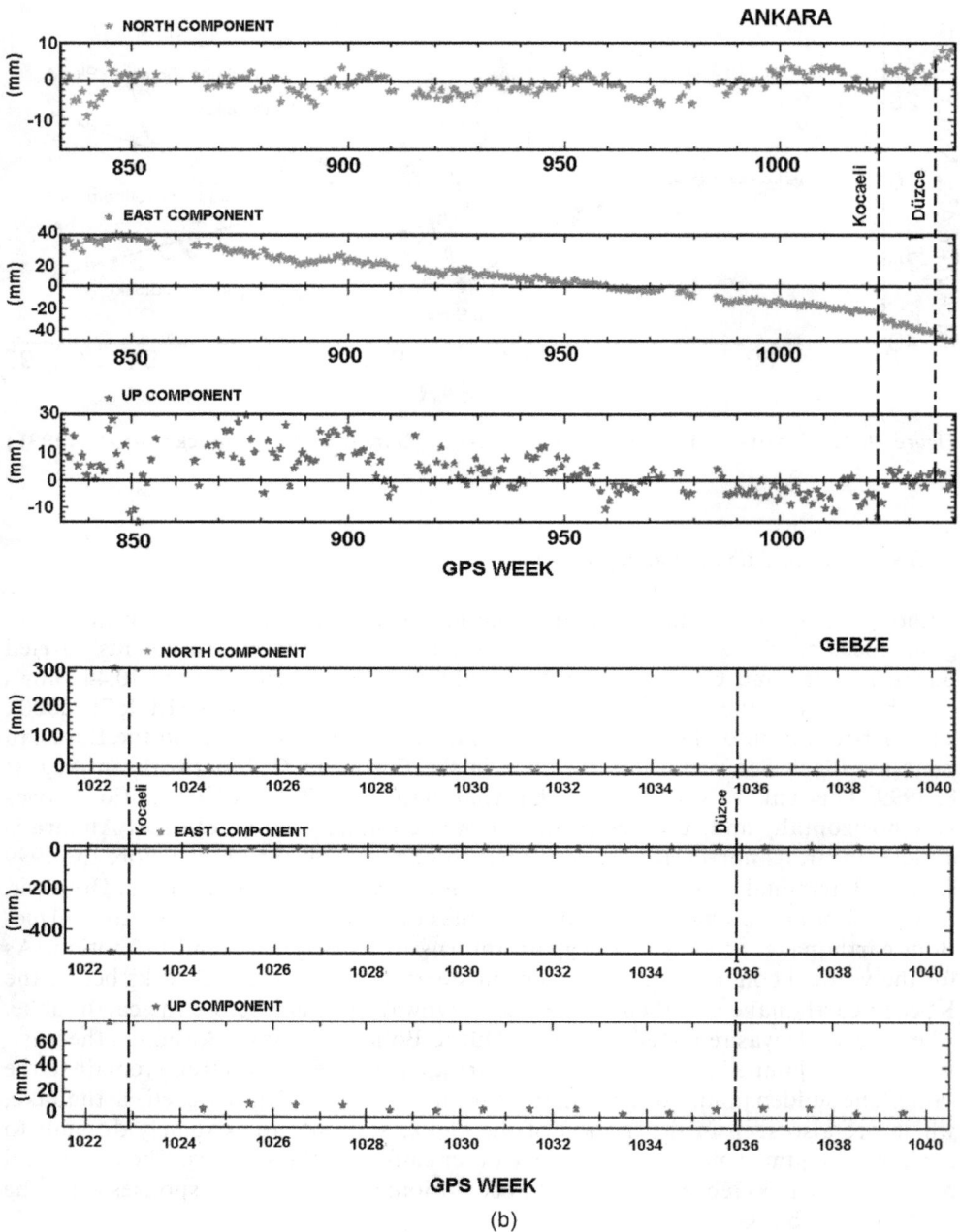

Figure 10.44 Measurements of deformation of Earth's crust in Ankara (a) and Gebze (b).

10.5.5   Anomalous animal behaviour

Although data is scarce, mass media have reported barking dogs and wildered cows and horses before the Düzce and Kocaeli earthquakes as well as other earthquakes investigated by the author. In addition, some dead fishes were scattered along the shores of Izmit Bay during the 1999 Kocaeli earthquake. When local people were interviewed in the epicentral area of the 1998 Adana-Ceyhan earthquake, they reported that dogs tried to receive the electromagnetic signals on their back while raising their feet in air. This observation was also confirmed by Ikeya et al. 1997, who carried out some experiments on various domestic animals.

10.5.6   Effects of the Sun and and the moon on earthquakes

It has been claimed that the Sun and the Moon have some effects on earthquake occurrence. Aydan et al. (2000b) presented some data on this. The sunspot index, which is defined as one of the parameters for solar activity, is said to have some effects on earthquake occurrence. The International Astronomy Union observes solar activity on a real-time base. Figure 10.45 shows the variation in sunspot index (Ssi) in the 20th century and earthquake occurrence in Turkey. The earthquake data come from a database developed by Aydan et al. (1996c) and NEIC. As seen from the figure, it seems that earthquakes with larger magnitudes occur when the sunspot index is close to its peak value. It is either on the ascending part or descending part of the Ssi and time response curve. On the other hand, earthquakes with smaller magnitudes generally occur in saddle points of the Ssi and time response curve. Next, the authors checked the relation between Ssi and earthquake magnitude M_s ($M_s > 4$) for 1999. Results are shown in Figure 10.46. The annual tendency of the relation between Ssi and earthquake magnitude in the 20th century is essentially repeated during 1999. Although this relation is still intriguing, it reserves further detailed studies.

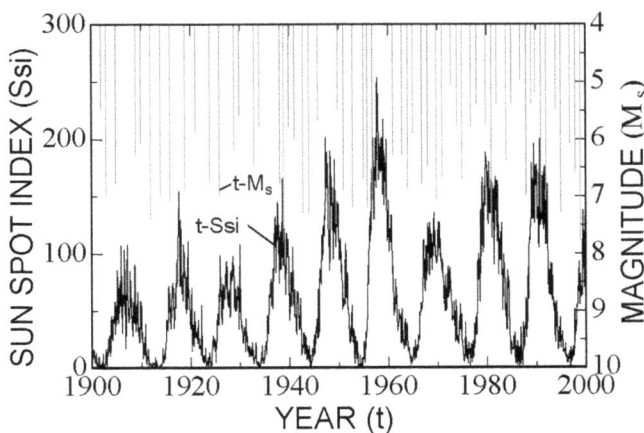

Figure 10.45 Relation between sunspot index and earthquake magnitude for Turkey in the twentieth century.

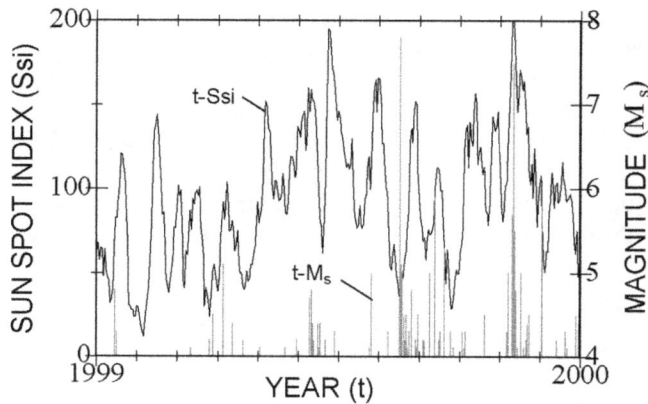

Figure 10.46 Relation between sunspot index and earthquake magnitude for Turkey during 1999.

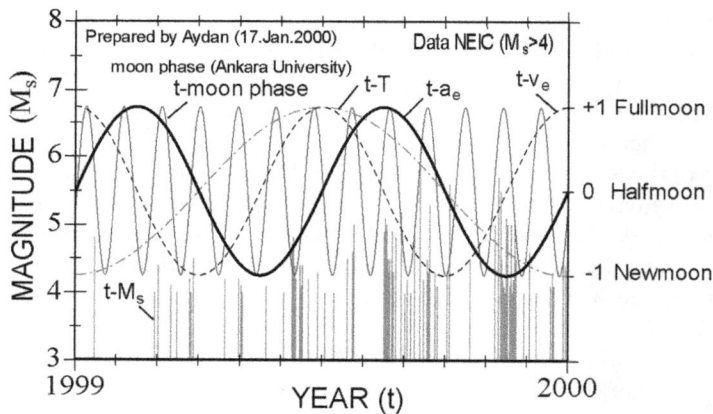

Figure 10.47 Relations between earthquake magnitude and moon phase, heat flux input and orbital acceleration and velocity.

The final investigation is concerned with the effect of moon phases, the variation in yearly orbital acceleration of Earth around the Sun and the variation in heat input from the Sun. The values of functions for each parameter are subtracted from the average values and the resulting value is normalized by the average value as shown in Figure 10.47. Although the weighting of each parameter may be different, the total effect should be the sum of these parameters in addition to the Ssi. A first glance at the figure implies that some correlations exist between earthquake occurrence and moon phases. It seems that earthquake occurrences, particularly those with greater magnitudes, are more likely during the new-moon and full-moon periods. In addition to the effect of moon phases, the variation in orbital acceleration of Earth and thermal stresses induced by the variation in heat flux input from the Sun should have some effects.

The effect of the variation of orbital acceleration of the Earth around the Sun should be maximum during summer and winter and minimum during spring and autumn periods. The effect of the heat flux will have an effect on temperature distributions in the atmosphere and the crust, which should induce some cyclic thermal stresses. Although time variation in temperature distribution may shift from the theoretical heat input function due to the heat conduction characteristics of the atmosphere, the resulting thermally induced stresses must be compressive during hot periods and tensile during cold periods. Although some good correlations exist among the parameters listed here, the weighting of each parameter on the overall stress state of the crust is still unclear. Therefore, this topic must be further studied before drawing any decisive conclusions.

10.6   APPLICATION OF THE MULTI-PARAMETER MONITORING SYSTEM TO EARTHQUAKES IN DENIZLI BASIN AND SUMATRA ISLAND OF INDONESIA

The author and his group established several multi-parameter monitoring systems (MPMS) in Denizli Basin (Aydan et al. 2005d, 2011; Kumsar et al. 2003) and they have obtained some experiences during past earthquakes in Denizli and nearby regions (Figure 10.48). The outcomes of the observations in Denizli Basin in relation to the 2003 Buldan earthquakes as an application of MPMS in practice are described herein.

There was an earthquake activity in Buldan and in surrounding Denizli on July 23, 2003 (Figure 10.48). An earthquake with a magnitude of 5.2 occurred at 07:56 a.m.

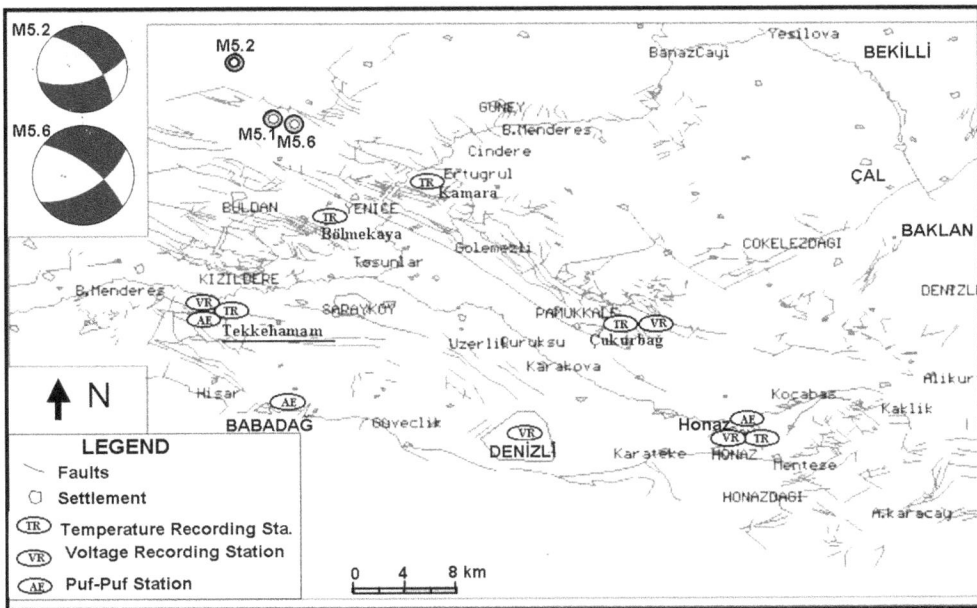

Figure 10.48 Locations of stations of multi-parameter monitoring system and epicentres of 2003 Buldan earthquakes.

on July 23, 2003. There were aftershocks with magnitudes of 4.1 on the following days. On July 26 another earthquake with a magnitude of 5.6 occurred near Buldan at 11:26 a.m. local time.

Tekkehamam puf-puf count is defined as the acoustic emission (AE) count in a unit time of gas pressure of thermal spring mud bubbles. There was a significant increase in the puf-puf numbers 2 days before the M5.2 earthquake in Buldan. The activity of the puf-puf count continued as the earthquake activity continued. After 3 days, an earthquake with a magnitude of 5.6 occurred. When the aftershocks became less than a magnitude of 4.0, the puf-puf counts decreased to a low level (Figure 10.49). This shows that there is a relation between earthquake activity of the region and puf-puf count changes.

The electric potential data variations of Honaz, Tekkehamam and Çukurbağ stations appear like the change of tidal wave height. During the measurements there were some unwanted and sudden artificial voltage changes due to external actions. At Çukurbağ station there were electric potential changes, which were thought to be related to the M5.2 and M5.6 earthquakes (Figure 10.49). At Tekkehamam station there were sudden changes.

The changes at the EW direction of the Honaz station and the NS direction of the Çukurbağ station started at the same time 2 days before the magnitude 5.2 earthquake. When the magnitude 5.6 of earthquake occurred, the Honaz station NS direction value was down to the minimum voltage value and then the graph started to rise (Figure 10.49). This type of change was also obtained by Aydan et.al. (2001b, 2002b, 2003) from the laboratory uniaxial tests. The changes at the Honaz and the Çukurbağ stations point out the possible changes in regional stress.

A devastating earthquake occurred with a magnitude of 9.3 (Mw) in Sumatra Islands of Indonesia on December 26, 2004 (Figure 10.50). Hundreds of thousands of people died as a result of the following tsunami. Although the earthquake epicentres were too far from the multi-parameter measurement stations in Denizli (Turkey) and in Japan, puf-puf count and electric potential variations were recorded at the stations in Denizli and at the multi-measurement stations established in Japan.

Figure 10.50 shows the responses of puf-puf counts in Tekkehamam station in Denizli (Turkey) and Koseto station in Shizuoka (Japan) during December 2004. It is of great interest that the December 26, 2004, earthquake occurred when the puf-puf counts became minimum following a peak. This was a common pattern observed in earthquakes in Turkey and Japan previously (Kumsar et al. 2003; Aydan and Tano 2003; Aydan et al. 2005d). The puf-puf count activity remarkably increased at Tekkehamam station thereafter, which is in accordance with the seismic activity increase in Turkey following the December 26, 2004, Northern Sumatra earthquake while the variations in puf-puf counts at Koseto (Shizuoka) station in Japan were not very remarkable.

Figure 10.50 shows the electric potential variations at the stations in Turkey and Japan. It is of great interest that there is a remarkable fluctuation ~1 week before the December 26, 2004, earthquake at Tekkehamam station in Denizli Basin, while the variations at Koseto station of Shizuoka in Japan were remarkable after the earthquake. The variations at Mitake station located in an abandoned lignite mine were also remarkable following the earthquake. The signal time becomes longer with increase of the volume involved with the magnitude of earthquake on the basis of outcomes from

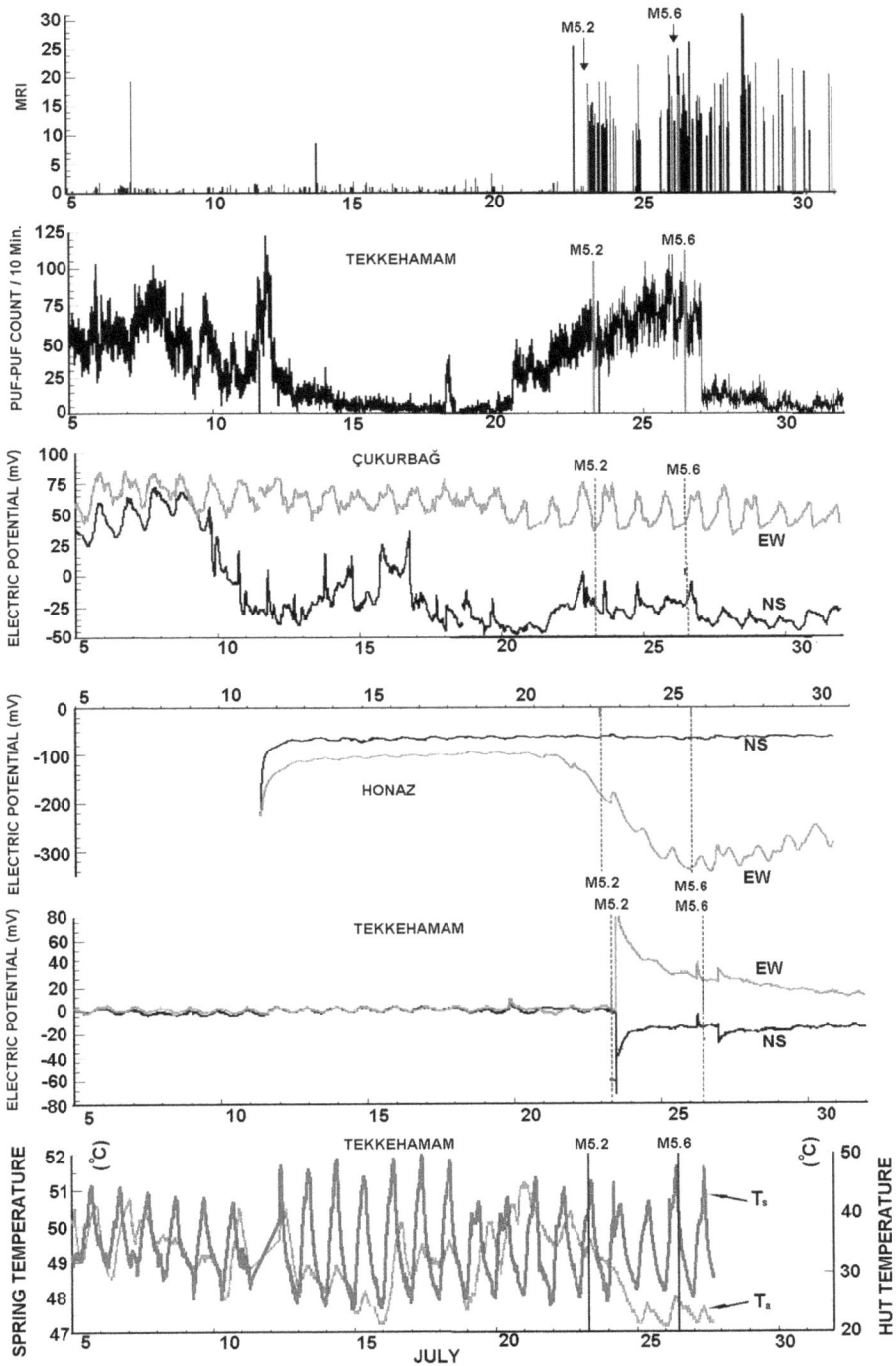

Figure 10.49 Acoustic emission (AE) count variations at monitoring stations in Denizli Basin associated with the July 2003 Buldan earthquakes.

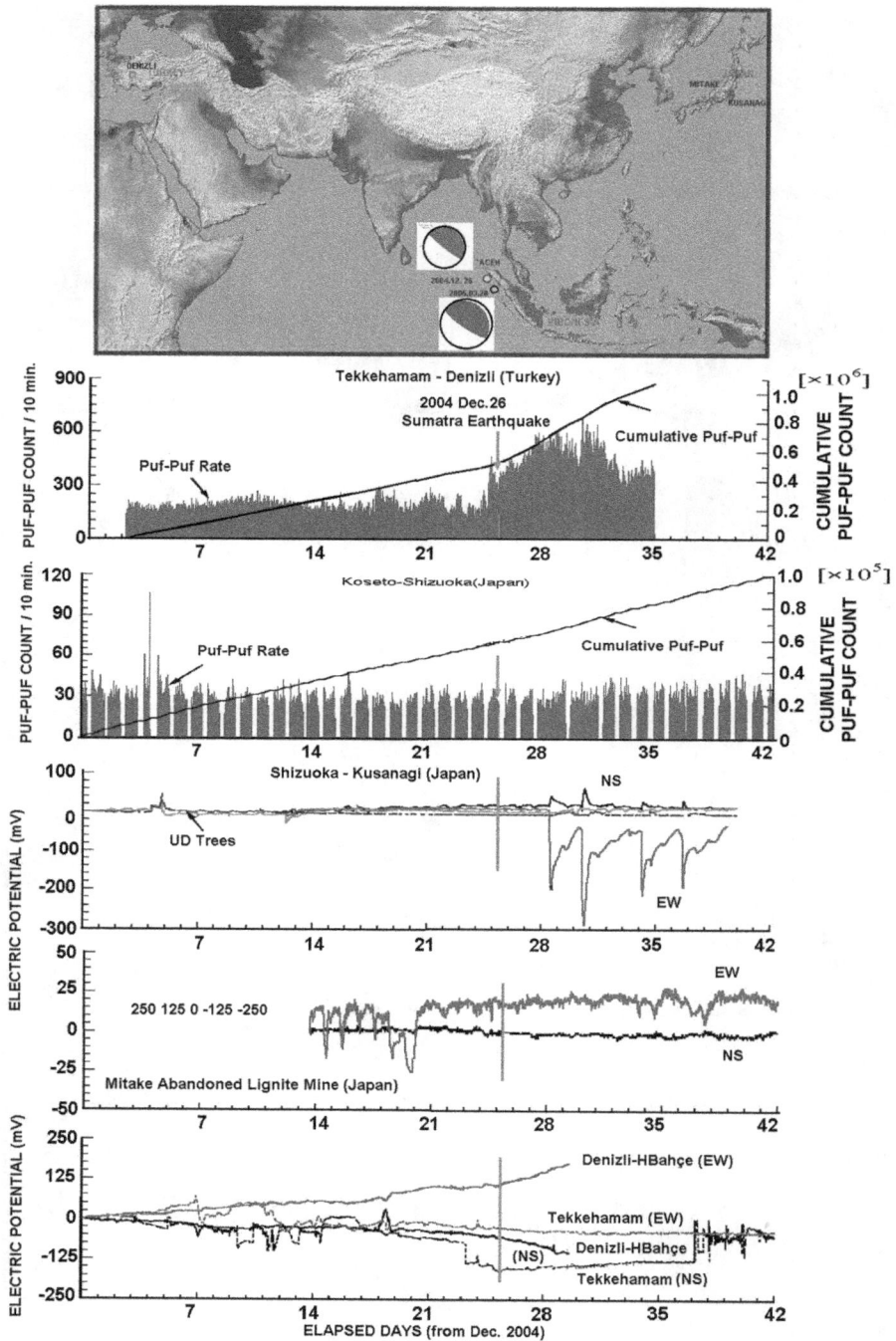

Figure 10.50 Puf-puf counts and electric potential variations at monitoring stations in Denizli Basin and in Shizuoka and Mitake in Japan associated with the 2004 and 2005 off-Sumatra earthquakes.

observations in the laboratory experiments, which indicated that precursor signals depend upon the size of the samples. Therefore, the period between the precursory electric potential variations (seismic electric signals – SES) and the earthquake are expected to be longer for large events.

Although the number of earthquakes and measurement period are still limited, a fairly good correlation exists between the crustal multi-parameter observations and earthquake activities. These preliminary results are very promising and it is expected that the system should be useful for short-term earthquake prediction if it is utilized together with other techniques such as GPS.

References

Abe, K. (1979) Size of great earthquakes of 1837–1974 inferred from tsunami data. *Journal of Geophysical Research* 84(B4), 1561–1568.

Abrahamson, N.A., and Somerville, P.G. (1996) Effects of the hanging wall and footwall on ground motions recorded during the Northridge earthquake. *Bulletin of the Seismological Society of America* 86(1B), 593–599.

AFAD (2007) 2007 Turkish seismic code, Ankara, Turkey.

Aggarwall, Y.P., Sykes, L.R., Armbruster, J., and Sbar, M.L. (1973) Premonitory changes in seismic velocities and prediction of earthquakes. *Science* 180, 632–635.

Aida, I. (1969) Numerical experiments for the tsunami propagation: The 1964 Niigata tsunami and the 1968 Tokachi-Oki tsunami tsunami. *Bulletin of the Earthquake Research Institute* 47, 673–700.

Aki, K. (1966) Generation and propagation of G waves from the Niigata earthquake of June 14, 1964. Part 2. Estimation of earthquake moment, released energy and stress-strain drop from G wave spectrum. *Bulletin of the Earthquake Research Institute* 44, 73–88.

Alpar, B., Altinok, Y., Gazioglu, C., and Yucel, Z.Y. (2003) Tsunami hazard assessment in Istanbul. *Turkish Journal of Maritime and Marine Sciences* 9(1), 3–29.

Altinok, Y., Tinti, S., Alpar, B., Yalciner, A.C., Ersoy, S., Bortolucci, E., and Armigliato, A. (2001) The tsunami of August 17, 1999 in Izmit Bay. *Turkey Natural Hazards* 24, 133–146.

Ambraseys, N.N. (1988) Engineering seismology. *Earthquake Engineering and Structural Dynamics* 17, 1–105.

Anderson, L.D. and Hart, R.S. (1976) An earth model based on free oscillations and body waves. *Journal of Geophysical Research* 81, 1461–1475.

Ando, M., Tsunoda, F., Hayakawa, Y., Hirahara, K., and Fujita, Y. (1996) *Earthquake and Volcanoes (Jishin to Kazan - in Japanese)*. Tokai University Publishing Co., Tokyo, p. 191.

Ariga, Y., Tsunoda, S., and Asaka, A. (2000) Determination of dynamic properties of existing concrete dam based on actual earthquake motions. 12WCEE, Paper No. 334, 8 p, Christcurch.

Asakura, T., and Sato, Y. (1998) Mountain tunnels damage in the 1995 in Hyogo-ken Nanbu Earthquake. Railway Technical Research Institute (RTRI), 39(1), pp. 9–16.

Asakura, T., Shiba, Y., Sato, Y., and Iwatate, T. (1996) Mountain tunnels performance in 1995 Hyogo-ken Nanbu Earthquake. Special Report of the 1995 Hyogo-ken nanbu Earthquake, Committee of Earthquake Engineering, JSCE.

Atak, O., Aksu, O., Aydan, Ö., Önder, M., and Toz, G. (2004) Measurement of ground deformation induced by liquefaction and faulting in the 1999 Kocaeli earthquake area. *Proceedings of the XXth ISPRS Congress*, 12–23 July 2004, Istanbul, Turkey, Paper No. 648 (on CD).

Aydan, Ö. (1986) Stability of slope and shallow underground openings in discontinuous rock mass. Nagoya University, Department of Geotechnical Engineering, Interim Report, 16 p. (unpublished).

Aydan, Ö. (1989) The stabilization of rock engineering structures by rockbolts. Doctorate Thesis, Nagoya University, 204 p.

Aydan, Ö. (1993) A consideration on the stress state of the earth due to the gravitational pull. *The 7th Annual Symposium on Computational Mechanics*, Tokyo, pp. 243–252.

Aydan, Ö. (1994) The dynamic shear response of an infinitely long visco-elastic layer under gravitational loading. *Soil Dynamics and Earthquake Engineering* 13, 181–186.

Aydan, Ö. (1995a) The stress state of the earth and the earth's crust due to the gravitational pull. *The 35th US Rock Mechanics Symposium*, Lake Tahoe, pp. 237–243.

Aydan, Ö. (1995b) Mechanical and numerical modelling of lateral spreading of liquefied soil. *International Conference on Earthquake Geotechnical Engineering*, Tokyo, pp. 881–886.

Aydan, Ö. (1996) Faulting and characteristics of earthquake waves in Hyogo-ken Nanbu Earthquake of January 17, 1995 (in Turkish). *Jeoloji Mühendisliği* 48, 63–77.

Aydan, Ö. (1997) The seismic characteristics and the occurence pattern of Turkish earthquakes. Turkish Earthquake Foundation, Report No.: TDV/TR 97-007, 41 p.

Aydan, Ö. (1998a) Analysis of masonary structures by finite element method. *Prof. Dr. Rifat Yarar Symposium*, Istanbul, Turkish Earthquake Foundation, TDV, pp. 141–150.

Aydan, Ö. (1998b) A simplified finite element approach for modelling the lateral spreading of liquefied ground. *The 2nd Japan-Turkey Workshop on Earthquake Engineering*, Istanbul.

Aydan, Ö. (2000a) A stress inference method based on structural geological features for the full-stress components in the earth's crust. *Yerbilimleri* 22, 223–236.

Aydan, Ö. (2000b) A stress inference method based on GPS measurements for the directions and rate of stresses in the earth' crust and their variation with time. *Yerbilimleri* 22, 21–32.

Aydan, Ö. (2001a) Comparison of suitability of submerged tunnel and shield tunnel for subsea passage of Bosphorus. *Engineering Geology Journal* 25(1), 1–17.

Aydan, Ö. (2001b) A finite element method for fully coupled hydro-thermo-diffusion problems and its applications to geo-science and geo-engineering. *The 10th IACMAG Conference*, Austin, pp. 781–786.

Aydan, Ö. (2001c) Modelling and analysis of fully coupled thermo-hydro-diffusion phenomena. IS-SHIZUOKA, Shizuoka, pp. 353–360.

Aydan, Ö. (2002) The inference of the earthquake fault and strong motions for Kutch earthquake of January 26, 2001. *A Symposium on the Records and Lessons from the Recent Large Domestic and Overseas Earthquakes*. JSCE, Earthquake Engineering Committee, Tokyo, pp. 135–140.

Aydan, Ö. (2003a) The earthquake prediction and earthquake risk in Turkey and the applicability of Global Positioning System (GPS) for these purposes. Turkish Earthquake Foundation, TDV/KT 024-87, 1–73 (in Turkish).

Aydan, Ö. (2003b) Actual observations and numerical simulations of surface fault ruptures and their effects engineering structures. The Eight U.S.-Japan Workshop on Earthquake Resistant Design of Lifeline Facilities and Countermeasures Against Liquefaction. Technical Report, MCEER-03-0003, pp. 227–237.

Aydan, Ö. (2004a) An experimental study on the dynamic responses of geomaterials during fracturing. *Journal of School of Marine Science and Technology, Tokai University* 1(2), 1–7.

Aydan, Ö. (2004b) Implications of GPS-derived displacement, strain and stress rates on the 2003 Miyagi-Hokubu earthquakes. *Yerbilimleri* 30, 91–102.

Aydan, Ö. (2004c) A reconnaissance report on Niigata Chuetsu earthquake of October 23, 2004. http://scc.u-tokai.ac.jp/jishin/chuetsu/, 57 p.

Aydan, Ö. (2006a) Strong motions, ground liquefaction and slope instabilities caused by the Kashmir earthquake of October 8, 2005 and their effects on settlements and buildings. *Bulletin of Engineering Geology, 23,* 15–34.

Aydan, Ö. (2006b) Geological and seismological aspects of Kashmir Earthquake of October 8, 2005 and geotechnical evaluation of induced failures of natural and cut slopes. *Journal of Marine Science and Technology, Tokai University* 4(1), 25–44.

Aydan, Ö. (2006c) The possibility of earthquake prediction by Global Positioning System (GPS). *Journal of School of Marine Science and Technology* 4(3), 77–89.

Aydan, Ö. (2007a) Inference of seismic characteristics of possible earthquakes and liquefaction and landslide risks from active faults (in Turkish). *The 6th National Conference on Earthquake Engineering of Turkey*, Istanbul, vol. 1, pp. 563–574.

Aydan, Ö. (2007b) The response of tunnels in liquefiable ground during earthquakes with special emphasis on bosphorus immersed tunnel. *Proceedings of 2nd Symposium on Underground Excavations for Transportation*, Istanbul, pp. 273–282.

Aydan, Ö. (2008a) Investigation of the seismic damage to the cave of Gunung Sitoli (Tögi-Ndrawa) by the 2005 Great Nias earthquake. *Yerbilimleri* 29(1), 1–16.

Aydan, Ö. (2008b) Some thoughts on seismic and tsunami hazard potentials in Indonesia with a special emphasis on Sumatra Island. *Journal of the School of Marine Science and Technology* 6(3), 19–38.

Aydan, Ö. (2009) The 13 March 1992 Erzincan earthquake and some thoughts and evaluations on its surface effects. *Bulletin of Engineering Geology, Turkey* 28–29, 33–56.

Aydan, Ö. (2012a) Ground motions and deformations associated with earthquake faulting and their effects on the safety of engineering structures. In: *Encyclopedia of Sustainability Science and Technology*, R. Meyers (Ed.). Springer, Berlin, Heidelberg, pp. 3233–3253.

Aydan, Ö. (2012b) The inference of physico-mechanical properties of soft rocks and the evaluation of the effect of water content and weathering on their mechanical properties from needle penetration tests. *The 46th US Rock Mechanics/Geomechanics Symposium, ARMA 12-639*, Chicago, Paper No. 639, 10 p. (on CD).

Aydan, Ö. (2013) The effects of earthquakes on rock slopes. *The 47th US Rock Mechanics/Geomechanics Symposium*, ARMA 13-378, San Francisco, 10 p.

Aydan, Ö. (2015a) Some considerations on a large landslide at the left bank of the Aratozawa Dam caused by the 2008 Iwate-Miyagi intraplate earthquake. *Rock Mechanics and Rock Engineering* 49(6), 2525–2539.

Aydan, Ö. (2015b) Large rock slope failures induced by recent earthquakes. *Rock Mechanics and Rock Engineering* 49(6), 2503–2524.

Aydan, Ö. (2015c) Crustal stress changes and characteristics of damage to geo-engineering structures induced by the Great East Japan Earthquake of 2011. *Bulletin of Engineering Geology and the Environment* 74(3), 1057–1070.

Aydan, Ö. (2015d) A critical testing of the applicability of some empirical relations used in the science and engineering of earthquakes through the 2011 great East Japan earthquake. *Bulletin of Engineering Geology and the Environment* 74(4), 1243–1254.

Aydan, Ö. (2016a) An integrated approach for the evaluation of measurements and inferences of in-situ stresses. *RS2016 Symposium, The 7th International Symposium on In-situ Rock Stress*, Tampere, pp. 38–57.

Aydan, Ö. (2016b) *Time Dependency in Rock Mechanics and Rock Engineering*. CRC Press, Taylor & Francis Group, New York, 241 p.

Aydan, Ö. (2017) *Rock Dynamics*. CRC Press, Taylor & Francis Group, New York, 462 p.

Aydan, Ö. (2018a) Infrared thermographic imaging in geoengineering and geoscience. In: *Encyclopedia of Sustainability Science and Technology*, R.A. Meyer (Ed.). Springer, Berlin, Heidelberg, pp. 413–438.

Aydan, Ö. (2018b) Multi-parameter and infrared monitoring systems for real-time rockburst susceptibility evaluation and their applications to practice in Japan, Section 8.4. In: *Rockburst: Mechanisms, Monitoring, Warning and Mitigation*, X.-T. Feng (Ed.). Elsevier, Amsterdam, Netherlands, pp. 282–299.

Aydan, Ö. (2018c) Some thoughts on the risk of natural disasters in Ryukyu Archipelago. *International Journal of Environmental Science and Technology* 9(10), 282–289.

Aydan, Ö. (2019a) *Rock Mechanics and Rock Engineering: Fundamentals*, Vol. 1. CRC Press, Taylor & Francis Group, New York, 406 p.

Aydan, Ö. (2019b) *Rock Mechanics and Rock Engineering: Applications.* CRC Press, Taylor & Francis Group, New York, 410 p.

Aydan, Ö. (2019c) A revisiting of photo-elasticity technique and its renewed potential use in Rock Mechanics and Rock Engineering. *The 47th Japan Rock Mechanics Symposium, JSCE,* Tokyo, pp. 299–304.

Aydan, Ö. (2021) *Continuum and Computational Mechanics for Geomechanical Engineers.* CRC Press, Taylor & Francis Group, New York, ISRM Book Series No 7, 324 p.

Aydan, Ö., and Amini, M.G. (2009) An experimental study on rock slopes against flexural toppling failure under dynamic loading and some theoretical considerations for its stability assessment. *Journal of Marine Science and Technology, Tokai University* 7(2), 25–40.

Aydan, Ö. and Daido, M. (2002) An experimental study on the seepage induced geo-electric potential in porous media. *Journal of the School of Marine Science and Technology, Tokai University* 55, 53–66.

Aydan, Ö. and Geniş, M. (2008a) Assessment of dynamic stability of an abandoned room and pillar underground lignite mine (in Turkish). *Turkish National Bulletin of Rock Mechanics,* TNSRM, Ankara 16, 23–44.

Aydan, Ö., and Geniş, M. (2008b) The seismic effect on the Bukittinggi WWII underground shelter by 2007 Singkarak (Solok) earthquake. *The 5th Asian Rock Mechanics Symposium (ARMS5),* Tehran, pp. 917–924.

Aydan, Ö., and Geniş, M. (2014) A numerical study on the response and stability of abandoned lignite mines in relation to the excavation of a large underground opening below. *The 14th International Conference of International Association for Computer Methods and Advances in Geomechanics (IACMAG),* Kyoto, Japan.

Aydan, Ö., and Hamada M. (2006) Damage to civil engineering structures by October 8, 2005 Kashmir Earthquake and recommendations for recovery and reconstruction. *Journal of Disaster Research* 1(3), 1–9.

Aydan, Ö., and Hamada, M. (1992) The site investigation of the Erzincan (Turkey) Earthquake of March 13, 1992. *The 4th Japan-US Workshop on earthquake Resistant Design of Lifeline Facilities and Countermeasures Against Soil Liquefaction,* Honolulu, May, 17–34.

Aydan, Ö., and Kawamoto, T. (1992) The stability of slopes and underground openings against flexural toppling and their stabilisation. *Rock Mechanics and Rock Engineering* 25(3), 143–165.

Aydan, Ö., and Kawamoto, T. (1997) The general characteristics of the stress state in the various parts of the earth's crust. *International Symposium of Rock Stress,* Kumamoto, pp. 369–373.

Aydan, Ö., and Kawamoto, T. (2004) The damage to abandoned lignite mines caused by the 2003 Miyagi-Hokubu earthquake and some considerations on its causes. *The 3rd Asian Rock Mechanics Symposium,* Kyoto, pp. 525–530.

Aydan, Ö., and Kim, Y. (2002) The inference of crustal stresses and possible earthquake faulting mechanism in Shizuoka Prefecture from the striations of faults. *Journal of the School of Marine Science and Technology, Tokai University* 54, 21–35.

Aydan, Ö., and Kumsar, H. (1997a) A site investigation of October 1, 1995 Dinar Earthquake. Turkish Earthquake Foundation, TDV/DR 97-003, 115 p.

Aydan, Ö., and Kumsar, H. (1997b) A new liquefaction assessment method and its applications. *The 2nd Symposium on Geotechnical and Earthquake Problems of Izmir and its Environment,* Izmir, Turkey, pp. 1–10.

Aydan, Ö., and Kumsar, H. (2002) Dinamik Yükler altında şevlerin kama yenilmesi sırasında mekanik davranışlarının modellenmesi üzerine deneysel ve kuramsal bir yaklaşım. KAYAMEK'2002, VI. Bölgesel Kaya Mekaniği Sempozyumu, pp. 235–243.

Aydan, Ö., and Kumsar, H. (2010) An experimental and theoretical approach on the modeling of sliding response of rock wedges under dynamic loading. *Rock Mechanics and Rock Engineering* 43(6), 821–830.

Aydan, Ö., and Nawrocki, P. (1998) Rate-dependent deformability and strength characteristics of rocks. *International Symposium on the Geotechnics of Hard Soils-Soft Rocks*, Napoli, vol. 1, pp. 403–411.

Aydan, Ö., and Ohta, Y. (2011) A new proposal for strong ground motion estimations with the consideration of characteristics of earthquake fault. *Seventh National Conference on Earthquake Engineering*, Istanbul, Turkey Paper No 65, 1–10 pp.

Aydan, Ö., and Paşamehmetoğlu, A.G. (1994) Dünyanın çeşitli yörelerinde ölçülmüş yerinde gerilimler ve yatay gerilim katsayısı. *Kaya Mekaniği Bülteni* 10, 1–17.

Aydan, Ö., and Tano, H. (2003) Multi-parameter measurement system at Koseto (Shizuoka) and its response during the M8.3 Tokachi-oki earthquake of 2003. *International Colloquium on Instrumentation and Monitoring of Landslides and Earthquakes*, Japan, Turkey, Koriyama, (Edited by H. Tano and Ö. Aydan), pp. 101–110.

Aydan, Ö., and Tano, H. (2012) The observations on abandoned mines and quarries by the Great East Japan Earthquake on March 11, 2011 and their implications. *Journal of Japan Association on Earthquake Engineering* 12(4), 229–248.

Aydan, Ö., and Tokashiki, N. (2012) The failure of retaining walls and their evaluation. *A Symposium on the Seismic Evaluation of Retaining Walls and Their Reinforcement*, Tokyo, JSCE, pp. 28–33.

Aydan, Ö., and Tokashiki, N. (2019) Tsunami boulders and their implications on the mega earthquake potential along Ryukyu Archipelago, Japan. *Bulletin of Engineering Geology and Environment* 78, 3917–3925.

Aydan, Ö., and Ulusay, R. (2002) A back analysis of the failure of a highway embankment at Bakacak during the 1999 Düzce-Bolu earthquake. *Environmental Geology* 42, 621–631.

Aydan, Ö., Shimizu, Y., and Ichikawa, Y. (1989) The effective failure modes and stability of slopes in rock mass with two discontinuity sets. *Rock Mechanics and Rock Engineering* 22(3), 163–188.

Aydan, Ö., Ichikawa, Y., Murata, K., and Shimizu, Y. (1991) An integrated systems for the stability of rock slopes. *Proceedings of International Conference on Computer Methods and Advances in Geomechanics*, Cairns, vol. 1, pp. 469–474.

Aydan, Ö., Shimizu, Y., and Kawamoto, T. (1992a) The stability of rock slopes against combined shearing and sliding failures and their stabilisation. *International Symposium on Rock Slopes*, New Delhi, pp. 203–210.

Aydan, Ö., Shimizu, Y., and Kawamoto, T. (1992b) The reach of slope failures. *The 6th International Symposium on Slope failures, ISL 92*, Christchurch, vol. 1, pp. 301–306.

Aydan, Ö., Shimizu, Y., and Karaca, M. (1994) The dynamic and static stability of shallow underground openings in jointed rock masses. *The 3rd International Symposium on Mine Planning and Equipment Selection*, Istanbul, pp. 851–858.

Aydan, Ö., Üçpırtı, H., and Kumsar, H. (1996a) Stability of slopes having an inclined sliding surface with a visco-plastic behaviour. *Proceedings of the Korea-Japan Joint Symposium on Rock Engineering*, Seoul , Korea, pp. 437–444.

Aydan, Ö., Üçpırtı, H., and Kumsar, H. (1996b) The stability of a rock slope having a visco-plastic sliding surface. *Kaya Mekaniği Bülteni* 12, 39–49.

Aydan, Ö., Sezaki, M., and Yarar, R. (1996c) The seismic characteristics of Turkish Earthquakes. *The 11th World Conference on Earthquake Engineering*, Acapulco, Mexico, CD-2, Paper No 1025.

Aydan, Ö., Ulusay, R., Kumsar, H., and Ersen, A. (1996e) Buckling failure at an open-pit coal mine. *EUROCK'96*, Torino, 641–648.

Aydan, Ö., Mamaghani, I.H.P., and Kawamoto, T. (1996d) Application of discrete finite element method (DFEM) to rock engineering structures. *NARMS'96*, Montreal, pp. 2039–2046.

Aydan, Ö., Kumsar, H., and Utsumi, H. (1997a) A comparative study on empirical liquefaction assessment methods. *The 8th International Symposium on Earthquake Engineering and Soil Dynamics*, Istanbul.

Aydan, Ö., Kumsar, H., and Sakoda, S. (1997b) Model wedge tests and re-assessment of limiting equilibrium methods for wedge sliding. *Proccedings of Rock Mechanics and Evironmental Geotechology RMEG*, Chonging University Press, China, pp. 261–266.

Aydan, Ö., Üçpırtı, H., and ve Ulusay, R. (1997c) Theoretical formulation of Darcy's law for fluid flow through porous and/or jointed rock and its validity (in Turkish). *Rock Mechanics Bulletin, TRMS* 13, 1–18.

Aydan, Ö., Ulusay, R., Kumsar, H., Sönmez, H., and Tuncay, E. (1998) A site investigation of Adana-Ceyhan earthquake of June 27, 1998. Turkish Earthquake Foundation, Report No. TDV/DR 006-30, 131 p.

Aydan, Ö., Ulusay, R., Hasgür, Z., and Taşkın, B. (1999a) A site investigation of Kocaeli earthquake of August 17, 1999. Turkish Earthquake Foundation, TDV/DR 08-49, 180 p.

Aydan, Ö., Ulusay, R., Hasgür, Z., and Hamada, M. (1999b) The behaviour of structures built in active fault zones in view of actual examples from the 1999 Kocaeli and Chi-chi Earthquakes. *ITU-IAHS International Conference on the Kocaeli Earthquake 17 August 1999: A Scientific Assessment and Recommendations for Re-Building*, Istanbul, pp. 131–142.

Aydan, Ö., Ulusay, R., and Kumsar, H. (2000a) Liquefaction phenomenon in the earthquakes of Turkey, including recent Erzincan, Dinar and Adana-Ceyhan earthquakes. *The 12th WCEE*, Auckland.

Aydan, Ö., Ulusay, R., Kumsar, H., and Tuncay, E. (2000b) Site investigation and engineering evaluation of the Düzce-Bolu Earthquake of November 12, 1999. Turkish Earthquake Foundation, TDV/DR 095-51, 307 pp.

Aydan, Ö., Dalgıç, S., and Kawamoto, T. (2000c) Prediction of squeezing potential of rocks in tunnelling through a combination of an analytical method and rock mass classifications. *Italian Geotechnical Journal* 34(1), 41–45.

Aydan, Ö., Seiki, T., Shimizu, Y., and Hamada, M. (2000d). Some considerations on rock slope failures due to earthquakes. Chonquing-Waseda Joint Seminar on Chi-Chi Earthquake.

Aydan, Ö., Kumsar, H., and Ulusay, R. (2000e) The implications of crustal strain-stress rate variations computed from GPS measurements on the earthquake potential of Turkey. *International Conference of GIS on Earth Science and Applications, ICGESA'2000*, CD, Menemen (Turkey).

Aydan, Ö., Minato, T., and Fukue, M. (2001) An experimental study on the electrical potential of geomaterials during deformation and its implications in Geomechanics. The 38th US Rock Mechanics Symposium, Washington, vol. 2, pp. 1199–1206.

Aydan, Ö., Kumsar, H., and Ulusay, R. (2002a) How to infer the possible mechanism and characteristics of from the striations and ground surface traces of existing faults. *JSCE, Earthquake and Structural Engineering Division I* 19(2), 199–208.

Aydan, Ö., Tokashiki, N., and Harada, A. (2002b) An experimental study on the dynamic responses and stability of the retaining walls of masonry type. *Proceedings of ISRM Regional Symposium on Rock Engineering Problems and Approaches in Underground Construction*, Seoul, Korea, vol. 2, pp. 769–776.

Aydan, Ö., Tokashiki, N., Ito, T., Akagi, T., Ulusay, R., and Bilgin, H.A. (2003) An experimental study on the electrical potential of non-piezoelectric geomaterials during fracturing and sliding. *The 9th ISRM Congress*, South Africa, pp. 73–78.

Aydan, Ö., Hamada, M., and Suzuki, Y. (2005a) Some observations and considerations on the damage induced by the tsunami of the 2004 Sumatra earthquake on structures and coast. *Journal of the School of Marine Science and Technology*, Tokai University 3(1), 79–94.

Aydan, Ö., Miwa, S., Kodama, H., and Suzuki, T. (2005b) The characteristics of M8.7 Nias earthquake of March 28, 2005 and induced tsunami and structural damages. *Journal of the School of Marine Science and Technlogy*, Tokai University 3(2), 66–83.

Aydan, Ö., Atak, V.O., Ulusay, R., Hamada, H., and Bardet, J.P. (2005c) Ground deformations and lateral spreading around the shore of Sapanca Lake induced by the 1999 Kocaeli earthquake. *Proceedings of the Geotechnical Earthquake Engineering Satellite Conference, Performance Based Design in Earthquake Geotechnical Engineering: Concepts and Research*, Osaka, Japan, pp. 54–61.

Aydan, Ö., Kumsar, H., and Tano, H. (2005d) Multi-parameter changes in the earth's crust and its relation to earthquakes in Denizli region of Turkey. *World Geothermal Congress 2005*, Antalya.

Aydan, Ö., Daido, M., Tano, H., Tokashiki, N., and Ohkubo, K. (2005e) A real-time multi-parameter monitoring system for assessing the stability of tunnels during excavation. *ITA Conference*, Istanbul, pp. 1253–1259.

Aydan, Ö., Hamada, H., and Konagai, K. (2006a) An evaluation of strong ground motions and failures of natural and cut slopes induced by Kashmir Earthquake of October 8, 2005. *The 1st European Conference on Earthquake Engineering and Seismology*, Geneva, Paper No. 1352 (on CD).

Aydan, Ö., Daido, M., Ito, T., Tano, H., and Kawamoto, T. (2006b) Instability modes of abandoned lignite mines and the assessment of their stability in long-term and during earthquakes. *The 3rd Asian Rock Mechanics Symposium*, Singapore (on CD).

Aydan, Ö., Hamada, M., and Sakamoto, A. (2007a) Characteristics of Noto Peninsula earthquakes with an emphasis on induced geotechnical damage. *Journal of the School of Marine Science and Technology* 5(2), 13–31.

Aydan, Ö., Miwa, S., Kodama, K., Suzuki, T., and Hamada, M. (2007b) Support activities of JSCE and EWoB-Japan for Nias Island following the Great Nias earthquake of 2005. *The International Symposium on Disaster in Indonesia (ISDI): Problems and Solutions, MS-3*, Padang, pp. 1–20.

Aydan, Ö., Daido, M., Tokashiki, N., Bilgin, A., and Kawamoto, T. (2007c) Acceleration response of rocks during fracturing and its implications in earthquake engineering. *The 11th ISRM Congress*, Lisbon, vol. 2, pp. 1095–1100.

Aydan, Ö., Tokashiki, N., and Sugiura, K. (2008a) Characteristics of the 2007 Kameyama earthquake with some emphasis on unusually strong ground motions and the collapse of Kameyama Castle. *Journal of the School of Marine Science and Technology* 6(1), 83–105.

Aydan, Ö., Ulusay R., and Atak, V.O. (2008b) Evaluation of ground deformations induced by the 1999 Kocaeli earthquake (Turkey) at selected sites on shorelines. In: *Environmental Geology*. Springer Verlag, Berlin, Heidelberg, vol. 54, pp. 165–182.

Aydan, Ö., Kumsar, H., and Toprak, S. (2009a) The 2009 L'Aquila earthquake (Italy): Its characteristics and implications for earthquake science and earthquake engineering. *Yerbilimleri* 30 (3), 235–257.

Aydan, Ö., Kumsar, H., Toprak, S., and Barla, G. (2009b) Characteristics of 2009 l'Aquila earthquake with an emphasis on earthquake prediction and geotechnical damage. *Journal Marine Science and Technology*, Tokai University 9(3), 23–51.

Aydan, Ö., Ohta, Y., Hamada, M., Ito, J., and Ohkubo, K. (2009c) The response and damage of structures along the fault rupture traces of the 2008 Wenchuan Earthquake. *International Conference on Earthquake Engineering: The 1st Anniversary of Wenchuan Earthquake*, Chengdu, pp. 625–633.

Aydan, Ö., Ohta, Y., Hamada, M., Ito, J., and Ohkubo, K. (2009d) The characteristics of the 2008 Wenchuan Earthquake disaster with a special emphasis on rock slope failures, quake lakes and damage to tunnels. *Journal of the School of Marine Science and Technology*, Tokai University 7(2), 1–23.

Aydan, Ö., Hamada, M., Itoh, J, and Ohkubo, K. (2009e) Damage to civil engineering structures with an emphasis on rock slope failures and tunnel damage induced by the 2008 Wenchuan Earthquake. *Journal of Disaster Research* 4(2), 153–164.

Aydan, Ö., Ohta, Y., and Hamada, M. (2009f) Geotechnical evaluation of slope and ground failures during the 8 October 2005 Muzaffarabad earthquake in Pakistan. *Journal Seismology* 13(3), 399–413.

Aydan, Ö., Ohta, Y., Geniş, M., Tokashiki, N., and Ohkubo, K. (2010a) Response and Earthquake induced damage of underground structures in rock mass. *Journal of Rock Mechanics and Tunneling Technology* 16(1), 19–45.

Aydan, Ö., Ohta, Y., Geniş, M., Tokashiki, N., and Ohkubo, K. (2010b) Response and stability of underground structures in rock mass during earthquakes. *Rock Mechanics and Rock Engineering* 43(6), 857–875.

Aydan, Ö., Kumsar, H., Toprak, S., and Barla, G. (2010c) Characteristics of 2009 L'Aquila earthquake with an emphasis on earthquake prediction and geotechnical damage. *Journal Marine Science and Technology*, Tokai University, 7(3), 23–51.

Aydan, Ö., Tano, H., and Ohta, Y. (2010d) A multi-parameter measurement system at Koseto (Shizuoka, Japan) and its responses during the large earthquakes since 2003. *Proceedings World Geothermal Congress 2010*, Bali, Indonesia, 25–29 April 2010.

Aydan, Ö., Ohta, Y., and Tano, H. (2010e) Multi-parameter response of soft rocks during deformation and fracturing with an emphasis on electrical potential variations and its implications in geomechanics and geoengineering. *The 39th Rock Mechanics Symposium of Japan*, Tokyo, pp. 116–121.

Aydan, Ö., Ohta, Y., Daido, M., Kumsar, H., Genis, M., Tokashiki, N., Ito, T., and Amini, M. (2011) Chapter 15: Earthquakes as a rock dynamic problem and their effects on rock engineering structures. In: *Advances in Rock Dynamics and Applications*, Y. Zhou and J. Zhao (Eds.). CRC Press, Taylor & Francis Group, New York, pp. 341–422.

Aydan, Ö., Ulusay, R., Hamada, M., and Beetham, D. (2012a) Geotechnical aspects of the 2010 Darfield and 2011 Christchurch earthquakes of New Zealand and geotechnical damage to structures and lifelines. *Bulletin of Engineering Geology and Environment* 71(4), 637–662.

Aydan, Ö., Uehara, F., and Kawamoto, T. (2012b) Numerical study of the long-term performance of an underground powerhouse subjected to varying initial stress states, cyclic water heads, and temperature variations. *International Journal of Geomechanics, ASCE* 12(1), 14–26.

Aydan, Ö., Fuse, T., and Ito, T. (2015) An experimental study on thermal response of rock discontinuities during cyclic shearing by Infrared (IR) thermography. *Proceedings of 43rd Symposium on Rock Mechanics, JSCE*, Tokyo, pp. 123–128.

Aydan, Ö., Nasiry, N.Z., Ohta, Y., and Ulusay, R. (2018a) Effects of earthquake faulting on civil engineering structures. *Journal of Earthquake and Tsunami* 12(4), 1841001.

Aydan, Ö., Tomiyama, J., Matsubara, H., Tokashiki, N., and Iwata, N. (2018b) Damage to rock engineering structures induced by the 2016 Kumamoto earthquakes. *The 3rd International Symposium on Rock Dynamics, RocDyn3*, Trondheim, pp. 525–531.

Aydan, Ö., Ohta, Y., Iwata, N., and Kiyota, R. (2019a) The evaluation of static and dynamic frictional properties of rock discontinuities from tilting and stick-slip tests. *Proceedings of 46th Japan Rock Mechanics Symposium*, Iwate, pp. 105–110.

Aydan, Ö., Ohta, Y., Amini, M., and Shimizu, Y. (2019b) The dynamic response and stability of discontinuous rock slopes. *Proceedings of 2019 Rock Dynamics Summit*, ISRM, Okinawa, Japan, 7–11 May, 2019. (Editors: Aydan, Ö., Ito, T., Seiki T., Kamemura, K., Iwata, N.), pp. 519–524.

Aydan, Ö., Tomiyama, J., Horiuchi, K., Aasim, B., Karimi, A.K., and Nasiry, N.Z. (2020) A comparative study on stress distributions in physical models using photo-elasticity and FEM. *IACMAG 2021*, Torino, pp. 3–21.

Aydan, Ö., Kiyota, R., and Iwata, N. (2022) Tilting and stick-slip tests for evaluating static and dynamic frictional properties of rock discontinuities. *Rock Mechanics and Rock Engineering* (in press).

Aytun, A. (1973) Creep measurements in the vicinity of Ismetpaşa Railway Station (in Turkish). *Proceedings of Symposium on the North Anatolian Fault and Earthquake Belt*, Ankara, March 29–31; published by the Mineral Research and Exploration Institute, MTA, Ankara.

Aytun, A. (1982) Creep measurements in the Ismetpaşa region of the North Anatolian fault zone. In: *Proceedings, Multidisplinary Approach to Earthquake Prediction, Friedr*, A.M. Isikara and A. Vogel (Eds.) Vieweg and Sohn, Braunshweig/Wiesbaden, pp. 279–292.

Bardet, J.P., Mace, N., Tobita, T., and Hu, J. (1999) Large-scale modeling of liquefaction-induced ground deformation Part I: A four parameter MLR model. *Proceedings of the 7th U.S.-Japan Workshop on Earthquake Resistant Design of Lifeline Facilities and Countermeasures against Soil Liquefaction*, Washington, pp. 155–173.

Barka, A. (1996) Slip distribution along the North Anatolian faultassociated with the large earthquakes of the period 1939–1967. *Bulletin of the Seismological Society of America* 86, 1238–1254.

Barrientos, S.E., and Ward, S.N. (1990) The 1960 Chile earthquake; inversion for slip distribution from surface deformation. *Geophysical Journal International* 103, 589–598.

Barsukov, O., and Sorokin, O.N. (1973) Variations in apparent resistivity of rocks in the seismically active Garm region. *Physics of Solid Earth* 10, 685.

Bartlett, S.F., and Youd, T.L. (1992) Empirical analysis of horizontal ground displacement generated by liquefaction-induced lateral spreads. Technical Report, NCEER 92-0021, National Center for Earthquake Engineering Research, Buffalo, New York, 5–15.

Bartlett, S.F. and Youd, T.L. (1995) Empirical prediction of liquefaction induced lateral spreading. *Journal of Geotechnical Engineering* 121(4), 316–329.

Biot, M. (1956) Theory of propagation of elastic waves in a fluid-saturated porous solid. *Journal of the Acoustical Society of America* 28, 168–191.

Bishop, A.W. (1955) The use of the slip-circle in the stability analysis of slopes. *Geotechnique* 5(1), 7–17.

Bowden, F.P. and Leben, L. (1939) The nature of sliding and the analysis of friction. *Proceedings of the Royal Society, London A* 169, 371–391.

Bowden, F.P., and Tabor, D. (1950) *The Friction and Lubrication of Solids*. The Clarendon Press, Oxford.

Brace, W.F., and Byerlee, J.D. (1966) Stick-slip as a mechanism for earthquakes. *Science* 153, 990–992.

Brown, E.T., and Hoek, E. (1978) Trends in relationships between measured in-situ stresses and depth. *International Journal of Rock Mechanics and Mining Sciences* 15, 211–215.

Buskirk, R.E., Frohlich, C., and Latham, G.V. (1981) Unusual animal behaviour before earthquakes: A review of possible sensory mechanisms. *Reviews of Geophysics and Space Physics* 19(2), 247–270.

Büyüksaraç, A., Reiprich, S., and Ateş, A. (1998) Three-dimensional magnetic model of amphibolite complex in Taskesti area, Mudurnu valley, North-West Turkey. *Journal of the Balkan Geophysical Society* 1(3), 44–52.

Byerlee, J.D. (1970) The mechanics of stick-slip. *Tectonophysics* 9, 475–486.

Byerlee, J. (1978) Friction of rocks. *Pure and Applied Geophysics* 116, 615–626.

Cakir, Z., Akoglu, A.M., Belabbes, S., Ergintav, S., and Meghraoui, M. (2005) Creeping along the İsmetpaşa section of the North Anatolian fault (Western Turkey): Rate and extent from InSAR. *Earth and Planetary Science Letters* 238, 225–234.

Campbell, K.W. (1981) Near source attenuation of peak horizontal acceleration. *Bulletin of the Seismological Society of America* 79(3), 549–580.

Chandrasekaran, A.R., and Krishna, J. (1954) Water towers in seismic zones. *Proceedings of the Third World Conference on Earthquake Engineering*, New Zealand, vol. IV, pp. 161–171.

Chang, T.-Y., Cotton, F., and Angelier, J. (2001) Seismic attenuation and peak ground acceleration in Taiwan. *Bulletin of the Seismological Society of America* 91(5), 1229–1246.

Chopra, A.K. (2020) *Earthquake Engineering for Concrete Dams: Analysis, Design, and Evaluation*. John Wiley & Sons, New York, 297 p.

Clough, R.W. and Penzien, J. (1975) *Dynamics of Structures*. McGraw, Hill, Inc., New York, 634 p.

Cohen, B.P., Somerville, P.G., and Abrahamson, N.A. (1991) Simulated ground motions for hypothesized $MW=8$ subduction earthquake in Washington and Oregon. *Bulletin of the Seismological Society of America* 81, 28–56.

CWB (1999) Free field strong-motion data from the 921 Chi-Chi earthquake. Seismological Center, Central Weather Bureau, Taipei, Taiwan.

DAD-ERD. (recently named AFAD) Earthquake Research Department. Earthquake and strong motion data bases of Turkey. http://www.deprem.gov.tr.

Dames & Moore. (1980) *Design and Construction Standards for Residential Construction in Tsunami-Prone Areas in Hawaii.* Prepared for the Federal Emergency Management Agency, Washington, DC.

Dean, R.G., and Dalrymple, R.A. (1998) *Water Wave Mechanics for Engineers and Scientists.* World Scientific Publishing Co., Singapore, 363 p.

Derr, J.S. (1973) Earthquake lights: A review of observations and present theories. *Bulletin of the Seismological Society of America* 63(6), 2177–2187.

Dobry, R., and Baziar, M.H. (1992) Modeling of lateral spreads in silty sands by sliding soil blocks. *A Specialty Conference on Stability and Performance of Slopes and Embankments*, University of California, Berkeley, 1–28.

Dobry, R., Idriss, I.M., and Ng, E. (1978) Duration characteristics of horizontal components of strong motion earthquake records. *Bulletin of the Seismological Society of America* 68(5), 1487–1520.

Dowding, C.H., and Rozen, A. (1978) Damage to rock tunnels from earthquake shaking. *Journal of the Geotechnical Engineering Division* ASCE, GT2, 175–191.

Ducellier, A., and Aochi, H. (2012) Effects of the interaction of topographic irregularities on seismic ground motion, investigated by the hybrid FD-FE method. Bulletin of Earthquake Engineering 10, 773–792.

EERI (1986) Reducing earthquake hazards: Lessons learned from past earthquakes. Earthquake Engineering Research Institute, Publication No: 86-02, El Cerito CA.

Egger, P. (1979) A new development in the base friction technique. *Colloquium on Geomechanical Models, ISMES*, Bergamo, 67–81.

Emre, Ö., Duman, T.Y., Doğan, A., Ateş, S., Keçer, Mç, Erkal, T., Özalp, S., Yıldırım, N., and Güner, N. (1999) 12 Kasım 1999 Düzce Depremi saha gözlemleri ve ön değerlendirme raporu. MTA.

Ergin, K., Güçlü, U., and ve Uz, Z. (1967) The catalogue of earthquakes of Turkey - Türkiye ve civarının deprem kataloğu (MS. 11–1964). İTÜ Maden Fakültesi Arz Fiziği Enstitüsü Yayınları, No 28 (in Turkish and English).

Eringen, A.C. (1980) *Mechanics of Continua*, R.E. Krieger Pub. Co., New York.

EUREF. (2000) The coordinate time series of Ankara and Gebze GPS stations. http://homepage.oma.be/euref/eurefhome.html

Eyidoğan, H., Güçlü, U., Utku, Z., and Değirmenci, E. (1991) A macro-seismic catalog of great earthquakes of Turkey for 1900–1988 (Türkiye büyük depremleri makrosismik rehberi), İTÜ, 199 p.

FEMA P-646. 2012. *Guidelines for Design of Structures for Vertical Evacuation from Tsunamis* (2nd Edn). Federal Emergency Management Agency, Washington, DC, 174 p.

Finkelstein, D., Hill, R.D., and Powell, J.R. (1973) The piezoelectric theory of earthquake lights. *Journal of Geophysical Research* 78(6), 992–993.

Fleeger, G.M., and Goode, D.J. (1999) Hydrologic effects of the Pymatuning Earthquake of September 25, 1998, in Northwestern Pennsylvania. U.S. Department of Interior, USGS, Water-Resources Investigations Report 99-4170.

Fowler, C.M.R. (1990) *The Solid Earth: An Introduction to Global Geophysics.* Cambridge University Press, Cambridge.

Fritz, H.M., Hager, W.H., and Minor, H.E. (2001) Lituya Bay case: Rockslide impact and wave run-up. *Science of Tsunami Hazards* 19, 3–22.

Friedman, H., Arıç, K., King, C.Y., Altay, C., and Sau, H. (1988) Radon measurements for earthquake prediction along the North Anatolian Fault Zone; a progress report. *Tectonophysics* 152, 209–214.

Fukushima, K., Yuji Kanaori, K., and Fusanori Miura, F. (2010) Influence of fault process zone on ground shaking of inland earthquakes: Verification of Mj=7.3 Western Tottori Prefecture and Mj=7.0 West Off Fukuoka Prefecture earthquakes, southwest Japan. *Engineering Geology* 116, 157–165.

Geller, R. (1997) Earthquakes cannot be predicted. *Science* 275, 161.

Gençoğlu, S., İnan, I., and Güler, H (1990) Türkiyenin Deprem Tehlikesi. (Earthquake danger of Turkey) Publication of the Chamber of Geophysical Engineers of Turkey, Ankara, 701 p.

Geniş, M. (2002) Evaluation of dynamic response and stability of shallow underground openings in discontinuous rock masses using model tests. Doctorate Thesis, Zonguldak Karaelmas University.

Geniş, M., and Aydan, Ö. (2002) Evaluation of dynamic response and stability of shallow underground openings in discontinuous rock masses using model tests. *Korea-Japan Joint Symposium on Rock Engineering*, Seoul, Korea, July, pp. 787–794.

Geniş, M., and Aydan, Ö. (2007) Static and dynamic stability of a large underground opening. *Proceedings of 2nd Symposium on Underground Excavations for Transportation*, Istanbul, pp. 317–326.

Geniş, M., and Aydan, Ö. (2008) Assessment of dynamic response and stability of an abandoned room and pillar underground lignite mine. *The 12th International Conference of International Association for Computer Methods and Advances in Geomechanics (IACMAG)*, Goa, India, pp. 3899–3906.

Geniş, M. and Gerçek, H. (2003) A numerical study of seismic damage to deep underground openings. *ISRM 2003-Technology Roadmap for Rock Mechanics, 10th ISRM Congress*, South African Institute of Mining and Metallurgy, South Africa, pp. 351–355.

Gerstenecker, C., Akın, D., and Demirel, H. (1999) Gravity changes along the western part of the North Anatolian Fault. http://www.gfz-potsdam.de/pb2/pb21/Mudurnu/.

Goto, K., Kawana, T., and Imamura, F. (2010) Historical and geological evidence of boulders deposited by tsunamis, southern Ryukyu islands, Japan. *Earth-Science Review* 102, 77–99.

Graves, R. (1996) Simulating seismic wave propagation in 3D elastic media using staggered-grid finite differences. *Bulletin of the Seismological Society of America* 86(4), 1091–1106.

GSI (Geographical Information Authority of Japan). Special web sites for selected earthquakes. http://www.gsi.go.jp/BOUSAI/.

GSI (Geographical Survey Institute of Japan). (2011) Disaster events (2004 Chuetsu earthquake, 2011 Tohoku Earthquake. http://www.gsi.go.jp/.

Gürpinar, A., Erdik, M., Yücemen, S., and Öner, M. (1979) Risk analysis of northern anatolia based on intensity attenuation. *Proceedings of the 2nd U.S. National Conference*, Stanford, California, pp. 72–81.

Hamada, M. (1999) Similitude law for liquefied-ground flow. *Proceedings of the 7th U.S-Japan Workshop on Earthquake Resistant Design of Lifeline Facilities and Countermeasures against Soil Liquefaction*, Salt Lake City, pp. 191–205.

Hamada, M. (2011) Lessons learned from the Great East Japan Earthquake and strategies for disaster reduction in Tokyo Metropolitan Area (in Japanese). *Journal of Japan Society for Safety Engineering* 50(6), 474–478.

Hamada, M. and Aydan, Ö. (1992) The site investigation of the March 13 Earthquake of Erzincan, Turkey. Reconnaissance Report for Association for the Development of Earthquake Prediction (ADEP), 86 p.

Hamada, M. and Kuno, M. (2014) *Earthquke Engineering for Nuclear Facilities*. Springer, Berlin, Heidelberg, 303 p.

Hamada, M. and Wakamatsu, K. (1998) Astudy on ground displacement caused by soil liquefaction. *Geotechnical Journal JSCE* 596 (III-43), 189–208.

Hamada, M., Yasuda, S., Isoyama, R., and Emoto, K. (1986) Study on liquefaction induced permanent ground displacements. Association for the Development of Earthquake Prediction in Japan, 1–87.

Hamada, M., O'Rouke, T.D., and Yoshida, N. (1993) Liquefaction-induced large ground displacement. Performance of Ground and Soil Structures during Earthquakes. *The 13th International Conference on Soil Mechanics and Foundation Engineering, Japanese Society for Soil Mechanics and Foundation Engineering*, New Delhi, 1–16.

Hanks, T.C., and Kanamori, H. (1979) A moment magnitude scale. *Journal of Geophysical Research* 84(B5), 2348–2350.

Haroun, M.A. (1983) Vibration studies and test of liquid storage tanks. *Earthquake Engineering and Structural Dynamics* 11, 179–206.

Hartzell, S.H. (1978) Earthquake aftershocks as Green's functions. *Geophysical Research Letters* 5, 1–4.

Hartzell, S.H., and Heaton, T.H. (1983) Inversion of strong ground motion and teleseismic waveform data for fault rupture history of the 1979 Imperial Valley, California earthquake. *Bulletin of the Seismological Society of America* 73, 1553–1583.

HARVARD (2011) Focal mechanism catalogue. http://www.seismology.harvard.edu/.

Hashimoto, S., Miwa, K., Ohashi, M., and Fuse, K. (1999) Surface soil deformation and tunnel deformation caused by the September 3, 1998, Mid-North Iwate Earthquake. *The 7th Tohoku Regional Convention*, Japan Society of Engineering Geology, Sendai.

Hestholm, S. (1999) Three-dimensional finite difference viscoelastic wave modelling including surface topography. *Geophysical Journal International* 139, 852–878.

Hirth, G., and Tullis, J. (1994) The brittle-plastic transition in experimentally deformed quartz aggregates. *Journal of Geophysical Research* 99, 11731–11747.

Hoek, E., and Brown, E.T. (1980) *Underground Excavations in Rock*. Inst. Mining, London.

Horiuchi, K., Aydan, Ö., and Tokashiki, N. (2018) Recent failures of limestone cliffs in Ryukyu Archipelago and their analyses. *Proceedings of 45th Rock Mechanics Symposium of Japan*, JSCE, Tokyo, pp. 131–136.

Housner, G.W. (1957) Dynamic pressures on accelerated fluid containers. *Bulletin of the Seismological Society of America* 47, 15–35.

Housner, G.W. (1961) Vibration of structures induced by seismic waves. In: *Shock and Vibration Handbook*, C.M. Harris and C.E. Crede (Eds.), McGraw Hill, New York, 501–532.

Housner, G.W. (1963) Dynamic behavior of water tanks. *Bulletin of the Seismological Society of the America* 53, 381–387.

Huber, A. and Hager, W.H. (1997) Forecasting impulse waves in reservoirs. *Dix-neuvième Congrès des Grands Barrages,* Florence, Commission Internationale des Grands Barrages, pp. 993–1005.

Hutchings, L. and Viegas, G. (2012) Application of empirical green's functions in earthquake source, wave propagation and strong ground motion studies, Chapter 3. In: *Earthquake Research and Analysis, New Frontiers in Seismology*, D'Amico, S. (Ed.). InTech Publishing, London, pp. 87–140.

Igarashi, G., Saeki, S., Takahata, N., Sumikawa, K., Tasaka, S., Sasaki, Y., Takahashi, M., and Sano, Y. (1995) Groundwater radon anomaly before the Kobe earthquake in Japan. *Science* 269, 60–61.

Iida, K. (1963) Magnitude, energy and generation mechanisms of tsunamis and a catalogue of earthquakes associated with tsunamis. *Proceedings of Tsunami Meetings Associated with the Tenth Pacific Science Congress*, The International Union of Geodesy and Geophysics, Paris, pp. 7–18.

Ikeda, T., Konagai, K., Kamae, K., Sato, T., Takase, Y. (2016) Damage investigation and source characterization of the 2014 Northern Part of Nagano Prefecture earthquake. *Journal of Structural Mechanics and Earthquake Engineering* 72(4), I_975-I_983.

Ikeya, M. and Matsumoto, H. (1997) Reproduced earthquake precursor legends using a Van de Graaff electrostatic generator: Candle flame and dropped nails. *Naturwissenschaften* 84, 539–541.

Ikeya, M., Komatsu, T., Kinoshita, Y., Teramoto, K., Inoue, K., Gondou, M., Yamamoto, T. (1997) Pulsed electric field before Kobe and Izu earthquakes from Seismically-induced Anomalous Animal Behaviour (SAAB). *Episodes* 20(4), 253–260.

Ikeya, M., Matsumoto, H., and Huang, Q.H. (1998) Alignment of silkworms as seismic animal anomalous behaviour (SAAB) and electro-magnetic model of a fault: A theory and laboratory experiment. *Acta Seismologica Sinica* 11(3), 365–374.

Imamura, F., Hashi, K., and Imteaz, M.A. (2001) Modeling for tsunamis generated by landsliding and debris flow. In: *Tsunami Research at the End of a Critical Decade*, G.T. Hebenstreit (Ed.). Kluwer Academic Publishers, New York, pp. 209–228.

International Atomic Energy Agency (IAEA) (2003) Seismic design and qualification for nuclear power plants. SAFETY GUIDE No. NS-G-1.6, IAEA Safety Standards Series, 59 p.

Irikura, K. (1983) Semi-empirical estimation of strong ground motions during large earthquakes. *Bulleting Disaster Prevention Research Institute (Kyoto University)* 33, 63–104.

Ishibashi, K. (2004) Status of historical seismology in Japan. Annals of Geophysics 47(2/3), 339–368.

Ishihara, K. (1985) Stability of natural deposits during earthquakes. *The 11th Soil Mechanics and Foundation Engineering Conference,* San Francisco, vol. 1, pp. 321–376.

Ishihara, K. (1993) Liquefaction and flow failure during earthquakes. *Geotechnique* 43(3), 351–415.

İspir, Y., Uyar, O., Güngörmüş, Y., Orbay, N., and Çağlayan, B. (1976) Some results from studies on tectonomagnatic effect in NW Turkey. *Journal of Geomagnetism and Geoelectricity* 28, 123–135.

ITASCA (2005) FLAC3D-Fast Lagrangian analysis of continua-user manual (dynamic option) (version 2.1). ITASCA Consulting Group Inc, Minneapolis.

Iwata, N., Adachi, K., Takahashi, Y., Aydan, Ö., Tokashiki, N., and Miura, F. (2016) Fault rupture simulation of the 2014 Kamishiro Fault Nagano Prefecture Earthquake using 2D and 3D-FEM. EUROCK2016, Ürgüp, pp. 803–808.

Jacobsen, L.S., and Ayre, R.S. (1951) Hydrodynamic experiments with rigid cylindrical tanks subjected to transient motions. *Bulletin of the Seismological Society of America* 41, 19.

Jaeger, J.C., and Cook, N.G.W. (1976) *Fundamentals of Rock Mechanics* (3rd Edn). Chapman and Hall Ltd, London, 593 p.

Japan Meteorological Agency (JMA) (2011) http://www.jma.go.jp/, Tokyo, Japan.

Japan Roadway and Bridges Society (1996) Roadway and Bridges Design Standard, Chapter V: Earthquake Resistance Design.

Japan Society of Civil Engineers (1923) Archives of structural damage by the 1923 Great Kanto Earthquake. http://www.jsce.or.jp.

Jeffreys, H., and Bullen, K.E. (1940) *Seismological Tables.* British Association for the Advancement of Science, London.

Joyner, W.B. and Boore, D.M. (1981) Peak horizontal acceleration and velocity from strong motion records including records from the 1979 Imperial Valley, California Earthquake. *Bulletin of the Seismological Society of America* 71(6), 2011–2038.

JSCE-EEC (2000) Earthquake resistant design codes in Japan. Japan Soceity of Civil Engineers (JSCE) – Earthquake Engineering Committee (EEC), Tokyo.

Kanai, K., and Tanaka, T. (1951) Observations of earthquake motion at different depths of the earth. *Bulletin of the Earthquake Research Institute*, Tokyo University 28, 107–113.

Kanamori, H., (1983) Global seismicity (LXXXV Corso), in Kanamori, H., and Boschi, E., eds., Earthquakes: Observation, Theory and Interpretation: Amsterdam, North-Holland, p. 596–608.

Kanibir, A., Ulusay, R., and Aydan, Ö. (2006) Assessment of liquefaction and lateral spreading on the shore of Lake Sapanca during the Kocaeli (Turkey) earthquake. *Engineering Geology* 83, 307–331.

Kawakami, H. (1984) Evaluation of deformation of tunnel structure due to Izu-Oshima Kinkai earthquake of 1978. *Earthquake Engineering & Structural Dynamics* 12(3), 369–383.

Kawamoto, T., and Aydan, Ö. (1999) A review of numerical analysis of tunnels in discontinuous rock masses. *International Journal of Numerical and Analytical Methods in Geomechanics* 23, 1377–1391.

Kawashima, K. (1999) Seismic design of underground structures in soft ground, a review. *Proceedings of the International Symposium on Tunneling in Difficult Ground Conditions*, Tokyo, Japan, pp. 3–20.

Kawashima, K. (2018) *Earthquake Engineering (in Japanese)*. Kajima Publishing Co., Tokyo, 317 p.

Kawashima, K., Aydan, Ö., Aoki, T., Kishimoto, I., Konagai, K., Matsui, T., Sakuta, J., Takahashi, N., Teodori, S.P., Yashima, A. (2010) reconnaissance investigation on the damage of the 2009 L'Aquila, Central Italy Earthquake. *Journal of Earthquake Engineering* 14, 817–841.

Keefer, D.K. (1984) Slope failures caused by earthquakes. *Geological Society of American Bulletin* 95, 406–421.

Ketin, İ. (1973) Umumi Jeoloji (General Geology). Published by İ. T. Ü., No: 30. Istanbul.

KiK-Net. (2007, 2008) Digital acceleration records of earthquakes since 1998. http://www.kik.bosai.go.jp/.

KiK-Net. (2011) Bedrock strong motion database. http://www.kik.bosai.go.jp/kik/.

Kikuchi, M. (1999, 2003) Focal mechanism and rupture mechanisms of 1999 Kocaeli, Chi-chi and Tokachi earthquakes. Seismological Notes. Tokyo University.

King, G.C.P., Stein, R.S., and Lin, J. (1994) Static stress changes and the triggering of earthquakes. *Bulletin of the Seismological Society of America* 84(3), 935–953.

Kiyomiya, O. (1995) Earthquake-resistant design features of immersed tunnels in Japan. *Tunnelling and Underground Space Technology* 10(4), 463–475.

Kiyota, R., Iwata, N., Aydan, Ö., and Tokashiki, N. (2019) Experimental study of scale effect in rock discontinuities on stick-slip behavior. *Proceedings of 2019 Rock Dynamics Summit*, ISRM, Okinawa, Japan, 7–11 May 2019. (Editors: Aydan, Ö., Ito, T., Seiki, T., Kamemura, K., and Iwata, N.), pp. 119–123.

K-NET. (2004) Surface strong motion database. http://www.kyoshin.bosai.go.jp/kyoshin/.

KOERI (Kandilli Ulusal Deprem İzleme Merkezi – Kandilli Observation and Earthquake Institute). http://www.koeri.boun.edu.tr/sismo/.

Koketsu, K. (2012) Seismological and geodetic aspects of the 2011 Tohoku earthquake and Great East Japan Disaster. *International Symposium on Engineering Lessons Learned from the 2011 Great East Japan Earthquake*, Tokyo, pp. 1–8.

Komada, H., and Hayashi, M. (1980) Earthquake observation around the site of undergound power station. CRIEPI Report, E379003, Central Research Institute of Electric Power Industry, Japan, pp. 1–34.

Kondo, G. (1968) The variation of the atmospheric field at the time of earthquake. *Memoirs of the Kakioka Magnetic Observatory*. 13, 11–23.

Konishi, I. and Yamada, Y. (1960) Studies on the earthquake resistant design of suspension bridge tower and pier systems. *II WCEE* 2, 107–118.

Kovari, K., and Fritz, P. (1975) Stability analysis of rock slopes for plane and wedge failure with the aid of a programmable pocket calculator. *The 16th US Rock Mechanics Symposium*, Minneapolis, USA, pp. 25–33.

Kowalik, Z. (2001) Basic relations between tsunamis calculation and their physics. *Science of Tsunami Hazards* 19, 99–116.

Kreyszig, E. (1983) *Advanced Engineering Mathematics*. John-Wiley & Sons, New York.

Kudo, K. (1983) Seismic source characteristics of recent major earthquakes in Turkey. In: *A Comprehensive Study on Earthquake Disasters in Turkey in View of Seismic Risk Reduction*, Y. Ohta (Ed.). Hokkaido University, Sapporo, Japan, pp. 23–66.

Kumsar, H., Aydan, Ö., and Ulusay, R. (2000) Dynamic and static stability of rock slopes against wedge failures. *Rock Mechanics and Rock Engineering* 33(1), 31–51.

Kumsar, H., Aydan, Ö., Tano, H., and Ulusay, R. (2003) Multi-parameter measurement system in Denizli and its response during July 2003 Buldan earthquakes. *International Colloquium on Instrumentation and Monitoring of Landslides and Earthquakes*, Japan and Turkey, Koriyama (Edited by H. Tano and Ö. Aydan), pp. 69–81.

Kuno, H. (1935) The geologic section along the Tanna Tunnel. *Bulletin of the Earthquake Research Institute*, University of Tokyo 14, 92–103.

Kuribayashi, E. and Tatsuoka, F. (1975) Brief review of soil liquefaction during earthquakes in Japan. *Soils and Foundations* 15(4), 81–92.

Lamb, H. (1932) *Hydrodynamics*. Cambridge University Press, Cambridge.

Liao, S.S.C and Whitman, R.V. (1986) Overburden correction factors for SPT in sand. *Journal of Geotechnical Engineering Division, ASCE* 112(3), 373–377.

Livaoğlu R., and Dogangün, A. (2006) Simplified seismic analysis procedures for elevated tanks considering fluid-structure-soil interaction. *Journal of Fluids and Structures* 22, 421–439.

Lockner, D.A., Johnston, M.J.S., and Byerlee, J.D. (1983) A mechanism to explain the generation of earthquake lights. *Nature* 302, 28–33.

Mader, C.L., and Gittings, M.L. (2002) Modeling the 1958 Lituya Bay mega-tsunami. *Science of Tsunami Hazards* 20, 241–250.

Mamaghani, I.H.P., Baba, S., Aydan, Ö., and Shimizu, S. (1994) Discrete finite element method for blocky systems. *Computer Methods and Advances in Geomechanics, IACMAG, Morgantown* 1, 843–850.

Mamaghani, I.H.P., Aydan, Ö., and Kajikawa, Y. (1999) Analysis of masonry structures under static and dynamic loading by discrete finite element method. *Journal of Structural Mechanics and Earthquake Engineering*. Japan Society of Civil Engineers, JSCE, No. 626/I-48, 1–12.

Mase, G. (1970) *Theory and Problems of Continuum Mechanics, Schaum Outline Series*. McGraw Hill Co., New York, 230 p.

Matsuda, T. (1975) The magnitude and periodicity of earthquakes from active faults. JISHIN, 28, 269–283 (in Japanese).

Matsuda, T. (1981) Active faults and damaging earthquakes in Japan. In: *Macroseismic Zoning and Precaution Fault Zones*. Maurice Ewing Series, 4, American Geophysics Union, Washington, pp. 271–289

Matsumoto, H., Ikeya, M., and Yamanaka, C. (1998) Analysis of Barber-Pole color and speckle noises recorded 6 and a half hours before the Kobe earthquake. *Japanese Journal of Applied Physics* 37, L1409–L1411.

Matsumoto, N., Sasaki, T. and Ohmachi, T. (2011) The 2011 Tohoku earthquake and dams. ICOLD 2011, Lucerne.

Meirovitch, L. (1986) *Elements of Vibration Analysis*. McGraw-Hill Int. Ed., New York, 560 p.

Melan, J. (1906) *Theory of Arches and Suspension Bridges*. Myron Clark Publ. Comp., London (1913) (German original third edition: Handbuch der Ingenieurwissenschaften, Vol. 2).

Milne, J. (1898) *Seismology, Kegan Paul, Trench*. Trubner & Co., London, 320 p.

Misawa, Y., Aydan, Ö., and Hamada, M. (1993) A consideration of the failure of Mt. Mayuyama in 1792 from rock mechanistic view point. *International Symposium on Assessment and Prevention of Failure Pheomena in Rock Engineering*, Istanbul, pp. 871–877.

Miyako, K. (2010) Numerical simulation of landslide movement and Unzen - Mayuyama Disaster in 1792, Japan. *Journal of Disaster Research* 5(3), 280–287.

Mizumoto, T., Tsuboi, T., and Miura, F. (2005) Fundamental study on fault rupture process and earthquake motions on and near a fault by 3D FEM (in Japanese). *Journal and Earthquake and Structure Division, JSCE* 780(I-70), 27–40.

Mizutani, H., Ishido, T., Yokokura, T., and Ohnishi, S. (1976) Electrokinetic phenomena associated with earthquakes. *Geophysical Research Letters* 3(2), 365–368.

Mogi, K. (1985) *Earthquake Prediction*. Academic Press, Orlando, FL, 355 p.

Morgernstern, N.R. and Price, V.E. (1965) The analysis of the tability of general slip surfaces. *Geotechnique* 15(1), 79–83.

Nadai, A.L. (1950) *Theory of Flow and Fracture of Solids*, Vol. II. McGraw-Hill, New York, pp. 623–624.

Nakamura, M. (2006) Source fault model of the 1771 Yaeyama tsunami, Southern Ryukyu Islands, Japan, inferred from numerical simulation. *Pure and Applied Geophys* 163, 41–54.

Nakano, T. and Ohta, Y. (2008) Non-linear dynamic response analysis of bridge crossing earthquake fault rupture plane. *The 14th World Conference on Earthquake Engineering*, October 12–17, 2008, Beijing, China.

Nakaza, E., Tokuyama, R., and Inagaki, K. (2015) Original locations of huge boulders moved by tsunami and its generation mechanism (in Japanese). *Journal of Coastal Engineering*, B2, JSCE 71(2), 193–198.

Nasu, N. (1931) Comparative studies of earthquake motions above ground and in a tunnel. *Bulletin of the Earthquake Research Institute*, Tokyo University 9, 454–472.

Newmark, N.M. (1965) Effects of earthquakes on dams and embankments. *Geotechnique* 15(2), 139–160.

Newmark, N.M. (1968) Problems in wave propagation in soils and rock. *Symposium on Wave Propagation and Dynamic Properties of Earth Materials*, University of New Mexico, Albuquerque, pp. 7–26.

Nistor, I., Saatcioglu, M., and Ghobarah, A. (2005) The 26 December 2004 Earthquake and Tsunami: Hydrodynamic forces on physical infrastructure in Thailand and Indonesia. *Proceedings 2005 Canadian Coastal Engineering Conference*, Halifax, Canada, CD-ROM, pp. 1–15.

NOAA (National Ocean and Atmosphere Administration). NCEI/WDS Global Historical Tsunami Database: 1883 Krakatau, 1960 Valdivia, 2011 Great East Japan Tsunami events. https://data.noaa.gov/.

Nur, A. (1972) Dilatancy, pore fluids and premonitory variations of ts/tp travel times. *Bulletin of the Seismological Society of America* 62(5), 1217–1222.

Nyman, D.J., and Kennedy, R.P. (1987) Seismic design of oil and gas pipeline systems. FEMA 139, *Earthquake Hazard Reduction Series* 30, 51–66.

Ohta, Y. (2011) A fundamental research on the effects of ground motions and permanent ground deformations neighborhoud earthquake faults on civil engineering structures (in Japanese). Doctorate Thesis, Graduate School of Science and Technology, Tokai University, 272 p.

Ohta, Y., and Aydan, Ö. (2004) An experimental study on ground motions and permanent deformation nearby faults. *Journal of the School of Marine Science and Technology* 2(3), 1–12.

Ohta, Y., and Aydan, Ö. (2007) Integration of ground displacement from acceleration records. *JSCE Earthquake Engineering Symposium*, Tokyo, pp. 1046–1051.

Ohta, Y., and Aydan, Ö. (2010) The dynamic responses of geomaterials during fracturing and slippage. *Rock Mechanics and Rock Engineering* 43(6), 727–740.

Ohta, Y., and Aydan, Ö. (2011) The characteristics of ground liquefaction caused by the August 11, 2009 Suruga Bay Earthquake (in Japanese). *Journal of the School of Marine Science and Technology*, Tokai University 9(2), 1–9.

Ohta, Y., Aydan, Ö., and Tokashiki, N. (2008) The dynamic response of rocks during fracturing and its implications in geo-engineering and earth science. *The 5th Asian Rock Mechanics Symposium (ARMS5)*, Tehran, pp. 965–972.

Ohta, Y., Aydan, Ö., and Yagi, M. (2014) Laboratory model experiments and case history surveys on response and failure process of rock engineering structures subjected to earthquake. *Proceedings of the 8th Asian Rock Mechanics Symposium*, Sapporo, pp. 843–852.

Okada, Y. (1992) Internal deformation due to shear and tensile faults in a half-space. *Bulletin of the Seismological Society of America* 82(2), 1018–1040.

Okamoto, S. (1973) *Introduction to Earthquake Engineering*. University of Tokyo Press, Tokyo.

Okawa, I. (2005) Strong earthquake motion recordings during the Pakistan, 2005/10/8, http://www.bri.go.jp/.

Omachi, T., Kojima, N., Murakami, A., and Komaba, N. (2003) Near field effects of hidden seismic faulting on a concrete dam. *Journal of Natural Disaster Science* 25(1), 7–15.

Omori, F. (1894) On the aftershocks of earthquake. *Journal of Colloid and Interface Science*, University of Tokyo 7, 111–200.

Osaki, Y. (1984) *Introduction to Spectral Analysis of Strong Motions*. Kajima Publishing Co., Tokyo, 260 p.

Owen, D.R.J., and Hinton, E. (1980) *Finite Elements in Plasticity: Theory and Practice*. Pineridge Press Limited, Swansea, UK.

Palermo, D., and Nistor, I. (2008) Tsunami-induced loading on structures: Beyond Hollywood's sSenarios. *Structure Magazine*, 10–13, March 2008.

Pavez, A., Sepúlveda, S.A., and Aguilera, R. (2007) Remote sensing analysis of landslides and coastal changes after the 2007 Aysén Mw 6.2 earthquake. *Geosur 2007*, Santiago, 120 p.

Pierce, E.T. (1976) Atmospheric electricity and earthquake prediction. *Geophysical Research Letters* 383, 185–188.

Pitarka, A. (1999) 3D elastic finite-differences modeling of seismic motion using staggered grids with nonuniform spacing. *Bulletin of the Seismological Society of America* 89(1), 54–68.

Port and Harbour Research Institute of Japan (1997) *Handbook on Liquefaction Remediation of Reclaimed Land*. A.A. Balkema, Rotterdam.

Port Authority Research Institute (PARI) (2011) Special web site for East Japan mega earthquake. http://www.pari.go.jp/info/tohoku-eq/.

Prentice, C., and Ponti, D. (1997) Coseismic deformation of the Wrights tunnel during the 1906 San Francisco earthquake: A key to understanding 1906 fault slip and 1989 surface ruptures in the southern Santa Cruz Mountains, California. *Journal of Geophysical Research* 102, 635–648.

Rahman, M.M., Schaab, C., and Nakaza, E. (2016) Experimental and numerical modeling of tsunami mitigation by canals. *Journal of Waterway, Port, Coastal, and Ocean Engineering* 143, 1–11.

Rathje, E.M., Karatas, I., Stephen, G.W., and Bachhuber, J. (2004) Coastal failures during the 1999 Kocaeli earthquake of Turkey. *Soil Dynamics and Earthquake Engineering* 24, 699–712.

Reddy, J.N. (1993) *Introduction to the Finite Element Method*. McGraw-Hill, New York.

Reilinger, R.E., Ergintav, S., Burgmann, R., McClusky, S., Lenk, O., Barka, A., Gürkan, O., Hearn, L., Feigl, K.L., Çakmak, R. Aktug, B., Özener, H., and Toksöz, M.N. (2000) Coseismic and postseismic fault slip for the 17 August 1999, M = 7.5, Izmit, Turkey Earthquake. *Science* 289, 1519–1524.

Reilinger, R.E., McClusky, S.C., Oral, M.B., King, R.W., Toksöz, M.N., Barka, A.A., Kınık, I., Lenk, O., and Şanlı, I. (1997) Global positioning system measurements of present-day crustal movements in the Arabia-Africa-Euroasia plate collision zone. *Journal of Geophysical Research* 102(B5), 9983–9999.

Richter, C.F. (1935) An instrumental earthquake magnitude scale. *Bulletin of Seismological Society of America* 25(1), 1–32.

Richter, C.F. (1958) *Elementary Seismology*. W.H. Freeman and Company, San Francisco.

Rodgers, J.E., Sanli, A.K., and Çelebi, M. (2004) Seismic response analysis of a 13-story steel moment-framed building in Alhambra, California. Open-File Report 2004-1338, USGS, 143 p.

Romano, F., Trasatti, E., Lorito, S., Piromallo, C., Piatanesi, A., Ito, Y., Zhao, D., Hirata, K., Lanucara, P., and Cocco, M. (2014) Structural control on the Tohoku earthquake rupture process investigated by 3D FEM, tsunami and geodetic data. *Scientific Reports* 4, 5631. DOI: 10.1038/srep05631.

Rouse, H. (1978) *Elementary Mechanics of Fluids*, Dover Publication, New York.

Rozen, A. (1976) Response of rock tunnels to earthquake shaking. M.Sc Thesis in Civil Engineering. Massachusetts Institute of Technology.

Sakurai, T. (1999) A report on the earthquake fault appearing in the Tanna tunnel under construction by North-Izu Earthquake 1930 (in Japanese). *The Journal of the Geological Society of Japan* 39(6), 540–544.

Satake, K. (2005) December 26 Tsunami in Indian Ocean. Geological Survey of Japan, National Institute of Advanced Industrial Science and Technology. http://staff.aist.go.jp/kenji.satake/Sumatra-E.html (see also Aydan et al. 2005a).

Sato, R. (1989) *Handbook on Parameters of Earthquake Faults in Japan*. Kajima Pub. Co, Tokyo (in Japanese).

Sato, T., Takahashi, M., Matsumoto, N., and Tsukuda, E. (1995) Anomalous discharge of groundwater after the 1995 Hyogo-ken nanbu earthquake in Awaji Island, Japan. (in Japanese) *Chishitsu News* 496, 61–66.

Sato, T., Aoyagi, K., Miyara, N., Aydan, Ö., Tomiyama, J., and Morita, T. (2019) The dynamic response of Horonobe Underground Research Center during the 2018 June 20 earthquake. *Proceedings of 2019 Rock Dynamics Summit*, ISRM, Okinawa, Japan, 7–11 May, 2019. (Editors: Aydan, Ö., Ito, T., Seiki T., Kamemura, K., Iwata, N.), pp. 640–645.

Sayısal Grafik (2004) http://www.sayisalgrafik.com.tr/deprem/.

Seed, H.B. (1979) Soil liquefaction and cyclic mobility evaluating for level ground during earthquakes. *Journal of Geotechnical Engineering Division ASCE* 109(3), 458–482.

Seed, H., and DeAlba, P. (1986) Use of SPT and CPT tests for evaluating the liquefaction resistance of sands. *In Use of In-situ Tests in Geotechnical Engineering*, ASCE Geotechnical Special Publication, vol. 6, pp. 281–302.

Seed, H.B., and Idriss, I.M. (1971) Simplified procedure for evaluating soil liquefaction procedure. *Journal of Soil Mechanics Foundation Division ASCE* 97(SM9), 1249–1273.

Seed, H.B. and Idriss, I.M. (1981) Evaluation of liquefaction potential of sand deposits based on observation. *In-situ Testing to Evaluate Liquefaction Susceptibility*, St. Louis, MO, 81–544, 45–65.

Seed, H.B., Tokimatsu, K., Harder, L.F., and Chung, R.M. (1985) The influence of SPT procedures in soil liquefaction resistance evaluations. *Journal of the Geotechnical Engineering Division ASCE* 111(12), 1425–1445.

Semenza, E. and Ghirotti, M. (2000) History of the 1963 Vaiont slide: The importance of geological factors. *Bulletin of Engineering Geology and the Environment* 59, 87–97.

Shamoto, Y., Zhang, J., and Tokimatsu, K. (1998) New charts for predicting large residual post-liquefaction ground deformations. *Soil Dynamics and Earthquake Engineering* 17, 427–438.

Shareq, A. (1981) Geological observations and geophysical investigations carried out in Afghanistan over the period of 1972–1979. In: *Zagros Hindu Kush Himalaya Geodynamic Evolution*, F.M. Delany, H.K. Gupta, Inter-Union Commission on Geodynamics (Eds.), American Geophysical Union, Washington, DC, pp. 75–86.

Sharma, S. and Judd, W.R. (1991) Underground opening damage to undergound facilities. *Engineering Geology* 30, 263–276.

Shimazaki, K. and Nakata, T. (1980) Time-predictable recurrence model for large earthquakes. *Geophysical Research Letters* 7(4), 279–282.

Shimohira, K., Aydan, Ö., Tokashiki, N., Watanabe, K., and Yokoyama, Y. (2019). An experimental study on the formation mechanism of tsunami boulders. *Proceedings of 2019 Rock Dynamics Summit*, ISRM, Okinawa, Japan, 7–11 May, 2019. (Editors: Aydan, Ö., Ito, T., Seiki T., Kamemura, K., and Iwata, N.), 195–200.

Şimşek, Ş., and ve Yıldırım, N. (2000a) 17 Ağustos ve 12 Kasım 1999 deprem bölgelerinde yeralan jeotermal kaynaklarda gözlenen değişimler ve önemi. TÜBİTAK Bilim ve Teknik Dergisi Şubat 2000 sayısı.

Slingerland, R.L., and Voight, B. (1979) Occurrences, properties, and predictive models of landslide-generated water waves. In: *Developments in Geotechnical Engineering 14B: Rockslides and Avalanches, 2, Engineering Sites*, Voight, B. (Ed.). Elsevier Scientific Publishing Company, Amsterdam, pp. 317–397.

Somerville, P.G., Sen, M., and Cohee, B. (1991) Simulation of strong ground motion recorded during the 1985 Michoacan, Mexico an Valparaiso, Chile earthquakes. *Bulletin of the Seismological Society of America* 81, 1–27.

Somerville, P.G., Smith, N.F., Graves, R.W., and Abrahamson, N.A. (1997) Modification of empirical strong ground motion attenuation relations to include the amplitude and duration effects of rupture directivity. *Seismological Research Letters* 68(1), 199–222.

Spencer, E. (1967) A method of analysis of the stability of embankments assuming parallel interslice forces. *Geotechnique* 17(1), 11–26.

Stein, R.S. (2003) Earthquake conversations. *Scientific American* 288 (1), 72–79.

Steinman, D.B. (1922) *A Practical Treatise on Suspension Bridges: Their Design, Construction and Erection*. John Wiley & Sons Inc., New York.

Sugito, M., Furumoto, Y., and Sugiyama, T. (2000) Strong motion prediction on rock surface by superposed evolutionary spectra. *The 12th World Conference on Earthquake Engineering*, 2111/4/A, Auckland, CD-ROM.

Sultankhodzhaev, A.N. (1984) Hydrogeoseismic precursors to earthquakes. *International Symposium on Earthquake Prediction*, Paris, pp. 181–191.

TEC-JSCE (Tunnel Engineering Committee, JSCE) (2005) Report of the 2004 Mid Niigata Prefecture Earthquake. JSCE (In Japanese).

Tezcan, L., Doğdu, N., and ve Kırmızıtaş, H. (2000) Sismik aktivitelere bağlı yeraltısuyunun değişimi. 53 Türkiye Jeoloji Kurultayı, 21–25, Şubat, Ankara.

Tinti, S, Armigliato, A., Manucci, A., Pagnoni, G., Zaniboni, F., Ahmet Cevdet Yalçıner, A.C., and Altinok, Y. (2006) The generating mechanisms of the August 17, 1999 Izmit bay (Turkey) tsunami: Regional (tectonic) and local (mass instabilities) causes. *Marine Geology* 225, 311–330.

Toda, S., Stein, R.S., and Sagiya, T. (2002) Evidence from the 2000 Izu Islands swarm that seismicity is governed by stressing rate. *Nature* 419, 58–61.

Tokashiki, N. (2011) Study on the engineering properties of Ryukyu limestone and the evaluation of the stability of its rock mass and masonry structures. PhD Thesis, Waseda University, Engineering and Science Graduate School, 221 p.

Tokashiki, N., and Aydan, Ö. (2010) Kita-Uebaru natural rock slope failure and its back analysis. *Environmental Earth Sciences* 62(1), 25–31.

Tokashiki, N., Aydan, Ö., Mamaghani, I.H.P., and Kawamoto, T. (1997) The stability of a rock block on an incline by discrete finite element method (DFEM). *Proceedings of the Ninth International Conference on Computer Methods and Advances in Geomechanics*, Wuhan, China, Vol. 1, pp. 523–528.

Tokashiki, N., Aydan, Ö., Shimizu, Y., and Mamaghani, I.H.P. (2001) A stability analysis of masonry walls by discrete finite element method(DFEM). *Computer Methods and Advances in Geomechanics* 2, 17.25–17.28.

Tokashiki, N., Aydan, Ö., Daido, M., and Akagi, T. (2005) An experimental and numerical study on the dynamic stability of masonary retaining walls. *The 35th Japan Rock Mechanics Symposium*, Tokyo, pp. 115–120.

Toki, K., and Miura, F. (1985) Simulation of a fault rupture mechanism by a two-dimensional finite element method. Journal of Physics of the Earth 33, 485–511.

Tokimatsu, K., and Yoshimi, Y. (1983) Empirical correlation of soil liquefaction based on SPT N-value and fines content. *Soil and Foundations* 26(4), 127–138.

Toksöz, M.N. (1977) Earthquake Prediction Research in the United States. Predicting Earthquakes, Panel fon Earthquake Prediction of the Committee on Seismology, NRC, pp. 37–50.

Toksöz, M.N., Shakal, A.F., and Michael, A.J. (1975) Space-time migration of earthquakes along North Anatolian Fault zone and seismic gaps. *Pure Applied Geophysics* (*Pageoph*) 17, 1258–1270.

Towhata, I. (2008) *Geotechnical Earthquake Engineering.* Springer Series in Geomechanics and Geoengineering, Springer, Berlin, Heidelberg, 684 p.

Towhata, I., and Matsumoto, H. (1992) Analysis of development of permanent displacement with time in liquefied ground. *The 4th US-Japan Workshop on Earthquake Resistant Design of Lifeline Facilities and Countermeasures for Soil Liquefaction,* Honolulu, vol. 2, pp. 335–349.

Towhata, I., Sasaki, Y., Tokida, K., Matsumoto, H., Tamari, Y., and Yamada, K. (1992) Prediction of permanent displacement of liquefied ground by means of minimum energy principle. *Soils and Foundations* 32(3), 97–116.

Trifunac, M.D., and Brady, A.G. (1975) On the correlation of seismic intensity scales with peaks of recorded ground motion. *Bulletin of the Seismological Society of America* 88, 139–162.

Tsuboi, T., and Miura, F. (2000) Simulation of near field seismic ground motions by FEM considering source rupture mechanism. *Proceeding of the 12th World Conference on Earthquake Engineering* (CD-ROM), Paper 271/4/A.

Tsuneishi, Y., Ito, T., and Kano, K. (1978) Surface faulting associated with the 1978 Izu-Oshima-Kinkai earthquake. *Bulletin of the Earthquake Research Institute,* University of Tokyo 53, 649–674.

Tsunogai, U., and Wakita, H. (1995) Precursory chemical changes in ground water: Kober earthquake, Japan. *Science* 269, 61–63.

Udias, A. (1999) *Principles of Seismology.* Cambridge University Press, Cambridge, 475 p.

Ueta, K., Miyakoshi, K., and Inoue, D. (2001) Left-lateral deformation of head-race tunnel associated with the 2000 western Tottori earthquake. *Journal of the Seismological Society of Japan* 54(2), 547–556.

Ulusay, R., and Aydan, Ö. (2005) Characteristics and geo-engineering aspects of the 2003 Bingöl (Turkey) earthquake. *Journal of Geodynamics* 40(2–3), 334–346.

Ulusay, R., and Aydan, Ö. (2011) Lateral spreading induced by ground liquefaction and associated damage to structures and lifelines during the Darfield (New Zealand) earthquake). *Seventh National Conference on Earthquake Engineering,* Istanbul, Turkey, 30 May to 3 June 2011, Paper No: 69, 1–10 pp.

Ulusay, R., Aydan, Ö., and Hamada, M. (2001) The behaviour of structures built on active fault zones: Examples from the recent earthquakes of Turkey. *A Workshop on Seismic-Fault Induced Failures – Possible Remedies for Damage to Urban Facilities,* Tokyo, Japan, pp. 1–26.

Ulusay, R., Aydan Ö., and Hamada, M. (2002) The behavior of structures built on active fault zones: Examples from the recent earthquakes of Turkey. *Structural Engineering/Earthquake Engineering, JSCE* 19(2), Special Issue, 149–167.

Ulusay, R., Aydan, Ö., Erken, E., Kumsar, H., Tuncay, E., and Kaya, Z. (2003a) Site investigation and engineering evaluation of the Cay-Eber Earthquake of February 3, 2002. Turkish Earthquake Foundation, TDV/DR 012-79, 213 p.

Ulusay, R., Aydan, Ö., Erken, E., Kumsar, H., Tuncay, E., and Kaya, Z. (2003b) Characteristics of the Cay-Eber earthquake and its evaluation in terms of local ground conditions. *Fifth National Conference on Earthquake Engineering,* Istanbul, Turkey, 26–30 May 2003, Paper No: AE-115.

Ulusay, R., Tuncay, E., Sonmez, H., and Gokceoglu, C. (2004) An attenuation relationship based on Turkish strong motion data and ISO-acceleration map of Turkey. *Engineering Geology* 74, 265–291.

US Army Corps of Engineers (1994) Engineering and design: Rock Foundations. EM1110-1-2908, 121 p.

Usami, T. (1980) *Earthquake Engineering for Architects*. Ichigaya Publishing Co., Tokyo, 267 p.

USBR (US Bureau of Reclamation) (1977) *Design of Arch Dams*. U.S. Department of the Interior, Denver, CO.

USGS (United States Geological Survey) (1999a) Crustal deformation measurements at Parkfield. http://quake.wr.us-gs.gov/.

USGS (United States Geological Survey) (1999b) Water level observations at Parkfield. http://quake.wr.usgs.gov/.

Uyeda, S., Al-Damegh, K.H., Dologlou, E., and Nagao, T. (1999) Some relationship between VAN seismic electric signals (SES) and earthquake parameters. *Tectonophysics* 304, 41–55.

Varotsos, P., and Alexopolous, K. (1984) Physical properties of the variations of the electric field of the earth preceding earthquakes. *Tectonophysics* 110, 73–98.

Wakamatsu, K. (1991) *Maps for Historic Liquefaction Sites in Japan*. Tokai University Press, Tokyo (in Japanese with English abstract).

Wakamatsu, K. (1993) History of soil liquefaction in Japan and assessment of liquefaction potential based on geomorphology. A Thesis in the Department of Architecture Presented in Partial Fulfillment of the Requirements for the Degree of Doctor of Engineering, Waseda University of Tokyo.

Wald, D., Somerville, P.G., Eeri, M., and Burdick, L.J. (1998) The Whitter Narrows, California earthquake of October 1, 1997: Simulation of recorded accelerations. *Earthquake Spectra* 4, 139–156.

Wald, D.J., Quitoriano, V., Heaton, T.H., and Kanamori, H. (1999) Relations between peak ground acceleration and Modified Mercalli Intensity in California. Earthquake Spectra, EERI.

Walsh, J.B. (1975) An analysis of local changes in gravity due to deformation. *Pure Applied Geophysics* 113, 97–106.

Wang, W.L., Wang, T.T., Su, J.J., Lin, C.H., Seng, C.R., and Huang, T.H., (2001) Assessment of damage in mountain tunnels due to the Taiwan Chi-Chi earthquake. *Tunneling and Underground Space Technology* 16, 133–150.

Ward, S.N. (2001) Landslide tsunami. *Journal of Geophysical Research* 106(6), 11201–11215.

Ward, S.N., and Asphaug, E. (2000) Asteroid impact tsunami: A probabilistic hazard assessment. *Icarus* 145, 64–78.

Wegener, A. (1912) Die Entstehung der Kontinente. *Geologische Rundschau (in German)* 3(4), 276–292.

Weiss, R., Wünnemann, K., and Bahlburg, H. (2006) Numerical modeling of generation, propagation and run-up of tsunamis caused by oceanic impacts: Model strategy and technical solutions. *Geophysical Journal International* 167, 77–88.

Wells, D.L., and Coppersmith, K.J. (1994) New empirical relationship among magnitude, rupture length, rupture width, rupture area, and surface displacement. *Bulletin of the Seismological Society of America* 84(4), 974–1002.

Westergaard, H.M. (1933) Water pressure on dams during earthquakes. *Transactions of ASCE* 95, 418–433.

Wünnemann, K., Collins, G.S., and Weiss, R. (2010) Impact of a cosmic body into earth's ocean and the generation of large tsunami waves: Insight from numerical modeling. *Reviews of Geophysics* 48(4), RG4006.

Yamamoto, T., Tateishi, A., and Tsuchiya, M. (2014) Seismic design of immersed tube tunnel and its connection with TBM tunnel in Marmaray Project. *The 2nd European Conference on Earthquake Engineering and Seismology*, Istanbul, 1–12.

Yarar, R., Ergunay, O., Erdik, M., and Gülkan, P. (1980) A preliminary probabilistic Assessment of the Seismic Hazard in Turkey. *Proceedings of the 7WCEE*, , Istanbul, vol. 1, pp 309–316.

Yashiro, K., Kojima, Y., and Shimizu, M. (2007) Historical earthquake damage to tunnels in Japan and case studies of railways tunnels in the 2004 Niigata-ken Chuetsu earthquake. *QR of RTRI* 48(3), 136–141.

Yasuda, S., Nagase, H., Kiku, H., and Uchida, Y. (1990) A simplified procedure for the analysis of a permanent displacement. *The 3rd US: Japan Workshop on Earthquake Resistant Design of Lifeline Facilities and Countermeasures for Soil Liquefaction,* San Francisco, pp. 225–236.

Yasuda, S., Irisawa, T., and Kazami, K. (2001) Liquefaction-induced settlements of buildings and damages in coastal areas during Kocaeli and other earthquakes. *The 15th International Conference on Soil Mechanics and Geotechnical Engineering*, İstanbul, Turkey, A.M. Ansal (Ed.), pp. 33–42.

Yasui, Y. (1973) A study of luminous phenomena accompanied with earthquake. *Memoirs Kakioka Magnetic Observatory* 13, 25–61.

Yeh, H.H. (2007) Design tsunami forces for onshore structures. *Journal of Disaster Research* 2(6), 531–536.

Youd, T.L., Idriss, I.M., Andrus, R.D., Arango, I., Castro, G., Christian, J.T., Dobry, R., Finn, W.D.L., Harder, L.F., Hynes, M.E., Ishihara, K., Koester, J.P., Liao, S.S.C., Marcuson, W.F., Martin, G.R., Mitchell, J.K., Moriwaki, Y., Power, M.S., Robertson, P.K., Seed, R.B., and Stokoe, K.H. (2001) Liquefaction resistance of soils: Summary report from the 1996 NCEER and 1998 NCEER/NSF workshops on evaluations of liquefaction resistance of soils. *Journal of Geotechnical and Geoenvironmental Engineering ASCE* 127(10), 817–833.

Youd, T.L., Hansen, C.M., and Bartlett, S.F. (2002) Revised multi linear regression equations for prediction of lateral spread displacement. *Journal of Geotechnical and Geoenvironmental Engineering ASCE* 128(12), 1007–1017.

Yu, S.B., Chen, H.Y., and Kuo, L.C. (1997) Velocity field of GPS stations in the Taiwan area. *Tectonophysics* 274, 41–59.

Zeyn, D., Stead, D., and Bornhold, B. (2006) Fjord rock slope failures and tsunamis. *International Symposium on Sea to Sky Geotechnique* 2006, 440–447.

Zienkiewicz, O.C., and Shiomi, T. (1984) Dynamical behavior of saturated porous media: the generalized Biot formulation and its numerical solution. *International Journal for Numerical and Analytical Methods in Geomechanics* 8, 71–96.

Index

For Product Safety Concerns and Information please contact our EU
representative GPSR@taylorandfrancis.com
Taylor & Francis Verlag GmbH, Kaufingerstraße 24, 80331 München, Germany